Titles in This Series

Titles in This Series

Titles in This Series

Volume

Mladen Luksic, Clyde Martin, and William Shadwick, Editors

69 **Methods and applications of mathematical logic,** Walter A. Carnielli and Luiz Paulo de Alcantara, Editors

70 **Index theory of elliptic operators, foliations, and operator algebras,** Jerome Kaminker, Kenneth C. Millett, and Claude Schochet, Editors

71 **Mathematics and general relativity,** James A. Isenberg, Editor

72 **Fixed point theory and its applications,** R. F. Brown, Editor

73 **Geometry of random motion,** Rick Durrett and Mark A. Pinsky, Editors

74 **Geometry of group representations,** William M. Goldman and Andy R. Magid, Editors

75 **The finite calculus associated with Bessel functions,** Frank M. Cholewinski

76 **The structure of finite algebras,** David C. Hobby and Ralph Mckenzie

77 **Number theory and its applications in China,** Wang Yuan, Yang Chung-chun, and Pan Chengbiao, Editors

78 **Braids,** Joan S. Birman and Anatoly Libgober, Editors

79 **Regular differential forms,** Ernst Kunz and Rolf Waldi

80 **Statistical inference from stochastic processes,** N. U. Prabhu, Editor

81 **Hamiltonian dynamical systems,** Kenneth R. Meyer and Donald G. Saari, Editors

82 **Classical groups and related topics,** Alexander J. Hahn, Donald G. James, and Zhe-xian Wan, Editors

83 **Algebraic K-theory and algebraic number theory,** Michael R. Stein and R. Keith Dennis, Editors

84 **Partition problems in topology,** Stevo Todorcevic

85 **Banach space theory,** Bor-Luh Lin, Editor

86 **Representation theory and number theory in connection with the local Langlands conjecture,** J. Ritter, Editor

87 **Abelian group theory,** Laszlo Fuchs, Rüdiger Göbel, and Phillip Schultz, Editors

88 **Invariant theory,** R. Fossum, W. Haboush, M. Hochster, and V. Lakshmibai, Editors

89 **Graphs and algorithms,** R. Bruce Richter, Editor

90 **Singularities,** Richard Randell, Editor

91 **Commutative harmonic analysis,** David Colella, Editor

92 **Categories in computer science and logic,** John W. Gray and Andre Scedrov, Editors

CONTEMPORARY MATHEMATICS

Volume 92

Categories in Computer Science and Logic

Proceedings of the AMS-IMS-SIAM Joint Summer Research Conference held June 14–20, 1987 with support from the National Science Foundation

John W. Gray and
Andre Scedrov, Editors

AMERICAN MATHEMATICAL SOCIETY
Providence · Rhode Island

The AMS-IMS-SIAM Joint Summer Research Conference in the Mathematical Sciences on Categories in Computer Science and Logic was held at University of Colorado, Boulder, Colorado, on June 14–20, 1987 with support from the National Science Foundation, Grant DMS-8613199.

1980 *Mathematics Subject Classification* (1985 *Revision*). Primary 18-06, 18D99, 68F20, 03B15.

Library of Congress Cataloging-in-Publication Data

AMS-IMS-SIAM Joint Summer Research Conference in the Mathematical Sciences on Categories in Computer Science and Logic (1987: University of Colorado, Boulder)

Categories in computer science and logic: proceedings of the AMS-IMS-SIAM Joint Summer Research Conference held June 14–20, 1987 with support from the National Science Foundation/John W. Gray and Andre Scedrov, editors.

p. cm. – (Contemporary mathematics, ISSN 0271-4132; v. 92) "The AMS-IMS-SIAM Joint Summer Research Conference in the Mathematical Sciences on Categories in Computer Science and Logic was held at University of Colorado, Boulder, Colo., on June 14–20, 1987 with support from the National Science Foundation" – T.p. verso.

Bibliography: p.

ISBN 0-8218-5100-4 (alk. paper)

1. Electronic data processing – Mathematics – Congresses. 2. Categories (Mathematics) – Congresses. I. Gray, John W. (John Walker), 1931-. II. Ščedrov, Andrej, 1955-. III. National Science Foundation. IV. Title. V. Series: Contemporary mathematics (American Mathematical Society); v. 92.

QA76.9.M35A47 1987 89-32893

004′.01′5–dc20 CIP

CONTENTS

PREFACE

John W. Gray

The conference on Categories in Computer Science and Logic was held from June 14 to June 20, 1987, at the University of Colorado, Boulder, Colorado. It was a joint Summer Research Conference in the Mathematical Sciences, sponsored by AMS-IMS-SIAM, and funded by the NSF. The organizers were:

J. W. Gray	University of Illinois at Urbana-Champaign, Chairman
A. Blass	University of Michigan
M. Makkai	McGill University
A. Pitts	University of Sussex
A. Scedrov	University of Pennsylvania
D. Scott	Carnegie-Mellon University

The program consisted of invited lectures plus contributed talks. The invited speakers were:

H. P. Barendregt	G. Huet
R. L. Constable	M. Hyland
P. J. Freyd	J. Lambek
J.-Y. Girard	F. W. Lawvere
J. Goguen	J. Reynolds

Furthermore, three evenings were devoted to animated discussions and presentations concerning various aspects of the syntax and semantics of polymorphism. These evenings were organized and led by P. Freyd, A. Scedrov, and D. Scott.

Category theory has had important uses in logic since the invention of topos theory seventeen years ago, and logic has always been an important component of theoretical computer science. What was new was the increasing direct interaction between category theory and computer science. The aim of this conference was to bring together researchers who were working on the interconnections between category theory and computer science or between category theory and logic. The conference emphasized how the general machinery developed in category theory could be applied to specific questions and utilized for category-theoretic studies of concrete issues. It was certainly not the first conference with this goal, but it happened at an especially propitious time since many people both in category theory and in computer science were coming to the realization that meaningful collaboration was in fact possible.

The papers in this collection reflect only part of the activities of the conference since the contributions of several of the participants are being published elsewhere. All the papers published here have been refereed. The papers that are here represent several aspects of the conference:

i) There is a kernel of topics relevant to all three fields. It includes, for example, Horn logic (Barr), lambda calculus (Freyd), normal form reductions (Scedrov), algebraic theories (Gray) and categorical models for computability theory (Hyland and Pitts, Lamarche, Mitchell and Scott, Mulry, Roman, Seely). Such topics were the central focus of the conference, but there was also time for category-theoretic topics related just to logic or to computer science.

ii) On the logic side, the topics include semantical (algebraic or topos-theoretic) approaches to proof-theoretic questions (Blass, de Paiva, Girard), problems concerning the internal properties of specific objects in (pre-) topoi and their representations, and categorical sharpening of model-theoretic notions (Cockett, Lambek). Category theory is useful in studying the proof theory and model theory (Lawvere, Paré, Power) of various non-classical logics as well as classical first order logic.

iii) On the computer science side, it has recently been recognized that category theory is appropriate for formalizing many aspects of computer programming and program design (Latch). One reason for this is that in computer science it is necessary to consider many different structures at the same time. These structures must be viewed from different aspects and the interactions between them are a central component of program design. Category theory is specifically designed to deal with this kind of a situation on an abstract level. Specific areas where active research is going on include: semantics of programming languages, data type specifications, categorical programming, and categorical logic.

Contemporary Mathematics
Volume **92**, 1989

MODELS OF HORN THEORIES

Michael Barr

Abstract

This paper explores the connection between categories of models of Horn theories and models of finite limit theories. The first is a proper subclass of the second. The question of which categories modelable by FL theories are also models of a Horn theory is related to the existence of enough projectives.

1 Introduction

Since both logical theories and sketches are used to present types of categories, the question naturally arises as to what classes of logical theories are equivalent to what kinds of sketches. In general, they give such different ways of organizing things that direct comparisons are not possible. It has been assumed, for example, that universal Horn theories classify the same kinds of theories as finite limit (also known as left exact) sketches. It turns out that this is not quite the case. In fact, all categories classified by universal Horn theories can also be classified by FL sketches, but not conversely. The basic difference is that a left exact theory allows one to define an operation whose domain is a limit, while a Horn theory allows one only to say that one limit is a subset of another.

Results similar to these have been obtained, using logical methods, by Volger [1987] and Makowsky [1985].

The author of this paper was supported by the Ministère de l'Education du Québec through FCAR grants to the Groupe Interuniversitaire en Etudes Catégoriques and by an individual operating grant from the National Science and Engineering Research Council.

2 Equational Horn theories

By an **equational Horn theory** we mean an equational theory augmented by a set of conditions of the form

$$[\phi_1(x) = \psi_1(x)] \wedge \cdots \wedge [\phi_n(x) = \psi_n(x)] \Rightarrow [\phi(x) = \psi(x)]$$

where ϕ, ψ, ϕ_i and ψ_i are operations in the theory and x stands for an element of a product of sorts to which they all apply. Note that since we are using the whole theory (that is

0271-4132/89 $1.00 + $.25 per page

the full clone), irrelevant arguments may be added to operations so that the operations in the above clause have the same arguments. Of course, if a sort is empty, any equational sentence stated in terms of that sort is satisfied.

By a **generalized equational Horn theory** we mean an equational theory augmented by a set of conditions of the form

$$[\bigwedge(\phi_i(x) = \psi_i(x)] \Rightarrow [\phi(x) = \psi(x)],$$

allowing a possibly infinite conjunction.

We begin with,

2.1 Theorem. *The category of models of a (generalized) equational Horn theory is sketchable by a finite limit (resp. limit) sketch.*

Proof. We give the proof in the finite case. It is well known that the models of a multi-sorted equational theory can be sketched by an FP sketch, that is one that uses only the predicates of finite products. For any equational Horn sentence

$$(\bigwedge(\phi_i = \psi_i)) \Rightarrow (\phi = \psi)$$

the operations ϕ_i, ψ_i, ϕ and ψ are all operations on some object in the theory. Next we observe that it is only conventional notation that puts a meet in the antecedent. In fact, that antecedent is the same as the single equation

$$\langle \phi_1, \phi_2, \ldots, \phi_n \rangle = \langle \psi_1, \psi_2, \ldots, \psi_n \rangle.$$

In a sketch, there are no distinguished generating sorts anyway. Thus a Horn sentence can be given by $(\phi = \psi) \Rightarrow (\phi' = \psi')$. Now in any model, the set of solutions to $\phi = \psi$ is simply the equalizer of two arrows. If we add those two arrows (along with their source and target) to the sketch, we can also add their equalizer. We can do the same for $\phi' = \psi'$. We can then add an operation from the one equalizer to the other along with commutative diagrams that ensure that this operation is merely an inclusion between two subobjects of the same object. □

The following theorem appears to be closely related to the main result of [Banaschewski & Herrlich, 1976], except that the condition on filtered colimits is replaced by closure under ultraproducts, a trivial modification. See also the paper of Shafaat [1969].

2.2 Theorem. *Let C be an equational category and $D \subseteq C$ be a full subcategory. A necessary and sufficient condition that D be the category of models of a generalized equational Horn theory based on the operations of C is that D is closed under subobjects and products. If the theory of C is finitary, then D is the category of models for a Horn theory if and only if it is also closed under filtered colimits.*

Of very similar import is,

2.3 Theorem. *Let C be the category of models of an FL sketch (or of an FL theory). Then C is the category of models of a generalized equational Horn theory if and only if C*

has a regular generating class of regular projectives. The Horn theory can be taken to be finitary if and only if this generating set can be taken to be finite.

Proof. It is left to the reader to show that if $\mathcal{A}$ is the category of algebra for a multi-sorted equational theory and $\mathcal{C}$ is the full subcategory defined by an equational Horn theory, then $\mathcal{A}$ has enough projectives and their reflections in $\mathcal{C}$ are a sufficient set of projectives there.

Let I be a sufficient set of projectives. There is a functor $U : \mathcal{C} \to \mathbf{Set}^I$ that sends an object C of $\mathcal{C}$ to the I-indexed set $\mathrm{Hom}(P, C)$, $P \in I$. Since $\mathcal{C}$ is the category of models of an FL theory, it is complete and cocomplete, which implies that this functor has a left adjoint F, which takes the I-indexed set $\{S_P\}$ to the algebra $\sum_P S_P \cdot P$ where we let $S \cdot P$ denote the sum of S copies of P. The adjoint pair induces a triple on the category $\mathbf{Set}^I$ whose category of algebras we denote by $\mathcal{A}$. We have the standard structure/semantics comparison $\Phi : \mathcal{C} \to \mathcal{A}$ with left adjoint Ψ. Since I includes a generating family, the embedding is faithful. Also it is a full, by a theorem of Beck's which asserts that Φ is full and faithful if and only if every object of $\mathcal{C}$ is a regular quotient of a free algebra. (See [Barr & Wells, 1985], Section 3.3, Theorem 9 for single-sorted theories; the general case offers no additional problem.)

Now we have $\mathcal{C}$ embedded as a full reflective subcategory of a multi-sorted equational category; moreover, $\mathcal{C}$ contains the free algebras. What we will do is show that given an algebra A which is not (isomorphic to an) algebra of the subcategory $\mathcal{C}$, then there is a generalized Horn clause that is not satisfied by A and is satisfied by every object of $\mathcal{C}$.

Consider a presentation of A,

$$F_1 \underset{d^1}{\overset{d^0}{\rightrightarrows}} F_0 \xrightarrow{d} A$$

with F_0 and F_1 free, say F_0 free on the I-indexed set X and F_1 free on the I-indexed set Y. The pair of arrows d^0 and d^1 induces one arrow $\langle d^0, d^1\rangle : F_1 \to F_0 \times F_0$. Each generator y of F_1 is taken by this map to a pair $\langle \phi_y(x), \psi_y(x)\rangle$ of elements of F_0. These elements are words in the free algebra generated by the elements $x \in X$. Let K be the kernel pair of the induced arrow $F_0 \to \Phi\Psi(A)$. Then since A is not in the image of Φ, there is an element $\langle \phi(x), \psi(x)\rangle \in K$, which is not in the image of $\langle d^0, d^1\rangle$. Thus A does not satisfy the generalized Horn clause

$$\bigwedge_y (\phi_y(x) = \psi_y(x)) \Rightarrow (\phi(x) = \psi(x)). \tag{*}$$

I claim that this clause is satisfied by every object of $\mathcal{C}$. In fact, let C be an object of $\mathcal{C}$. Choose a set of elements $c_x \in C$ in such a way that c_x has the same sort as x and so that every one of the equations $\phi_y(c_x) = \psi_y(c_x)$ is satisfied. There is a homomorphism $f : F_0 \to C$ that takes x to c_x by the property of free algebras. The fact that $\phi_y(c_x) = \psi_y(c_x)$ is equivalent to $(f \circ d^0)(y) = (f \circ d^1(y)$. Since this is true for all the generators y of F_1, it follows that $f \circ d^0 = f \circ d^1$ and so f factors through A. But the adjunction property of Ψ implies that f factors through $\Phi\Psi(A)$ and hence that $f(\phi(x) = f(\psi(x)$ and hence that $\phi(c_x) = \psi(c_x)$. Thus any tuple of elements that satisfies all $\phi_y(x) = \psi_y(x)$ also satisfies $\phi(x) = \psi(x)$, so that (*) is satisfied by every object of $\mathcal{C}$.

If the projectives are small, $\mathcal{C}$ is closed in $\mathcal{A}$ under filtered colimits. To see this, we first note that the projectives are small in $\mathcal{C}$ and so the functors they represent commute with filtered colimits. But then the triple induce also commute with filtered colimits and then the underlying functors they represent on $\mathcal{A}$ also commute with filtered colimits and hence the representing objects are still small. Now if $C = \operatorname{colim} C_i$ is a filtered colimit in $\mathcal{C}$ and $A = \operatorname{colim} \Phi C_i$ is the colimit in $\mathcal{A}$ then for any $P \in X$, we have

$$\begin{aligned}\operatorname{Hom}(\Phi P, A) &\cong \operatorname{Hom}(\Phi P, \operatorname{colim} \Phi C_i) \cong \operatorname{colim} \operatorname{Hom}(\Phi P, \Phi C_i) \cong \operatorname{colim} \operatorname{Hom}(P, C_i) \cong \\ &\operatorname{Hom}(P, \operatorname{colim} C_i) \cong \operatorname{Hom}(P, C) \cong \operatorname{Hom}(\Phi P, \Phi C)\end{aligned}$$

whence $\Phi C \cong A$.

This implies that Φ commutes with filtered colimits, while Ψ, being a left adjoint, commutes with all colimits.

Repeat the proof using only finitely presented algebras A with the free algebras all free on finite sets. We will find a set of finite (since the elements y range over a finite set) Horn clauses that are satisfied by a finitely presented algebra if and only if the algebra is isomorphic to an algebra coming from $\mathcal{C}$.

Now let A be an arbitrary object of $\mathcal{A}$ that satisfies all the Horn sentences. Like any object of $\mathcal{A}$, A is a filtered colimit of finitely presented algebras. Write $A = \operatorname{colim} A_i$ with each A_i finitely presented. Exactly as in the proof of the first part, that fact that A satisfies all those Horn clauses implies that each $A_i \to A$ factors $A_i \to \Phi\Psi(A) \to A$. Then we have a diagram for each i

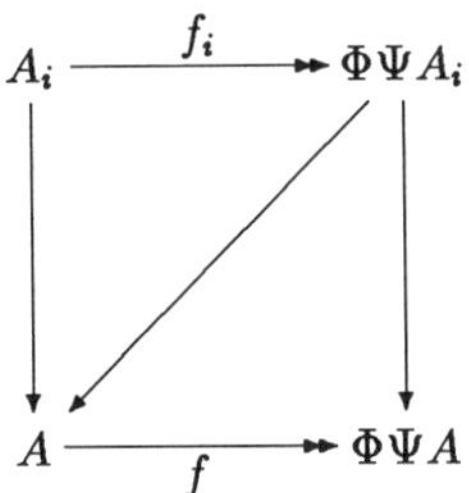

where the horizontal arrows are epic. The upper triangle commutes and the lower one does too because the upper arrow is epic. Taking colimits, we get

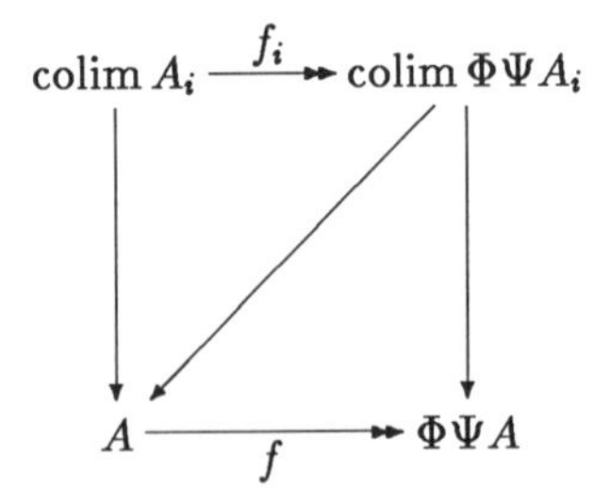

From the fact that both vertical arrows are isomorphisms, it follows by an easy diagram chase that the diagonal is too. Thus A is a filtered colimit of objects of $\mathcal{C}$ and hence is one too. □

As a consequence, we can conclude that neither the category of posets nor of small categories can be the category of models of a generalized equational Horn theory. In fact, in each case it is easy to see that the only projectives are discrete (posets or categories). For if P is a projective, let a be an object of P that is the target of a non-identity arrow. Construct a new category Q by adding a new object b and a single arrow $a \to b$, subject to no equations so that for any object c of P, the induced $\mathrm{Hom}(c,a) \to \mathrm{Hom}(c,b)$ is an isomorphism, while $\mathrm{Hom}(b,c) = \emptyset$. If $\mathit{2}$ is the category with two objects and one arrow between and no other non-identity arrow, then Q is the regular quotient of $P+\mathit{2}$ gotten by identifying a with the head of the non-identity arrow of $\mathit{2}$. The functor $P \to Q$ that takes the object a to b and everything else to itself does not lift to an arrow $P \to P+\mathit{2}$. On the other hand, any regular quotient of a discrete is discrete and so the regular projectives are not a regular generating family.

2.4 For the record, here is an example of a finitary equational theory and a full subcategory closed under products and subobjects not closed under filtered colimits and hence not the category of models of an ordinary Horn theory, at least not one based on the given operations.

The equational theory is the category of commutative rings with countably many constants $a_1, a_2, \ldots$ adjoined. Consider the generalized Horn theory

$$(\bigwedge a_i x = 0) \Rightarrow (x = 0).$$

Let R be the free ring on one variable, which may be thought of as the ring of polynomials $R = Z[x, a_1, a_2, \ldots]$. In the filtered sequence of rings

$$R/(a_1 x) \to R/(a_1 x, a_2 x) \to \cdots$$

each of these rings is a model of the theory. It satisfies neither the antecedent nor the consequent of the sentence. The colimit, on the other hand, satisfies the antecedent, but continues not to satisfy the consequent and is thus not a model.

3 Relational Horn theories

We saw in the last section that the discrete posets (resp. categories) are a generating family (in the sense of representing a faithful family of functors), but not a regular generating family, for the category of posets, but not for categories. Since posets are the models of a Horn theory, this suggests it may be possible to characterize the categories of models of a Horn theory by the existence of a family of regular projectives that generate. This may be so, but until now I have been able to verify only one half of this conjecture (but it is the half that establishes that **Cat** is not the category of models of a Horn theory and demonstrates that FL theories have more expressive power).

In general, a Horn theory allows relations and predicates involving those relations. I am unaware of any intrinsic characterization (one that is independent of particular choices of sorts) of such categories analogous to those for equational categories. That is the main reason the results are only partial. C. Lair claims that the answer to every interesting question involving sketches in a series of articles appearing in the "publication" **Diagrammes**, so all the answers can no doubt be found there.

3.1 Theorem. *The category of models of a universal Horn theory has a generating family of regular projectives.*

Proof. Let $\mathcal{C}$ be the category of models of a universal Horn theory. Let $\mathcal{A}$ and $\mathcal{B}$ be the categories of models of the corresponding equational and relational theories, resp. That is, the algebras in $\mathcal{A}$ have all the operations and satisfy all the equations that of the theory and those of $\mathcal{B}$ have additionally been equipped with the relations. We note that we have $\mathcal{C} \subseteq \mathcal{B} \subseteq \mathcal{A}$, with the first inclusion being full. Now let $C_1 \twoheadrightarrow C_2$ be a regular epi among objects of $\mathcal{C}$. I claim that this is also a regular epi in $\mathcal{A}$. In fact, for each sort s of the theory, let $C_3(s)$ be the image of the arrow $C_1(s) \to C_2(s)$. This image is, as is well known, a model of the underlying equational theory. It can be made into a model of the relational theory by letting, for a relation $\rho \subseteq s_1 \times \cdots \times s_n$,

$$C_3(\rho) = C_1(\rho) \cap [C_3(s_1) \times \ldots \times C_3(s_n)].$$

Call such a subalgebra of an algebra in $\mathcal{B}$ *full*. Then it is easy to see that every full subalgebra of an algebra in $\mathcal{C}$ is still in $\mathcal{C}$. Moreover the arrow $C_1 \to C_2$ factors through C_3 and since it was assumed a regular epi, it follows that $C_3 = C_2$.

Next we observe that the underlying functors for the sorts of the operations are representable (and in fact, have adjoints). Since these functors preserve regular epis, the representing objects are regular projectives.

Finally we note that an arrow is monic if and only if it is injective on all sorts. In other words that the inclusion of $\mathcal{C}$ into $\mathcal{A}$ reflects monos. It certainly preserves them, in fact has a left adjoint. But since arrows in $\mathcal{C}$ are functions of the sorts, a non-mono is also not injective on a sort. But then the kernel pair (which is preserved by the inclusion) provides two distinct arrows that are identified by the given arrow. Thus the evaluations at sorts, represented by the regular projectives, are collectively faithful and so the representing objects are a generating family. □

3.2 As observed above, the only projectives in the category of small categories are the discrete ones, which represent the set of objects functor (and its powers) and these are not faithful since two functors can agree on objects and not on arrows. A homomorphism on a poset is determined by its value on elements.

The converse could be verified if I could show that given a category of models of a coherent theory and homomorphisms that preserve positive sentences and given a faithful set of representable functors, there is a coherent theory *on that given set of functors* for which the given category is equivalent to the category of models. The point is that in model theory, the sorts are generally prescribed beforehand, whereas we are looking into an intrinsic characterization.

References

B. Banaschewski and H. Herrlich, *Subcategories defined by implication*, **Houston J. Math. 2** (1976), 149-171.

M. Barr, C. Wells, **Toposes, Triples and Theories.** Springer-Verlag, 1985.

J. Makowsky, *Why Horn formulas matter in computer science: initial structures and generic examples.* Technical Report #329, Department of Computer Science, Technion, Haifa, Israel, 1984.

A. Shafaat, *On implicationally defined classes of algebras*, **J. London Math. Soc. 44** (1969), 137-140.

H. Volger, *On theories which admit initial structures.* Preprint: Universität Passau, 1987.

MICHAEL BARR
DEPARTMENT OF MATHEMATICS & STATISTICS
MCGILL UNIVERSITY
805 SHERBROOKE ST. WEST
MONTREAL, QUEBEC
CANADA H3A 2K6

Contemporary Mathematics
Volume 92, 1989

GEOMETRIC INVARIANCE OF EXISTENTIAL FIXED-POINT LOGIC

Andreas Blass[1]

ABSTRACT. We prove that the truth values of existential fixed-point formulas are preserved by the inverse image parts of geometric morphisms of topoi. We apply this result to establish several other properties of existential fixed-point logic.

Let L be a first-order language, and let φ be a formula of higher-order logic over L, such that only the first-order variables $x_1, \cdots, x_n$ are free in φ. We say that φ is <u>geometrically invariant</u> if, whenever $\mathfrak{A}$ is an L-structure in an elementary topos $\mathcal{E}$ and $f:\mathcal{F} \longrightarrow \mathcal{E}$ is a geometric morphism of topoi, then $\|\varphi\|_{f^*(\mathfrak{A})} = f^*(\|\varphi\|_{\mathfrak{A}})$. Here $\|\varphi\|_{\mathfrak{A}}$ is the subobject of A^n defined by φ in the internal logic of $\mathcal{E}$ (so the higher-order variables in φ are interpreted as ranging over the full power sets as defined in $\mathcal{E}$) and $f^*(\mathfrak{A})$ is the L-structure in $\mathcal{F}$ obtained by applying the left adjoint part f^* of f to the structure $\mathfrak{A}$, i.e., to its universe A and to all its predicates and operations.

We shall show that all formulas of existential fixed-point logic, a fragment of second-order logic that has arisen naturally in connection with certain aspects of theoretical computer science [5,7], are geometrically invariant. Combining this result with general facts about geometrically invariant formulas, we deduce some additional properties of existential fixed-point formulas.

1980 Mathematics Subject Classifications 03C80, 03G30, 18B25.
[1]Partially supported by NSF grant DMS8501752

The syntax of existential fixed-point logic has atomic formulas, conjunction, disjunction, and existential quantification, all defined as in ordinary first-order logic, plus the least-fixed-point construction

(1) $\quad$ LET $Q_1(\underset{\sim}{x}^1) \longleftarrow \delta_1, \dots, Q_r(\underset{\sim}{x}^r) \longleftarrow \delta_r$ THEN φ,

where each Q_i is a new n_i-place predicate symbol, each $\underset{\sim}{x}^i$ is an n_i-tuple of distinct variables, and the δ_i's and φ are formulas of the language including the Q_i's. A variable is free in (1) if it is either free in some δ_i and absent from the corresponding $\underset{\sim}{x}^i$ or free in φ. The meaning of (1) is that φ becomes true if the Q_i's are interpreted as the least fixed-point of the simultaneous recursion defined by the δ_i's. For brevity, we give details only for the case $r = 1$, referring to [5] for the general case.

Let an L-structure $\mathfrak{A}$ and values in its universe A for all the free variables of

(2) $\quad$ LET $Q(\underset{\sim}{x}) \longleftarrow \delta$ THEN φ

be given. Define a monotone operator Δ on n-ary relations $X \subseteq A^n$ by

$$\Delta(X) = \{\underset{\sim}{a} \in A^n \mid \delta \text{ holds in } \mathfrak{A} \text{ when } Q \text{ is interpreted as } X \text{ and } \underset{\sim}{x} \text{ is interpreted as } \underset{\sim}{a}\},$$

and let $\Delta^\infty \subseteq A^n$ be the least fixed-point of Δ. Then (2) holds in $\mathfrak{A}$ (with the given values of its free variables) if and only if φ holds in $\mathfrak{A}$ when Q is interpreted as Δ^∞.

As in the preceding paragraph, we shall often consider only those LET...THEN... constructions where only a single predicate Q is defined recursively, and indeed we shall often assume that Q is unary. All this is only for notational simplicity; our arguments apply equally to the general LET...THEN... construction (1).

In [5], a limited use of negation was allowed in existential fixed-point formulas. For our present purposes, it is more convenient to exclude negation, but we shall discuss later how to extend our results to the framework of [5].

Since we have not allowed negation, all existential fixed-point formulas are monotone with respect to all their predicate symbols. This implies that (2) is equivalent to the second-order formula

$$\forall Q \; [\forall \underset{\sim}{x}(\delta \Rightarrow Q(\underset{\sim}{x})) \Rightarrow \varphi],$$

and similarly for (1). So existential fixed-point logic is a fragment of second-order logic and can therefore be interpreted in any elementary topos [8, 11].

THEOREM 1. _Every existential fixed-point formula is geometrically invariant._

Proof. We proceed by induction on the total number of connectives, quantifiers, and fixed-point operators in the formula. The cases of atomic formulas, conjunctions, disjunctions, and existential quantifications are well known; see [11] or [13]. So we need only consider a formula $\psi(\underset{\sim}{z})$ of the form

$$\text{LET } Q(x) \longleftarrow \delta(Q,\underset{\sim}{z},x) \text{ THEN } \varphi(Q,\underset{\sim}{z}),$$

where we have explicitly exhibited the presence of Q and the n free variables $\underset{\sim}{z}$ of ψ in δ and φ. The general idea for proving that ψ is geometrically invariant is to invoke Theorem 6 of [4], which says roughly that the left-adjoint parts of geometric morphisms preserve inductively constructed objects. The present context does not exactly match that of [4], and in fact [4] will give us only an inclusion, $f^*(\|\psi\|_{\mathfrak{A}}) \subseteq \|\psi\|_{f^*(\mathfrak{A})}$, where we want equality. Fortunately, the reverse inclusion will be relatively easy to verify.

Let an L-structure $\mathfrak{A}$ in the topos $\mathcal{E}$ be given. We begin by describing $\|\psi\|_{\mathfrak{A}}$, using the internal logic of $\mathcal{E}$. Until further notice, we work in the internal logic.

By definition, $\psi(\underset{\sim}{z})$ holds (in $\mathfrak{A}$, for specified $\underset{\sim}{z}$) if and only if

$$\forall Q \subseteq A \; [\text{Closed}(Q,\underset{\sim}{z}) \Longrightarrow \varphi(Q,\underset{\sim}{z})],$$

where $\text{Closed}(Q,\underset{\sim}{z})$ means that

$$\forall x \in A \; [\delta(Q,\underset{\sim}{z},x) \Longrightarrow Q(x)],$$

i.e., that Q is closed under the operator Δ defined earlier. Because δ is monotone, there is, for each $\underset{\sim}{z}$, a smallest such Q, namely the intersection of all such Q's. (The usual proof of this works in intuitionistic type theory, hence in the internal logic of $\mathcal{E}$.) Thus, there is an $(n+1)$-place predicate $C \subseteq A^{n+1}$ such that, for all $\underset{\sim}{z}$, $\{x | C(\underset{\sim}{z}, x)\}$ is the smallest Q such that Closed$(Q, \underset{\sim}{z})$. Therefore, by monotonicity of φ,

$$(3) \qquad \psi(\underset{\sim}{z}) \iff \varphi[C(\underset{\sim}{z}, -), \underset{\sim}{z}]$$

where the right side means the result of replacing every $Q(t)$ in $\varphi(Q, \underset{\sim}{z})$ with $C(\underset{\sim}{z}, t)$, renaming bound variables if necessary to avoid clashes.

We record for future reference that C is the smallest subset of A^{n+1} that satisfies, for all x and $\underset{\sim}{z}$

$$(4) \qquad \delta[C(\underset{\sim}{z}, -), \underset{\sim}{z}, x] \Longrightarrow C(\underset{\sim}{z}, x).$$

In order to compare the phenomena just discussed in $\mathcal{E}$ with their analogues in $\mathcal{F}$, via a geometric morphism $f: \mathcal{F} \longrightarrow \mathcal{E}$, we leave the internal logic. From the external point of view, we have, according to (3),

$$(5) \qquad \|\psi(\underset{\sim}{z})\|_{\mathfrak{A}} = \|\varphi[Q'(\underset{\sim}{z}, -), \underset{\sim}{z}]\|_{\mathfrak{A}, C},$$

where Q' is a new $(n+1)$-place predicate symbol and "$\mathfrak{A}, C$" means $\mathfrak{A}$ with Q' interpreted as C, and where C is the smallest subobject of A^{n+1} such that

$$(6) \qquad \|\delta[Q'(\underset{\sim}{z}, -), \underset{\sim}{z}, x]\|_{\mathfrak{A}, C} \leq C.$$

(Note that C, having been explicitly defined in the internal logic, exists externally as a subobject of A^{n+1}.) Of course, in the structure $f^*(\mathfrak{A})$ in $\mathcal{F}$, there is a smallest subobject C' of $f^*(A)^{n+1}$ such that

$$(6') \qquad \|\delta[Q'(\underset{\sim}{z}, -), \underset{\sim}{z}, x]\|_{f^*(\mathfrak{A}), C'} \leq C',$$

and we have

$$(5') \qquad \|\psi(\underset{\sim}{z})\|_{f^*(\mathfrak{A})} = \|\varphi[Q'(\underset{\sim}{z}, -), \underset{\sim}{z}]\|_{f^*(\mathfrak{A}), C'}$$

Since $\varphi[Q'(\underset{\sim}{z}, -), \underset{\sim}{z}]$ and $\delta[Q'(\underset{\sim}{z}, -), \underset{\sim}{z}, x]$ have lower complexity than $\psi(\underset{\sim}{z})$, they are geometrically invariant by induction hypothesis. Thus, we infer from (5) that

$$f^*(\|\psi(\underset{\sim}{z})\|_{\mathfrak{A}}) = \|\varphi[Q'(\underset{\sim}{z}, -), \underset{\sim}{z}]\|_{f^*(\mathfrak{A}), f^*(C)}.$$

Comparing this with (5′), we see that the theorem will be proved if we establish that $f^*(C) = C'$.

Half of this is easy, since (6) and the geometric invariance of $\delta[Q'(\underset{\sim}{z},-),\underset{\sim}{z},x]$ give

$$\|\delta[Q'(\underset{\sim}{z},-),\underset{\sim}{z},x]\|_{f^*(\mathfrak{A}),f^*(C)} \le f^*C.$$

Since C' is the smallest subobject of $f^*(A)^{n+1}$ satisfying (6′), it follows that $C' \le f^*(C)$.

To prove the reverse inclusion, we shall use Theorem 6 of [4]. To apply this theorem, we define an inductive construction (in the sense of [4]) (C,S,P,R) as follows, working in the internal logic of $\mathcal{E}$. The ambient set C has already been defined. The set S of construction steps consists of all $(X,\underset{\sim}{z},x) \in \mathcal{P}(C) \times C$ such that $\delta[X(\underset{\sim}{z},-),\underset{\sim}{z},x]$ holds. The prerequisites of $(X,\underset{\sim}{z},x)$ are the members of X, and its result is $(\underset{\sim}{z},x)$, i.e.,

$$P = \{[(\underset{\sim}{w},y),\ (X,\underset{\sim}{z},x)] \mid (\underset{\sim}{w},y) \in X\}, \text{ and}$$

$$R(X,\underset{\sim}{z},x) = (\underset{\sim}{z},x).$$

An easy calculation shows that a subobject K of C is closed for this inductive construction if and only if

$$\|\delta[Q'(\underset{\sim}{z},-),\underset{\sim}{z},x]\|_{\mathfrak{A},K} \le K.$$

So C is the only closed set, i.e., the inductive construction (C,S,P,R) is total. By Theorem 6 of [4], $(f^*(C),\ f^*(S),\ f^*(P),\ f^*(R))$ is also total. So in order to show that $C' = f^*(C)$ and thus finish the proof of the theorem, we need only show that C' is closed for $(f^*(C),\ f^*(S),\ f^*(P),\ f^*(R))$.

Consider the two-sorted structure $\mathfrak{A}^+$ obtained by adding to $\mathfrak{A}$:

S as the second domain,

$P \subseteq A^{n+1} \times S$ as an $(n+2)$-place relation, and

the $n+1$ projections $p_1,\cdots,p_n$, q sending $(X,\underset{\sim}{z},x) \in S$ to $\underset{\sim}{z}$ and x.

By definition of S, the formula (with s of sort S free)

$$\delta[P(\underset{\sim}{p}(s),-,s),\ \underset{\sim}{p}(s),\ q(s)]$$

has truth value in $\mathfrak{A}^+$ equal to all of S. Since this formula has lower complexity than ψ, the induction hypothesis ensures that it has truth value $f^*(S)$ in $f^*(\mathfrak{A}^+)$. That is, it is internally valid in $\mathcal{F}$ that

$$(7) \qquad (\forall s \in S)\ \delta[f^*(P)(\underset{\sim}{p}'(s),-,s),\ \underset{\sim}{p}'(s),\ q'(s)],$$

where we have written $\underset{\sim}{p}'$ and q' for the projections in $\mathcal{F}$ (since f^* preserves products).

We can now verify that C' is closed for $(f^*(C), f^*(S), f^*(P), f^*(R))$ by arguing as follows in the internal logic of $\mathcal{F}$. Suppose $s \in f^*(S)$ and

$$(8) \qquad \forall c \in f^*(C)[(c,s) \in f^*(P) \Longrightarrow c \in C'];$$

we must show $f^*(R)(s) \in C'$. Writing s as $(X,\underset{\sim}{z},x)$, we have, by (7),

$$\delta[f^*(P)(\underset{\sim}{z},-,s),\ \underset{\sim}{z},x],$$

and therefore by (8) and the monotonicity of δ,

$$\delta[C'(\underset{\sim}{z},-),\ \underset{\sim}{z},x\ ].$$

By (6′), we have $(\underset{\sim}{z},x) \in C'$, i.e., $f^*(R)(s) \in C'$, as desired. □

REMARK. If one restricts attention to complete topoi, then it is easier to prove Theorem 1 by reducing it to the known geometric invariance of infinitary geometric formulas. Given an existential fixed-point construct LET $Q(x) \longleftarrow \delta(Q,x)$ THEN $\varphi(Q)$, we define approximations $\delta^\alpha(x)$ to the recursively defined predicate $Q(x)$ by

$\delta^0(x)$ is false

$\delta^{\alpha+1}(x)$ is $\delta(\delta^\alpha(-),x)$

$\delta^\lambda(x)$ is $\bigvee_{\alpha<\lambda} \delta^\alpha(x)$ for limit ordinals λ.

Then, for all sufficiently large α, LET $Q(x) \longleftarrow \delta(Q,x)$ THEN $\varphi(Q)$ holds in $\mathfrak{A}$ (with specified values for its free variables) if and only if $\varphi(\delta^\alpha(-))$ holds. Applying this observation to all of the LET...THEN... constructions in a formula ψ, we obtain infinitary geometric formulas ψ_α such that each structure $\mathfrak{A}$, in any topos, satisfies $\psi \Longleftrightarrow \psi_\alpha$ for all sufficiently large α. (_A priori_, "sufficiently large" could depend on $\mathfrak{A}$, but see [5], Theorem 9.) Since each ψ_α is geometrically invariant, so is ψ.

Theorem 1 establishes for existential fixed-point formulas a property that was well-known for so-called _geometric_ or _coherent_ formulas, i.e., first-order formulas built from atomic formulas using only conjunction, disjunction, and existential quantification [11,13]. It allows us to extend to existential fixed-point formulas certain other known properties of geometric formulas. For example, let us define (in analogy to geometric sequents [11,13]) an _existential fixed-point sequent_ to be a formula of the form

$$\forall \underset{\sim}{x}(\varphi \Longrightarrow \psi)$$

where φ and ψ are existential fixed-point formulas and where the tuple $\underset{\sim}{x}$ includes all the free variables of φ and ψ. An _existential fixed-point theory_ is a theory axiomatized by existential fixed-point sequents.

COROLLARY 2. _If_ $f:\mathcal{F} \longrightarrow \mathcal{E}$ _is a geometric morphism and if_ $\mathfrak{A}$ _is a model in_ $\mathcal{E}$ _of an existential fixed-point theory_ T, _then_ $f^*(\mathfrak{A})$ _is a model (in_ $\mathcal{F}$_) of_ T.

Proof. For each axiom $\forall x(\varphi(x) \Longrightarrow \psi(x))$ of T, we have, since the axiom holds in $\mathfrak{A}$, $\|\varphi(\underset{\sim}{x})\|_{\mathfrak{A}} \leq \|\psi(\underset{\sim}{x})\|_{\mathfrak{A}}$. As f^* preserves $\leq$, Theorem 1 gives us that $\|\varphi(\underset{\sim}{x})\|_{f^*(\mathfrak{A})} \leq \|\psi(\underset{\sim}{x})\|_{f^*(\mathfrak{A})}$, so the axiom in question holds in $f^*(\mathfrak{A})$. □

COROLLARY 3. _If an existential fixed-point sequent_ $\forall \underset{\sim}{x}(\varphi \Longrightarrow \psi)$ _is deducible from an existential fixed point theory_ T _in classical higher-order logic, then it is also deducible from_ T _in intuitionistic higher-order logic._

Proof. By [8] it suffices to show that $\|\varphi\|_{\mathfrak{A}} \leq \|\psi\|_{\mathfrak{A}}$ whenever $\mathfrak{A}$ is a model of T in any topos $\mathcal{E}$. Let $\mathcal{E}$ and $\mathfrak{A}$ be given, and, by Barr's theorem [2, 11(Prop. 7.54)], let $f:\mathcal{F} \longrightarrow \mathcal{E}$ be a surjective geometric morphism (which means that f^* reflects isomorphisms) with $\mathcal{F}$ a Boolean topos. By Corollary 2, $f^*(\mathfrak{A})$ is a model of T. As classical logic is internally valid in $\mathcal{F}$ and $\forall \underset{\sim}{x}(\varphi \Longrightarrow \psi)$ is classically deducible from T, $f^*(\mathfrak{A})$ must satisfy $\forall \underset{\sim}{x}(\varphi \Longrightarrow \psi)$. This means that the inclusion of $\|\varphi \wedge \psi\|_{f^*(\mathfrak{A})}$ in $\|\varphi\|_{f^*(\mathfrak{A})}$ is an isomorphism. But this inclusion is, according to Theorem 1, the image under f^* of the inclusion $\|\phi \wedge \psi\|_{\mathfrak{A}}$ in $\|\varphi\|_{\mathfrak{A}}$. As f^* reflects isomorphisms, the latter inclusion is also an isomorphism. So $\mathfrak{A}$ satisfies $\forall \underset{\sim}{x}(\varphi \Longrightarrow \psi)$. □

The corollary just proved is applicable to a version of Hoare's logic of asserted programs [1,10]. An asserted program has the form {p}S{q} where p and q are formulas and S is a program; its meaning is that if p is true of the initial state of a run of S and if that run terminates, then q is true of the final state. We let S range over while-programs with (parameterless) recursive procedures, as in Section 3 of [1]. As suggested in [5], we take the formulas p and q to be existential fixed-point formulas. (One of the main points in [5] is that this choice leads to a smoother theory than the usual choice, first-order formulas.) It then follows from Theorem 1 of [5] that the asserted program {p}S{q} can be expressed by an existential fixed-point sequent. Thus, by Corollary 3 above, the law of the excluded middle is conservative for deductions of asserted programs from any existential fixed-point theory; that is, if such a deduction can be carried out using the law of the excluded middle, then it can also be carried out without this law.

Corollary 3 provides further evidence for the thesis that existential fixed-point logic is intimately tied to computation, for it says that, in a certain sense, whatever can be proved at all can be proved constructively.

COROLLARY 4. *Let* T *be an existential fixed-point theory with finitely many axioms, and let* $\mathcal{S}$ *be any elementary topos with a natural numbers object. Then there exists a clasifying topos for* T *over* $\mathcal{S}$.

Recall [3,11,15] that a classifying topos for T over $\mathcal{S}$ is a topos $\mathcal{E}$ defined over $\mathcal{S}$ (i.e., equipped with a geometric morphism $\mathcal{E} \longrightarrow \mathcal{S}$) such that, for any topos $\mathcal{F}$ defined over $\mathcal{S}$, the category of geometric morphisms over $\mathcal{S}$ from $\mathcal{F}$ to $\mathcal{E}$ (and natural transformations between their left adjoint parts) is equivalent to the category of models of T in $\mathcal{F}$. We omit the proof of Corollary 4, since it is exactly like one of the standard proofs [3,11,15] for the existence of classifying topoi of finitely axiomatized geometric theories. (It may, however, be worth remarking that this argument uses Theorem 1, not merely Corollary 2. There exist properties of structures such that (i) if $\mathfrak{A}$ has the property, then so does $f^*(\mathfrak{A})$ for all geometric morphisms f, but (ii) there is no classifying topos for structures having the property. Examples of such properties are K-finiteness and well-foundedness; see [14] and [4].)

It was shown in [5], Theorem 7, that, whenever an existential fixed-point formula holds of some elements in a structure $\mathfrak{A}$, then this is so "because of" some finite subset of $\mathfrak{A}$. We shall show that this property of existential fixed-point formulas is a consequence of their geometric invariance and that it holds in all topoi with natural numbers objects.

THEOREM 5. Let $\mathfrak{A}$ be a structure in a topos $\mathcal{S}$ with a natural numbers object, and let $\varphi(\underset{\sim}{z})$ be a geometrically invariant formula. Then, in the internal logic of $\mathcal{S}$, the following assertions (about elements $\underset{\sim}{z}$ of $\mathfrak{A}$) are equivalent.

(a) $\varphi(\underset{\sim}{z})$ holds in $\mathfrak{A}$.

(b) There exist a finitely presented structure $\mathfrak{B}$, elements $\underset{\sim}{y}$ satisfying $\varphi(\underset{\sim}{y})$ in $\mathfrak{B}$, and a homomorphism $\mathfrak{B} \longrightarrow \mathfrak{A}$ sending $\underset{\sim}{y}$ to $\underset{\sim}{z}$.

Before proving the theorem, we make some remarks about it.

(1) We assume that the language L of our structures consists only of the non-logical symbols actually occuring in $\varphi(\underset{\sim}{z})$, or at least that L is finite. Then the concept of a finitely presented structure makes sense in the internal logic of a topos, as explained in [12]. In fact, there is an object of finite presentations of L-structures, so the quantification in (b) is legitimate.

(2) The theorem implies that, if $h:\mathfrak{A} \longrightarrow \mathfrak{A}'$ is a homomorphism, then $\|\varphi(\underset{\sim}{z})\|_{\mathfrak{A}} \le \|\varphi(h(\underset{\sim}{z}))\|_{\mathfrak{A}'}$. Indeed, (b) is obviously preserved by homomorphisms, and therefore so is (a).

Proof of Theorem 5. Since $\mathcal{S}$ has a natural numbers object, the classifying topos $\mathcal{S}[\mathfrak{G}]$ for L-structures can be obtained as the category of internal $\mathcal{S}$-valued funtors on the internal category $\mathbb{C}$ of finitely presented L-structures. The generic L-structure $\mathfrak{G}$ is the underlying set functor on $\mathbb{C}$. The usual proof [12] that this functor category is the classifying topos provides the following explicit construction of the left-adjoint part f^* of the geometric morphism $f:\mathcal{S} \longrightarrow \mathcal{S}[\mathfrak{G}]$ classifying an L-structure $\mathfrak{A}$ in $\mathcal{S}$. For any $X \in \mathcal{S}[\mathfrak{G}]$, $f^*(X)$ is obtained from the internal disjoint union

$$\coprod_{\mathfrak{B} \text{ object of } \mathbb{C}} X(\mathfrak{B}) \times \mathrm{Hom}(\mathfrak{B},\mathfrak{A})$$

(where Hom denotes the set of L-structure homomorphisms) by

identifying (x, hg) with $(X(g)(x), h)$ whenever $g: \mathfrak{B}_1 \longrightarrow \mathfrak{B}_2$ is a morphism of $\mathbb{C}$, $x \in X(\mathfrak{B}_1)$, and $h \in \mathrm{Hom}(\mathfrak{B}_2, \mathfrak{A})$.

We shall apply this with $X = \|\varphi(z)\|_{\mathfrak{G}}$. In this case, $X(\mathfrak{B})$ admits a simple description. (The following incorrect argument, to be corrected below, gives the essential idea. For each object $\mathfrak{B}$ of $\mathbb{C}$, the evaluation functor $\mathcal{S}[\mathfrak{G}] \longrightarrow \mathcal{S}$ sending every C to $X(\mathfrak{B})$ is the left adjoint part b^* of a geometric morphism b, because it has adjoints (Kan extensions) on both sides. So

$$\|\varphi(\underset{\sim}{z})\|_{\mathfrak{G}}(\mathfrak{B}) = b^*\left(\|\varphi(\underset{\sim}{z})\|_{\mathfrak{G}}\right) = \|\varphi(\underset{\sim}{z})\|_{b^*(\mathfrak{G})} = \|\varphi(\underset{\sim}{z})\|_{\mathfrak{B}}.$$

The error in this argument lies in treating $\mathfrak{B}$, a variable (of type $\mathrm{Ob}(\mathbb{C})$) in the internal logic, as designating actual structures and thus giving rise to an (external) geometric morphism b. To correct the error, we "internalize" an appropriate part of the preceding argument.) Let $\mathbb{D}$ be the internal discrete category with the same objects as $\mathbb{C}$. The inclusion $\mathbb{D} \longrightarrow \mathbb{C}$ induces a geometric morphism b from $\mathcal{S}^{\mathbb{D}}$ to $\mathcal{S}[\mathfrak{G}] = \mathcal{S}^{\mathbb{C}}$; its left-adjoint part b^* acts on functors X in $\mathcal{S}^{\mathbb{C}}$ by restricting them to $\mathbb{D}$. By Theorem 1, $b^*\left(\|\varphi(\underset{\sim}{z})\|_{\mathfrak{G}}\right) = \|\varphi(\underset{\sim}{z})\|_{b^*(\mathfrak{G})}$. This equation, when expressed in the internal logic of $\mathcal{S}$, says that $\|\varphi(\underset{\sim}{z})\|_{\mathfrak{G}}(\mathfrak{B}) = \|\varphi(\underset{\sim}{z})\|_{\mathfrak{B}}$ for all objects $\mathfrak{B}$ of $\mathbb{D}$ (i.e., of $\mathbb{C}$).

Returning to our description of $f^*(X)$, where $f: \mathcal{S} \longrightarrow \mathcal{S}[\mathfrak{G}]$ classfies $\mathfrak{A}$, we find that $f^*\left(\|\varphi(\underset{\sim}{z})\|_{\mathfrak{G}}\right)$ which equals $\|\varphi(\underset{\sim}{z})\|_{\mathfrak{A}}$ by Theorem 1, is obtainable from

$$\coprod_{\mathfrak{B} \text{ object of } \mathbb{C}} \|\varphi(\underset{\sim}{z})\|_{\mathfrak{B}} \times \mathrm{Hom}(\mathfrak{B}, \mathfrak{A})$$

by making identifications as described above. This $\|\varphi(\underset{\sim}{z})\|_{\mathfrak{A}}$ is a subobject of A^n via f^* of the inclusion of $\|\varphi(\underset{\sim}{z})\|_{\mathfrak{G}}$ in G^n and the identification of $f^*(G^n)$, a quotient of

$$\coprod_{\mathfrak{B} \text{ object of } \mathbb{C}} B^n \times \mathrm{Hom}(\mathfrak{B}, \mathfrak{A}),$$

with A^n via the correspondence that sends $(\underset{\sim}{y}, h) \in B^n \times \mathrm{Hom}(\mathfrak{B}, \mathfrak{A})$ to $h(\underset{\sim}{y}) \in A^n$. Thus, as a subobject of A^n, $\|\varphi(\underset{\sim}{z})\|_{\mathfrak{A}}$ is the set of those $h(\underset{\sim}{y})$ such that $\underset{\sim}{y} \in \|\varphi(\underset{\sim}{z})\|_{\mathfrak{B}}$ for some finitely presented $\mathfrak{B}$ and h is a homomorphism from $\mathfrak{B}$ to $\mathfrak{A}$. □

If $\mathcal{S}$ is the category of sets (or, more generally, a Boolean topos with a natural numbers object) and if the (finite) language L has only relation and constant symbols, then the finitely presented L-structures are just the finite ones, and all their homomorphic images are also finite. So, in this situation, we can replace (b) of Theorem 2 with

(b′) $\varphi(\underset{\sim}{z})$ <u>holds in a finite substructure of</u> $\mathfrak{A}$ (<u>containing</u> $\underset{\sim}{z}$).

This replacement is clearly not legitimate if L has function symbols (other than constants), for then $\mathfrak{A}$ need not have any finite substructures. Its legitimacy can, however, be restored by simply replacing n-place function symbols by (n+1)-place relation symbols. Such a relation symbol would, in general, denote partial functions in the finite substructures considered in (b′), even though it denotes a total function on $\mathfrak{A}$. In this situation, the equivalence of (a) and (b′) for existential fixed-point formulas is essentially Theorem 7 of [5], except that we have not allowed any negations in our formulas.

To conclude this paper, we comment on the extension of our results to the case where certain predicate symbols of L are declared to be <u>negatable</u>, meaning that negations of atomic formulas built from these predicate symbols are permitted in existential fixed-point formulas. As in [5], we do not admit negations of compound formulas, and we do not permit negatable predicate symbols as the recursively defined symbols Q_i in the LET...THEN... construction (1). These restrictions ensure that universal quantification is not surreptitiously introduced (as negated existential quantification) and that the recursive definitions in LET...THEN... are monotone and therefore have least fixed-points.

To interpret this extended version of existential fixed-point logic in the intuitionistic environment of a topos, we require, as part of the definition of the concept of L-structure, that the interpretation in a structure of a negatable predicate be a complemented subobject of the domain of the structure. This convention ensures that negated atomic formulas, being interpreted as complements, are geometrically invariant, for the left-adjoint parts of geometric morphisms, though they need not preserve negations in general, must preserve complements. With this observation, the proofs of Theorem 1 and its corollaries immediately extend to the present situation, provided that, in Corollary 3, we include in intuitionistic logic the assumption that the negatable predicates are decidable, $\forall \underset{\sim}{x}(P(\underset{\sim}{x}) \vee \neg P(\underset{\sim}{x}))$.

The proof of Theorem 5 does not quite generalize to the new situation, because the classifying topos $\mathcal{S}[\mathcal{G}]$ for L-structures, in the new sense, need not be a presheaf topos. An L-structure in the new sense can be viewed as an $\overline{L}$-structure in the ordinary sense, where $\overline{L}$ consists of L plus a new predicate symbol $\overline{P}$ for each negatable P of L, subject to the axioms

(9) $\forall \underset{\sim}{x}\ (P(\underset{\sim}{x}) \wedge \overline{P}(\underset{\sim}{x}) \longrightarrow \text{false})$ and

(10) $\forall \underset{\sim}{x}\ (\text{true} \longrightarrow P(\underset{\sim}{x}) \vee \overline{P}(\underset{\sim}{x}))$.

The axioms (9) are universal Horn sentences, so their classifying topos is, by [6], the topos $\mathcal{E}$ of presheaves on the category of finitely presented models. But the classifying topos for (9) and (10) is the sheaf subtopos of $\mathcal{E}$ with respect to a topology in which, for each family presented structure $\mathfrak{B}$, each n-ary negatable P, and each n-tuple $\underset{\sim}{b}$ from B, the homomorphisms h from $\mathfrak{B}$ to structures in which $P(h(\underset{\sim}{b})) \vee \overline{P}(h(\underset{\sim}{b}))$ holds constitute a cover of $\mathfrak{B}$. If $\mathcal{S}$ is Boolean and L has no function symbols except constants (so the finitely presented structures are finite), then every finitely presented model is covered by ones in which the interpretations of P and $\overline{P}$ are complementary, and these latter structures have no non-trivial covers. In this situation, the comparison lemma [9] implies that the sheaf subtopos is a presheaf topos, and then the proof of Theorem 5 still works. If function symbols are present, we can replace them with predicate symbols, and we can add to (9) and (10) the universal Horn sentences saying that these relations are partial functions, but then the hypothesis of Theorem 5 must be strengthened to require that φ be geometrically invariant even when structures are admitted in which the function symbols are interpreted by partial functions.

This strengthened hypothesis holds when φ is an existential fixed-point formula, since its natural translation into the language with relation symbols in place of function symbols is again an existential fixed-point formula. Notice, however, that for this special case of existential fixed-point formulas, we could avoid (10) and the consequent need to pass to a sheaf topos. Replacing each negated atomic formula $\neg P(\underset{\sim}{t})$ in φ with $\overline{P}(t)$, we obtain an existential fixed-point formula $\overline{\varphi}$

of $\bar{L}$, without negation, and we can apply the original version of Theorem 5. The $\mathfrak{B}$ in part (b) of the theorem will, in general, have interpretations P and $\bar{P}$ disjoint (as we have kept (9)) but not necessarily complementary, even if they are complementary in $\mathfrak{A}$. In this situation, we can replace $\mathfrak{B}$ with a structure $\mathfrak{B}'$ in which P and $\bar{P}$ are complementary, by pulling back the P and $\bar{P}$ of $\mathfrak{A}$ along the homomorphism; $\bar{\varphi}(\underline{y})$ remains true, as $\bar{\varphi}$ is monotone in P and $\bar{P}$, but $\mathfrak{B}'$ need not be finitely presented.

Summarizing the preceding discussion, we have that Theorem 5 remains true when negatable predicate symbols are admitted, provided that, in the finitely presented structure $\mathfrak{B}$, we either (if $\mathcal{S}$ is Boolean) allow the function symbols other than constants to be interpreted as partial functions or allow the negation of P to be interpreted as a predicate disjoint from, but not necessarily complementary to, the interpretation of P.

REFERENCES

1. K. Apt, "Ten years of Hoare's logic: a survey--Part I," A.C.M. Trans. Prog. Lang. Syst. 3(1981), 431-483.

2. M. Barr, "Toposes without points," J. Pure Appl. Alg., 5 (1974), 265-280.

3. J. Bénabou, "Théories relatives à un corpus," C.R. Acad. Sci. Paris, 281 (1975), A831-834.

4. A. Blass, "Well-ordering and induction in intuitionistic logic and topoi," in Mathematical Logic and Theoretical Computer Science, eds. D.W. Kueker, E.G.K. Lopez-Escobar, and C.H. Smith, Marcel Dekker, Inc. (1987), 29-48.

5. A. Blass and Y. Gurevich, "Existential fixed-point logic," in Computation Theory and Logic, ed. E. Börger, Springer Lecture Notes in Computer Science 270 (1987), 20-36.

6. A. Blass and A. Scedrov, "Classifying topoi and finite forcing," J. Pure Appl. Alg., 28 (1983), 111-140.

7. A.K. Chandra and D. Harel, "Horn clause queries and generalizations," J. Logic Programming, 1985:1:1-15.

8. M. Fourman, "The logic of topoi," in Handbook of Mathematical Logic, ed. J. Barwise, North-Holland (1977), 1053-1090.

9. A. Grothendieck and J. Verdier, Séminaire de Géometrie Algébrique IV, Tome 1, Springer Lecture Notes in Mathematics, 269 (1972).

10. C.A.R. Hoare, "An axiomatic basis for computer programming," Comm. A.C.M., 12 (1969), 576-580, 583.

11. P.T. Johnstone, Topos Theory, Academic Press (1977).

12. P.T. Johnstone and G.C. Wraith, "Algebraic theories in toposes," in Indexed Categories and their Applications, eds. P.T. Johnstone and R. Paré, Springer Lecture Notes in Mathematics, 661 (1978), 141-242.

13. A. Kock and G. Reyes, "Doctrines in categorical logic," in Handbook of Mathematical Logic, ed. J. Barwise, North-Holland, (1977), 283-313.

14. F.W. Lawvere, "Variable quantities and variable structures in topoi," in Algebra, Topology and Category Theory, eds. A. Heller and M. Tierney, Academic Press, (1976), 101-131.

15. M. Tierney, "Forcing topologies and classifying topoi," in Algebra, Topology and Category Theory, eds. A. Heller and M. Tierney, Academic Press, (1976) 211-219.

MATHEMATICS DEPARTMENT
UNIVERSITY OF MICHIGAN
ANN ARBOR, MI 48109

Contemporary Mathematics
Volume 92, 1989

On the Decidability of Objects in a Locos

J.R.B. Cockett *
Department of Computer Science,
University of Tennessee,
Knoxville, TN 37996-1301, U.S.A.

August 19, 1988

A **locos** is a an open coherent category with arithmetic given by a parameterized list constructor. A *coherent* category is a finitely complete regular category with stable joins of subobjects. It is *open* if it has disjoint coproducts. When a parameterized list constructor is present it becomes possible to do primitive recursive arithmetic: $list(1)$ is the natural number object.

There are well-known (full and faithful) embedding theorems for weak first order logics into stronger first order logics and into higher order logics. However, when arithmetic is added to these weaker settings less is known about how well-behaved the corresponding embeddings are. That they are not *as* well-behaved is known. In fact, the embeddings are not full or faithful into higher order logics.

The main results of this paper are concerned with the structure of a locos constructed from decidable objects. It is shown that in such a locos every object is decidable. This is of practical significance when the construction of type theories over a locos is considered. This also helps to elucidate some of the properties of the initial locos.

1 Introduction

The main results of this paper are concerned with the structure of a locos constructed from *decidable objects*. The results rely on theorems described in section 4, and proved in [5,11,10]. Before proving these results the concept of a locos is introduced in section 2 and the relationship of the logic of a locos to other weak arithmetic settings is discussed in section 3.

One of the motivations behind developing locos theory (see [5]) was a curiosity over the question of how much of mathematics can be obtained from the

*This work was done while the author was visiting the "Groupe Interuniversitaire en Étude Catégorique", McGill University, Montreal, Canada

presence of a *list* constructor instead of a *powerset* constructor. Of course, as a computer scientist, this is not totally idle curiosity as one cannot easily compute with the powerset constructor, while one clearly can with the list constructor. Indeed I introduced the setting of a locos because it seems to have the minimal amount of structure that is required in order to *program* in a reasonable fashion.

However, it should not therefore be thought that locos theory is a minimalist approach to the semantics of programming. A much meaner approach is to simply provide the natural numbers and require that everything be coded as a number. This has the effect of making everything almost incomprehensible!

As usual the problem, both for developing theory and programming languages, is to try and capture the structure which makes what one wants to do with the setting as natural as possible while not overloading the formulation with clearly unnecessary axioms and constructs. If it later turns out that one can encode everything as numbers anyway (which actually happens in a very strong sense in the initial locos) then one can simply pass this off as an interesting fact and be thankful that one has neatly avoided the explicit necessity of coding!

The logic of a locos is, excepting the arithmetic, strictly first order. In general, a locos will not be cartesian closed and will not have a subobject classifier. The specification of a locos is *elementary*: thus, no infinite limits or colimits are involved. Despite this a locos always has solutions to the "simple" fixed point constructions. For example, free term algebras can always be constructed [5]. As a locos does not have arbitrary quotients of objects, unlike an arithmetic universe, it is not possible to construct arbitrary free algebras. Indeed an obvious constraint on constructions, which is a consequence of the results in this paper, is that the constructed objects must be decidable. Mathematically this may seem like an unnecessary handicap. However, if one is designing a programming language, this is a highly desirable feature.

The fact that an initial locos exists can be seen from the way in which a locos can be specified as a model of a finite limit (**FL**) sketch [3,2]. The properties of the initial locos are obviously of special interest, and a simple consequence of the results of this paper is that every object in the initial locos is decidable.

Since I gave the talk considerably more has come to light about the initial locos, for which I am in debt to André Joyal and the thesis of his student Stanislas Roland, [10], dating from 1976! It is impossible not to discuss these results as they have particular relevance to the question of how these weak arithmetic logics are related. This is done in section 3.

It is easy to dismiss the study of weak arithmetic logics as irrelevant. However, the failure to understand the relationships between these settings probably lies at the heart of many of the knottiest philosophical problems in theoretical computer science. The relationship between **NP** and **P**, for example, can be viewed in this light.

Indeed, it is possible to take this further and to parade all sorts of practical reasons why computer programmers, and thus computer scientists, should be *primarily* concerned with such weak settings. For example, for obvious reasons, programs should terminate within a reasonable time. Ensuring that this happens implies that the program must be a map of an appropriate weak arithmetic logic. Also, it is questionable whether a programmer should use data structures for which he may not be able to decide in any reasonable length of time whether two instances are really the same!

2 What is a Locos?

The quick answer to this question is that a locos is an open coherent category with parameterized recursion.

A *coherent category* is a finitely complete regular category which has stable joins of subobjects. The term *stable* here means that when subobjects are pulled back over a map the join operation is preserved. Coherent categories are essentially equivalent to coherent logic [9].

A category is said to be *open* if it has stable disjoint coproducts. In a coherent category this is equivalent to requiring that any two objects can be embedded into a third disjointly, which is the motivation for the word "open".

Finally, it is necessary to explain what is meant by *parameterized* recursion: let $\mathbf{X}$ be a category with products then by $act(A \times Id_{\mathbf{X}}, Id_{\mathbf{X}})$ shall be denoted the category of *actions* (or *dynamorphisms* in the terminology of Arbib and Manes [1]) of $\mathbf{X}$. This means that the objects of $act(A \times Id_{\mathbf{X}}, Id_{\mathbf{X}})$ are arrows:

$$x : A \times X \longrightarrow X$$

while the morphisms are commutative squares:

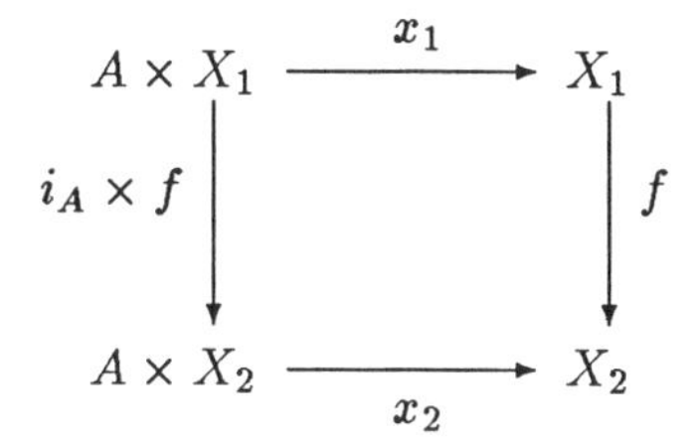

One may regard the objects of this category as state-changing devices on an input alphabet A. The map which gives the state change gives an *action* of A on the states. The maps are then the natural notion of homomorphism for these devices.

There is an obvious underlying functor, U_A: $act(A \times Id_{\mathbf{X}}, Id_{\mathbf{X}}) \longrightarrow \mathbf{X}$, which simply forgets the action. The category shall be said to have *enough recursive objects* in case for each object A this U_A has a left adjoint F_A. When the various adjunction formulae are unwound this means that for every pair of objects A and B there is an object $rec(A, B)$ with the two maps:

$$r_0(A, B) : B \longrightarrow rec(A, B)$$

$$r_1(A, B) : A \times rec(A, B) \longrightarrow rec(A, B)$$

which have the universal property for actions. Explicitly given an action

$$x : A \times X \longrightarrow X$$

and a map

$$y : B \longrightarrow X,$$

there is a unique $act(x, y)$ which makes the following diagrams commute:

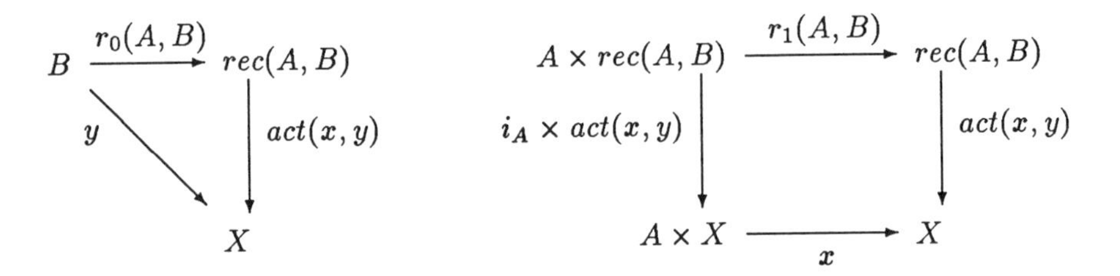

The property of having enough recursive objects can be expressed in various ways. The most common alternative ways are given in the following proposition:

Proposition 2.1 *For a category* $\mathbf{X}$ *with products the following are equivalent:*

(i) $\mathbf{X}$ *has enough recursive objects,*

(ii) Each endofunctor $A \times Id_{\mathbf{X}}$ *has an associated free triple,*

(iii) Each endofunctor $B + A \times Id_{\mathbf{X}}$ *has a least fixed-point* $rec(A, B)$.

However, the form of the recursion generated by any of these specifications does not seem to have sufficient structure to give arithmetic. [1] Indeed, there seems to be no reason why the recursive objects, $rec(A, B)$, should be at all well-behaved. What is lacking is the ability to do *parameterized* recursion from which primitive arithmetic arises.

To obtain parameterized recursion it suffices to demand that the recursive objects in the slice category are closely related to those in the overlying category. It is certainly not obvious that the slice category of a category with enough recursive objects need itself have enough recursive objects. However, what is clear is that the slice category inherits many of the recursive objects.

Let $\mathbf{X}$ have products and let Y be an object of $\mathbf{X}$ then we shall denote the overlying functor:

$$V_Y : \mathbf{X}/Y \longrightarrow \mathbf{X}; [F : X \longrightarrow Y] \mapsto X,$$

and the right adjoint to this by:

$$I_Y : \mathbf{X} \longrightarrow \mathbf{X}/Y; X \mapsto [p_1 : X \times Y \longrightarrow X].$$

[1] It is interesting to note that the initial models do have a well-behaved arithmetic [4]

Lemma 2.2

$$rec(I_Y(A), I_Y(B)) = [act(p_1(A \times Y), p_1(B \times Y)) : rec(A, B \times Y) \longrightarrow Y].$$

Now there is a unique comparison map

$$act(\langle\langle p_0, p_1.p_0\rangle.r_1, p_1.p_1\rangle, r_0 \times i_Y) : rec(A, B \times Y) \longrightarrow rec(A, B) \times Y$$

and it is clear that demanding that this map is an isomorphism is equivalent to requiring that the functor I_Y preserves the recursion.

Definition 2.3 *The recursion is said to be* parameterized *in case the functor, I_Y, preserves recursion in the above sense.*

Of course this immediately makes the *list constructor* extremely important in categories with parameterized recursion as the following series of isomorphisms shows:

$$rec(A, B) \longrightarrow rec(A, 1 \times B) \longrightarrow rec(A, 1) \times B \longrightarrow list(A) \times B$$

where it is clear that $list(A)$ is $rec(A, 1)$. The object B is now clearly identifiable as the *parameter* of the recursion. Furthermore it is not hard to show that $list(A)$ is the free monoid on A, [10,5].

The following notions are useful to distinguish the level of recursion which is present in a category:

Definition 2.4 *A category* $\mathbf{X}$*, which has products, is said to be:*

(i) number-arithmetic *if for every B the recursive object $rec(1, B)$ exists and is parameterized,*

(ii) list-arithmetic *if for every A and B the recursive object $rec(A, B)$ exists and is parameterized,*

(iii) fully-arithmetic *if* $\mathbf{X}$ *has finite limits and every internal graph has a free internal category which is parameterized.*

One point of generalizing number-arithmetic is to try to capture some useful aspect of the *logic* of arithmetic. The notion of fully-arithmetic is derived from André Joyal's use of the word arithmetic in his *arithmetic universes.* In this setting one can build free internal categories from internal graphs and, in particular, therefore one can then generate equivalence relations from arbitrary graphs. In fact the exact completion of a locos will be finitely cocomplete, fully-arithmetic and actually an arithmetic universe.

We shall not discuss the notion of fully-arithmetic and how it differs from list-arithmetic in any depth in this paper. However, it is useful to realize that the notion of fully-arithmetic is obviously *local*, in the usual sense that the slice category is fully-arithmetic. Similarly, being number-arithmetic is obviously a local property. However, it is not obvious that being list-arithmetic is always a local property: indeed, it appears that the assumption that coproducts are stable is necessary.

Obviously, if a category is fully-arithmetic then it is list-arithmetic. Similarly if a category is list-arithmetic then it is number-arithmetic. An important question concerns when a weaker form of arithmetic implies a stronger form. For example, the algebraic theory of primitive recursive functions is only cartesian, thus it cannot be fully-arithmetic. However, it is certainly number-arithmetic and is, in fact, list-arithmetic. Similarly, although a locos is only list-arithmetic by definition it is actually fully-arithmetic.

3 Arithmetic Logics

From the above development it is clear that one can have list-arithmetic categories which have considerably less structure than that of a locos. It would therefore be instructive to consider the relations between the logics of these categories.

The overall situation is depicted in figure 1 where each arrow represents a psuedo left-adjoint between the 2-categories of categories which belong to the level of logic indicated at the domain and codomain of the arrow. The unit of the two adjunction is known to be a full faithful embedding if the arrow is thick. The following abbreviations have been used:

- **FP** is the 2-category of categories with finite products, final object, and functors which preserve these,
- **FL** is the 2-category of categories with finite limits and functors which preserve these limits,
- **REG** is the 2-category of (finitely complete) regular categories with functors which preserve regularity,
- **COH** is the 2-category of coherent categories with coherent functors,
- **OPCOH** is the 2-category of open coherent categories with coherent functors (preservation of the open property is automatic),
- **PTOPOS** is the 2-category of pretoposes with coherent functors,
- **TOPOS** is the 2-category of toposes with logical functors.

To indicate that list-arithmetic is present these abbreviations have been prefixed by **A**. There are two exceptions: a list-arithmetic open coherent category is a locos, thus **OPCOH** becomes **LOCOS**, and at the pretopos level **PTOPOS** becomes **AU**, an arithmetic universe.

On the right of the diagram, where arithmetic is present, considerably less is known about the form of the embeddings. It is well-known that coherent categories embed fully and faithfully into toposes [3]. However, an embedding which preserves the arithmetic as well is much harder to secure. In fact such an embedding, cannot be full as the recursion in a locos need only be primitive. However, the situation is worse than that: not only can the embedding not be conservative (reflect isomorphisms), it need not even be faithful.

One indication of the behavior of an embedding is given by looking at the initial objects of the 2-categories concerned. The initial object of **AFP** is the algebraic theory (using the terminology of Lambek ,[8]) of primitive recursive functions,[2] $\mathbf{P}_0$.

It is important to emphasize that there is every reason why this theory should not simply be the primitive recursive functions in **Sets**, or indeed in any other topos or cartesian closed category. In particular, the maps cannot be determined by their points. This may be readily seen as there is a complete logic for the initial category. This together with an element by element evaluation would give a decision procedure which would condradict the undecidability of primitive recursive arithmetic.

The category, however, does satisfy almost all the theorems one would expect if the situation had been that simple. For example $N \times N$, $N + N$ and $1 + N$ are all present as isomorphs of N, and all the (basic) primitive recursive functions behave just as one might expect. In fact, the difficulty is rather to find the respects in which $\mathbf{P}_0$ *differs* from primitive recursive functions. Unfortunately, establishing the basic results in an elementary manner can be difficult enough and these look singularly unimpressive to those outside the field.

To obtain the initial category of **AFL**, which we shall call $\mathbf{L}_0$, one must add equalizers to the algebraic theory of primitive recursive functions, $\mathbf{P}_0$. However, while doing this one must also preserve the list arithmetic. It is certainly not obvious that one can do this while maintaining $\mathbf{P}_0$ as a full and faithful subcategory.

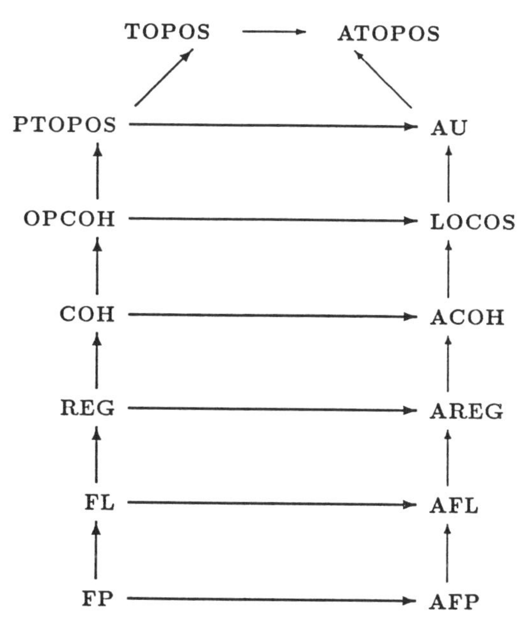

Figure 1: Embedding Theorems for Logics

[2]Burroni calls them the "formal primitive recursive functions" [4]. Regarding the category, $\mathbf{P}_0$, as an algebraic theory begs the question of what the category of algebras in **Sets** (or any topos) looks like.

In fact, Leopaldo Román, [12], has shown that in any finitely complete number-arithmetic category that the *graph* of every (partial or total) recursive function is present. Of course, the presence of the graph of a total recursive function does not mean that it is present as a *map* from the natural number object even when it is known to be total in other settings. However, it certainly does lead one to wonder whether more than primitive recursive maps may be present in $\mathbf{L}_0$.

The structure of $\mathbf{L}_0$ is of central importance. In [6] it is shown that $\mathbf{L}_0$ is exactly the free equalizer completion of $\mathbf{P}_0$. This means that one can forget entirely about the arithmetic structure in this completion step: it is all automatically transmitted. As the equalizer completion always gives a full faithful embedding this has the consequence that $\mathbf{P}_0$ embeds fully and faithfully into $\mathbf{L}_0$. But much more is true!

In a remarkable masters thesis under the direction of André Joyal, Stanislas Roland, [10], constructed this category. Although he did not prove that it was the equalizer completion, undoubtedly Joyal was aware of this fact. The construction, in the eyes of Joyal, was simply one step in the creation of the initial arithmetic universe, $\mathbf{A}_0$, which is the exact completion of this category, $\mathbf{L}_0$. I was surprised to learn thereby that arithmetic universes are also a primitive recursive setting.[3]

In his thesis Stanislas Rolland proves, in effect, that $\mathbf{L}_0$ is an open coherent category. From this it is possible to conclude that it is a locos, indeed the initial locos. In fact Joyal has pointed out that much more is true: given any theory of primitive recursive functions one can always complete in this fashion to obtain a locos in which that theory is fully and faithfully embedded and this can be extended by forming the exact completion to an arithmetic universe.

This has the rather important consequence that there is little point in studying the theory of primitive recursive functions in the weakest setting, $\mathbf{P}_0$. The theory will be exactly the same as the theory in $\mathbf{L}_0$. The advantage of using the stronger logic is that the effort involved in obtaining results will be much reduced. One can, of course, push this even further to the arithmetic universe level if the proof is helped by exactness.

AU is the 2-category of arithmetic universes: an arithmetic universe is a fully-arithmetic pretopos. A pretopos is a balanced category. Thus, in the initial arithmetic universe $\mathbf{A}_0$ the map which includes the domain of Ackermann's function into $N \times N$, which is known to be monic [12], cannot also be epic without making Ackermann's function a map. Thus this inclusion is not epic which means that there are two maps which distinguish the domain of Ackermann's function. So if Ackermann's function is added, by forcing this inclusion to be

[3] In my talk I incorrectly stated that an arithmetic universe has all recursive functions. The confusion arose due to the meaning of *bijective* in the following result (attributed to Joyal): in a number-arithmetic finitely complete category if all bijectives are isomorphisms then all total recursive maps are present as maps from N to N. The meaning of bijective here is very disappointing: it is in terms of the (**Sets** map) of global sections or elements of the category. Of course, I interpreted it incorrectly to be epic and monic (*bijic*?). It is unfortunately that the theory of arithmetic universes has still not been published even though most of the work dates from the mid 70's.

an isomorphism and thus increasing the power of the arithmetic, these two maps must be identified. This destroys any hope that there might be a *faithful* embedding from these settings into higher order settings. Furthermore this argument works for any theory of recursion which does not contain all total functions.

The initial categories are very special and, while this tell us little about the form of the embeddings between the first order logics with arithmetic, they do provide some negative results concerning embeddings into higher order settings. The embeddings between the regular, coherent and open coherent settings with arithmetic remain unexplored. Are they full and faithful? Unfortunately, again, the answer is complex.

It is certainly not the case that an arbitrary arithmetic coherent category can be embedded into a locos. An example of an arithmetic coherent category is a distributive lattice. The natural number object is the top, $\top$, and $list(x)$ for any x is also the top, $\top$. In a locos coproducts must be disjoint. The preservation of the arithmetic forces the isomorphism $1 + N \cong N$ to be preserved which clearly cannot be while embedding a distributive lattice into a locos. It is an open question as to whether this is essentially the only aberration.

The following is a conjecture concerning the state of affairs between the logics which are weaker than that of a locos.

Conjecture 3.1 *The units of the 2-adjunctions*

$$\mathbf{AFP} \longrightarrow \mathbf{AFL}$$

$$\mathbf{AFL} \longrightarrow \mathbf{AREG}$$

$$\mathbf{AREG} \longrightarrow \mathbf{ACOH}$$

are all full and faithful.

4 Arithmetic Theorems

It is common knowledge that there are various ways of primitive recursively encoding $N + N$ and $N \times N$ in terms of N in **Sets**. It is less well-known that these encodings work in all number-arithmetic categories.

Proposition 4.1 *In any number-arithmetic cartesian category,* $\mathbf{C}$*, the following objects exist by the following isomorphisms:*

(i) $\langle 0; s\rangle : 1 + N \longrightarrow N$,

(ii) $\langle n(0, s.s); n(0.s, s.s)\rangle : N + N \longrightarrow N$,

(iii) $act(n(0, s.s), n(0.s, s.s)) : N \times N \longrightarrow N$.

Here the map, $n(0, s.s) : N \longrightarrow N$, simply doubles each number, while $n(0.s, s.s)$ doubles each number and adds one. There are various ways to encode $N \times N$, the manner I have chosen is somewhat different from that in [10,11] where the usual diagonal counting technique is used. However, this encoding does seem to follow most naturally from the encoding of the coproduct.

It is well-known that there is a total order on N it is described in the following proposition, which is is proven in [10,11]:

Proposition 4.2 *In any cartesian number-arithmetic category the following coproduct exists and is given by the isomorphism*

$$\langle\langle p_0, \langle p_0, p_1\rangle.add.s\rangle; \langle i_N, i_N\rangle; \langle\langle p_0, p_1\rangle.add.s, p_1\rangle\rangle : N\times N + N + N\times N \longrightarrow N\times N.$$

This is a useful observation as it immediately establishes the decidability of N in a locos:

Corollary 4.3 *In a locos N is decidable.*

In a locos there are much stronger results characterizing the recursion which are proved in [5]. These results lead to a generalization of Freyd's characterization of the natural number in a topos [7] as an object X such that $\langle 0; s\rangle : 1 + X \longrightarrow X$ is an isomorphism and the coequalizer of s and i_X is the final object. The generalization we discuss below both extends the result to weaker arithmetic settings and extends the scope to include arbitrary recursive objects.

The proof of the generalization of Freyd's characterization relies on the rather useful fact that the recursion construction preserves connected limits. This fact is the "hammer" in the toolkit for proving general results about recursive objects.

Proposition 4.4 *For any locos, $\mathbf{L}$, the functor, $rec : \mathbf{L}\times\mathbf{L} \longrightarrow \mathbf{L}$, preserves finite connected limits.*

This result appears to require almost all of the power of the logic of a locos. Thus, it is likely that it cannot be applied to weaker settings. The use of the *hammer* in a proof, therefore, may condemn that proof as a basis for generalization to even weaker logical settings.

An immediate corollary is that connected limits of free monoids with free maps is a free monoid, or more simply:

Corollary 4.5 *In any locos, $\mathbf{L}$, the list constructor $list : \mathbf{L} \longrightarrow \mathbf{L}$ preserves connected limits.*

If the above result is a hammer, the generalization of Freyd's result is a sledge hammer in the toolkit. Before stating it, however, it is necessary to discuss what is meant by a coequalizer being *precise*.

For some coequalizer pairs in a list-arithmetic category we can always *generate* an equivalence relation. It is *not* the case, in general, that we can do this for arbitrary pairs. However, for the pair in the theorem below this is the case. In a locos in fact we can generate all equivalence relations. This means that we can compare the equivalence relation induced by the coequalizing map with the generated one. The coequalizer is *precise* if these two are the same. More formally:

Definition 4.6 *A coequalizer in a locos is said to be* precise *if the equivalence relation generated by the pair (exists and) is the same as the induced equivalence relation.*

A rather different way of stating this is that it must remain a coequalizer under all functors in **LOC**.

Theorem 4.7 *In any locos,* $\mathbf{L}$, *an object* X *such that*

(i) there is an isomorphism $\langle r_0; r_1\rangle : B + A \times X \longrightarrow X$,

(ii) there is a map $c : X \longrightarrow B$ *such that*

$$A \times X \overset{p_1}{\underset{r_1}{\rightrightarrows}} X \overset{c}{\longrightarrow} B$$

is a precise *coequalizer,*

is isomorphic to $rec(B, A)$ *and therefore a recursive object on* B *and* A.

As $rec(B, A) \cong list(A) \times B$ we can further simplify the statement to be about lists. In a fully-arithmetic setting all equivalence relations can be generated. If in addition the setting is exact then every coequalizer is precise. This means that in an arithmetic universe or a topos the preciseness is automatically satisfied by all coequalizers. This gives the following form of the result for arithmetic universes:

Corollary 4.8 *In an arithmetic universe,* $\mathbf{U}$, *and object,* X, *is equivalent to the free monoid on* A, $list(A)$, *if and only if there is an isomorphism*

$$\langle r_0, r_1\rangle : 1 + A \times X \longrightarrow X$$

and

$$A \times X \overset{r_1}{\underset{p_1}{\rightrightarrows}} X \overset{1_X}{\longrightarrow} 1$$

is a coequalizer.

This means that when a functor is a pretopos functor it will preserve the list arithmetic. In the case of a locos this must be weakened: a functor is a locos functor if and only if it is an open coherent functor which preserves the number arithmetic.

5 Decidability Properties

In order to develop the elementary properties of decidability in a locos it is necessary to agree on some terminology.

Definition 5.1 *Let* **X** *be a locos then:*

(i) A map $y : Y \longrightarrow X$ *is* recognizable *if there is a characteristic map* $\chi_y : Y \longrightarrow 1 + 1$ *such that*

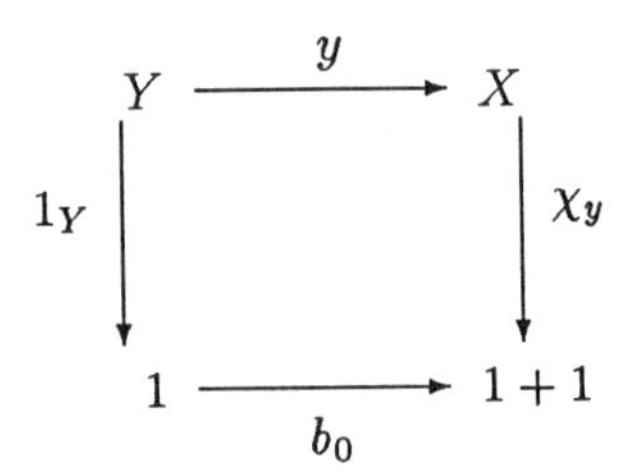

commutes and is a pullback.

(ii) An object, X, is decidable *if the diagonal map,*

$$\langle i_X, i_X \rangle : X \longrightarrow X \times X$$

is recognizable.

The word *recognizable* has been imported from pattern recognition where one has to build a *recognizer*, χ_y, to distinguish the relevant simuli from the irrelevant. It is more usual to use the word decidable for both the above concepts and to let the fact that they refer respectively to maps and objects give the distinction. This has the minor problem that ambiguity can arise when referring to a decidable subobject: is the map or the object decidable?

Of course, a subobject is recognizable precisely when it has a complement. In fact there are various equivalent ways of characterizing recognizable maps. A possibly less familiar way uses the concept of a decision. A decision is an idempotent cooperation, that is a map $q : A \longrightarrow A+A$ such that $q.\langle i_X ; i_X \rangle = i_X$.

Lemma 5.2 *In any open coherent category the following are equivalent:*

(i) $y : Y \longrightarrow X$ *is recognizable,*

(ii) $y : Y \longrightarrow X$ *has a complement in X, that is there is a subobject* $\neg y : \neg Y \longrightarrow X$ *such that* $y \wedge \neg y$ *is empty and yet* $y + \neg y$ *covers X,*

(iii) There is a decision, $q : X \longrightarrow X + X$, *such that*

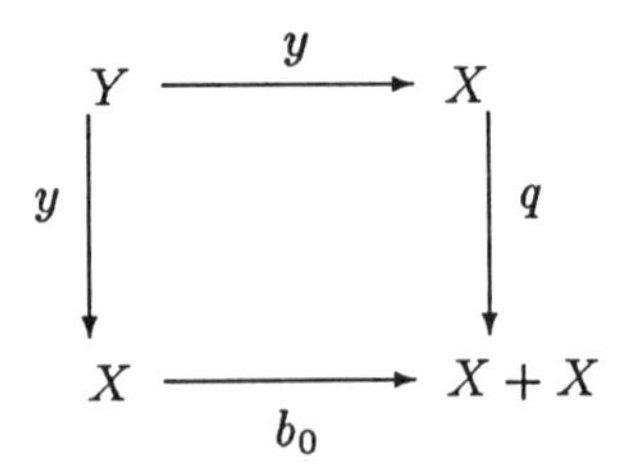

is a pullback.

Proof. We shall prove $(i) \Rightarrow (ii) \Rightarrow (iii) \Rightarrow (i)$:

$(i) \Rightarrow (ii)$ $\neg y$ is given by the pullback

$$\begin{array}{ccc} \neg Y & \xrightarrow{\neg y} & Y \\ {\scriptstyle 1_{\neg Y}}\downarrow & & \downarrow{\scriptstyle \chi_y} \\ 1 & \xrightarrow[b_1]{} & 1+1 \end{array}$$

$(ii) \Rightarrow (iii)$ Let $q = \langle y; \neg y\rangle^{-1}.\langle y + \neg y\rangle$,

$(iii) \Rightarrow (i)$ $\chi_y = q.\langle 1_Y + 1_Y\rangle$.

□

Recognizable maps can be composed, they are stable under pulling back and under many other standard operations. Here then is a summary of the operations which preserve recognizability.

Proposition 5.3 *In any open coherent category:*

(i) the complement of a recognizable subobject is recognizable,

(ii) i_X and 0_X are always recognizable,

(iii) if $x.g$ is recognizable then x is recognizable,

(iv) the composition of recognizable maps are recognizable,

(v) the pullback of a recognizable map is recognizable,

(vi) if $x : X \longrightarrow Z_1$ and $y : Y \longrightarrow Z_2$ are recognizable maps then $x \times y : X \times Y \longrightarrow Z_1 \times Z_2$ and $x + y : X + Y \longrightarrow Z_1 + Z_2$ are recognizable maps,

(vii) the join and meet of two recognizable subobjects is recognizable,

Proof. The proofs are straightforward, however, they introduce some useful terminology and maps.

(i) Clearly $\neg a$ is recognized by $\chi_a.not$ where

$$not := \langle b_1; b_0\rangle,$$

(ii) i_X has 0_X as complement,

(iii) The recognizer for x is $g.\chi_{x.g}$,

(iv) Let $x : X \longrightarrow Y$ and $y : Y \longrightarrow Z$ both be recognizable then

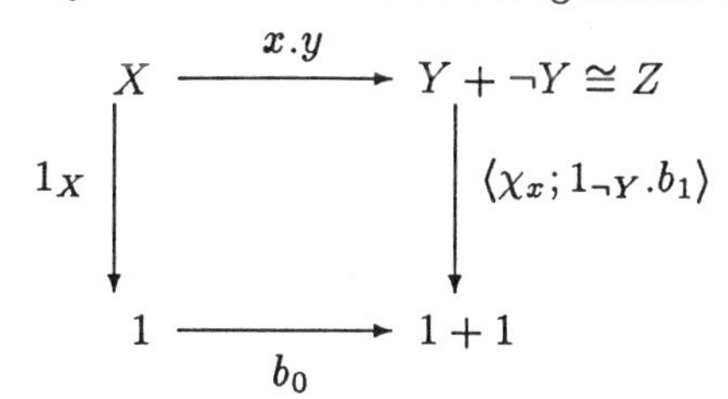

is a pullback.

(v) Consider the pullback

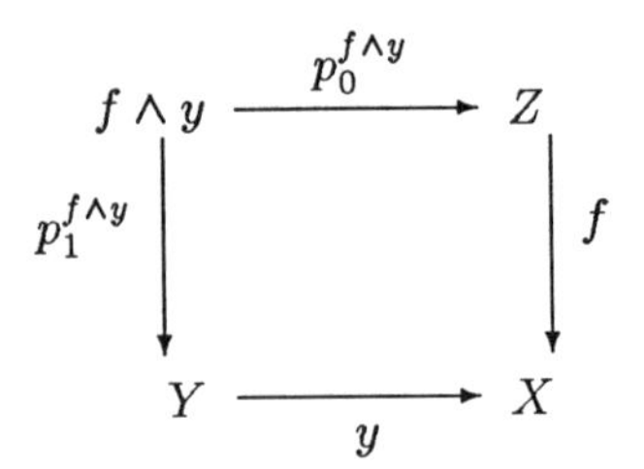

where y is recognizable, that is

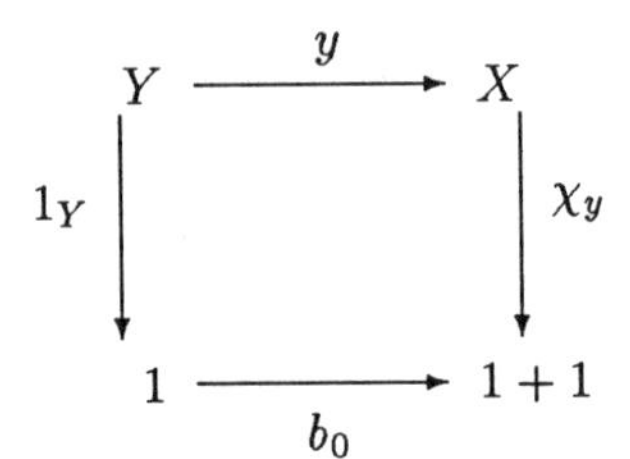

is a pullback. Now stacking these squares shows that $p_0^{f \wedge y}$ is recognized by $f.\chi_y$.

(vi) We have the following stacked pullback squares:

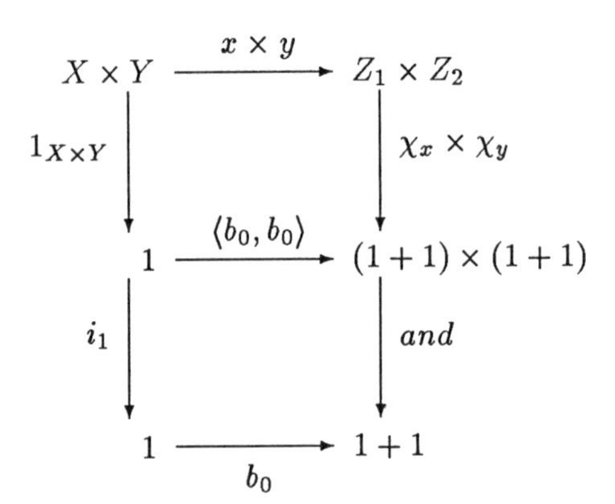

It is easy to check that the lower square is a pullback given the following definition for *and*:

$$and := d_0.\langle p_1, 1_{1+1}.b_1 \rangle = d_1.\langle p_0, 1_{1+1}.b_1 \rangle$$

where $d_0 : (1+1) \times (1+1) \longrightarrow 1 \times (1+1) + 1 \times (1+1)$ and $d_1 : (1+1) \times (1+1) \longrightarrow (1+1) \times 1 + (1+1) \times 1$ are the distributors. This gives the recognizer for $x \times y$.

For $x + y$ we have the following stacked pullback squares:

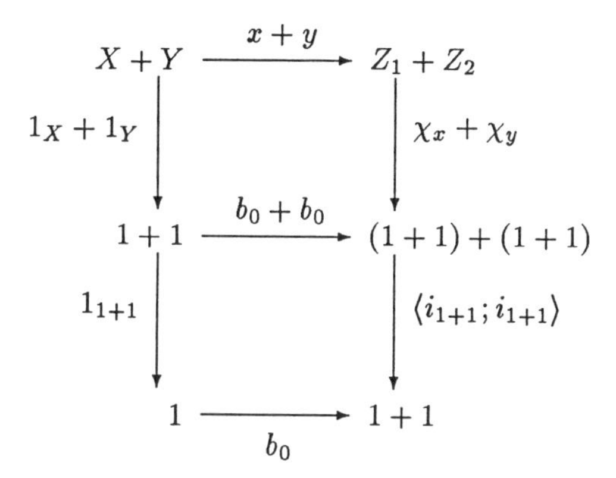

which gives the recognizer for $x + y$.

(vii) Let $a : A \longrightarrow X$ and $b : B \longrightarrow X$ be recognizable then we wish to show that $a \cup b$ is the monic factor of $\langle a; b\rangle$. From the previous result we know that $\langle a + b\rangle$ is recognized by $\langle \chi_a; \chi_b\rangle$ but

$$\langle a; b\rangle.\langle \chi_a, \chi_b\rangle.or = 1.b_0$$

where

$$or := d_0.\langle 1.b_0; p_1\rangle = d_1.\langle 1.b_0; p_0\rangle$$

To verify this it suffices to check the effect of these maps on the components.

The above equation gives the following commuting squares where the bottom square is a pullback but the top square is not necessarily a pullback. The map $e(\langle a; b\rangle)$ is similarly given by the comparison map to the pullback.

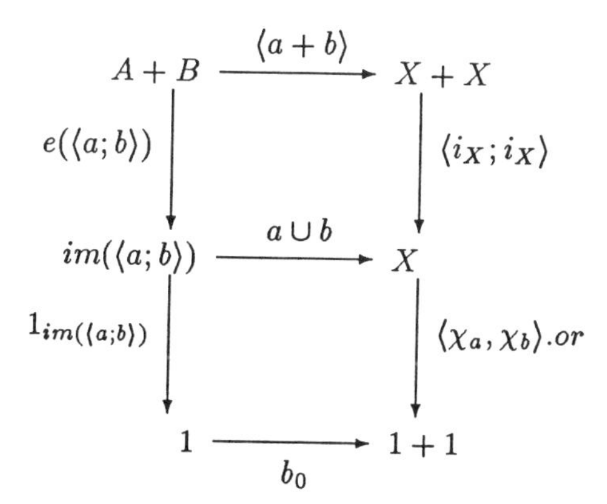

We know that $\langle i_X; i_X\rangle$ is regular epic as it is a retract, we wish to show that $e(\langle a; b\rangle)$ is regular epic by showing that it is also a retract. To show this we use the following double pullback to give a form for $a \cup b$:

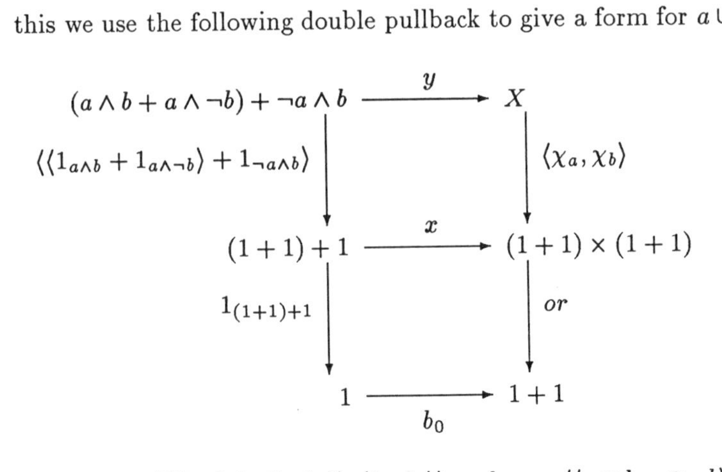

where $x := \langle\langle\langle b_0, b_0\rangle; \langle b_0, b_1\rangle\rangle; \langle b_1, b_0\rangle\rangle$ and $y := \langle\langle a \cap b; a \cap \neg b\rangle; \neg a \cap b\rangle$. It is clear now that $\langle\langle p_0^{a\wedge b}; p_0^{a\wedge\neg b}\rangle + p_1^{\neg a\wedge b}\rangle : (a\wedge b + a\wedge\neg b) + \neg a\wedge b \longrightarrow A+B$ and that this is a left inverse for $e(\langle a; b\rangle)$.
We already know that the meet is recognizable using the fact that pullbacks and compositions of recognizables are recognizable. However we should like to describe another approach to this which involves the *and* map. We know that the following is a pullback:

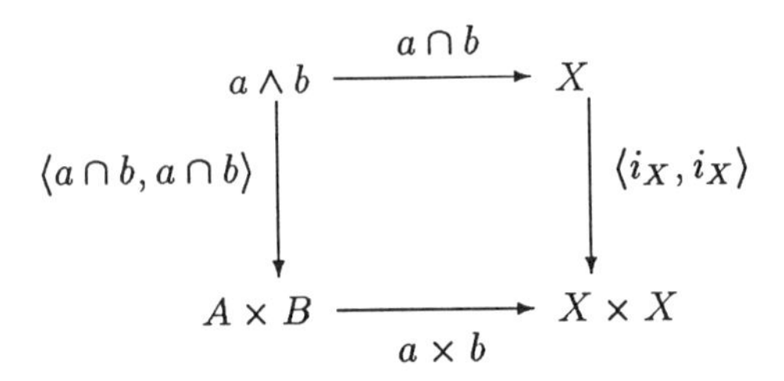

and this can be stacked with the recognizer for $a \times b$ above to give the recognizer of $a \cap b$.

□

The way in which we set up the definitions means that $1+1$ is the recognizable subobject classifier. In the course of proving the above result we have shown that *and*, *or*, and *not* behave in the expected manner on the recognized subobjects. This means that *and*, for example, is an associative, commutative and idempotent operation with identity, b_0, which acts as the top, and zero, b_1, which acts as bottom. It is also clear that $not.not = i_{1+1}$ and de Morgan's identities hold, that is $\langle not \times not\rangle.or.not = and$ and $\langle not \times not\rangle.and.not = or$. Furthermore it is easy to check that absorption holds, that is $\langle p_0, \langle p_0, p_1\rangle.or\rangle.and = p_0$ and that *and* distributes over *or*. This means that $1+1$ is as expected a boolean algebra.

Proposition 5.4 *In any open coherent category the recognizable subobject classifier* $1+1$ *together with* b_0, b_1, *not, and, and or forms a boolean algebra such that:*

- $\langle \chi_x, \chi_y \rangle.and = \chi_{x \cap y}$,
- $\langle \chi_x, \chi_y \rangle.or = \chi_{x \cup y}$,
- $\chi_x.not = \chi_{\neg x}$,
- $1_X.b_0 = \chi_{i_X}$,
- $1_X.b_1 = \chi_{0_X}$.

We shall now prove that, in a locos, the recognizable maps are preserved by the functor $rec(-,-)$. Thus if $b : B \longrightarrow B_0$ and $a : A \longrightarrow A_0$ are recognizable maps then $rec(b,a) : rec(B,A) \longrightarrow rec(B_0,A_0)$ is recognizable. This result is intuitively obvious: we expect the map $\chi_{rec(b,a)}$ given by the following recursion scheme to be the recognizer.

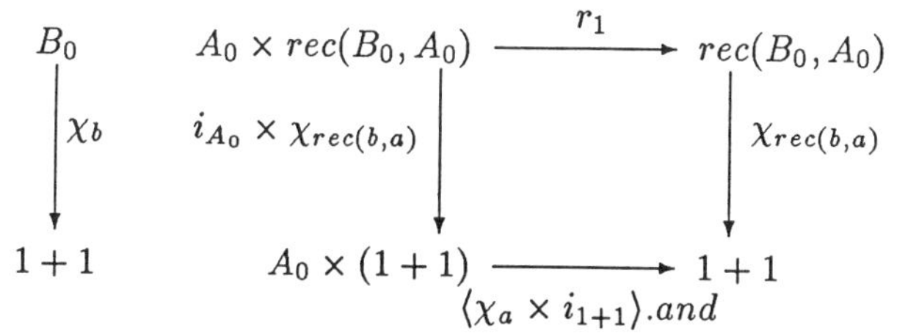

$$\begin{array}{ccccc}
B_0 & & A_0 \times rec(B_0,A_0) & \xrightarrow{\ r_1\ } & rec(B_0,A_0) \\
\Big\downarrow \chi_b & & \Big\downarrow i_{A_0} \times \chi_{rec(b,a)} & & \Big\downarrow \chi_{rec(b,a)} \\
1+1 & & A_0 \times (1+1) & \xrightarrow[\langle \chi_a \times i_{1+1} \rangle.and]{} & 1+1
\end{array}$$

Of course, proving that this intuition is correct is somewhat harder.

Proposition 5.5 *In any locos if* b *and* a *are recognizable then* $rec(a,b)$ *is recognizable. Furthermore, the map* $\chi_{rec(b,a)}$ *is the recognizer.*

Proof. We shall prove this in a very elementary fashion by showing that we can construct the recursive object on two recognizable subobjects as a recognizable subobject of the recursive object on the objects themselves.

Let $g : B \longrightarrow X$ and $f : A \times X \longrightarrow X$ define the initial conditions and the action for a recursion scheme on B and A. Suppose $b : B \longrightarrow B_0$ and $a : A \longrightarrow A_0$ are recognizable. We define an extension of the above recursion scheme as $g_0 : B_0 \longrightarrow X+1$ and $f_0 : A_0 \times (X+1) \longrightarrow X+1$, where

$$g_0 := \langle b; \neg b \rangle^{-1}.\langle g + 1_{\neg B} \rangle$$

$$f_0 := \langle \langle a; \neg a \rangle^{-1} \times i_{X+1} \rangle.d_0.\langle d_1 + d_1 \rangle.\langle \langle f.b_0; 1.b_1 \rangle; \langle 1.b_1; 1.b_1 \rangle \rangle$$

The idea is to show that extending a recursive scheme in this manner produces a unique map with the correct properties from an appropriate recognizable subobject of $rec(B_0, A_0)$. However we shall need some technical lemmas to complete the proof.

□

Lemma 5.6 *For the maps described above*

(i) the square

$$\begin{array}{ccc} A_0 \times (X+1) & \xrightarrow{f_0} & X+1 \\ {\scriptstyle \chi_a \times \langle 1_X + i_1 \rangle} \downarrow & & \downarrow {\scriptstyle \langle 1_X + i_1 \rangle} \\ (1+1) \times (1+1) & \xrightarrow[and]{} & 1+1 \end{array}$$

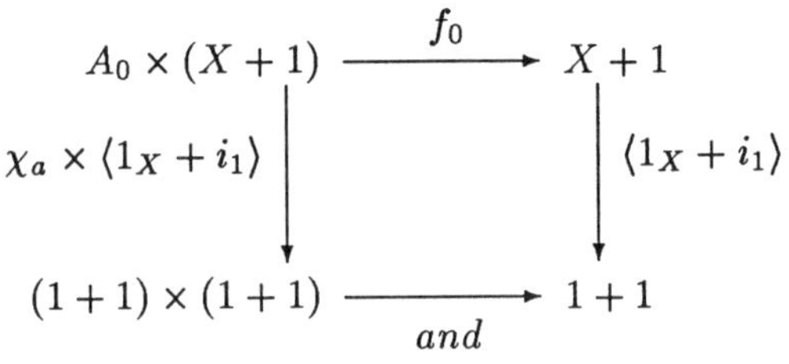

commutes,

(ii) $g_0.\langle 1_X + i_1 \rangle = \chi_b$.

Proof.

(i) $f_0.\langle 1_X + i_1 \rangle = \langle \chi_a \times \langle 1_X + i_1 \rangle.and$ if and only if the equation holds when each side is prefixed in turn by $a \times b_0$, $a \times b_1$, $\neg a \times b_0$, and $\neg a \times b_1$. Although the checking is tedious it works!

(ii) This time we must prefix the equation by b and $\neg b$ and provided these prefixed equations hold the original must. Again it works here is a sample for the prefix with b:

$$\begin{aligned} & b.\langle b; \neg b \rangle^{-1}.\langle g + 1_{\neg B} \rangle.\langle 1_X + i_1 \rangle \\ = \;& b_0.\langle g + 1_{\neg B} \rangle.\langle 1_X + i_1 \rangle \\ = \;& g.b_0.\langle 1_X + i_1 \rangle \\ = \;& g.1_X.b_0 \\ = \;& 1_B.b_0 \\ = \;& b.\chi_b \end{aligned}$$

□

Lemma 5.7 *For the maps described above:*

(i) the square

$$\begin{array}{ccc} A \times X & \xrightarrow{f} & X \\ {\scriptstyle x \times b_0} \downarrow & & \downarrow {\scriptstyle b_0} \\ A_0 \times (X+1) & \xrightarrow[f_0]{} & X+1 \end{array}$$

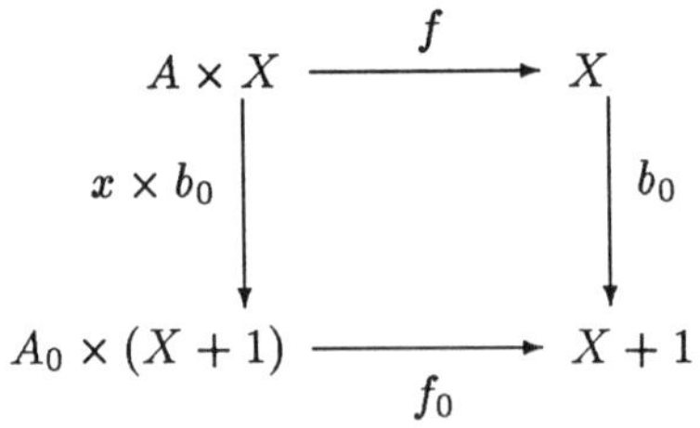

is a pullback,

(ii) the square

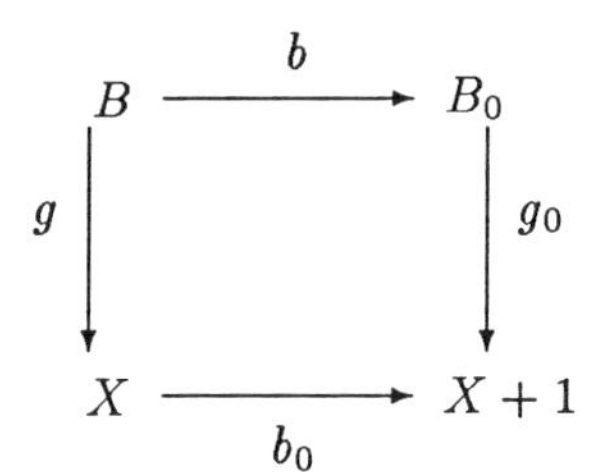

is a pullback.

Proof.

(i) It is clear that

$$\langle\langle a;\neg a\rangle^{-1}\times i_{X+1}\rangle.d_0.\langle d_1+d_1\rangle.\langle\langle b_0;b_1.b_0\rangle;\langle b_1.b_1.b_0;b_1.b_1.b_1\rangle\rangle$$

is an isomorphism and that

$$\begin{array}{ccc} A\times X & \xrightarrow{\quad f\quad} & X \\ \downarrow b_0 & & \downarrow b_0 \\ A\times X+(A\times 1+(\neg A\times X+\neg A\times 1)) & \xrightarrow[\langle f.b_0;\langle 1.b_1;\langle 1.b_1;1.b_1\rangle\rangle\rangle]{} & X+1 \end{array}$$

is a pullback. This makes the diagram in question a pullback.

(ii) As $b.g_0 = b.\langle b;\neg b\rangle^{-1}.\langle g+1_{\neg B}\rangle = b_0.\langle g+1_{\neg B}\rangle$ this square is equivalent to the following square

$$\begin{array}{ccc} B & \xrightarrow{\quad b_0\quad} & B+\neg B \\ g\downarrow & & \downarrow g+1_{\neg B} \\ X & \xrightarrow[\quad b_0\quad]{} & X+1 \end{array}$$

which is clearly a pullback.

□

Proof(5.5 continued). We have the following commutative cube:

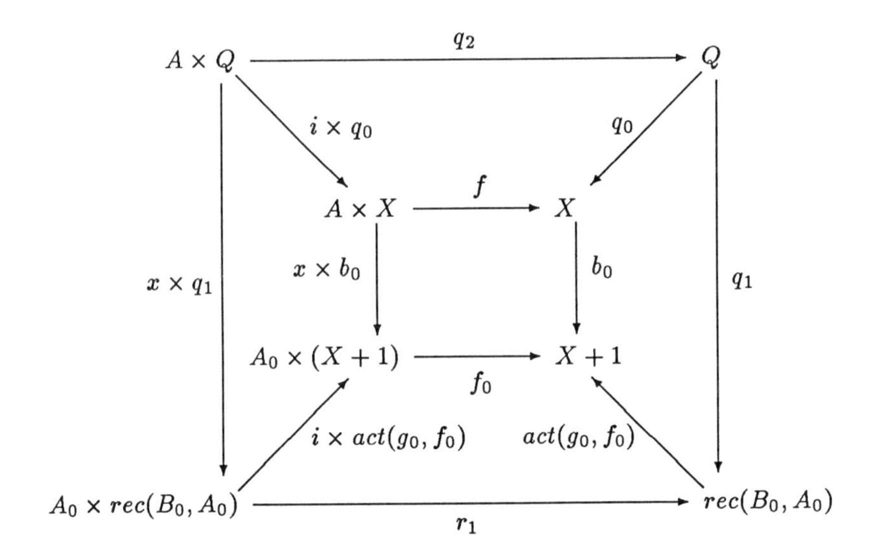

where q_0, q_1, and q_2 are filled in by making Q be the pullback of b_0 and $act(g_0, f_0)$. This makes all the commuting squares except the top and bottom trapezoids pullbacks. In particular this means that q_1 is recognizable.

To complete the proof we must show that Q is independent of the choice of f and g in which case we will be able to argue that it is the recursive object on A and B.

Letting $X = 1$ in the above diagram we obtain the following commutative cube:

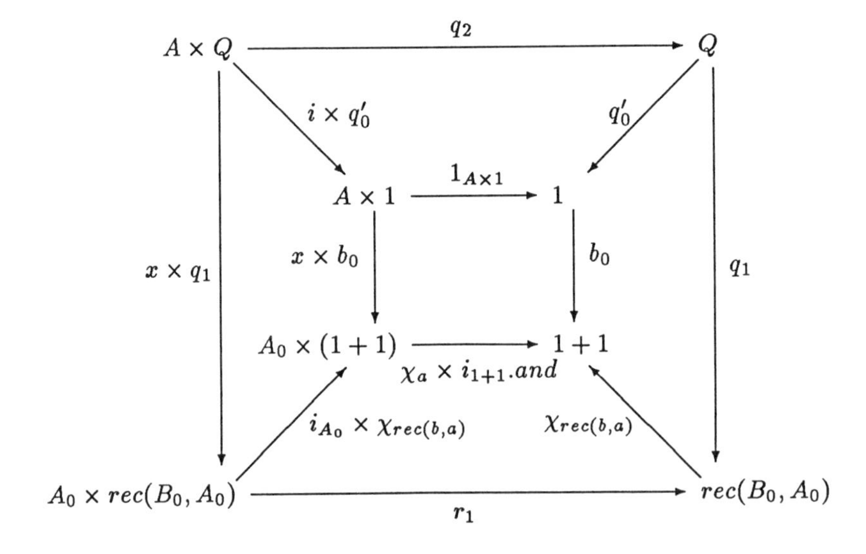

but using lemma 5.6

$$\chi_{rec(b,a)} = act(\chi_a, \langle \chi_a \times i_{1+1} \rangle.and) = act(g_0, f_0).\langle 1_X + i_1 \rangle$$

as

$$r_0.act(g_0, f_0).\langle 1_X + i_1 \rangle = g_0.\langle 1_X + i_1 \rangle$$

and

$$\begin{aligned} & r_1.act(g_0, f_0).\langle 1_X + i_1 \rangle \\ = \; & \langle i_{A_0} \times act(g_0, f_0) \rangle.f_0.\langle 1_X + i_1 \rangle \\ = \; & \langle i_{A_0} \times act(g_0, f_0) \rangle.\langle \chi_a \times \langle 1_X + i_1 \rangle \rangle.and \\ = \; & \langle i_{A_0} \times act(g_0, f_0).\langle 1_X + i_1 \rangle \rangle.\langle \chi_a \times i_{1+1} \rangle.and \end{aligned}$$

so that as

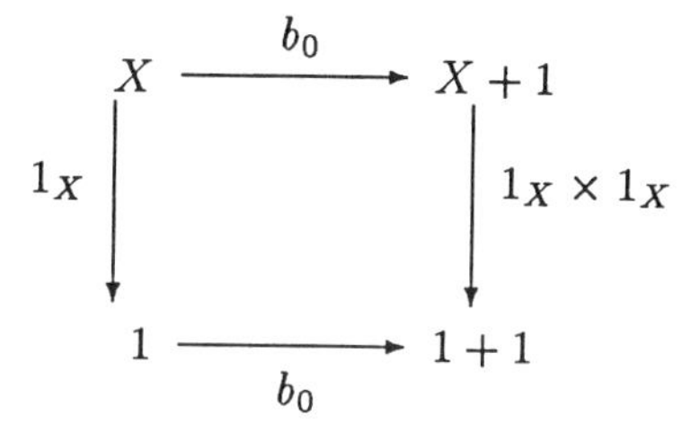

is a pullback, the following square must also be a pullback

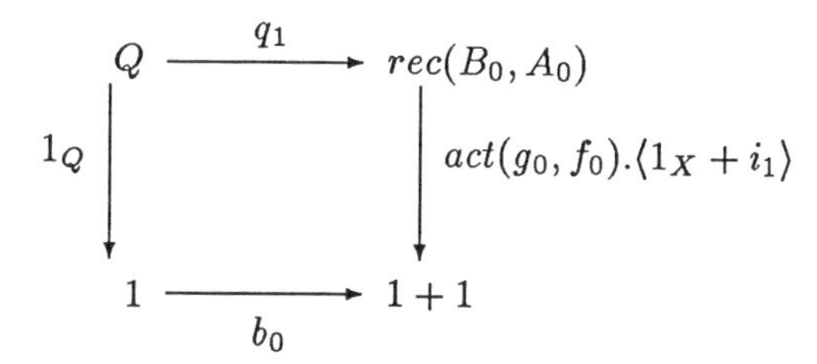

Thus Q, q_2, and q_1 are independent of the particular recursion scheme. It remains to show there is an independent q_3 such that

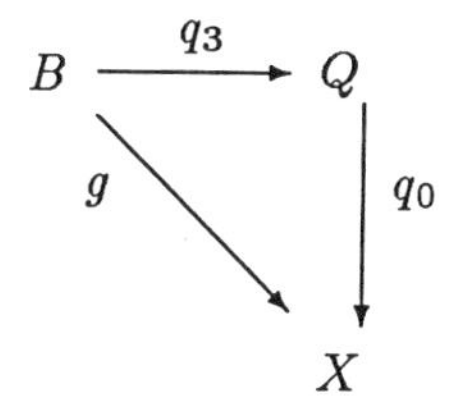

commutes. This is given by the following commuting diagram

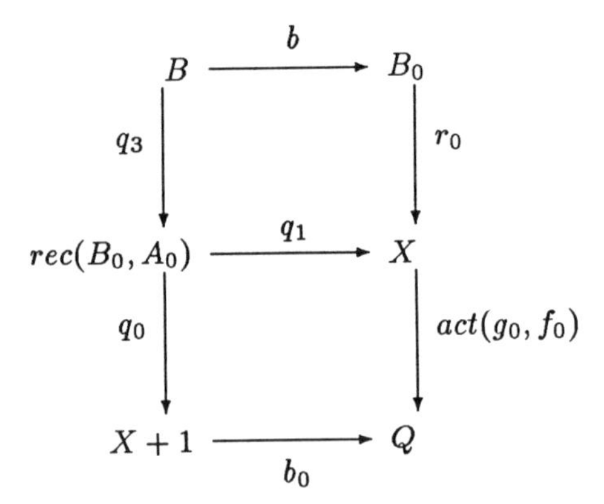

in which, as $r_0.act(g_0, f_0) = g_0$, the outer square is a pullback and lower square is a pullback, thus the upper square must also be a pullback. Clearly this is independent of the recursion scheme again. Thus Q behaves like a recursive object on B and A and so it must be a recursive object.

Clearly q_1 is equivalent to $rec(b, a)$ which is thus recognizable.

□

Decidable objects are easily seen to be closed under all the defining constructions of an open coherent category:

Proposition 5.8 *In any open coherent category if X, Y, and Z are decidable then the following are decidable:*

(i) the final object, 1, and the initial object, 0,

(ii) $X \times Y$ and $X + Y$,

(iii) W where $z : W \longrightarrow X$ is monic.

Proof.

(i) $i_{1\times 1}$ and $i_{0\times 0}$ are the diagonal maps,

(ii) let $\delta_X : X \times X \longrightarrow 1+1$ decide X and δ_Y decide Y, then it is easy to check that

$$\langle\langle p_0.p_0, p_1, p_0\rangle.\delta_X, \langle p_0.p_1, p_1.p_1\rangle.\delta_Y\rangle.and : (X \times Y) \times (X \times Y) \longrightarrow 1+1$$

decides $X \times Y$. To decide $X + Y$ it is necessary to distribute out the product and, ignoring the cross terms decide the diagonal. Thus

$$d_0.\langle d_1 + d_1\rangle.\langle\langle \delta_X; 1_{X\times Y}.b_1\rangle; \langle 1_{Y\times X}.b_1; \delta_Y\rangle\rangle : (X+Y) \times (X+Y) \longrightarrow 1+1$$

decides $X + Y$.

(iii) $\langle z \times z\rangle.\delta_X$ decides W.

□

To ensure that decidability transmits through all the constructions of a locos it is now sufficient to check that the recursive object on decidable objects is decidable.

Proposition 5.9 *In any locos if A and B are decidable then $rec(A, B)$ is decidable.*

Proof. $rec(A, B) \cong list(A) \times B$ thus it suffices to prove that $list(A)$ is decidable. First we observe that, using 4.4, the square

$$\begin{array}{ccc} list(A \times A) & \xrightarrow{\langle list(p_0), list(p_1)\rangle} & list(A) \times list(A) \\ {\scriptstyle list(1_{A \times A})} \downarrow & & \downarrow {\scriptstyle list(1_A) \times list(1_A)} \\ list(1) & \xrightarrow[\langle list(i_1), list(i_1)\rangle]{} & list(1) \times list(1) \end{array}$$

is a pullback as separating the product show that there is an underlying connected diagram involved. Notice that $list(1_A)$ is the length function on a list as $list(1) \cong N$. Now, by 4.2, the lower map is recognizable and, thus, the upper map is recognizable.

However $list(\langle i_A, i_A\rangle) : list(A) \longrightarrow list(A \times A)$ is recognizable as A is decidable so that the composition

$$list(\langle i_A, i_A\rangle).\langle list(p_0), list(p_1)\rangle = \langle list(i_A), list(i_A)\rangle$$

is recognizable. Thus, $list(A)$ is decidable.

□

We now have the result we were searching for:

Theorem 5.10 *If a locos, **L**, is constructed from decidable objects then every object in **L** is decidable.*

In particular, we may apply this to the initial locos to obtain a result, which, given that we now know a great deal about the initial locos, is to be expected:

Corollary 5.11 *The initial locos has every object decidable.*

References

[1] M.A. Arbib, E.G. Manes, *Arrows, structures, and functors: the categorical imperative,* (Academic Press, 1975).

[2] M.Barr *Models of Sketches.* Cahier de Topologie et Géométie Différentielle Catégorique, Vol XXVII, pp93-107, 1986.

[3] M. Barr, C. Wells *Toposes, Triples and Theories,* (Springer Verlag, 1984).

[4] A. Burroni, *Recurivite graphique (premier partie): catégorie des fonctions récursive primitive formelles.* Cahier de Topologie et Geométrie Différentielle Catégorique, Vol XXVII - 1 (1986) 49-79.

[5] J. R. B. Cockett *Categories with list-arithmetic: the locos,* Technical Report of the Computer Science Dept., Univ. of Tennessee, Knoxville, TN 37996-1301 (to appear). (Previous title *Locally recursive categories* Aug. '86).

[6] J. R. B. Cockett *Notes on Constructions over Canonical Equalizer Extensions,* Technical Report of the Computer Science Dept., University of Tennessee, Knoxville, TN 37996-1301 (to appear). (Currently notes in room 920, Math. Dept., Burside Hall, McGill Univ.)

[7] P.J. Freyd, *Aspects of Topoi,* Bull. Austral. Math. Soc. 7 (1972) pp1-76.

[8] J. Lambek *Cartesian Close Categories and type λ-calculi,* in Combinators and functional programming languages. Eds. C. Cousineau, P.L. Curien and B. Robinet, Lecture Notes in Computer Science, 242, (Springer Verlag, 1986).

[9] M. Makkai, G. Reyes *First order categorical logic,* Lecture Notes in Math. 611 (Springer Verlag, 1977).

[10] S. Roland *Essai sur les Mathematiques Algorithmiques,* Memoire de Maitise Univerisité du Quebec á Montreal, 1976.

[11] L. Román *Cartesian Categories with Natural Numbers Object,* Dept. Mathematics, McGill Univ., (Submitted to J. of Pure Appl. Algebra, 1986).

[12] L. Román *Categories with Finite Limits and a Natural Numbers Object,* Dept. Mathematics, McGill Univ., (Submitted to J. of Pure Appl. Algebra, 1987).

Contemporary Mathematics
Volume **92**, 1989

The Dialectica categories

V.C.V de Paiva[1]

Abstract. This paper is a resumé of my work on a categorical version of Godel's "Dialectica interpretation" of higher-order arithmetic. The idea is to analyse the Dialectica interpretation using a category **DC** where objects are relations on objects of a basic category **C** and maps are pairs of maps of **C** satisfying a certain pullback condition. If **C** has finite limits then **DC** exists and has a symmetric monoidal structure. If **C** is cartesian closed, **DC** is monoidal closed; if **C** has stable coproducts, **DC** has products and weak- coproducts. Moreover, if **C** has free monoids then **DC** has cofree comonoids and we define an endofunctor ! on **DC** which is a comonad. Using the structure above, **DC** is a categorical model for intuitionistic linear logic. The category of !-coalgebras is isomorphic to the category of comonoids in **DC** and the!-Kleisli category corresponds to the Diller-Nahm variant of the Dialectica interpretation.

Introduction

These notes contain a resumé of work I have been doing on a categorical version of the "Dialectica Interpretation" of higher order arithmetic.They form, more or less, an extended abstract of the first few sections of the Ph.D. thesis I am currently preparing under the supervision of Martin Hyland.

The original idea, as suggested to me by Hyland, was to consider the interpretation in a way now familiar from the "propositions as types" school of categorical proof-theory. As always the objects of the category are well-determined, in our case they represent essentially the Φ^D, where Φ is a formula in higher-order arithmetic and $(\ )^D$ is the Dialectica translation, see [T]. The maps however are more problematic - looked at from the proof-theoretic point of view they should represent normalisation classes of proofs, but more abstractly a map from Φ^D to Ψ^D can be taken to be some kind of realisation of the formula "$\Phi^D \to \Psi^D$". Hyland's observation was that in the case of the Dialectica Interpretation this realisation could be given very abstractly, leading to the notion of a Dialectica category **DC** for an arbitrary category **C** with limits. The objects of the category are relations in the base category **C**, which we write as

[1] Supported by CNPq,Brazil

$U \xleftarrow{\alpha} X$ and the maps from an object $U \xleftarrow{\alpha} X$ to another $V \xleftarrow{\beta} Y$ are pairs of maps $f:U \to V$ and $F:U \times Y \to X$ in **C**, satisfying a certain condition. The motivation behind such an odd definition of maps can be found in the Dialectica translation of implication. Implication, by far the most interesting rule in the Dialectica translation, is described by Troelstra [p.231] as:

$$
\begin{aligned}
(A \Rightarrow B)^D &\equiv (\exists u \forall x A_D \Rightarrow \exists v \forall y B_D)^D \\
&\equiv [\forall u (\forall x A_D \Rightarrow \exists v \forall y B_D)]^D \\
&\equiv [\forall u\, \exists v\, (\forall x A_D \Rightarrow \forall y B_D)]^D \\
&\equiv [\forall u\, \exists v \forall y\, (\forall x A_D \Rightarrow B_D)]^D \\
&\equiv [\forall u\, \exists v \forall y\, \exists x (A_D \Rightarrow B_D)]^D \\
&\equiv \exists \mathfrak{U} \mathfrak{X} \forall uy\, (A_D(u, \mathfrak{X}(u,y)) \Rightarrow B_D(\mathfrak{U}u, y))
\end{aligned}
$$

So to translate implication we need the functionals $\mathfrak{U}:U \to V$ and $\mathfrak{X}:U \times Y \to X$, which correspond to (f,F) in our definition.

A Dialectica category, however, differs from conventional proof-theoretic categories in that it is not cartesian closed. In fact usually in a Dialectica category, we have two constructions that seem to correspond to the interpretation of conjunction and it is the construction which is not a product, just a tensor, that provides us with a "good" categorical structure - the interpretation of implication gives us a monoidal closed category.

A new input came when we received accounts of Girard's work on linear logic (cf.[G]), and realised that there were many aspects of his work that seemed close to the categorical behaviour of the Dialectica categories. Indeed it became clear that we had a categorical version of the intuitionistic fragment of linear logic (cf.[G-L]). The main remaining problem was the interpretation of the operator ! (pronounced "of course"). That was solved by looking at co-free comonoid structures in DC and as a spin-off from this categorical setting we got another category $DC_!$, the !-Kleisli category, which corresponds to the variant of the Dialectica Interpretation described by Diller and Nahm. These notes are divided into two sections, the first constructs the Dialectica category DC and establishes its categorical properties, the second describes the connector "of course".

Section 1 The Dialectica category DC

1.1 Basic definitions

In this section we describe the general Dialectica category DC associated with a basic category **C** with finite limits. Martin Hyland's idea was to build a category on relations between objects of the basic category. So if U and X are objects of **C**, a typical object of DC would be $\alpha:A \rightarrowtail U \times X$, a subobject of the product $U \times X$. A map between two such objects $\alpha:A \rightarrowtail U \times X$ and $\beta:B \rightarrowtail V \times Y$ would be a pair of maps of **C**, (f,F) $f:U \to V$, $F:U \times Y \to X$ such that a non-trivial condition is satisfied, namely pulling back $A \rightarrowtail U \times X$ and $B \rightarrowtail V \times Y$ as the diagram shows, the first subobject A' is smaller than the second B', i.e there is a map making the triangle commute:

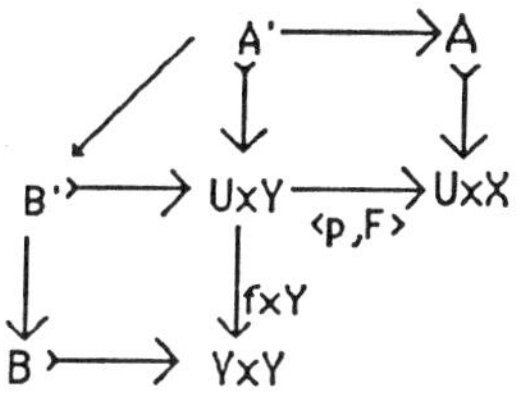

$$
\begin{array}{ccccc}
 & \swarrow & A' & \longrightarrow & A \\
 & & \downarrow & & \downarrow \\
B' & \rightarrowtail & U\times Y & \xrightarrow[\langle p,F\rangle]{} & U\times X \\
\downarrow & & \downarrow{\scriptstyle f\times Y} & & \\
B & \rightarrowtail & V\times Y & &
\end{array}
$$

Given this presentation it is not obvious that DC is indeed a category. To simplify notation we write $(U \xleftarrow{\alpha} X)$ for $\alpha: A \rightarrowtail U\times X$. Then a map in DC can be represented as the pair (f,F) in the diagram below

$$
\begin{array}{c}
U \xleftarrow{\alpha} X \\
f\downarrow \nwarrow F \\
V \xleftarrow{\beta} Y
\end{array}
\qquad \text{where } f: U \to V,\quad F: U\times Y \to X
$$
$$
\text{satisfy } \langle p,F\rangle^{-1}(\alpha) \le (f\times Y)^{-1}(\beta) \qquad [*]
$$

If, intuitevely, one thinks of α and β as usual set-theoretic relations, condition [*] says that if $u\alpha F(u,y)$ then $f(u)\beta y$. Since [*] is not a straightforward categorical condition we will show that DC is a category. Given two maps $(f,F):\alpha\to\beta$ and $(g,G):\beta\to\gamma$ their composition $(g,G)\circ(f,F)$ is $gf: U\to W$ in the first coordinate and $G\circ F: U\times Z\to X$ given by:

$$
U\times Z \xrightarrow{\delta\times Z} U\times U\times Z \xrightarrow{U\times f\times Z} U\times V\times Z \xrightarrow{U\times G} U\times Y \xrightarrow{F} X
$$

in the second. To check that the new map $(gf, G\circ F):\alpha\to\gamma$ satisfies condition [*] we use pullback patching. It is easy to see that composition is associative and that identities are (id_U, π_1) where $id_U: U\to U$ is the identity and $\pi_1: U\times X\to X$ is the canonical second projection in C.

1.2 The monoidal structure in DC

DC has a natural symmetric monoidal structure. For objects $U \xleftarrow{\alpha} X$ and $V \xleftarrow{\beta} Y$, we take their tensor product $\alpha\otimes\beta$ to be $U\times V \xleftarrow{\alpha\times\beta} X\times Y$. Note that the tensor product above does not define a product, since we in general do not have projections. The operation $\otimes$ is a bifunctor and one can easily check that it defines a symmetric monoidal structure with $I=(1\leftarrow 1)$ as a unit.

1.3 The category DC is monoidal closed

Assuming C is finitely complete, we defined DC and verified that it has a monoidal structure. Now if C is also a cartesian closed category we want to show that the monoidal structure on DC is closed. To show that we have to define internal homs in DC (cf. [K]) and check the adjunction $(-)\otimes\beta \dashv [\beta\to(-)]$. Define the internal homs in DC by recalling that intuitively $[\beta\to\gamma]$ should represent "the set of pairs of maps in C, $f: V\to W$, $F: V\times Z\to Y$

satisfying the [*] condition". So it is fairly obvious that $[\beta\to\gamma]$ should be a subobject of $W^V\times Y^{V\times Z}\times V\times Z$. Consider the following diagram:

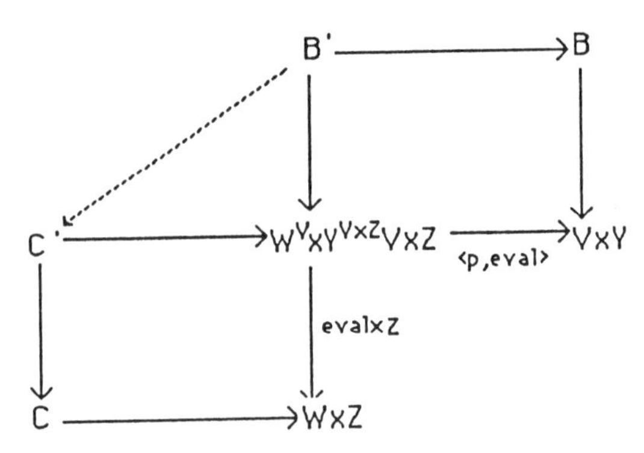

where B' and C' are defined by the pullbacks.

Finally we define $[\beta\to\gamma]$ to be the greatest subobject $A\rightarrowtail W^V\times Y^{V\times Z}\times V\times Z$ such that $A\wedge B'\leq C'$ ($\wedge$ here means pullback over $W^V\times Y^{V\times Z}\times V\times Z$). Note that this is the usual categorical translation of Heyting implication.To garantee the existence of the greatest subobject, we ask for C locally cartesian closed as well as cartesian closed. Having defined $[\beta\to\gamma]$ we check that $DC(\alpha\otimes\beta,\gamma)$ is naturally isomorphic to $DC(\alpha,[\beta\to\gamma])$. To see that $DC(\alpha\otimes\beta,\gamma)$ is in bijective correspondence with $DC(\alpha,[\beta\to\gamma])$ take (f,F) in $DC(\alpha\otimes\beta,\gamma)$, it is of the form $(f,\langle F_1,F_2\rangle)$ where $f:U\times V\to W$, $F_1:U\times V\times Z\to X$ and $F_2:U\times V\times Z\to Y$. The map f is bijectively associated (by exponential transpose in C) to $f:U\to W^V$ and analogously F_2 is associated to $F_2:U\to Y^{V\times Z}$. So the mapping $(f,\langle F_1,F_2\rangle)\to(\langle f,F_1\rangle,F_2)$ has appropriate domain and codomain and is clearly bijective. Also a long and tedious arrow-chasing proves that $(f,\langle F_1,F_2\rangle)$ is a map in DC if and only if so is $(\langle f,F_1\rangle,F_2)$.

1.4 Products and weak-coproducts in DC

In this subsection we consider **C** a cartesian closed category with stable and disjoint coproducts. Then we shall define categorical products and weak-coproducts in DC and show a weak form of distributivity of product over weak-coproducts.

The product $U\times V\xleftarrow{\alpha\&\beta}X+Y$ of two objects $U\xleftarrow{\alpha}X$ and $V\xleftarrow{\beta}Y$ of DC is obtained by considering the subobjects $U\times B\rightarrowtail^{1\times\beta}U\times V\times Y$ and $A\times V\rightarrowtail^{\alpha\times 1}U\times X\times V$ and adding them up. Thus $\alpha\&\beta=A\times V+U\times B\rightarrowtail^{1\times\beta+\alpha\times 1}U\times X\times V+U\times V\times Y\cong U\times V\times(X+Y)$. To check that '$\alpha\&\beta$' gives a categorical product, note that:

1. There are canonical projections given by $(\pi_0,i\pi_1)$ and $(\pi_1,j\pi_2)$, i and j canonical injections in **C**.
2. The object $\alpha\&\beta$ has the universal property, i.e given maps $(f,F):\delta\to\alpha$ and $(g,G):\delta\to\beta$

there is a unique map in DC, namely $(\langle f,g\rangle,F+G)$, making the diagram

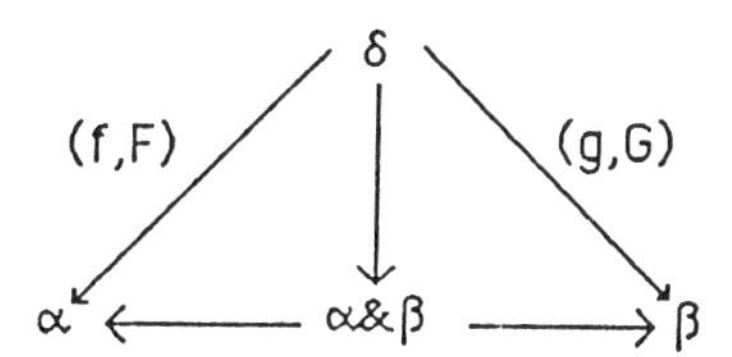

commute. Again some pullback patching is needed to show $(\langle f,g\rangle,F+G)$ is a map in DC.

We do not seem to have all coproducts in DC, but we do have some special ones, e.g for $U\xleftarrow{\alpha}X$ and $V\xleftarrow{\beta}X$ the object $U+V\xleftarrow{\alpha+\beta}X$ is a coproduct, where the relation '$\alpha+\beta$' is defined in the obvious way.

More importantly we always have weak-coproducts, i.e there is an operation $\oplus$, not a bifunctor, which satisfies:

1. There are canonical injections i_1, i_2.
2. If there are maps $(f,F):\alpha\to\delta$ and $(g,G):\beta\to\delta$, then there is a map $\alpha\oplus\beta\to\delta$, but it is not necessarily unique. To define $\oplus$ first take the pullback of $A\rightarrowtail^{\alpha} U\times X$ along $U\times X^U\xrightarrow{\langle p,ev\rangle}U\times X$, multiply the new α' by Y^V, and add it to the correspondent new β. So that $\alpha\oplus\beta=U+V\xleftarrow{\alpha+\beta}X^U\times Y^V$. For the canonical injections we just use canonical injections in C and evaluation. For the map $\alpha\oplus\beta\to\delta$ we use the natural map $U+V\to W$ and any of the possible maps in the second coordinate.

Another point to mention is that we do not have distributivity of tensor product over weak-coproduct, nor of categorical product over weak-coproduct. But we do have a map (i,I) going from $\alpha\otimes(\beta\oplus\gamma)$ to $(\alpha\otimes\beta)\oplus(\alpha\otimes\gamma)$ and a map (j,J) coming back and these maps $(i,I),(j,J)$ form a retraction. The map (i,I) will be necessary in **1.5.** It consists of the usual iso $i:U\times(V+W)\to U\times V+U\times W$ in the first coordinate ; $I:U\times(V+W)\times(X\times Y)^{U\times V}\times(X\times Z)^{U\times W}\to X\times Y^V\times Z^W$ in the second coordinate can be decomposed as (H,M,N) where $H=H_1+H_2$, all of them consisting of evaluations and projections.

Recapitulating:

If C	then DC
has products, 1, pbs	exists with $\otimes$
is cartesian closed	is monoidal closed
has stable, disjoint coproducts	has products, weak coproducts weak-distributivity

1.5 Linear Categories

Linear categories as described in [G-L] are categories with enough structure to model at least part of Intuitionistic Linear Logic, which means that they are symmetric monoidal closed categories with categorical products and coproducts. That implies the existence of units for product (**t**), for tensor (**I**) and for coproduct (**0**).

The aim of this section is to show that DC, despite not being a linear category, (not all coproducts) can be considered as a model for linear logic. That is, the constructions in DC, even the weak-coproduct, satisfy the rules of the Gentzen style system for propositional intuitionistic linear logic. We recall those rules:

Structural Rules:

$$1.\ \frac{}{A \vdash A} \qquad 2.\ \frac{\Gamma \vdash A \qquad \Delta, A \vdash B}{\Gamma, \Delta \vdash B} \qquad 3.\ \frac{\Gamma, \Delta, A, B \vdash C}{\Gamma, \Delta, B, A \vdash C}$$

Logical Rules:

$$4.\ \frac{}{\vdash 1} \qquad 5.\ \frac{\Gamma \vdash A}{\Gamma, 1 \vdash A} \qquad 6.\ \frac{\Gamma \vdash A \qquad \Delta \vdash B}{\Gamma, \Delta \vdash A \otimes B}$$

$$7.\ \frac{}{\Gamma \vdash t} \qquad 8.\ \frac{}{\Gamma, 0 \vdash A} \qquad 9.\ \frac{\Gamma, A, B \vdash C}{\Gamma, A \otimes B \vdash C}$$

$$10.\ \frac{\Gamma, A \vdash C}{\Gamma, A \& B \vdash C} \qquad 11.\ \frac{\Gamma, B \vdash C}{\Gamma, A \& B \vdash C} \qquad 12.\ \frac{\Gamma \vdash A \qquad \Gamma \vdash B}{\Gamma \vdash A \& B}$$

$$13.\ \frac{\Gamma \vdash A}{\Gamma \vdash A \oplus B} \qquad 14.\ \frac{\Gamma \vdash B}{\Gamma \vdash A \oplus B} \qquad 15.\ \frac{\Gamma, A \vdash C \qquad \Gamma, B \vdash C}{\Gamma, (A \oplus B) \vdash C}$$

$$16.\ \frac{\Gamma, A \vdash B}{\Gamma \vdash A \multimap B} \qquad 17.\ \frac{\Gamma \vdash A \qquad \Delta, B \vdash C}{\Gamma, \Delta, A \multimap B \vdash C}$$

where $\Gamma = \langle G_1, \ldots, G_n \rangle$ and Δ are strings of formulae and Γ, Δ is juxtaposition.

We define an interpretation of (the propositional part of) intuitionistic linear logic into a category **D**, as a map $|-|_0$ which associates to each atomic formula A an object of the category **D**. That interpretation can be extended to all the formulae of propositional intuitionistic linear logic, via $|-|$:Form i.L.L$\to$**D**, if we can associate constructions in the category **D** to the logical connectives in linear logic. We say that **D** is a categorical model for (propositional) intuitionistic linear logic, if using the interpretation $|-|$ we can define an appropriate categorical notion of $\vdash_D$ such that the following holds: $\Gamma\vdash_{Lin} A \Rightarrow |\Gamma|\vdash_D |A|$

In our case we want to read $\vdash_D$ as "there exists a map in **D** from $|\Gamma|$ to $|A|$". So to show now that DC is a model of propositional intuitionistic linear logic we suppose we are given an interpretation of atomic formulae as objects of DC, $|A| = U\xleftarrow{\alpha}X$, we extend that interpretation to the sets of formulae by setting $|\Gamma| = |G_1,\ldots,G_n| = |G_n|\otimes\ldots\otimes|G_1|$ (note the reverse order) and by interpreting the connectives $\otimes$, &, $\oplus$, $\multimap$ as the corresponding constructions in DC. Then it is straightforward to check that the structures defined for DC satisfy the rules above. Rule **1** is ensured by the existence of identities and rule **2** is obtained by tensoring and composing the maps. The symmetric monoidal structure in DC ensures **3** to **6** and **9**. Rules **7** and **10** to **12** are obtained by interpreting $|A\&B|$ as the categorical product of objects $|A|$ and $|B|$, $|A|\&|B|$. In addition, there are logical rules corresponding to the weak-coproduct **8** , **13** and **14**. Rule **15**, since we are reading the rules only downwards, corresponds to the weak-distributivity, i.e to the existence of the map $(i,I):\alpha\otimes(\beta\oplus\gamma)\to(\alpha\otimes\beta)\oplus(\alpha\otimes\gamma)$ mentioned before. Finally, rules **16** and **17** reflect the monoidal closed structure.

As a last remark in this section we note that we are amalgamating two steps, i.e we are considering "propositions as types" and "types as objects of a category", which gives us "propositions as objects of a (suitable) category". Suitable, at this level, only means that linear entailment corresponds to existence of a map in the category.

Section 2. The linear connective !

The logical idea behind the connective ! in linear logic is that it should give you the possibility of using the same hypothesis as many times as you wish. So, even if there is no diagonal map in DC we would like to have a natural map $!A\to!A\otimes!A$. From that to develop the idea that ! should be, not only an endofunctor, but a comonad in DC is not, perhaps the most natural thought, but it seems to work.

In this section we recall some basic facts about comonads and state a folklore proposition that we shall need later. Then assuming DC with all the structure described in section **1**, we discuss the comonad !, its basic properties and some logical consequences .

2.1 Basic facts about comonads

We recall some results from chapter 6 in [CWM], using comonads instead of monads. They shall be stated as facts, since their proofs are just dualizations of MacLane's.

FACT 1: Every adjunction $\langle F,G,\eta,\varepsilon\rangle:\mathbf{D}\to\mathbf{C}$ gives rise to a monad in the category $\mathbf{D}$ and a comonad in $\mathbf{C}$. The comonad is given by the endofunctor FG, the co-unit of the comonad by the co-unit of the adjunction $\varepsilon:FGU\to U$ and the unit of the adjunction $\eta:I\to GF$ yields by composition a natural transformation μ, where $\mu=F\eta G:FGU\to FGFGU$.

FACT 2: Every comonad $T:\mathbf{C}\to\mathbf{C}$ gives rise to two categories, the category $\mathbf{C}^T$ of T-coalgebras (or Eilenberg-Moore category) and the T-Kleisli category, $\mathbf{C}_T$. The category $\mathbf{C}^T$ has as objects T-coalgebras, that is pairs $\langle U,h:U\to TU\rangle$, where U is an object of $\mathbf{C}$ and h is a map, called the structure map of the algebra, which makes both diagrams commute:

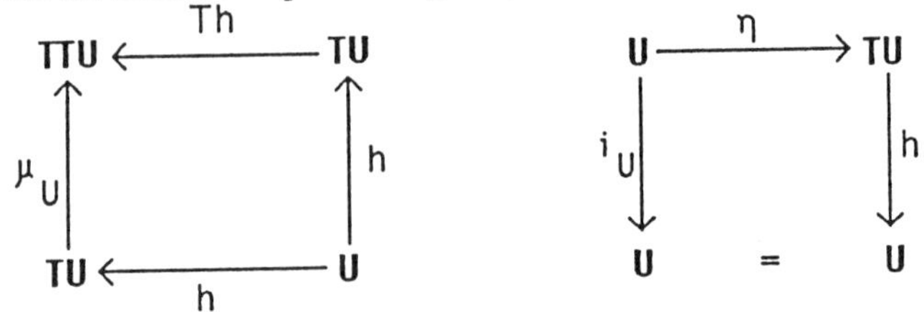

A morphism of T-algebras is an arrow $f:U\to U'$ of $\mathbf{C}$ which renders commutative the diagram:

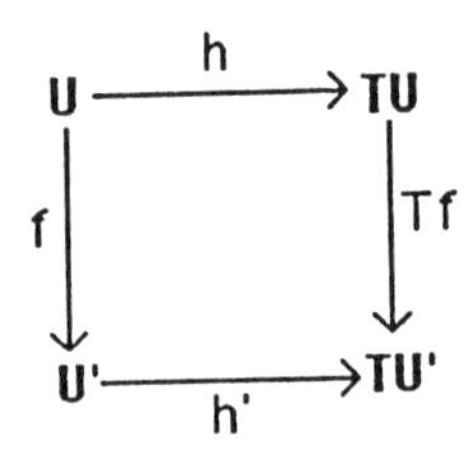

The category $\mathbf{C}_T$, the T-Kleisli category has the same objects as $\mathbf{C}$, but $\mathrm{Hom}\mathbf{C}_T(X,Y)$ corresponds to $\mathrm{Hom}\mathbf{C}(TX,Y)$. Composition of $f:TX\to Y$ and $g:TY\to Z$ is given by:

$$TX \xrightarrow{\ \mu\ } TTX \xrightarrow{\ Tf\ } TY \xrightarrow{\ g\ } Z$$

FACT 3: Let $\langle F,G,\eta,\varepsilon\rangle:\mathbf{D}\to\mathbf{C}$ be an adjunction, $T=\langle FG,\varepsilon,\mu\rangle$ the comonad it defines in $\mathbf{C}$. Then there are unique functors $\mathbf{K}:\mathbf{D}\to\mathbf{C}^T$ and $\mathbf{L}:\mathbf{C}_T\to\mathbf{D}$ making the following diagram commute:

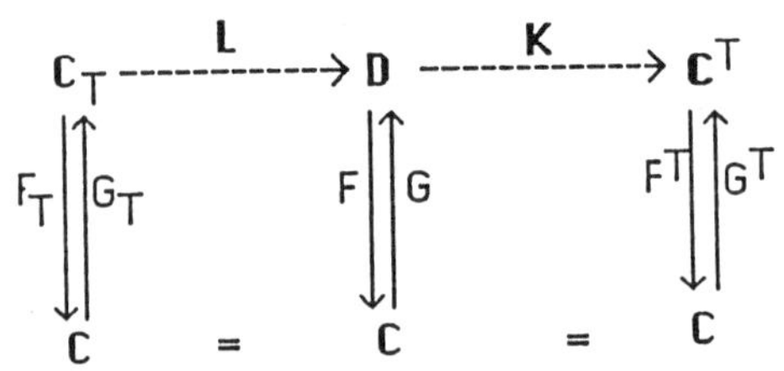

Under certain hypothesis the unique functor $\mathbf{K}:\mathbf{D}\to\mathbf{C}^T$ can be an equivalence as we shall discuss.

2.2

Recall that if **D** is any symmetric monoidal category we can consider the category **Mon D**, consisting of monoid objects in **D**, i.e triples $(M,\mu_M:M\otimes M\to M,\eta_M:I\to M)$, where M is an object of **D**, μ is its monoid multiplication and η its unit and these maps make the following diagrams commute:

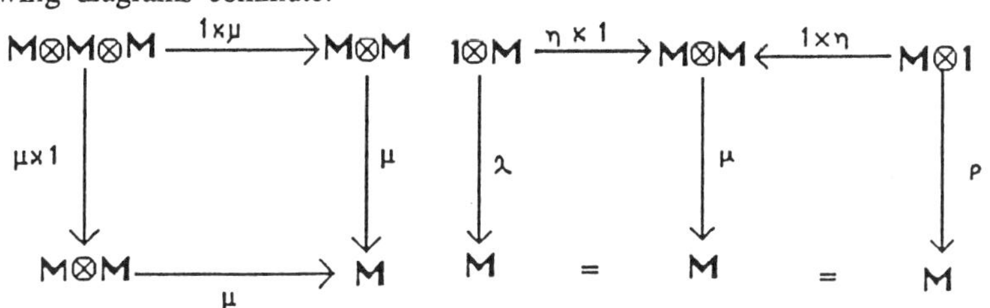

Maps $f:(M,\mu,\eta)\to(M',\mu',\eta')$ are maps $f:M\to M'$, which preserve the structure. Notice that as a monoidal structure is self-dual, one can dually define the category **Comon D** of comonoids on **D**, where objects are triples $(C,\gamma_C:C\to C\otimes C,\varepsilon_C:C\to I)$. In this subsection we want to show the following proposition, which is just an application of Beck's Theorem, as well as being part of the folklore of monoidal closed categories.

Proposition: If **D** is any monoidal category and U:**Comon D**→**D** has a right-adjoint R, then U is comonadic, i.e **Comon D**≅T-coalgebras, where T is the comonad defined in **D** by the adjunction $U\dashv R$.

Proof: To show the proposition we quote Beck's theorem in an appropriate form and verify that the forgetful functor U satisfies the condition required.

Theorem: Let $\langle U,R,\eta,\varepsilon\rangle$: **Comon D**→**D** be an adjunction, $T=\langle UR,\varepsilon,\mu\rangle$ the comonad it determines in **D**, $\mathbf{D}^T$ the category of coalgebras for this comonad. Then the following are equivalent:

1.The unique comparison functor K: **Comon D**→$\mathbf{D}^T$ is an equivalence.

2.The functor U:**Comon D**→**D** creates equalizers for those parallel pairs f,g in **D** for which Uf,Ug has an absolute equalizer in **Comon D.**

So to use the theorem we have to show condition 2 for U: **Comon D**→**D** above.

If the following diagram is an absolute equalizer in **D** we want it to be an equalizer in **Comon D.**

$$E \overset{e}{\rightarrowtail} A \underset{g}{\overset{f}{\rightrightarrows}} B$$

But since the equalizer is an absolute equalizer we have that

$$E\times E \overset{e\times e}{\rightarrowtail} UA\times UA \underset{Ug\times Ug}{\overset{Uf\times Uf}{\rightrightarrows}} UB\times UB$$

is an equalizer too. Using that we define a comultiplication in E, induced by the comultiplication in A. Similarly we can define a co-unit for E, E→I. Then E with the induced structure is a comonoid object, the map $E \rightarrowtail^{e} A$ is a comonoid homomorphism and it is easy to see that (E,e) is an equalizer in **Comon D**.

2.3

Now we want to apply the theory of sections **2.1** and **2.2** to the Dialectica categories DC. For that we consider C cartesian closed, with stable and disjoint coproducts and with free monoid structures. So there exists a functor $*$:**C→Mon C**, which is left-adjoint to the forgetful functor U:**Mon C→C.**

To define the endofunctor !:DC→DC, which is the promised comonad, we need first to define special maps $C_{(-,-)}$ for pairs of objects V,Y of **C**. Define $C_{V,Y}$ using the following transformations:

$$\frac{\dfrac{\dfrac{V\times Y \to (V\times Y)^*}{Y \to (V\times Y)^{*V}}}{Y^* \to (V\times Y)^{*V}}}{V\times Y^* \xrightarrow[C_{V,Y}]{} (V\times Y)^*} \qquad \begin{matrix}\eta \text{ co-unit of adjunction}\\ \\ * \dashv U\end{matrix}$$

Using the maps above, to define the endofunctor ! we recall that objects in DC are monos in C, i.e arrows of the form $A \rightarrowtail^{\alpha} U\times X$, so $U \leftarrow^{\alpha} X$ or $\alpha: A \rightarrowtail U\times X$ is taken by ! to $U \leftarrow^{!\alpha} X^*$, where the relation $!\alpha$ is given by the pullback:

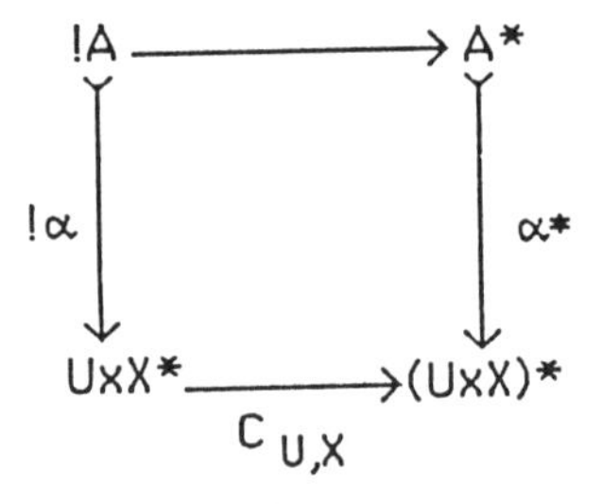

Intuitively the relation "$u\alpha x$" is transformed into "$u(!\alpha)[x_1\ldots x_k]$ iff $u\alpha x_1$ and $u\alpha x_2$ and...and $u\alpha x_k$". A map in DC $(f,F):\alpha\to\beta$ goes to $!(f,F)=(f,!F):!\alpha\to!\beta$, where $!F:U\times Y^*\to X^*$ is the composite

$$U\times Y^* \xrightarrow{C_{U,X}} (U\times Y)^* \xrightarrow{F^*} X^*$$

It is easy to check that (f,!F) is a map in DC and that ! does define an endofunctor. Now to show that ! is a comonad we have to exhibit two natural transformations, $\mu:!\to!!$ and $\varepsilon:!\to Id$ which make the usual diagrams commute. Specifically the natural transformation $\mu_\alpha:!\alpha\to!!\alpha$

is given by the identity in the first coordinate and "forgetting parenthesis" $P:U\times X^{**}\to X^*$ in the second coordinate, while $\varepsilon_\alpha:!\alpha\to\alpha$ is given by identity and "repetition of the argument" on the second coordinate, $R:U\times X\to X^*$. An easy manipulation shows that the two natural transformations satisfy the comonad equations:

1. $!\mu_\alpha \circ \mu_\alpha = \mu_{!\alpha} \circ \mu_\alpha$
2. $\varepsilon_{!\alpha} \circ \mu_\alpha = id_{!\alpha} = !\varepsilon \circ \mu_\alpha$

The next step is to show that ! provides us with good categorical structure, so:

Proposition: There exists a functor $F:DC\to$ **Comon** DC such that $U \dashv F$ and $UF\cong !$.

Proof: To define F we just check that objects $!\alpha$ admit a natural comonoid structure. Indeed, there is $\mu_{!\alpha}:!\alpha \to !\alpha \otimes !\alpha$, given by diagonal in **C**, $\Delta:U\to U\times U$ and concatenation of sequences $C:U\times X^*\times X^*\to X^*$.

Also $\varepsilon_{!\alpha}:!\alpha\to I$ is easily seen as the canonical unique map to terminal object, $t:U\to 1$ and canonical injection into coproduct, $U\times 1\to X^*$. So $!\alpha$ has a natural comonoid structure and we call F the functor from DC to **Comon** DC which takes α to $<!\alpha,\mu_{!\alpha}:!\alpha\to!\alpha\otimes!\alpha,\varepsilon_{!\alpha}:!\alpha\to 1>$.

To show $U\dashv F$, that is Hom DC $[U\alpha,\beta] \cong$ Hom ComonDC $[\alpha,F\beta]$, we have to describe another natural transformation $\tau_\alpha:\alpha\to!\alpha$, which exists only for α a comonoid. The natural transformation $\tau_\alpha:\alpha\to!\alpha$, is given by identity in the first coordinate and $T:U\times X^*\to X$ in the second, and T is obtained using the comonoid multiplication as many times as necessary to transform the sequence $[x_1x_2...x_k]$ into a single element of X, as in:

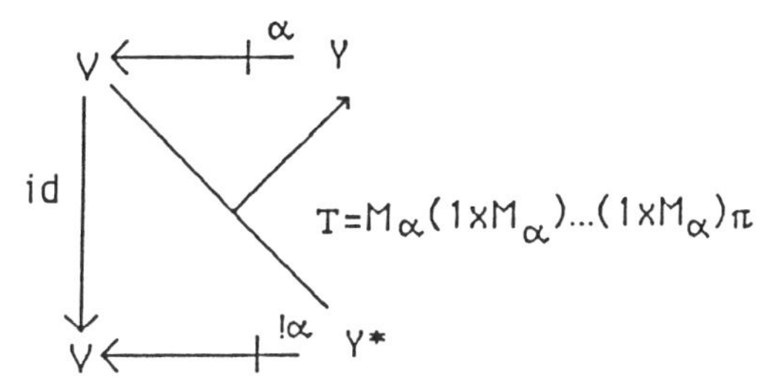

So given $(f,F):U\alpha\to\beta$ we get $(g,G):\alpha\to!\beta$ via composition, $(g,G)=!(f,F).\tau_\alpha$. Conversely given $(t,T):\alpha\to F\beta$ to get $(s,S):U\alpha\to\beta$ we simply compose (t,T) with the natural map $!\beta\to\beta$, given by identity and co-unit of the adjunction on **C**,

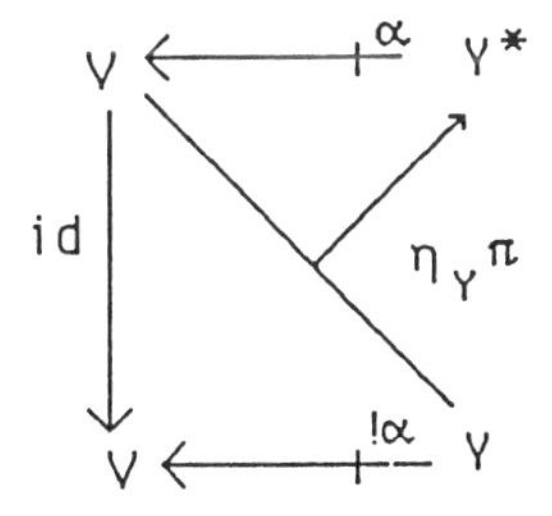

Obviously $UF\cong!$.

2.4

We apply the general theory in section **2.1** to the comonad ! in DC.

As we have shown $U \dashv F$:**Comon** DC$\to$DC, **FACT 1** only says we have a comonad in DC, namely !. FACT **2** says that the comonad $<!,\mu_!,\varepsilon_!>$ in DC gives rise to two categories, respectively, $DC^!$ the !-coalgebras and $DC_!$, the !-Kleisli category.

The objects of $DC^!$ are pairs $<\alpha, h:\alpha\to !\alpha>$, where h satisfies the !-coalgebra diagrams and morphisms are maps in DC, which preserve the coalgebra structure. $DC_!$, on the other hand, has the same objects as DC but $DC_![\alpha,\beta]=DC\,[!\alpha,\beta]$, that is a map in $DC_!$ from α to β corresponds to a map in DC from $!\alpha$ to β. It is easy to see that $DC_!$ inherits the cartesian products from DC, since $DC_!(\gamma,\alpha\&\beta)=DC(!\gamma,\alpha\&\beta)=DC(!\gamma,\alpha)\times DC(!\gamma,\beta)$ $=DC_!(\gamma,\alpha)\times DC_!(\gamma,\beta)$.

FACT 3 says we can draw the following diagram:

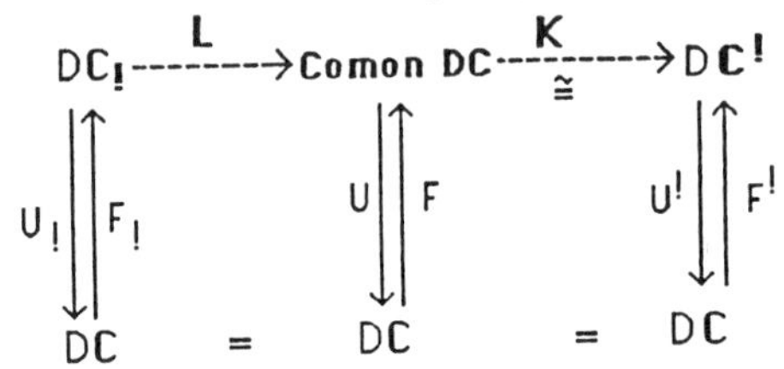

and proposition **2.2** tells us that **K** is an equivalence of categories, i.e that $DC^!\cong$**ComonDC.** We can also easily verify that the tensor product of two comonoids is a comonoid and has natural projections.

One could be tempted to say that the tensor product in DC becomes a cartesian product in **ComonDC**, but for that we need the additional hypothesis of commutativity of the comonoid structure (see [F]).If we assume commutativity of the comonoid structures then we can have **ComonDC** is cartesian, but not necessarily closed, since the internal hom of two comonoids $[\alpha\to\beta]$ need not be a comonoid. More importantly we can have the following isomorphism $!(\alpha\&\beta)\cong !\alpha\otimes !\beta$, which leads directly to the nice result that $DC_!$ is cartesian closed.

2.5

In this subsection we consider the functor ***:Mon C$\to$C** as giving us **free** commutative monoids in **C.** Then we can, as before, define the comonad !:DC$\to$DC and it naturally defines a functor F:DC$\to$**Comon**$_C$DC, where the subscript c only serves to remind us that we are considering now commutative comonoids. Everything goes as before, with **Comon**$_C$**DC**$\cong DC^!$ and all the previous adjunctions. We also have that **Comon**$_C$**DC** is cartesian, its product being $\alpha\otimes\beta$ and the mentioned **Lemma:** In **Comon**$_C$**DC** the isomorphism $!(\alpha\&\beta)\cong !\alpha\otimes !\beta$ holds.

Proof: Since F is a left-adjoint it preserves products, $(\alpha\&\beta)$ is the product in DC, so $F(\alpha\&\beta)=(!(\alpha\&\beta),\mu,\varepsilon)$ is isomorphic to the product of comonoids $!\alpha\otimes !\beta$.

That takes us to the last result in this short subsection, which has a one-line proof.

Proposition: The Kleisli category $DC_!$ associated with the comonad ! is cartesian closed.

Proof: To check that the internal hom $[\beta,\gamma]_{DC!}$ is $[!\beta,\gamma]_{DC}$ we look at:
$DC_!(\alpha\&\beta,\gamma)=DC(!(\alpha\&\beta),\gamma) \cong DC(!\alpha\otimes!\beta,\gamma) \cong DC(!\alpha,[!\beta,\gamma]_{DC}) = DC_!(\alpha,[!\beta,\gamma]_{DC}) = DC_!(\alpha,[\beta,\gamma]_{DC!})$.

2.6

The aim of this subsection is to tie up the linear logic aspects with the category theory presented in the last subsections. So we shall show that (DC,!) corresponds to a model of Intuitionistic Linear logic with modality and that $DC_!$, the Kleisli category of DC, which is cartesian closed, can be related to DC in a interesting way.

We start by reading off the rules for the modality from Girard's paper.These are:

$$\text{I. } \frac{\Gamma, A \vdash B}{\Gamma, !A \vdash B} \qquad \text{II. } \frac{\Gamma \vdash B}{\Gamma, !A \vdash B}$$

$$\text{III. } \frac{\Gamma, !A, !A \vdash B}{\Gamma, !A \vdash B} \qquad \text{IV. } \frac{!\Gamma \vdash A}{!\Gamma \vdash !A}$$

Then it is again very easy (cf **1.4**) to check that the category DC with the comonad $<!,\mu_!,\varepsilon_!>$ described in **2.3** is a model of intuitionistic linear logic with modality. To have rule I it is enough to have a map $!A \to A$ which we have since ! is a comonad.To have rules II and III it is enough to have maps $!A \to I$ and $!A \to !A \otimes !A$ which we have since !A is a comonoid. Finally, to have rule IV it is enough to have half of the adjunction $U \dashv F$ or

$DC(U!B,A) \cong DC_!(!B,!A)$, since we only need that given a map $!B \to A$ we can get a map $!B \to !A$ in a natural way.

The modality was introduced by Girard to recover the strenght of intuitionistic logic, by means of the following translation:

$$A^t = A \quad \text{for } A \text{ atomic}$$
$$(A \wedge B)^t = A^t \,\&\, B^t$$
$$(A \vee B)^t = !A^t \oplus !B^t$$
$$(A \to B)^t = !A^t \multimap B^t$$
$$(\neg A)^t = !A^t \multimap 0$$

So using that translation we want to show the following proposition, which is slightly stronger than the corresponding one in [G-L].

Proposition: $\Gamma \vdash_{Int} A$ iff $!\Gamma^t \vdash_{Lin} A^t$

Proof: We show the direct implication by structural induction on the deduction $\Gamma \vdash_{Int} A$, i.e we look at the last application of any of the rules of I.L and we check that if the premises have been translated then using L.L+modality rules, we can get a translation of the consequence. For instance, the CUT-rule i.e

$$\frac{\Gamma \vdash A \qquad \Delta, A \vdash B}{\Gamma, \Delta \vdash B}$$

becomes the following deduction:

$$\frac{\dfrac{!\Gamma^t \vdash_{Lin} A^t}{!\Gamma^t \vdash_{Lin} !A^t} \qquad !\Delta^t, !A^t \vdash_{Lin} B^t}{!\Gamma^t, \Delta^t \vdash_{Lin} B^t}$$

All the other structural rules are straightforward applications of the modality rules, but for the exchange rule which is simply linear exchange. For the logical rules we have to add some steps, in general very easy ones as in the introduction of conjuntion which becomes

$$\dfrac{\dfrac{\dfrac{!\Gamma^t, !A^t \vdash_{Lin} C^t}{!\Gamma^t, !A^t, !B^t \vdash_{Lin} C^t}}{!\Gamma^t, !A^t \otimes !B^t \vdash_{Lin} C^t}}{!\Gamma^t, !(A\&B)^t \vdash_{Lin} C^t}$$

where we use the isomorphism $!(A\&B) \cong !A \otimes !B$. The only slightly complicated case is the introduction of $\rightarrow$ on the left, which requires the lemma $!(!A \multimap B) \vdash_{Lin} !A \multimap !B$. The lemma is easily given by :

$$\dfrac{\dfrac{\dfrac{!A \multimap B, !A \vdash_{Lin} B}{!(!A \multimap B), !A \vdash_{Lin} B}}{!(!A \multimap B), !A \vdash_{Lin} !B}}{!(!A \multimap B) \vdash_{Lin} !A \multimap !B}$$

and all the other rules are similar to the ones above.

To show the converse we follow the suggestion in [G-L] again and look at the translation which takes linear logic into intuitionistic logic via:

$|!A| = A$ for A atomic

$|A\&B| = |A| \wedge |B|$

$|A \otimes B| = |A| \wedge |B|$

$|A \oplus B| = |A| \vee |B|$

$|A \multimap B| = |A| \rightarrow |B|$

Then it is trivial to check that for A an intuitionistic formula $|A^{\iota}| = A$. Moreover, if $\Delta \vdash_{Lin} B$ then $|\Delta| \vdash_{Int} |B|$, so applying this to $!\Gamma^{\iota} \vdash_{Lin} A^{\iota}$ we have $|!\Gamma^{\iota}| \vdash_{Int} |A^{\iota}|$ which implies $\Gamma \vdash_{Int} A$.

Conceptually the proposition above reflects the fact that in the same way as DC is a model for intuitionistic linear logic, its Kleisli category $DC_!$ is a model for Intuitionistic logic. It is interesting to note that the Kleisli category corresponds exactly to the Diller-Nahm variant of the Dialectica Interpretation. As a remark we recall that the original Dialectica assumed decidability of the atomic formulae. Decidability was essential to prove the consistency of $A \rightarrow A \wedge A$ and the soundness of the whole system depended upon it, cf. Troelstra's comments, page 230. The categorical model gives us a glimpse of why that happens. There we do not have, in general, maps $A \rightarrow A \otimes A$, but if A is decidable we can use the following trick to get a map in DC from a to $\alpha \otimes \alpha$:

$$\begin{array}{ccc} U & \xleftarrow{\ \alpha\ } & X \\ {\scriptstyle \Delta}\downarrow & \searrow\!\!\swarrow\ {\scriptstyle D} & \\ U \times U & \xleftarrow{\ \alpha \wedge \alpha\ } & X \times X \end{array}$$

$$\text{where } D(u,x,x') = \begin{cases} x' & \text{if } u \alpha x \\ x & \text{otherwise} \end{cases}$$

makes (Δ, D) a map of the category.

2.7 Conclusions and acknowledgements

In this paper we have had space to discuss just one of the relations between the Dialectica Interpretation and Linear Logic using categorical models. There is another which I am working on which is the result of a suggestion of Girard. It provides a way to model Classical Linear Logic via certain *-autonomous categories.

I would like to thank R.A.G Seely for the opportunity of presenting this paper, as well as the encouragement to do so. At a different, more personal level I would like to thank Edmund Robinson and Pino Rosolini for unending explanations and discussions. Also the whole "team" in Cambridge, including Peter Johnstone, Glynn Winskel, Thierry Coquand and Thomas Forster, provided a lively and estimulating atmosphere in which work was a pleasure.

But to conclude, I would like one day, perhaps, to be able to express the immense debt I owe my supervisor Martin Hyland. Any one who has ever worked with him, knows exactly what I mean, not only in terms of his fascinating mathematical insights, but also the feelings of friendship and cooperation that he communicates to his colleagues and students.

References

[F] T.FOX, *Coalgebras and cartesian categories*, Comm. Alg. (7) 4 (1976), 665-667.

[G] J.-Y.GIRARD, *Linear Logic*, Theo. Comp. S. 46, (1986), 1-102.

[G-L] J.-Y.GIRARD and Y.LAFONT, *Linear Logic and Lazy Computation*, preprint, 1986

[K] G.M. KELLY, *Basic Concepts of Enriched Category Theory,* Cambridge University Press,1982.

[CWM] S.MACLANE, *Categories for the Working Mathematician,* Springer-Verlag, 1971.

[T] A.S.TROELSTRA, *Metamathematical Investigation of Intuitionistic Arithmetic and Analysis,* Springer-Verlag, 1973.

Valeria C. V. de Paiva
University of Cambridge
Lucy Cavendish College and
Department of Pure Mathematics
16, Mill Lane Cambridge CB2 1SB - U.K.

Contemporary Mathematics
Volume **92**, 1989

COMBINATORS

by
Peter Freyd
University of Pennsylvania

Combinator Algebra has always struck me as a subject almost impossible to watch: it's a lousy spectator sport but has a reputation for being habit-forming as a participator sport. It is only with reluctance that I have been talked into writing this up: I am particularly dubious about the newness of the proof of the extensional principle.

Let $\mathbf{A}$ be a set with a binary operation denoted by catenation. Every element $a \in \mathbf{A}$ names a unary operation on $\mathbf{A}$, to wit, the operation that sends x to ax. The first goal of combinatorial logic is to obtain an $\mathbf{A}$ such that every operation is so nameable. In the usual foundations and with the usual interpretation of the word 'every' this is, of course, impossible. Let us, here, tinker with the notion of 'every': we seek an $\mathbf{A}$ such that every operation *that results from the structure* is nameable by an element of $\mathbf{A}$. Before we get to a formal definition we'll consider a few special cases: $\mathbf{A}$ must have an element I that names the identity operation i.e. that satisfies the equation:

$$Ix = x.$$

For each element $x \in \mathbf{A}$ we will need another element x' that names the *constant* operation whose unique value is x. But this is itself a unary operation (from x to x') and it needs a name. Hence $\mathbf{A}$ must have an element K such that Kx names the constant function valued x, i.e. an element that satisfies the equation:

$$(Kx)y = x.$$

We will henceforth use the standard convention that in absence of parentheses the binary operation is performed moving from left to right: $xyz = (xy)z$. Hence the above equation may be rewritten:

$$Kxy = x.$$

For the next example, note that for each element x we need an element that names the operation obtained by *evaluating at* x, and the unary operation that delivers this name is itself named by an element E, i.e. an element that satisfies the equation:

$$Exy = yx.$$

Our requirement for unary operations forces a requirement about binary operations. For example, the *composition* operation: given two elements x and y let $x \circ y$ denote the operation defined by $(x \circ y)z = x(yz)$. For each x and y there must be an element that names the operation $x \circ y$. For each x there must be a unary operation that sends y to the element that names $x \circ y$. There must be an element that names the operation that sends each x to the element that names the operation that sends each y to the element that names $x \circ y$. This last element is traditionally named B. All is summarized by the equation:

$$Bxyz = x(yz).$$

The binary operation that sends the pair x,y to $x \circ y$ is named by B since $Bxy = x \circ y$.

The standard notation for undoing the mess of words is, of course, the λ-calculus. $x \circ y = \lambda z.\ x(yz)$. $Bx = \lambda y.\ x \circ y$. $B = \lambda xyz.\ x(yz)$.

(It is worth noting that I,K and E are easily interpretable as binary operations: in reverse order, E names the binary operation that is the 'transpose' of the primitive binary operation; K names the binary operation usually called the *left projection* operation; I names the primitive binary operation. The *right projection* operation is named by KI: $KIxy = (KIx)y = Iy = y$.)

The first goal of combinatorial logic is now formalizable as follows: for any expression ϕ in the language and for any variable x we need an expression ψ which names the operation that carries x to ϕ. That is, $\psi x = \phi$ where x does not occur in the expression ψ. (In the language of λ-calculus, we are requiring the set of expressions to be closed under λ-abstraction.) Schönfinkel solved the existential side back in the early 20's: define a **COMBINATOR ALGEBRA** as a set with a binary operation denoted with catenation and constants K and I satisfying the equations:

$$Kxy = x,$$
$$Sxyz = xz(yz).$$

That's all. Note first that if we define I as SKK then we may easily verify the equation $Ix = x$. The proof of **functional completeness**, as it is usually called, proceeds as follows: if x does not occur in a term ϕ then we may take ψ to be $K\phi$; if x <u>is</u> ϕ then ψ may be taken to be I; in the remaining case ϕ is necessarily of the form $\phi'\phi''$ and by induction we may assume that ψ' and ψ'' are x-free terms such that $\psi'x = \phi'$ and $\psi''x = \phi''$ so that we may take ψ to be $S\psi'\psi''$.

(It is worth checking that B is definable as $S(KS)K$, and E as $B(SI)K$.)

The uniqueness condition is usually called the **extensional principle**:

If ψ and ψ' are x-free terms such that $\psi x = \psi' x$ then $\psi = \psi'$.

As just one example, note that our definition of I as SKK was a bit arbitrary: SKS would have done as well. The extensional principle easily implies that $SKK = SKS$. (Indeed, it implies that SKy is independent of y, or if preferred, that $SK = KI$.) The extensional principle implies a host of further equations. It is a remarkable fact, first discovered by Curry, that it is a consequence of just a finite number of those equations. (A combinator algebra that satisfies the extensional principle is often called a **CURRY ALGEBRA**.)

Our approach starts as follows: let $\mathbf{A}$ be a subalgebra of $\mathbf{B}$ and b an element of $\mathbf{B}$. *The subalgebra generated by* $\mathbf{A}$ *and* b *is constructable as* $\mathbf{A}b = \{\ ab : a \in \mathbf{A}\ \}$. The proof: $\mathbf{A}b$ is closed under the primitive binary operation because $(ab)(a'b) = (Saa')b$; it contains the constants because $(KK)b = K$ and $(KS)b = S$; it contains **A** because $(Ka)b = a$; it contains b because $Ib = b$.

The set $\mathbf{A}b$ is the image of the obvious function from **A**: this enlargement of **A** can be viewed as a quotient of **A**. We will specialize to the case that **B** is $\mathbf{A}[X]$, the result of freely adjoining a generator, X, to **A** and we will take the element b to be X. The 'obvious function' from **A** to $\mathbf{A}[X]$ sends a to aX. *The extensional principle says that this onto function is one-to-one.* It says, therefore, that we can construct $\mathbf{A}[X]$ using **A** as its underlying set. We do so as follows:

Given a combinator algebra **A** define $\bullet\mathbf{A}$ ('dot A') to be the combinator algebra obtained by taking the same set as that for **A** and defining $x \cdot y$ (the 'dot product') as Sxy; defining $K^\bullet$ as KK; defining $S^\bullet$ as KS. We need the equations:

$$1^\bullet) \quad K^\bullet \cdot x \cdot y = x,$$
$$2^\bullet) \quad S^\bullet \cdot x \cdot y \cdot z = x \cdot z \cdot (y \cdot z).$$

The undotted version of these equations is:

$$1) \quad S(S(KK)x)y = x,$$
$$2) \quad S(S(S(KS)x)y)z = S(Sxz)(Syz).$$

These are the first two of the four equations which we will use to imply the extensional principle. It is worth checking immediately that they are consequences of the extensional principle: for 1) compute $S(S(KK)x)yz = S(KK)xz(yz) = KKz(xz)(yz) = K(xz)(yz) = xz$; for 2) compute $S(S(S(KS)x)y)zt = S(S(KS)x)yt(zt) = S(KS)xt(yt)(zt) = KSt(xt)(yt)(zt) = S(xt)(yt)(zt) = xt(zt)(yt(zt))$ and $S(Sxz)(Syz)t = Sxzt(Syzt) = xt(zt)(yt(zt))$.

The inclusion map from **A** into $\mathbf{A}[X]$ corresponds to what we will call the **K-map**, the function that sends $a \in \mathbf{A}$ to $Ka \in \bullet\mathbf{A}$. We need an equation to say that this is a homomorphism of combinator algebras:

$$3^\bullet) \quad K(xy) = (Kx) \cdot (Ky).$$

The undotted version is:

$$3) \quad K(xy) = S(Kx)(Ky).$$

The fact that the K-map preserves the two constants is an immediate consequence of the definition of $K^\bullet$ and $S^\bullet$. The verification of 3) from the extensional principle is easier than it was for the previous cases: $K(xy)z = xy$ and $S(Kx)(Ky)z = Kxz(Kyz) = xy$.

We will not need the freeness of $\bullet\mathbf{A}$, but the verification of its freeness continues to serve as a discovery procedure. Given a homomorphism g from **A** to another algebra **B** and an element $b \in \mathbf{B}$ we seek a map f from $\bullet\mathbf{A}$ to **B** so that $f(Ka) = g(a)$ for all $a \in \mathbf{A}$ and $f(I) = b$. Define it by $f(x) = (gx)b$. The uniqueness condition for freeness says that if one choses **B** to be $\bullet\mathbf{A}$, g to be the K-map and X to be I, then f is the identity function. That becomes our final equation:

$$4^\bullet) \quad (Kx) \cdot I = x.$$

The undotted version is:

$$4) \qquad S(Kx)I \;=\; x.$$

The verification: $S(Kx)Iy \;=\; Kxy(Iy) \;=\; xy.$

The semantic argument is now easy: suppose that ψ and ψ' are x-free terms such that $\psi x \;=\; \psi' x$. Let $\mathbf{A}$ be the free algebra on the variables occurring in ψ and ψ'. If $\mathbf{B}$ is any combinator algebra and g any homomorphism from $\mathbf{A}$ to $\mathbf{B}$ then we are given that $(g\psi)b \;=\; (g\psi')b$ for any $b \in \mathbf{B}$. The first two equations say that we may take $\mathbf{B}$ to be $\bullet\mathbf{A}$ and the third equation that we may take g to be the K-map. No equations are needed to take b to be I and we obtain $(K\psi)\cdot I \;=\; (K\psi')\cdot I$. The fourth equation now yields $\psi \;=\; \psi'$.

Contemporary Mathematics
Volume **92**, 1989

POLYNAT IN PER

by
Peter Freyd
University of Pennsylvania

Preliminaries: $A \in \boldsymbol{PER}$ means that A is a partial equivalence relation on the natural numbers; $x \equiv y \bmod A$ means that xAy, i.e. A relates x to y; $x \in Dom(A)$ means that xAx. We need a recursive binary partial operation on the natural numbers, denoted by $x`y$, with the property that for any recursive partial map $f$ there exists $n$ such that $f(x) = n`x$ whenever $f(x)$ converges. If $A,B \in \boldsymbol{PER}$ then $A \to B$ is defined by: $m \equiv n \ mod\ A \to B$ iff

$$\forall xy.\ x \equiv y \bmod A \ \Rightarrow\ m`x \equiv n`y \bmod B.$$

Let p be an element of $\bigcap A.\ (A \to A) \to (A \to A)$. We wish to show:

THEOREM: *There is a formula of the form*

$$(p`x)`a \ = \ x`(x`(\cdots (x`(x`a)) \cdots))$$

that holds for all relevant x,a.

(The converse is straightforward.) By *relevant* x,a we mean that there is $A \in \boldsymbol{PER}$ such that $a \in Dom(A)$ and $x \in Dom(A \to A)$. Note that if x,a is relevant then so is the pair $x`(x`a)$. And so on. That is, there is a canonical choice for A given x,a, to wit, the identity relation on the set $x,\ x`a,\ x`(x`a),\cdots$. We will say that x,a is a **relevant pair** if the following succeeds in defining an infinite sequence:

$$a_n \ = \ if\ (n = 0)\ then\ a\ else\ x`a_{n-1}.$$

Let s name the successor function. i.e. $s`n = n + 1$. We will reserve the variable $k$ to denote $p`s`0$. (In the absence of parentheses we associate to the left: $p`s`n = (p`s)`n$.) We wish to show:

If x,a *is a relevant pair then* $p`x`a \ = \ a_k$.

Two more definitions: $Orb(x,a)$ is the *set* $\{a_0,\ a_1,\ \cdots\}$. The sequence is infinite but $Orb(x,a)$ can be finite. We will say that two relevant pairs x,a and y,a are **extensionally equivalent** if they induce the same sequence. It is clear that, in that case, $p`x`a = p`y`a$.

Our first task will be to prove that $p`s`m = m + k$ for all large m. For that purpose let t be such that $t`n = if\ (n = k + 1)\ then\ n\ else\ n + 1$. Let A be the PER that partitions the natural numbers as

$$\{0\},\ \{1\},\ \cdots\ \{k\},\ \{k + 1,\ k + 2,\ \cdots\}.$$

($k + 2$ cells, all but one of them being one-element sets.) Then $s \equiv t \ mod\ A \to A$ hence $p`s`0 \equiv p`t`0 \bmod A$. But the cell containing $p`s`0$ contains nothing other than k hence $p`t`0 = k$.

For any $m \geq k+2$ let B be the PER whose domain is $\{0,1, \cdots k, k+1\} \cup \{m, m+1, \cdots\}$ and which partitions its domain as
$\{0, m\}, \{1, m+1\}, \cdots \{k, m+k\}, \{k+1, m+k+1, m+k+2, \cdots\}$.
($k+2$ cells, all but one of them being two-element sets.) Then $t \in Dom(B \to B)$. Since $0 \equiv m \ mod\ B$ we have $p`t`0 \equiv p`t`m$, that is, $p`t`m \in \{k, m+k\}$. But clearly $p`t`m \in Orb(t,m)$ hence $p`t`m \geq m$ and we can infer that $p`t`m = m+k$. But s,m and t,m are extenstionally equivalent hence

$$p`s`m = m + k \quad for \quad m \geq k+2.$$

Next, we consider a relevant pair x,a such that $Orb(x,a)$ is finite. Let m be large enough so that $m \geq k+2$ and $m \geq a_n$ all n. Let y be such that $y`n = if\ (n < m)\ then\ x`n\ else\ n+1$. Since y,a and x,a are extensionally equivalent it suffices to show that $p`y`a = a_k$. Let B be the PER whose domain is $Orb(x,a) \cup \{m, m+1, \cdots\}$. In order to describe the identifications B makes on its domain we define a function f by $f(n) = if\ (n < m)\ then\ n\ else\ a_{n-m}$. For $n,n' \in Dom(B)$ define $n \equiv n'\ mod\ B$ iff $f(n) = f(n')$. B has been chosen so that $a \equiv m\ mod\ B$ and $y \in Dom(B \to B)$. Since y,m and s,m are extensionally equivalent we have $p`y`m = m + k$ hence $p`y`a \equiv m + k$. That is, $f(p`y`a) = f(m+k)$. Since $p`y`a < m$ the definition of f thus yields $p`y`a = a_k$.

We finish with the case that $Orb(x,a)$ is infinite. Let k' be defined by $p,x,a = a_{k'}$. We wish to show that $k = k'$. Let m be bigger than both k and k' and let y be such that $y`n = if\ (n = a_m)\ then\ n\ else\ x`n$. Let C be the PER whose domain is $Orb(x,a)$. C partitions its domain as

$$\{a_0\}, \{a_1\}, \cdots, \{a_{m-1}\}, \{a_m, a_{m+1}, \cdots\}.$$

(y,a and C are to x,a as $t,0$ and A were to $s,0$.) Then $x \equiv y\ mod(C \to C)$, hence $p`x`a \equiv p`y`a\ mod\ C$. But $Orb(y,a)$ is finite hence $p'y'a = a_k$ thus $a_{k'} \equiv a_k$. The cell that contains $a_{k'}$ has just one element, hence $k = k'$.

Contemporary Mathematics
Volume **92**, 1989

TOWARDS A GEOMETRY OF INTERACTION

Jean-Yves Girard

dedicated to dag prawitz

Abstract : the paper presents a program for proof-theory, inspired by its growing connections with computer science ; what follows can therefore be seen as an updating of the -now classical- paradigms of Hilbert and Brouwer. Such an illustrious company is a bit daunting, and requires at least some evidence for the program : such evidence is based on the author's recent work on **linear logic** *[7] and on further developments, e.g. [8], [9]. In two words, we start from the idea that both Hilbert and Brouwer have still something to say (and not as antagonistic things as one could expect) provided one reacts against the* **reductionism** *of the former, and the* **subjectivism** *of the latter. The program is essentially about the development of a logic of* **actions**, *i.e. of non-reusable facts (versus* **situations***) : we will accept the intuitionistic dogma that the meaning of a formula is in a proof of it (and not in its truth). The intuitionistic tradition understands proofs as subjectivistic entities, and develops an ideology of* **intensionality**, *which is often nothing more than an alibi for* **taxonomy**, *whereas one may reasonably advocate that proofs are the written trace of underlying geometrical structures. From formalism we shall keep the dogma of* **finitism**, *which was pushed to absurdity by Hilbert because he aimed at an absolute elimination of infinity : we now know that his proposed task, as carried out by Gentzen, involves a finite dynamics whose eventual behaviour is so unpredictible that only the reintroduction of infinite tools -more or less equivalent to the one under elimination- can master it. What must be kept of finitism is the idea of replacement of static infinite situations by finite dynamic actions ; this finite dynamics should lie in the geometrical structure of Gentzen's* **Hauptsatz** *which precisely eliminates the use of infinity in proofs.*

I. basic logical commitments

Before discussing the logical problems that will lead us to a drastic reformulation of logic, let us explain why such essential points have been overlooked by the whole logic tradition (including the author himself, who first found linear logic as a technical decomposition of intuitionistic logic, and only later on reconstructed a kind of commonplace justification for it). The reason has to be searched for in the obsession of *Grundlagen*, i.e. the furious reductionism under Hilbert's flag : since it was possible to *reduce* the formal core of any scientific activity to mathematics, it has been assumed that it was enough to analyze mathematics. Surely -in the spirit the Jivaro ideology- a reduction of mathematics would have induced a reduction of the formal core of other sciences. But reductionism in mathematics failed, and the reduction of -say the formal core of physics- to mathematics is simply a lemma in view of a wrong theorem. In fact, this reduction was a very awkward one, not taking care of the fact that the meaning of implication in real life or physical sciences has nothing to do with its familiar mathematical meaning : it is only through very heavy and *ad hoc* paraphrases that real implication may be put into mathematics (usually the paraphrase is done by adding a parameter for an extraneous time). The logical laws extracted from mathematics are only adapted to eternal truths ; the same principles applied in real life, easily lead to absurdity, because of the interactive (causal) nature of real implication.

I.1. platonism

One usually calls platonism the naive ideology shared by mathematicians about their field :

there is an external (ideal) world formulas are about.

The name "platonism" is a bit unfair to Plato who was not so simple-minded. Anyway, with this external stable infinite world, statements are true or false (independently of our ability to check them), and this justifies principles like

tertium non datur : $A \vee \neg A$

which is the core of *classical logic*, of universal use. Even if there is a lot of criticism to address to this logic, it is still our ultimate mathematical reference : even the most rabid constructivist must acknowledge that. The main problem with platonism is that it leaves no room for the very heart of mathematics, namely *proofs*. If "reality" is prior to anything else, proofs should be seen as a subjective process of understanding the real world. This is not a very satisfactory status, and we are lead to seek a less naive ontology for mathematics. The price to pay for that will be a farewell to the principles of classical logic. Although it is not excluded that classical logic could be compatible with the more elaborated viewpoints to be discussed below, there seems to be serious obstacles to its "constructivization" (however, the fact that linear logic is symmetric w.r.t. linear negation makes the situation a bit less hopeless).

I.2. intuitionism

The intuitionist paradigm is to dump the external world to focus on proofs. The fact that "reality" -which was anyway an abstraction- no longer finds a satisfactory status, has been interpreted in a subjectivistic way (leading to unbelievable nonsense, e.g. Brouwer's "creative subject"). But this relation of intuitionism with some kind of spiritualism is merely a historical accident : remember that at the beginning of the century, there was a lot of metaphysics about the actual nature of the world, matter or energy ; this was a hidden way to be for or against religion and the same kind of opposition existed among logicians between scientism (Hilbert) and mysticism (Brouwer). Brouwer's excessive ideological commitments should be interpreted as a defensive attitude against the spirit of the times -the Absolute Triumph of Science-. Just to say that subjectivism is far from being the only possible reading of Brouwer. One of the greatest ideas in logic (which has not yet received the mathematical treatment it deserves) is Heyting's *semantics of proofs*, which can be summarized by the slogan

proofs as functions.

Typically a proof of $A \Rightarrow B$ is a function mapping proofs of A into proofs of B. This explanation has been thought of as incomplete, since there is no way of deciding whether or not a given function maps proofs of A to proofs of B. But let us observe that :

a definite answer to the problem of "proving that a proof is a proof" would induce a *reduction* of intuitionism to a fixed formal system, which is absurd ;

the criticism to Heyting's paradigm relies on a subjectivistic attitude : I must recognize a proof when I see one of them. But, as we already mentioned, subjectivism is not the only issue : if a proof of A is a program enjoying specification A, we must accept the fact that, in most cases, a program will meet a given specification, but that there will be no way of checking it.

There is a more serious objection to Heyting's paradigm, namely the word "function". The standard acception of this term is "functional graph", while obviously Heyting meant something else. One usually speaks of *intensionality* with a clear subjective background : the function is given together with a description, and this widely opens the door for any kind of *taxonomy*. At that point, Heyting's semantics becomes a bubble-gum, since taxonomy allows us to distinguish between "a" and "A", 1 + 2 and 2 + 1, or even between "A" and "A", since they occur at different places.

I.3. actions

Since the paradigm "function & description" is clearly deficient, we shall propose another one :

proofs as actions.

The term action has to be understood with its familiar meaning (and not with a specific technical one like "group action"). Our program is essentially to try to give a precise mathematical contents to this expression. In the sequel, we shall try to give some hints as to the solution of this problem. The main difficulties will be :

to make a clear distinction between functions and actions

to try to get rid of taxonomy in the description of actions, i.e. to find out what is "beyond" syntax.

The logical twist from functions to actions leads us to linear logic.

II. Linear logic : a logic of action

II.1. actions versus situations

Classical and intuitionistic logics deal with *stable* truths :

If A and A $\Rightarrow$ B , then B, *but* A *still holds.*

This is perfect in mathematics, but wrong in real life, since real implication is *causal*. A causal implication cannot be iterated since the conditions are modified after its use ; this process of modification of the premises (conditions) is known in physics as *reaction*. For instance, if A is to spend $1 on a pack of cigarettes and B is to get them, you lose $1 in this process, and you cannot do it a second time. The reaction here was that $1 went out of your pocket. The first objection to that view is that there are in mathematics, in real life, cases where reaction does not exist or can be neglected. Such cases are *situations* in the sense of stable truths. Our logical refinements should not prevent us to cope with situations, and there will be a specific kind of connectives (*exponentials*, "!" and "?") which shall express the iterability of an action, i.e. the absence of any reaction ; typically !A means to spend as many $'s as one needs. If we use the symbol $\multimap$ (*linear implication*) for causal implication, a usual intuitionistic implication A $\Rightarrow$ B therefore appears as

$$A \Rightarrow B = (!A) \multimap B$$

i.e. A implies B exactly when B is caused by some iteration of A. As far as intuitionistic logic is concerned, the translation inside linear logic using essentially this principle, will be faithful, so nothing will be lost ; it remains to see what is gained.

II.2. the two conjunctions

In linear logic, two conjunctions $\otimes$ (*times*) and & (*with*) coexist. They correspond to two radically different uses of the word "and". Both conjunctions express the availability of two actions ; but in the case of $\otimes$, both will be done, whereas in the case of &, only one of them will be performed (but we shall decide which one). To understand the distinction consider A,B,C :

A :	to spend $1
B :	to get a pack of Camels
C :	to get a pack of Marlboro.

An action of type A will be a way of taking $1 out of one's pocket (there may be several actions of this type since we own several notes). Similarly, there are several packs of Camels at the dealer's, hence there are several actions of type B. An action of type $A \multimap B$ is a way of replacing any specific $ by a specific pack of Camels.

Now, given an action of type $A \multimap B$ and an action of type $A \multimap C$, there will be no way of forming an action of type $A \multimap B \otimes C$, since for $1 you will never get what costs $2 (there will be an action of type $A \otimes A \multimap B \otimes C$, namely getting two packs for $2). However, there will be an action of type $A \multimap B \& C$, namely the superposition of both actions. In order to perform this action, we have first to choose which among the two possible actions we want to perform, and then to do the one selected. This is an exact analogue of the computer instruction IF...THEN...ELSE... : in this familiar case, the parts THEN... and ELSE... are available, but only one of them will be done. A typical misconception is to view "&" as a disjunction : this is wrong since the formulas $A \& B \multimap A$ and $A \& B \multimap B$ will both be provable. By the way, there are two disjunctions in linear logic :

"$\oplus$" (*plus*) which is the dual of "&", which expresses the choice of one action between two possible types ; typically an action of type $A \multimap B \oplus C$ will be to get one pack of Marlboro for the $, another one is to get the pack of Camels. In that case, we can no longer decide which brand of cigarettes we shall get. In terms of computer science, the distinction &/$\oplus$ is reminiscent of the distinction outer/inner non determinism.

"$⅋$" (*par*) which is the dual of "$\otimes$", which expresses a dependency between two types of actions ; the meaning of "$⅋$" is not that easy, let us just say that $A ⅋ B$ can either be read as $A^{\perp} \multimap B$ or as $B^{\perp} \multimap A$, i.e. "$⅋$" is a symmetric form of "$\multimap$"; if one prefers, "$⅋$" is the constructive contents of classical disjunction.

II.3. states and transitions

A typical consequence of the excessive focusing of logicians on mathematics is that the notion of *state* of a system has been overlooked. Let us give some examples to which the discussion applies :

(i) the current state of a Petri net, i.e. the repartition of tokens (a suggestion of Andrea Asperti and Carl Gunter, private communications)

(ii) the current state of a Turing machine (a suggestion of Vincent Danos, private communication)

(iii) the current position during a chessboard game

(iv) the list of molecules present before (or after) a chemical reaction

(v) the current list of beliefs of an expert system, etc.

Observe that all these cases are modelized according to precise protocols, hence can be formalized, so can eventually be written in mathematics : but in all cases, one will have to introduce an extraneous temporal parameter, and the formalization will explain, in classical logic, how to pass from the state S (modelized as (S,t)) to a new one (modelized as (S',t+1)). This is very awkward, and it would be preferable to ignore this *ad hoc* temporal parameter.

In fact, one would like to represent states by formulas, and transitions by means of implications of states, in such a way that S' is accessible from S exactly when $S \multimap S'$ is provable from the transitions, taken as axioms. But here we meet the problem that, with usual logic, the phenomenon of *updating* cannot be represented. For instance take the chemical equation

$$2H_2 + O_2 \Rightarrow 2H_2O$$

a paraphrasis of it in current language could be

"H_2 and H_2 and O_2 imply H_2O and H_2O".

The common sense knows how to manipulate this as a logical inference ; but this common sense knows that the sense of "and" here is non idempotent (because the proportions are crucial) and that once the starting state has been used to produce the final one, it cannot be reused. The features which are needed here are those of "$\otimes$" to represent "and" and "$\multimap$" to represent "imply"; a correct representation will therefore be

$$H_2 \otimes H_2 \otimes O_2 \multimap H_2O \otimes H_2O$$

and it turns out that if we take chemical equations written in this way as axioms, then the notion of linear consequence will correspond to the notion of accessible state from an initial one. On this example we see that it is crucial that the two following rules of classical logic

$$A \Rightarrow A \wedge A \qquad (1) \qquad\qquad A \wedge B \Rightarrow A \qquad (2)$$

become wrong when $\Rightarrow$ is replaced by $\multimap$ and $\wedge$ is replaced by $\otimes$ (principle (1) would say that the proportions do not matter, whereas principle (2) would enable us to add an atom of carbon to the left, that would not be present on the right). The first rule is a way of writing *contraction*, whereas the second rule is a way of writing *weakening*.

Let us now go to an example from computer science, which essentially simplifies what Danos did for Turing machines : take a formal grammar, generated by a finite alphabet $L = \langle a_1, \ldots, a_k \rangle$, and defined by means of transition rules

$$m_i \Rightarrow m'_i \qquad (i = 1, \ldots, n)$$

where the m_i's are words on L ; now, if we take $a_1, \ldots, a_k$ as propositional atoms

we can represent any word	$m = a_{i_1} \ldots a_{i_p}$
by means of the proposition	$m* = a_{i_1} \otimes \ldots \otimes a_{i_p}$
and any transition	$m \Rightarrow m'$
by means of the axiom	$m* \multimap m'*$

Then it is plain than a word m is accessible from a word m_0 just in case the linear implication

$$m_0 \multimap m$$

is provable from the axioms. Here together with the interdiction of the analogues of (1) and (2), a third principle from classical logic :

$$A \wedge B \Rightarrow B \wedge A \qquad (3)$$

becomes wrong if one replaces "$\wedge$" by "$\otimes$" and "$\Rightarrow$" by "$\multimap$". (3) is a possible way of writing *exchange*. In that case, we are not longer speaking of standard linear logic, but of *non-commutative* linear logic, which should play a prominent role in the future, but which is still very experimental.

The example of a formal grammar shows very clearly the close connection between linear logic and any kind of computation process, and it is therefore no wonder that linear logic finds natural applications in computer science. Observe also that the meaning of the arrow of state transition systems is now exactly the meaning of a logical implication, what it should be !

To sum up our discussion about states and transitions : the familiar notion of theory : classical logic + axioms, should therefore be replaced by :

theory = linear logic + axioms + current state

The axioms are there forever; but the current state is available for a single use: hence once it has been used to prove another state, then the theory is updated, i.e. this other state becomes the next current state.

II.4. expert systems

Let us say something more specific about expert systems : to cope with the inadequation of classical logic w.r.t. updating, incredible solutions have been proposed, namely to replace theories by models. The authors of such "ameliorations" of classical logic, e.g. the so called "default reasoning" -that would better be called "deficient reasoning"- seem to ignore that processes of the kind "if A cannot be proved, add ¬A" have been known for more than 50 years to be non-effective at all. By the way, even the idea of thinking of a decision as a completion process is a nonsense : are we interested to add

"Mary is a student"

to our knowledge on the basis it is consistent ? The answer is negative, since there maybe 300 applicants like Mary (all consistently students) but we have limited *resources*, and we have to select 30 of them by means of some reasonable criteria, money, base-ball etc. Here the real decision problem is of *quantitative* nature (minimal inputs, maximal outputs) and cannot be handled by means of classical logic or any of its "improvements". Here linear logic could be of essential use : for reasons already explained, this logic keeps an exact maintenance of resources ; hence, from the initial state S (representing our resources) it could deduce several possibilities, e.g.

$$S \multimap S'\&S''\&S'''$$

and it is up to us (external non-determinism) to select which among S', S'', and S''' will be our next state. We can even refine the logical axioms in such a way that the choice is directly made during the deduction process, but at no moment we have to go to the shame of completion ! It seems that people were led to such regressions simply because they were not able to give divergent transition rules, e.g. $S \Rightarrow S'$ together with $S \Rightarrow \neg S'$, which in classical logic entail the contradiction of S. In linear logic, one can write simultaneous transitions $S \multimap S'$ and $S \multimap S''$, with S', S'' contradictory, without S becoming contradictory ($S\otimes S$ becomes contradictory, but it has a meaning very far from S).

II.5. linear negation

The most important linear connective is *linear negation* $(-)^\perp$ (*nil*). Since linear implication will eventually be rewritten as $A^\perp \mathbin{⅋} B$, "nil" is the only negative operation of logic. Linear negation behaves like transposition in linear algebra ($A \multimap B$ will be the same as $B^\perp \multimap A^\perp$, at least in the commutative case), i.e. it expresses a *duality*, that is a change of standpoint :

action of type A = *reaction of type* $A^{\perp}$

(other aspects of the duality action/reaction are output/input, or answer/question). This change of standpoint is ultimately an inversion of causality, i.e. of the sense of time, but this aspect is still very mysterious, since it involves non-commutative linear logic, which is not yet ripe.

The main property of $(-)^{\perp}$ is that $A^{\perp\perp}$ can, without any problem, be identified with A, like in classical logic ; but, in classical logic, the price to pay was the loss of constructivity. In intuitionistic logic, it is well known that $\neg\neg A$ is not equivalent with A. But it is less known that the familiar equivalence between $\neg A$ and $\neg\neg\neg A$ is not an isomorphism w.r.t. proofs : we have indeed maps in both directions, which form a retraction pair (like in topological vector spaces, between a dual and a tridual). In fact the familiar Gödel $\neg\neg$-translation of classical logic into intuitionistic logic heavily uses identifications of the form $\neg\neg A = \neg\neg\neg\neg A$, which are wrong in terms of proofs, and this is why this translation is not enough to make classical logic constructive.

The involutive character of "nil" ensures De Morgan-like laws for all connectives and quantifiers, e.g.

$$\exists xA = (\forall xA^{\perp})^{\perp}$$

which may look surprising at first sight, especially if we keep in mind that the existential quantifier of linear logic is *effective* : typically, if one proves $\exists xA$, then one proves $A[t/x]$ for a certain term t. This exceptional behaviour of "nil" comes from the fact that $A^{\perp}$ negates (i.e. reacts to) a single action of type A, whereas usual negation only negates some (unspecified) iteration of A, what usually leads to a Herbrand disjunction of unspecified length, whereas the idea of linear negation is not connected to anything like a Herbrand disjunction. Linear negation is therefore more primitive, but also *stronger* than usual negation (i.e. more difficult to prove).

II.6. structural rules

In 1934 Gentzen introduced *sequent calculus*, which is the basic synthetic tool for studying the laws of logic. This calculus is not always convenient to build proofs, but it is essential to study their properties. (In the same way, Hamilton's equations in mechanics are not very useful to solve practical problems of motion, but they play an essential role when we want to discuss the very principles of mechanics.) Technically speaking, Gentzen introduced *sequents*, i.e. expressions $\Gamma \vdash \Delta$, where Γ $(= A_1,\dots,A_n)$ and Δ $(= B_1,\dots,B_m)$ are finite sequences of formulas. The intended meaning of $\Gamma \vdash \Delta$ is that

$$A_1 \text{ and } \ldots \text{ and } A_n \text{ imply } B_1 \text{ or } \ldots \text{ or } B_m$$

but the sense of "and", "imply", "or" has to be clarified. The calculus is divided into three groups of rules (**identity, structural, logical**), among which the structural block has been systematically overlooked. In fact, a close inspection shows that the actual meaning of the words "and", "imply", "or", is wholly in the structural group : in fact it is not too excessive to say that a logic is essentially a set of structural rules ! The three standard structural rules are all of the form

$$\frac{\Gamma \vdash \Delta}{\Gamma' \vdash \Delta'} \quad \text{, more precisely :}$$

α) *weakening* opens the door for fake dependencies : in that case Γ' and Δ' are just extensions of the sequences Γ, Δ. Typically, it speaks of causes without effect, e.g. spending $1 to get nothing -not even smoke-; but is an essential tool in mathematics (from B deduce $A \Rightarrow B$) since it allows us not to use all the hypotheses in a deduction. It will rightly be rejected from linear logic. It is to be remarked that this rule has been criticized a long time ago by philosophers in the tradition of Lewis's "strict implication", and has led to various "relevance logics", which belong to the philosophical side of logic, see [3] for instance. Technically speaking, the rule says that ⊗ is stronger than &, which is wrong, but not that absurd :

$$\text{(I)} \qquad \cfrac{\cfrac{\cfrac{A \vdash A}{A,B \vdash A}\,w \qquad \cfrac{B \vdash B}{A,B \vdash B}\,w}{A,B \vdash A\&B}\,\&}{A\otimes B \vdash A\&B}\,\otimes$$

β) *contraction* is the fingernail of infinity in propositional calculus : it says that what you have, you will always keep, no matter how you use it. The rule corresponds to the case where Γ' and Δ' come from Γ and Δ by identifying several occurences of the same formula (on the same side of "$\vdash$"). To convince yourself that the rule is about infinity (and in fact that without it there is no infinite at all in logic), take the formula INF : $\forall x \exists y\ x < y$ (together with others saying that $<$ is a strict order). This axiom has only infinite models, and we show this by exhibiting 1,2,3,4,... distinct elements ; but, if we want to exhibit 27 distinct elements, we are actually using INF 26 times, and without a principle saying that 26 INF can be contracted into one, we would never make it ! Another infinitary feature of the rule is that it is the only responsible for undecidability : Gentzen's subformula property yields a decision method for predicate calculus, provided we can bound the length of the sequents involved in a cut-free proof, and this is obviously the case in the absence of contraction. This fact has been observed by various people, see e.g. [13] (but there is a lot of reported literature on this subject in Japan and Russia, which may be anterior to the one just mentioned). In linear logic, both contraction and weakening will be forbidden *as structural rules* ; but it would be nonsense not to recover them in some way : we have introduced a new interpretation for the basic notions of logic (actions), but we do not want to abolish the old one (situations), and this is why special connectives (exponentials ! and ?) will be introduced, with the two missing structurals as their main rules. The main difference is that we now *control* in many cases the use of contraction, which, -one should not forget it- means controlling Herbrand disjunctions.

Intuitionistic logic accepts contraction (and weakening as well), but only to the left of sequents : this is done in a very hypocritical way, by restricting the sequents to the case where Δ consists of one formula, so that we are never actually in position to write a single right structural. So, when we have a cut-free proof of $\vdash A$, the last rule must be logical, and this has immediate consequences, e.g. if A is $\exists yB$, then $B[t]$ has been proved for some t etc. These features, that just come from the absence of right contraction, will therefore be present in linear logic, in spite of the presence of an involutive negation. It is perhaps the place to have a discussion on the fuzzy expression "constructive"; what is wrong with classical logic is not that we have *tertium non datur*, since the example of linear logic (namely $A \parr A^{\perp}$) shows that it can very well be interpreted as a communication : on the whole any interpretation of classical disjunction in terms of operations on proofs would be admissible. Such operations are in fact defined via cut-elimination, which precisely gives the meaning of proofs as functions. But, in the case of a cut

$$\dfrac{\dfrac{\Gamma \vdash A,A,\Delta}{\Gamma \vdash A,\Delta} \qquad \dfrac{\Gamma',A,A \vdash \Delta'}{\Gamma',A \vdash \Delta'}}{\Gamma,\Gamma' \vdash \Delta,\Delta'}$$

between a right and a left contraction, Gentzen's procedure yields two essentially different answers, depending on which side we take as the most important : in other terms a symmetric problem gets several asymetric answers. This example shows why classically speaking, all proofs must be identified, that is why classical logic is eventually about *provability*, and not about proofs (in terms of categories, there is a similar remark by Joyal, see [15], pp. 65-67, 126. Hence the ultimate reason why classical logic is not constructive is not because it uses contraction on the right, but because it uses it on *both* sides.

Relevance logic accepts contraction on both sides, and removes weakening ; the result of this cocktail of structural rules is very awkward since it seems that the good combinations are : C+W+E (classical), W+E (affine), E (linear), nothing (linear non commutative). The awkwardness of the logic is made even worse by the adjunction of *ad hoc* distributivity rules, which come from an attempt to stick -as much as possible in the absence of weakening- to classical logic. In terms of resources, relevantists correctly stated that the premise must be used in a causality, but their acceptation of contraction now says that resources may be used *ad libitum* : from two pieces of bread you will never go to one without eating, but from one, you can get 1000, like in Jesus's miracles... The fact that they kept contraction on both sides left the problem of constructivity untouched. In fact, if we clearly lose something with relevantism (namely the simplicity, the elegance of classical logic), it is hard to say what is gained since relevantism is not constructive, and just corresponds to vague philosophical motivations. However, to be fair, relevantists were the first people to distinguish between two conjunctions, two disjunctions, with exactly the rules that we later wrote for ⊗, ⅋, &, ⊕. In terms of formulation of the rules, the distinction ⊗/⅋ is already legitimate in classical logic, but has been overlooked, since these connectives are provably equivalent (this is the reason why the distinction is more natural in logics that do not contain weakening and/or contraction). The fact that "⊗" is stronger than "&" comes from weakening (see (I) above) ; using contraction, we get the reverse implication :

$$\text{(II)} \qquad \dfrac{\dfrac{\dfrac{\dfrac{A \vdash A \qquad B \vdash B}{A,B \vdash A\otimes B}\,\otimes}{A\&B,B \vdash A\otimes B}\,\&}{A\&B,A\&B \vdash A\otimes B}\,\&}{A\&B \vdash A\otimes B}\,c$$

This proof -available in relevance logic too- says that "&" is stronger than "⊗", which is much worse than the other side of the equivalence proved by (I) above. It makes the distinction between ⊗ and & very tiny, and on the whole useless. If we now look at classical logic, we see that the main meaning of "∧" and "∨" as *connectives* is additive (∧ = &, ∨ = ⊕), whereas the meaning of "∧","∨" as *commas* in sequents, is multiplicative (∧ = ⊗, ∨ = ⅋). Classical logic -and to a large extent relevance logic- operates a confusion between additive and multiplicative features, and these features dislike extremely to be confused.

γ) *exchange* expresses the commutativity of multiplicatives : Γ' and Δ' are obtained from Γ and Δ by inner permutations of formulas. It is only for reasons of expressive power that this rule is still present in the main version of linear logic : a certain amount of commutativity is needed in order to make a good use of exponentials. Here one has to mention the work of Lambek (1958) [14], which came out from linguistic considerations; his *syntactic calculus* is based on a non-commutative conjunction (corresponding to our $\otimes$) and two implications, $\multimap$ and $\multimapinv$, one to the right and one to the left. The general framework is that of intuitionistic sequents $\Gamma \vdash A$, with no structurals at all. In spite of its limited expressive power (only multiplicatives) and the artificial intuitionistic framework, this work must be acknowledged as a true ancestor to linear logic ; its connection to linguistics can be seen as the first serious evidence against the exclusive focus on mathematics. Moreover, its rejection of exchange seem to indicate that, eventually, linear logic should be non-commutative, i.e. without exchange. See II.9. for a discussion of non-commutativity.

As soon as weakening and contraction have been expelled, one can imagine other structural rules, among which the *mix* rule

$$\frac{\Gamma \vdash \Delta \qquad \Gamma' \vdash \Delta'}{\Gamma, \Gamma' \vdash \Delta, \Delta'}$$

has some interest. If one were to accept this rule, then good taste would require to add the void sequent $\vdash$ as an axiom (without weakening, this has no dramatic consequence). *Mix* is connected with the neutral multiplicative elements 1 and $\bot$, more precisely, it states that they are the same, which is of course a simplifying hypothesis. It can also be viewed as the fact that $\otimes$ is stronger than $⅋$, which means that the absence of interaction should be a form of interaction. As a matter of fact there is not enough material to make a definite judgement as to a possible inclusion of *mix*. For the reader acquainted with proof-nets [7], the inclusion of this rule would have an unpleasant feature, namely to abolish the idea of *cyclicity* which enabled us to forward the information from any part of the structures to any other part.

<u>II.7. linear sequent calculus</u>

In order to present the calculus, we shall adopt the following notational simplification : formulas are written from literals $p, q, r, p^\perp, q^\perp, r^\perp$ etc. and constants 1, $\bot$, $\top$, 0, by means of the connectives $\otimes$, $⅋$, $\&$, $\oplus$ (binary) $!$, $?$ (unary) and the quantifiers $\forall x$, $\exists x$. Negation is *defined* by De Morgan equations , and linear implication is also a defined connective :

$$1^\perp =_d \perp \qquad \perp^\perp =_d 1$$
$$\top^\perp =_d 0 \qquad 0^\perp =_d \top$$
$$p^\perp =_d p^\perp \qquad p^{\perp\perp} =_d p$$
$$(A \otimes B)^\perp =_d A^\perp ⅋ B^\perp \qquad (A ⅋ B)^\perp =_d A^\perp \otimes B^\perp$$
$$(A \& B)^\perp =_d A^\perp \oplus B^\perp \qquad (A \oplus B)^\perp =_d A^\perp \& B^\perp$$
$$(!A)^\perp =_d ?A^\perp \qquad (?A)^\perp =_d !A^\perp$$
$$(\forall x A)^\perp =_d \exists x A^\perp \qquad (\exists x A)^\perp =_d \forall x A^\perp$$

$$A \multimap B =_d A^\perp ⅋ B$$

Sequents are now of the form $\vdash \Delta$, i.e. the left hand side is void. General sequents $\Gamma \vdash \Delta$ can be mimicked as $\vdash \Gamma^\perp, \Delta$.

IDENTITY GROUP

$\vdash A, A^\perp$ (Identity axiom)

$$\frac{\vdash \Gamma, A \qquad \vdash A^\perp, \Delta}{\vdash \Gamma, \Delta}\ (Cut)$$ (Cut rule)

STRUCTURAL GROUP

$$\frac{\vdash \Gamma}{\vdash \Gamma'}\ (e)$$ (exchange)

in this rule Γ' is obtained from Γ by a permutation.

LOGICAL GROUP

multiplicatives

$$\frac{\vdash \Gamma, A \qquad \vdash B, \Delta}{\vdash \Gamma, A \otimes B, \Delta}\ (\otimes) \qquad \frac{\vdash \Gamma, A, B}{\vdash \Gamma, A ⅋ B}\ (⅋)$$

$\vdash 1$ (axiom)

$$\frac{\vdash \Gamma}{\vdash \Gamma, \perp}\ (\perp)$$

additives

$$\frac{\vdash \Gamma, A \qquad \vdash \Gamma, B}{\vdash \Gamma, A \& B}\ (\&) \qquad \frac{\vdash \Gamma, A}{\vdash \Gamma, A \oplus B}\ (\oplus^1) \qquad \frac{\vdash \Gamma, B}{\vdash \Gamma, A \oplus B}\ (\oplus^2)$$

$\vdash \Gamma, \top$ (axiom)

exponentials

$$\frac{\vdash ?\Gamma, A}{\vdash ?\Gamma, !A}\ (!) \qquad \frac{\vdash \Gamma, A}{\vdash \Gamma, ?A}\ (d?) \quad \text{(dereliction)}$$

$$\frac{\vdash \Gamma}{\vdash \Gamma, ?A}\ (w?) \quad \text{(weakening)}$$

$$\frac{\vdash \Gamma, ?A, ?A}{\vdash \Gamma, ?A}\ (c?) \quad \text{(contraction)}$$

quantifiers

$$\frac{\vdash \Gamma, A}{\vdash \Gamma, \forall x A}\ (\forall) \qquad \frac{\vdash \Gamma, A[t/x]}{\vdash \Gamma, \exists x A}\ (\exists)$$

(In rule ($\forall$), x must not be free in Γ)

II.8. comments

The rule for "⅋" shows that the comma behaves like a hypocritical "⅋" (on the left it would behave like "⊗"); "and", "or", "imply" are therefore read as "⊗", "⅋", "⊸".

identity group : the principles of this group express that "A is A", which is perhaps the ultimate meaning of logic ... In other terms, they say that an action (output, answer) of type A is a reaction (input, question) of type $A^\perp$. One can also view the identity axiom as the identity function from A to A, or from $A^\perp$ to $A^\perp$, and the symmetry of the axiom forbids us to choose between these two legitimate interpretations. But in fact, the interpretation as a function is wrong, since it forgets the dynamics. Let us try to understand this very important point : the only dynamical feature of the system is the cut-rule ; without cut, there would be no action performed. The cut puts together an action and a reaction of the same type (i.e. an action of type A and an action of type $A^\perp$) and something happens, namely "cut elimination". Let us take a very trivial analogy, namely DIN plugs for electronic equipment : we may think of the axiom $\vdash A, A^\perp$ as an extension wire between two complementary DIN plugs :

$A^\perp$ A

More generally, we can think of a sequent Γ as the interface of an electronic equipment, this interface being made of DIN plugs of various forms ; the negation corresponds to the complementarity between male and female plugs. Now a proof of Γ can be seen as any equipment with interface Γ. Now, the cut rule is well explained as a plugging :

Γ....——■ ■——....Δ

A $A^\perp$

the main property of the extension wire is that

Γ....——■ ■——■

can be replaced by

Γ....——■

It seems that the ultimate, deep meaning of cut-elimination is located there. Observe that commonsense would forbid self-plugging of an extension wire :

which would correspond, in terms of the proof-nets of [7] to the incestuous configuration :

$$\frac{\overline{A \qquad\qquad A^\perp}}{}\ Cut$$

which admits two shortrips.

structural group : see additional discussion in II.9. .

logical group :

multiplicatives and additives : notice the difference between the rule for ⊗ and the rule for & : ⊗ requires disjoint contexts (which will never be identified unless ? is heavily used) whereas & works with twice the same context. In a similar way, the two disjunctions are very different, since ⊕ requires one among the premises, whereas ⅋ requires both).

exponentials : ! and ? are modalities. Modalities are very special connectives. The rule for ! does not define this connective, since the context Γ must start with "?", which is the dual of "!", i.e. "!" is eventually defined in terms of itself. To understand the difference between exponentials and -say- additives, let us remark that, if we write additive rules for another pair &', ⊕', then the equivalence of this new pair with &, ⊕ is immediate, whereas, one can have another pair !', ?', with the same rules as !, ?, but non provably equivalent. The general symmetries of sequent crlculus (namely that a cut on a complex formula splits into simpler cuts, and the same for axioms) are such that, when

we know the (right) rule for &, then the rules for $\oplus$ are forced, and conversely. In the case of modalities, the rule for ! does not determine the rules for ? (except dereliction), in particular, the possibility of refining the exponentials is widely open for that reason. The connective "!" (*of course*) has the meaning of a storage. To be very precise, "!" indicates the *potentiality* of a duplication : an action of type !A consists of an action of type A on the slot of a copying machine. The rules for "?" (*why not*) enable us, via cut, to make this machine work : dereliction just takes back the original action, weakening destroys it, while contraction duplicates the data. In terms of computer, the rules for ! and ? correspond to storing, reading, erasing and duplicating. The importance of exponentials w.r.t. memory has been pointed out by Yves Lafont : see [11], in particular the idea of computing without garbage collector.
quantifiers : they are not very different from what they are in usual logic, if we except the disturbing fact that $\exists$ is now the exact dual of $\forall$. It is important to remark that $\forall$ is very close to & (and that $\exists$ is very close to $\oplus$). But are there quantifier analogues for "$\otimes$" and "$⅋$" ? Clearly such "multiplicative quantifiers" should differ from $\forall$ and $\exists$ in the sense that they would allow not a single instanciation, but several simultaneous instanciations, and therefore they would be very close to exponentials. In other terms, one cannot exclude the replacement of the awkward modals "!" and "?" by some more regular quantifiers. The eight logical operations can be written on a cubic pattern :

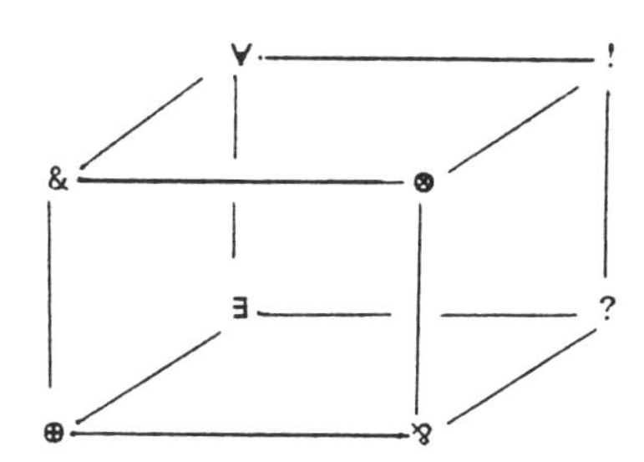

organized along the three oppositions

vertical : conjunctive/disjunctive

horizontal : one/all

oblique : binary/uniform.

II.9. non-commutativity

Non-commutative linear logic is still very experimental, and we shall here just discuss the most obvious system, which contains no exponentials. Compared to II.7., we must make the following adaptations :

$(A\otimes B)^{\perp} =_d B^{\perp}⅋A^{\perp}$ $\qquad\qquad$ $(A⅋B)^{\perp} =_d B^{\perp}\otimes A^{\perp}$

Besides linear implication

$$A \multimap B =_d A^\perp ⅋ B$$

there will coexist linear *retro*-implication,

$$B \multimapinv A =_d B ⅋ A^\perp$$

Transposition will no longer be the identification between $A \multimap B$ and $B^\perp \multimap A^\perp$ (which involved commutativity of ⅋), but the fact that $A \multimap B$ and $A^\perp \multimapinv B^\perp$ are literally the same. One of the most exciting (and problematic) intuitions about non-commutativity is that it should have a temporal meaning : $\multimap$ is usual causality (in the future), whereas $\multimapinv$ is past causality. Everybody will easily find examples of past causality in real life; however since we are not accustomed to think rigorously in those terms, the most basic evidences may be misleading, and the existence of some formal model (even very incomplete) might be of essential use. This is to say that in such matters, we cannot use common sense at all, and that we must follow some kind of mathematical pattern, even against intuition. Here comes a very interesting phenomenon (that can be understood by those who know proof-nets, see [7]) : the natural way of introducing non-commutativity is not to expell exchange, but to restrict it to circular permutations. In fact this corresponds exactly to the restriction to *planar* proof-nets, i.e. to proof-nets with no crossings between the axiom links. But if we start to forbid crossings of lines in the non-commutative case, do this mean that the commutative proofs are incorrect ? Or should we, as suggested by Freyd (private discussion) view them as *proof-braids* ? Surely linear logic has a lot of relations with monoidal categories (there are even categories like Barr's "*-Autonomous categories" [1] which have additives, multiplicatives and an involution), and proof-braids could be a good candidate to describe the various isomorphisms in a non-commutative monoidal category. But what could be the temporal meaning of the twistings between axiom links ? This seems to be an absolute mystery ...

The restriction to circular permutations (which means that we consider the sequents as written on a circle) makes a reasonable candidate for a logic (as long as we ignore exponentials). In view of the identification between

$$B \otimes A \multimap C \qquad \text{and} \qquad A \multimap (B \multimap C)$$

which is essentially associativity of "$\otimes$", we can even understand that in the product $B \otimes A$, the second component is done before the first one.

We can keep the rules as they have been written in II.7. ; if we want to have 2-sided sequents, it is natural to translate $A_1, \ldots, A_n \vdash B_1, \ldots, B_m$ as $\vdash A_n^\perp, \ldots, A_1^\perp, B_1, \ldots, B_m$. One of the obvious features of the calculus is that the two implications do not mingle, and this will be essential when we shall try to work with PROLOG.

II.10. logic programming

It seems that the exact relation of PROLOG (and more generally Robinson's resolution) to Gentzen's *Hauptsatz* has been overlooked in the full literature on logic programming, so let us explain this roughly : the famous result of Gentzen states that, if $\vdash A$ is a logical consequence (in the sense of classical logic) of axioms $\Gamma_i \vdash \Delta_i$, then there is a proof of A, with the following properties :

it uses as axioms instanciations $\Gamma'_i \vdash \Delta'_i$ of the original axioms

the cut rule is restricted to cut-formulas occuring in some instanciation of those axioms.

(For a textbook presentation of this result, see e.g. [6], p.123.)

This is not exactly Robinson's resolution, but it does not take much to get it from that result : assume now that the axioms are Horn clauses, i.e. Γ_i and Δ_i are made of atomic formulas, and Δ_i is just one formula ; assume also that A is atomic. Then since cut is restricted to atomic formulas, logical rules cannot be used (because logical symbols cannot disappear), and we are therefore left with the following list of principles :

i) instanciations of axioms (it is enough to consider non-logical axioms)

ii) cut

iii) weakening

iv) contraction

v) exchange

It is easily shown that one can restrict oneself to deductions made only of Horn clauses ; then one gets rid of :

iii) weakening : if we replace some clause $\Gamma \vdash C$ by $\Gamma, B \vdash C$, since our goal has no left part, the fate of B is to disappear by means of a cut, and we can assume that this will happen by proving at some later moment $\vdash B$: but this is masochism ! It would have been simpler not to weaken at all.

iv) contraction : if we replace some clause $\Gamma, B, B \vdash C$ by $\Gamma, B \vdash C$, since at some moment, our B must be cut with one proof of $\vdash B$ (as above), it would have been possible to cut twice at the level of $\Gamma, B, B \vdash C$ to get $\Gamma \vdash C$. Observe that this idea removes a rule, but the price to pay is a doubling of the task, since B will be searched for twice.

v) exchange : it is finally possible to do our cuts in such an order that exchange is never used.

Finally, we are left with steps i) and ii), and the familiar resolution method is nothing but automatic proving inside i) + ii). But this means that we are in fact trying to make proofs inside non-commutative linear logic ! Now it is necessary to make a very important -although commonplace- remark : so-called logic programming is logical at two levels

externally, since it is concerned with the notion of classical consequence

internally, since its operationality obeys to certain logical laws (which indeed are those of non-commutative linear logic)

The two logics, the internal and the external one, have no reason to coincide ; it is because people want them to coincide that they are lead to absolute nonsense. Let us take an example : in classical logic (which is the external logic of PROLOG), everybody knows that there are 16 binary connectives, period. A quick inspection of the list shows that only one of them looks like a conjunction : this is the familiar boolean conjunction which is desperately commutative and idempotent. But is the inner conjunction of PROLOG like that ? If we compare the clauses $A,A \vdash B$ and $A \vdash B$, it is very clear that the first one requires a duplication of the intermediate goal A, hence *internally* speaking, "A and A" does not mean "A", unless we decide to ignore problems of efficiency.

But the inner conjunction is not even commutative, since the clauses $A,B \vdash C$ and $B,A \vdash C$ have a very different behaviour, typically when B fails and A neither succeeds nor fails ; the failure of commutativity is by the way much more extreme than the failure of idempotency. Since classical logic implies the commutativity of conjunction, the only way to avoid an immediate contradiction in the example just given is to identify failure with the absence of success, i.e. to introduce unprovability, which leads to the deficient nonsense already met in section II.4. : this is the so-called "closed world assumption", which came from the furious attempt to identify inner and outer logic, and just succeeded in writing absurdities both from the viewpoint of classical logic and from the viewpoint of operationality.

But failure is not the absence of success : failure is a positive information, namely that we have followed a certain operational pattern *up to the end*, and that we are aware of it. This is definitely different from the absence of success which could for instance involve non terminating loops. So, if we can express exactly which operational pattern we have followed, it will be possible to internalize failure as the provability of some statement, and since provability is the kind of feature that can be mechanized, we are not led to computational nonsense, like in the case of the "closed world assumption". In particular, the idea of "negation as failure" is perfectly sound, provided we are prepared to accept that such a negation cannot at the same time be classical ! In fact we propose to identify negation in PROLOG with linear negation, then to use the usual implication "$\multimap$" to speak of success, and the reverse arrow "$\multimapinv$" to speak of failure. Let us give a very primitive example : we just consider propositional clauses A,B,C etc. and we assume that there are only two clauses ending with E, namely

(I) $A,B \vdash E$

(II) $C,D \vdash E$

Let us describe in familiar terms the part of the procedure involving the subgoal E : we try to prove E by means of (I), and in case attempt (I) fails, we try by means of (II); if (II) fails as well, then E fails. To say that attempt (I) succeeds is just $A \otimes B$ ("do B, then do A"), and the fact that success is then forwarded to E is therefore

(1) $A \otimes B \multimap E$

Failure of attempt (I) can be decomposed into two subcases : either B fails, or B succeeds and then A fails. This will be exactly represented by $B^{\perp} \oplus (A^{\perp} \otimes B)$; now, once attempt (I) has failed, a success of attempt (II) will be forwarded to E :

$C \otimes D \otimes (B^{\perp} \oplus (A^{\perp} \otimes B)) \multimap E$

which can be written as two axioms

(2) $C \otimes D \otimes B^{\perp} \multimap E$

(3) $C \otimes D \otimes A^{\perp} \otimes B \multimap E$

Finally, failure of both (I) and (II) will cause a failure of E :

$(D^{\perp} \oplus (C^{\perp} \otimes D)) \otimes (B^{\perp} \oplus (A^{\perp} \otimes B)) \multimap E^{\perp}$

which can be written as

(4) $D^{\perp} \otimes B^{\perp} \multimap E^{\perp}$

(5) $D^{\perp} \otimes A^{\perp} \otimes B \multimap E^{\perp}$

(6) $C^{\perp} \otimes D \otimes B^{\perp} \multimap E^{\perp}$

(7) $C^{\perp} \otimes D \otimes A^{\perp} \otimes B \multimap E^{\perp}$

Axioms (4)-(7) are indeed retro-causalities "in order to do E, one must do...", Typically (4) is $B \parr D \multimapinv E$, etc. In fact, (1)-(7) can be written as right-handed sequents made of literals, typically (5) becomes

(5') $\vdash B^{\perp}, A, D, E^{\perp}$ etc.

Then one should now compare this axiomatization with the operationality of PROLOG, and that task should better be done by specialists of logic programming (by the way, there is some work in preparation with Jean Gallier and Stan Raatz, whose goal is the study of any form of control in logic programming by means of a linear logic axiomatization).

Let us just mention some key points :

i) it is very important that the two implications do not mingle ; cyclic exchange may cause some problems, easily solved by adding two special constants **e**, **s**, with $\mathbf{e}^\perp =_d \mathbf{s}$, $\mathbf{s}^\perp =_d \mathbf{e}$, and writing, instead of (5') :

$$(5'') \qquad \vdash \mathbf{e}, B^\perp, A, D, E^\perp \otimes \mathbf{s}$$

and then looking at the provability of goals

$$\vdash \mathbf{e}, E \otimes \mathbf{s} \qquad \text{or} \qquad \vdash \mathbf{e}, E^\perp \otimes \mathbf{s}$$

to express success or failure.

ii) a careful look shows that we are indeed using several meanings of "and", "or", "implies". Typically, "$\oplus$" is used when we are listing several possibilities (e.g. for failure), whereas "$⅋$" occurs by means of retrocausality and indicates a complex interaction between subgoals.

II.11. relation with intuitionistic logic

There is a translation of intuitionistic logic inside linear logic : read

$A \Rightarrow B$	as	$!A \multimap B$
$A \wedge B$		$A \,\&\, B$
$A \vee B$		$!A \oplus\, !B$
$\forall x A$		$\forall x A$
$\exists x A$		$\exists x !A$
$\neg A$		$!A \multimap \mathbf{0}$

see [7], ch. 5 for more details. This translation is faithful not only w.r.t. provability, but also w.r.t. proofs. As a matter of fact, linear logic came from a denotational decomposition of disjunction which involved linearization processes, and later on gave rise to all the connectives of linear logic. But an essential concern has always been the possibility of a faithful translation of intuitionistic logic, since we were afraid of a possible loss of expressive power : this is why non-commutative linear logic was a bit overlooked in the beginning.

To give a demonstration of the improvements immediately caused by this translation, let us look at the familiar technique of *fake substitution* in typed λ-calculus : in order to improve implementations, people imagined to indicate substitutions, but not to make them : between $(\lambda x t)u$ and $t[u/x]$, they have introduced an intermediate step, namely $t\langle u/x\rangle$, where $.\langle./x\rangle$ stands for a fake substitution. This operation has no logical status, which makes its manipulation dangerous. In linear logic, this intermediate step will be built in : since $A \Rightarrow B$ is $!A \multimap B$, we need two steps to mimick a λ-abstraction, so to speak

(1) a "memory" step

(2) a "linear λ-abstraction"

when we normalize, usual β-conversion splits into two steps, one dealing with (2), the other with (1), and these two steps altogether mimick β-conversion. But if one stops at step (2), then one gets something that could be denoted by $t\langle u/x\rangle$, and this eventually gives a logical status to what was originally just control.

III. the main methodological conflicts

III.1. against reductionism

The first important methodological contradiction lies in the oppositions

dynamic	/	*static*
sense	/	*denotation*
finite	/	*infinite*

these three oppositions are different aspect of the same problem.

Let us start with Frege, who distinguished between *sense* and *denotation* : if we take the sentence

Erich von Stroheim is the author of *Greed*

"Erich von Stroheim" and "the author of *Greed*" have the same denotation, i.e. represent the same external object, but have not the same sense (otherwise it would be pointless to state such a sentence). Denotationally speaking the two expressions refer to the same thing, whereas one has to check something (look at a dictionary, make a proof, a computation) to relate their two distinct senses. This is why

denotation is static, sense is dynamic.

The only extant mathematical semantics for computation are denotational, i.e. static. This is the case for the original semantics of Scott [17], which dates back to 1969, and this remains true for the more recent *coherent semantics* of the author [7]. These semantics interpret proofs as functions, instead of actions. But computation is a dynamic process, analogous to -say- mechanics. The denotational approach to computation is to computer science what statics is to mechanics : a small part of the subject, but a relevant one. The fact that denotational semantics is kept constant during a computational process should be compared to the existence of static invariants like mass in classical mechanics. But the core of mechanics is dynamics, where other invariants of a dynamical nature, like energy, impulsion etc. play a prominent role. Trying to modelize programs as actions is therefore trying to fill the most obvious gap in the theory. There is no appropriate extant name for what we are aiming at : the name "operational semantics" has been already widely used to speak of step-by-step paraphrases of computational processes, while we are clearly aiming at a less *ad hoc* description. This is why we propose the name

geometry of interactions

for such a thing.

The inadequation of the denotational approach w.r.t. computation becomes conspicuous if we observe that such semantics will have a strong tendency to be infinite, whereas programs are finite dynamical processes. So to speak, the denotational approach exchanges a finite dynamical action for an infinite static situation. How is it possible ? Simply by considering not only the behaviour of the system in an actual run (which is very difficult to analyze), but taking at the same time all possible behaviours ; typically, if a program is functional, by listing, in front of every possible input, the corresponding output. This yields an inifinitary expansion, in which the results are flatly embedded. Let us take a very basic example : a program of type

$$(A_1\&B_1)\otimes\ldots\otimes(A_n\&B_n)$$

will be run by choosing between A_1 and B_1, ..., A_n and B_n. The denotational approach will consider altogether the 2^n possible runs, so to include the actual one. But only one of these runs is actual, and this interpretation is clearly wrong. Remark here that syntax is much less greedy, since it encodes the situation with only 2n data. Another familiar example of replacement of a finite dynamical process by an infinite listing is the well-known "ω-rule", prominent in German proof-theory : since the dynamics of induction (i.e. recurrence) is very difficult to handle, one introduces flat listings A[0],A[1],A[2],....; then one has to cope with infinitary syntax, which means that eventually one has to encode it by means of -say- Kleene brackets to come back to the finite. This would not be so bad if the dynamics of Kleene indices were not so *ad hoc*. One has eventually exchanged an intrinsic dynamics for an *ad hoc* one, and something essential has been lost. This is why there is little room for the ω-rule in computer science. It is not absurd to dream of a direct dynamical approach to induction, where the infinite proof-tree would only be an ideal (direct) limit that we never reach : but this is not that easy...

Hilbert's mistake, when he tried to express the infinite in terms of the finite was of a reductionist nature : he neglected the dynamics. The dynamics coming from the elimination of infinity is so complex that one can hardly see any reduction there. But once reductonism has been dumped, Hilbert's claim becomes reasonable : infinity is an undirect way to speak of the finite ; more precisely infinity is about finite dynamical processes.

These basic oppositions are at work when we try to understand "!" ; should we think of !A as something like an infinite tensor $\bigotimes_{\omega} A$? With some minor adjustments, e.g. $\bigotimes_{\omega}(1\&A)$, this is denotationally sound ; however, such an identification would be a complete dynamical nonsense. This is because the rule for "!" indicates unlimited *possibilities* of duplication, but not a concrete one : the duplication occurs during elimination of cuts with $?A^{\perp}$, and it is in this dual part that the information "how many copies do you want" is located. We must not confuse a copying machine that can produce 3000 copies of an original document with these 3000 copies : maybe we only need one. The clarification of this point could be of great importance : consider for instance *bounded exponentials* $!\alpha A, ?\alpha A$, that could be added to linear logic with the intuitive meaning of "iterate α times". They obviously obey to the following laws :

$$\frac{\vdash ?\tau\Gamma, A}{\vdash ?\alpha\tau\Gamma, !\alpha A}(!) \qquad \frac{\vdash \Gamma, A}{\vdash \Gamma, ?1A}(d?) \quad \text{(dereliction)}$$

$$\frac{\vdash \Gamma}{\vdash \Gamma, ?0A}(w?) \quad \text{(weakening)}$$

$$\frac{\vdash \Gamma, ?\alpha A, ?\beta A}{\vdash \Gamma, ?\alpha{+}\beta A}(c?) \quad \text{(contraction)}$$

and this shows that there is some underlying polynomial structure in the exponentials. Now, it is not always the case that we can associate polynomials to all exponentials occuring in a proof of standard linear logic, especially when we have to deal with cut ; hence the proofs admitting such polynomial indexings are very peculiar. In fact they admit a normalization in polynomial time : it has already been observed in [7], that, without contraction, the cut-elimination process is essentially shrinking (hence in linear time). Now, if we are with polynomial exponentials, what we can do is first replace all $!\alpha A$ (resp. $?\alpha A$) by something like $\bigotimes_{\alpha} A$ (resp. $\bigparr_{\alpha} A$), which induces a polynomial expansion of the proof, and then normalize. Some experimental refinements of linear logic -in order to cope with the converse problem- are under study (with André Scedrov and Phil Scott) and it seems likely that polynomial time functions are typable in such systems. Unfortunately what has so far been done is syntactically awkward, whereas the idea clearly deserves a natural syntax.

IV.2. against subjectivism

Finally, what is closest to the idea of dynamics is syntax, which makes all the necessary distinctions of sense and has the good taste of being finite. So why not contenting oneself with syntax ? This leads to our second opposition :

geometry / taxonomy

In fact this opposition is much more central than the former ones. The problem with syntax is that it is good in many respects but one : it contains irrelevant informations. These informations are very often of temporal nature, and induce an *ad hoc* temporality. Our problem will be to find out what is hidden behind syntax, without going to denotation, so to speak :

a non-subjectivistic approach to sense.

To understand how syntax may convey artificial information, imagine that I want to write something like $A \otimes B \otimes C$. Since $\otimes$ is binary, I must choose between $(A \otimes B) \otimes C$ and $A \otimes (B \otimes C)$, whereas I had in mind a ternary construction. So both solutions contain an irrelevant information, namely some *ad hoc* temporality in the building of $A \otimes B \otimes C$. For instance if I have chosen the first representation, then the subconfiguration $A \otimes B$ will be easier to handle than $B \otimes C$.

α) *what is taxonomy ?*

Taxonomy is the habit of classification by means of dictionaries, languages etc. It is an essential human activity, and corresponds to our need for putting some order into the reality we are dealing with. Let us mention :

i) the *classification of animals* into various species, subspecies etc.

ii) Mendelejeff's *periodic classification* of elements

iii) the use of *coordinates* on Earth

iv) *musical notation*

These four examples are very useful in real life, but not of the same quality. For instance, there is something arbitrary in iii) (e.g. the starting meridian, the units), but it stays reasonable as long as the coordinates are organized along the rotating axis of Earth ; iv) is clearly the product of historical accidents, which have eventually produced 10-odd keys, and is clearly an obstacle to musical practice ; i) is very *ad hoc* : one counts numbers of eyes, wings, testicles, but the discovery of Australia forced taxonomists to add new entries to the classification ; ii) is very good, because based on the number of protons of atoms ; it is so good that new elements were found (or even created) by looking at the gaps of the table, which would have been impossible if these elements had been classified in the style of i) by means of shapes, colours, odours etc. To understand the difference between the scientific quality of i) and ii) : nowadays, we can exclude the existence of a fabulous metal like orichalch, but not of a fabulous beast like a unicorn.

Taxonomy is the bureaucracy of science. This means that -like its State analogue- it is useful, but has a propensity to produce endless obstacles to the execution of the simplest task : its natural tendency is to live on its own and to expell the reality it is about. Taxonomy can be that bad because of the human need to name, to classify, to number, by means of anything whatsoever, Zodiacal signs, "psychological" tests etc.

β) *progress in science* is very often connected with a fight against taxonomy : typically, the discovery that the distinction parabola / hyperbola / ellipse is irrelevant from the algebraic viewpoint, and the subsequent introduction of the projective plane. Algebraically speaking the distinction between these three types of curves appears as a taxonomy, namely the choice of points at infinity. The usual way of getting rid of taxonomy is by exhibiting some kind of *invariants* w.r.t. a natural notion of equivalence, e.g. homotopy etc. The determination of the kind of equivalence depends on scientific choices, e.g. we decide to focus on algebraic properties, and to forget metric ones.

γ) *mathematical logic*, which deals by definition with language, is often taking the opposite viewpoint, namely that any mathematical object comes with an *extrinsic* description, which is claimed to be part of its structure. This viewpoint is widely spread under the name "intensionality". On the whole, the tendency of intensionality would be to distinguish between right and left-handed cups, because we have in mind different uses (which hand holds the cup), whereas mathematicians (and manufacturers) will identify the two things. However the positive aspect of intensionality is that it has kept alive distinctions like

functions as graphs / *functions as programs*

which are surely very important, and that have been overlooked by the main stream of mathematics -especially the Cantorian approach-. The negative aspect of this tradition lies in its implicit slogan "*the map is the territory*", which strongly opposes to any serious structural study.

δ) *recursive functions* are a typical example of this situation. We have been knowing for more than 50 years that a recursive function is not a graph, but a finite program. Kleene indices introduce a class of finite programs which is enough to make decent study of recursive functions, together with nice tools -like the recursion theorem, the S_{nm} theorem- which enable us to manipulate them in a uniform way. Kleene indices convey the necessary amount of finiteness, dynamicity etc. needed to *speak* of programs, *but they are not programs*. They are just some *ad hoc* taxonomy enabling us to reduce an actual program to something which -roughly speaking- does the same thing. The main stream of recursion theory treats Kleene brackets as a black box, enjoying some abstract enumeration property... *But who has ever seen the tail of an actual index ?*

There should be somewhere a purely geometrical notion of finite dynamical structure (not relying on *ad hoc* schemes and dirty encodings). The problem is to find tools sharp enough to isolate them. Our methodological hypothesis is that the problem is *implicitely* solved by Gentzen's *Hauptsatz*, which eliminates abstract notions in proofs (cuts) by introducing finite dynamics, and therefore looks like a good approximation to a universal dynamics. To solve the problem explicitly would mean to find out the geometrical meaning of the *Hauptsatz*, i.e. what is hidden behind the somewhat boring syntactical manipulations it involves.

On this precise point (to take the *Hauptsatz* as our ultimate reference) we shall certainly disagree with category-theorists ; besides personal taste, one must acknowledge that category-theory presents a very clean approach to many problems of computer-science. Unfortunately (if one forgets hypothetical uses of bicategories) it is purely denotational, i.e. modelizes computation by equalities. However one can distinguish between *dynamic* and *static* uses of equality in the categorical approach : typically when we formulate a universal problem, we write a commuting diagram, together with unicity requirements. The commuting diagrams are enough for computing, so the equality here has a dynamic meaning ; the unicity requirement is essentially about possible transpositions of rules, hence is just a change of description (taxonomy), and the equality there is static. See the introduction of [9] for a short discussion.

IV. the geometry of cut-elimination

IV.1. system F

This system, also known as polymorphic λ-calculus, is built as the Howard-isomorphic copy of the system of natural deduction for second order propositional logic. Due to the Howard isomorphism (which is a precise technical restatement of Heyting's paradigm) it will be enough to concentrate on the aspect "natural deduction" of the system. The main features of the system have been established in [5] :

i) normalization : one can execute the programs represented by the proofs by means of certain rewritings that do converge

ii) representation : any numerical algorithm that has a proof of termination inside usual mathematics (i.e. : second order arithmetic) can be represented inside the system. However the internalized algorithm will be slightly different from the original one. Therefore we expect that, when we express some provably terminating algorithm inside **F**, *some regularity is added.*

The first approximation to a universal geometry of computation could be the rewriting process i), but this is wrong : this process has in turn to be implemented. This implementation should not be left to engineering : the mathematical study of what is behind rewriting is a more technical expression of our program of "geometry of interaction".

In what follows, a complete familiarity with Howard's isomorphism [12] (see also [10]) is supposed. This isomorphism enables us to replace functional expressions by deductions. The gain is that, when we view functional terms as proofs, there are some tortures that we can inflict to them that we could not perform on functional expressions : in particular, the symmetrization I/O that will eventually lead to proof-nets. Many misconceptions concerning linear logic and proof-nets come from the fact that people stick too much to the idea of variable. But the use of variables must be seen as a taxonomy : we follow too much our old-fashioned intuitions about inputs and outputs and in particular the functional notation asymmetrizes situations which are perhaps symmetric (again think of right and left handed cups).

IV.2. natural deduction

Natural deduction has mainly been studied by Dag Prawitz in the 60's, see [16], [10]. It can be seen as an alternative formulation of sequent calculus : instead of sequents $\Gamma \vdash A$, one considers *deductions*

$$\begin{array}{c}\Gamma\\ \vdots\\ A\end{array}$$

with a tree-like form. Every proof in sequent calculus induces a unique natural deduction, and conversely, every natural deduction comes from a sequent calculus proof, *but this ancestor is far from being unique* ! To explain the importance of this fact let us quote Prawitz (approximation of a private discussion, June '82)

<< J.Y.G. : I prefer sequent calculus which is more synthetic...

D.P. : maybe you are right. But sequent calculus is just a system of *derived* rules about proofs, whereas natural deduction tries to represent the proofs themselves as primitive objects. >>

This point is very central : sequent calculus is *the* synthetic way of manipulating those mysterious hidden objects (that we call proofs, programs, and we would like to see as actions). But natural deduction has been the first serious attempt to find out what these objects could be. To give an analogy : we can manipulate synthetic units like "tuner", "amplifier", "loudspeaker", and plug them together when certain matchings are fulfilled. Now, the hifi unit we thus obtain works by itself, without any reference to our decomposition into several units that was so essential to us : the resulting object is a complex mixture of transistors, diods etc., in which, by the way, other relevant synthetic units could have been individualized. In other terms : to build something (a proof, a program) we must go step-by-step and produce bigger and bigger synthetic configurations. But the object produced should not remember our particular step by step decomposition, which is purely taxonomic. Sequent calculus is presumably the best possible taxonomic system for actions. But an action may come from several descriptions, typically when

transposition of rules

occurs. The situation becomes dramatic with the *Hauptsatz*, where 90% of one's energy is spent on bureaucratic problems of transposing rules. Let us give an example : when we meet a configuration

$$\dfrac{\dfrac{\vdash \Gamma, A}{\vdash \Gamma', A}\,R \qquad \dfrac{\vdash A^\perp, \Delta'}{\vdash A^\perp, \Delta'}\,S}{\vdash \Gamma', \Delta'}\,\text{cut}$$

there is no natural way to eliminate this cut, since the unspecified rules (R) and (S) do not act on A or $A^\perp$; then the idea is to forward the cut upwards :

$$\dfrac{\dfrac{\dfrac{\vdash \Gamma, A \qquad \vdash A^\perp, \Delta}{\vdash \Gamma, \Delta}\,\text{cut}}{\vdash \Gamma', \Delta}\,R}{\vdash \Gamma', \Delta'}\,S$$

But, in doing so, we have decided that rule (R) should now be rewritten *before* rule (S), whereas the other choice would have been legitimate too. Hence, from a symetrical problem, we are led to an asymetric solution : the taxonomical devices that force us to write (R) before (S) or (S) before (R) are not more respectable than the alphabetical order in the dictionary. One should try to get rid of them, or at least, ensure that their effect is limited. What natural deduction achieves is to identify intuitionistic sequent calculus proofs that are the same up to order of rules. (In classical logic, if we start with two proofs which are just variants w.r.t. transposition of rules, then the peculiarities of contraction on both sides are such that eventually the *Hauptsatz* will lead to two cut-free proofs which are not even variants.)

IV. 3 limitations of natural deduction

Natural deduction, which succeeds in identifying a terrific number of inversion-related sequent calculus proofs, is not free from serious defects :

α) Natural deduction is only satisfactory for $\Rightarrow$, $\forall$, $\wedge$; the connectives $\vee$

and ∃ receive a very *ad hoc* treatment : the elimination rule for "∨" is

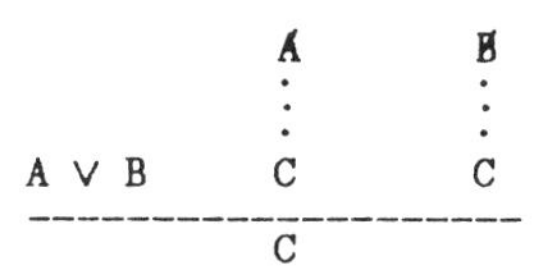

with the presence of an extraneous formula C. Then the problems of commutation of rules (i.e. changing the extraneous C) become prominent, and the solution of the literature (so-called "commutative conversions") is just *bricolage*.

β) In fact, there is a hidden taxonomy in natural deduction : in deduction

$$\begin{array}{c}\Gamma\\ \vdots\\ A\end{array}$$

one distinguishes *one* conclusion, and *several* hypotheses. Since the conclusion is always unambiguous, there is no problem to determine which is the last rule used etc. The connectives $\Rightarrow$, $\wedge$, and the quantifier $\forall$ accept this taxonomy without any apparent problem. But disjunction would rather prefer a rule of the form

$$\frac{A \vee B}{A \quad B}$$

with two conclusions. But this would go against the taxonomic requirement "one conclusion at a time".

γ) In natural deduction, one distinguishes between *introductions* and *eliminations*. Our claim is that eliminations rules are just introductions (for a dual connective), but *written upside down*. For instance, the elimination rule for implication is written as

$$\frac{A \quad A \Rightarrow B}{B}$$

and B is the official conclusion of the rule. But the ***hidden*** conclusion of the rule is $A \Rightarrow B$. This point should be familiar to specialists of natural deduction, where one introduces the "main hypothesis", which plays the role of the actual conclusion of a deduction. For instance, in order to cope with the lack of compositionality of "hexagons" in denotational semantics, this change of viewpoint becomes prominent, see e.g. [4] : in this paper, this shift of viewpoint yields the hexagon property for normal proofs (hence for all proofs, using normalization), whereas if one sticks to the usual taxonomy, one gets the impression that some compositionality is required.

δ) Natural deduction uses global rules, typically

$$\frac{\begin{array}{c}[A]\\ \vdots\\ B\end{array}}{A \Rightarrow B}$$

which apply to whole deductions, in contrast to rules like the elimination rule for $\Rightarrow$, which apply to formulas.

In the 70's, Rick Statman made an attempt [18] to study the geometry of natural deduction, and specially emphasized point δ), associating a "genus" to deductions. But since he had no way to restore symmetry, in order to cope with -say- point γ), the attempt eventually failed.

IV.4. proof-nets

Linear logic makes us hope that a final answer to the problem of inversion of rules might be found. This is because of the symmetrical nature of linear logic : the essential connective is now "$\multimap$", which conveys the purely implicative meaning of "$\Rightarrow$", putting aside the component "!", not of implicative nature. Hence, if one wants to study "$\Rightarrow$", we can study separately "$\multimap$" and "!". Here we shall concentrate on "$\multimap$", which in linear logic is defined from "$\wp$" by $A \multimap B =_d A^\perp \wp B$. The main idea will be to replace a natural deduction

Γ
⋮
A by a *proof-net* with several conclusions, $\Gamma^\perp$,A. Everytime a formula will be turned upside down, the negation symbol $(-)^\perp$ will be needed.

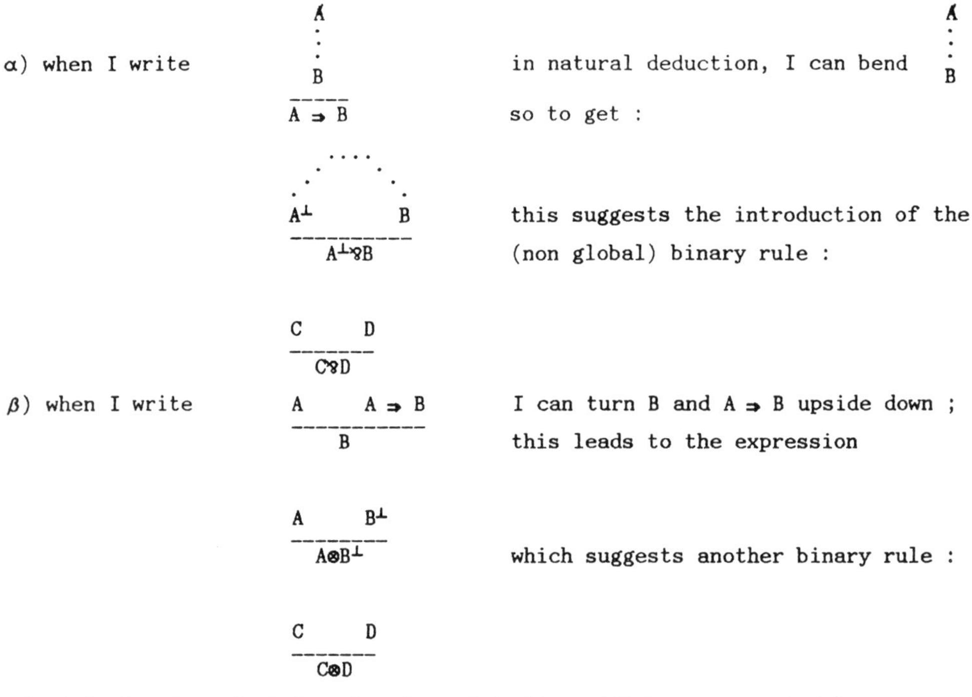

γ) minimal and maximal formulas in a deduction will be represented by means of configurations

$$\frac{}{A \qquad A^\perp} \quad \text{and} \quad \frac{A \qquad A^\perp}{Cut} \quad \text{respectively.}$$

The first case (*axiom link*) represents an identity axiom, whereas the second case (*cut link*) is the cut-rule ; the expression *Cut*, which is not a formula, is there for inessential reasons.

This representation of proofs is the ultimate possible identification of proofs of sequent calculus. Typically the net below represents two algorithms :

on one side, the algorithm which takes a binary function f of type $A \multimap (B \multimap C)$ into a function g of type $B \multimap (A \multimap C)$ by interchanging the inputs

on the other side, the algorithm which takes a ternary sequence $b \otimes a \otimes c$ of type $B \otimes (A \otimes C^{\perp})$ into $a \otimes b \otimes c$ of type $A \otimes (B \otimes C^{\perp})$.

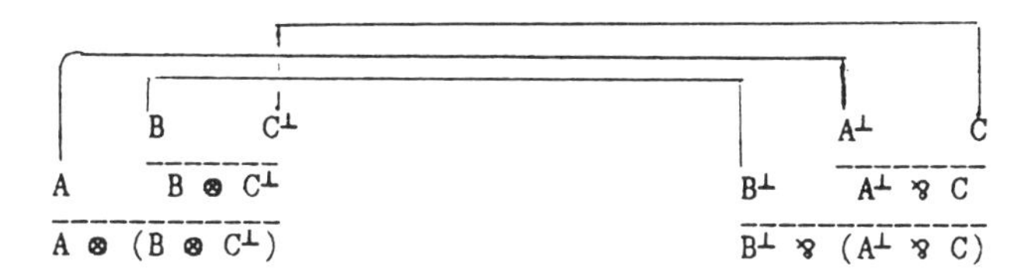

These two algorithms are physically the same because they correspond to the same pointer moves (represented by the axiom links) independently of the choice of the input side and of the output side. (By the way, observe that this proof-net is non-planar, i.e. wrong from the non-commutative standpoint.)

The main mathematical problem comes from the fact that we have expelled global (i.e. contextual) rules. Locally speaking the rules for $\otimes$ and $⅋$ are exactly the same. Hence one must find a global soundness criterion for the graphs written with such rules to be proof-nets, i.e. to represent (at least) one proof of sequent calculus. The answer is by means of the notion of a cyclic *trip*, which mimicks, in a formal way, the transportation of information during a cut-elimination process. For more details, one may consult [7] . However, this work is satisfactory only w.r.t. $(-)^{\perp}$, $\otimes$, $⅋$ (hence $\multimap$ as well). More recently, the notion of proof-net has been extended to quantifiers, see [9]. The solution found there could be the basis for a further extension to additives and exponentials, but what is known so far is only very partial. The weakening rules ($\perp$) and (w?) seem also to pose a very delicate problem, since they seem to contradict cyclicity.

IV.5. getting rid of syntax

This is the ultimate aim, which has been achieved only for multiplicatives, see [8], with essential improvements by Vincent Danos and Laurent Regnier [2]. In this case, the basic underlying structure turns out to be a permutation of pointers, and the notion of a cyclic permutation plays a prominent role (Danos and Regnier replaced cyclicity conditions by conditions involving connected acyclic graphs, i.e. trees). It seems that the general solution (not yet known, even for quantifiers) could be something like permutations of pointers, with variable addresses, so that some (simple form of) unification could be needed for the composition of moves : yet another connection with logic programming !

When one gets rid of syntax, an immediate question is

what is a type ?

Types have always been part of some rigid syntactic (i.e. taxonomic) discipline, while we would prefer a general answer, not depending on the choice of a particular system. Here we have to remember the familiar analogy between

"π is a proof of A" and
"π is a program enjoying specification A"

i.e. types are *specifications*. Specifications can be seen as *plugging instructions*. For instance one can plug something of type A with something of type $A \Rightarrow B$ and get something of type B. This is a particular case of the general paradigm of *plugging of complementary specifications*, which is the meaning of the cut-rule.

But let's assume that our program has been pushed to the end (!), so that we are without syntax. In the geometrical structure β representing a program π, let us individualize two arbitrary complementary parts, β' and β'', which communicate via a common border $\partial\beta' = \partial\beta''$. Does it make sense to "type" β' and β'' in such a way that these "typings" will ensure that, once plugged via their common border, β' and β'' yield a sound action ? The answer to this question is only known in the multiplicative case, and still open in the other cases, even for quantifiers. The short description below is taken from [8] (see also [2]), and we shall assume a complete familiarity with proof-nets :

any switching S of β can uniquely be decomposed as a disjoint sum S'+S" of independent switchings of β' and β''. Using S', we get a certain permutation $\sigma_{S'}$ of $\partial\beta'$, and similarly, S" yields a permutation $\tau_{S''}$ of $\partial\beta''$ (= $\partial\beta'$). The condition for β to be a proof-net is exactly the fact that $\sigma_{S'}\tau_{S''}$ is cyclic for all S',S". Now, if we introduce

the notation $\sigma \perp \tau$
to say that σ and τ are two permutations defined on the same set, and that $\sigma\tau$ is cyclic

the sets of permutations

$$\Sigma(\beta') = \{\sigma_{S'}; S' \text{ switching of } \beta'\}$$
$$\Sigma(\beta'') = \{\tau_{S''}; S'' \text{ switching of } \beta''\}$$

then β is a proof-net if and only if :

$$\Sigma(\beta') \perp \Sigma(\beta'').$$

If we define the *principal types* of β' and β'' by :

$$pT(\beta') = \Sigma(\beta')^{\perp\perp} \qquad pT(\beta'') = \Sigma(\beta'')^{\perp\perp}$$

then β' and β'' can be plugged together exactly when

$$pT(\beta') \perp pT(\beta'')$$

In other terms, the principal type of an algorithm should be a collection of "border behaviours", permutations in the case just considered (more generally, something like partial isometries in a **C***-algebra ?). There is a notion of orthogonal border behaviours, cyclicity in the multiplicative case (more generally some kind of ergodicity ?). The principal type of an algorithm is not the set of all its border behaviours (which has a bad structure), but a larger set, its biorthogonal. (The reason is that if we take an algorithm of type $A \otimes B \otimes C$, we shall not get the same border behaviour whether we write it using $A \otimes (B \otimes C)$ or $(A \otimes B) \otimes C$, but this difference, due to taxonomy, is swallowed by the use of biorthogonality.

Of course, it would be a nonsense to try to compute principal types, as defined above. But plugging requires orthogonality of the principal types, and not at all that each of them is the orthogonal of the other, so that there is room in between. In particular, preset systems of typing can be seen as convenient ways to get and manipulate majorizations of principal types. The practical way of showing that β' is pluggable with β'' is therefore to find convenient majorizations

$$pT(\beta') \subset B' \qquad pT(\beta'') \subset B'' \qquad \text{with } B' \perp B''.$$

By the way, the types we attribute to algorithms in the multiplicative fragment are always majorizations, and are seldom optimal.

There is still the problem of how to interpret cut-elimination. If we could see an action as something as a partial isometry p in -say- a **C***-algebra, then an action is performed (cut-free) when $p^2 = 0$, i.e. when its domain X and its codomain Y are orthogonal subspaces : $XY = 0$. For a general action p (corresponding to the idea of a proof with cuts), it would no longer be true that $XY = 0$, but X would commute with Y, so that it would make sense to speak of the projectors (subspaces) $X' = X - XY$ and $Y' = Y - XY$. The syntactic operation of cut-elimination should be geometrically translated as a construction leading from p, with domain X and codomain Y, to p', with domain X' and codomain Y', the idea being to start from X' and to iterate p the number of times necessary to exit through Y'. For instance the process of cut-elimination in the multiplicative case can perfectly be interpreted in this way. However the consideration of finite dimensional spaces would be enough in that case ; the introduction of more abstract spaces seems to be necessary in order to represent irreversible processes like erasing, or simply to make room for duplications when we shall interpret contraction. An attempted treatment of actions as partial isometries in Hilbert space was started in Spring '87. It turned out that this approach could not replace down to earth syntactic considerations, in other terms that not enough material had been accumulated for a decent conceptualization ; but we nevertheless still believe that Gentzen's Hauptsatz should eventually be reformulated using the language of functional analysis !

IV.6. time, space, communication

Linear logic is eventually about time, space, communication, but is not a temporal logic, or a kind of parallel language : such approaches try to develop preexisting conceptions about time, processes, etc.. In those matters, the general understanding is so low that one has good chances to produce systems whose aim is to *avoid* the study of their objects (remember the sentence of Clémenceau : <<Quand je veux enterrer un problème, je nomme une commission>>. Linear logic is not "la commission du temps" or "la commission de la communication". The main methodological commitment is to refuse any *a priori* intuition about these objects of study, and to assume that (at least part of) the temporal, the parallel features of computation are already in Gentzen's approach, but are simply hidden by taxonomy. We shall therefore search for the answers to these essential problems inside *refinements* of usual logic, and not in such and such *ad hoc* extensions.

Methodologically speaking we concentrate on the central issue of

unbracketing.

We are forced to program sequentially, by means of nested brackets. This bracketing is a particular temporality for a program : if we perform the operations in the order imposed by the brackets, then we shall eventually make it. However, there might be other temporalities which may for instance come from some unexpected property of the inputs, and which might be more interesting. The idea of removing taxonomy, i.e. *our* temporality, by means of something like proof-nets, is to give the maximum degree of freedom for the execution of a program. The extension of proof-nets to quantifiers yields some additional hints as to a possible nature of this temporality, namely the relative dependencies between variables in massive processes of unification. In general, one should see time as the partial order of causality ("I must compute this to get that"), the absence of causality being perhaps of spacial nature. Negation would therefore appear as the inversion of the sense of time, which is not an unpleasant idea. The main problem one is faced with is that we have at least three intuitions about time :

i) time is logic modulo the order of rules
ii) time is the cut-elimination process
iii) time is the contents of non-commutative linear logic

These three intuitions should be unified to some extent ; however, one of the immediate difficulties with ii) is that cut-elimination is, at it stands, an irreversible process (which is consistent with our current experience with time) whereas the logical rules of iii) are symmetric w.r.t. the exchange past/future. Moreover, technically speaking, there are problems to develop iii), namely to have simultaneously commutative (spacial) and non-commutative (temporal) features.

As to the symmetry of time, there is a new element coming in : in the study of quantifiers, the normalization process uses some global procedures, which seem to involve a global time. This is a computational nonsense (since the thing will anyway be implemented by a local procedure) which seems to come from the symmetry between past and future. If one drops this symmetry, i.e. if the implication

$$\forall x A \Rightarrow A[t/x]$$

is only kept as a retro-causality, but not as a causality :

$$A[t/x] \leftarrow \forall x A, \qquad \text{(but not } \forall x A \rightarrow A[t/x]\text{)}$$

it seems possible to keep local procedures, but it is very difficult to find any milestone on which we could test such possible refinements of rules. But the idea that starts to make its way is that

logical rules should not be symmetric w.r.t. time.

A last word about communication : on the basis of the work so far done, the following conception of communication between systems seems to be reasonable :

processes communicate without understanding each other.

The idea should be that -at a very abstract level-, what processes share is a common border, but that their inner instructions have nothing in common. So when A receives a message from B, he can only perform global operations on it

-erasing, duplicating, sending back to B-depending on which gate of the common border he received it through. When a message is sent back to B, then B receives again his own stuff, that he can read, but through an unexpected gate etc. Massive iterations of such incomprehensions can perhaps mimick comprehension ; by the way this corresponds to the way scientists use the patience of their colleagues to improve their intuitions about their own work.

Finally, we must confess that we keep an eye on physics, especially quantum mechanics. It is not excluded that strange linear connectives like "⅋" could be useful to interpret some basic phenomena of physics...
But this is science-fiction.

REFERENCES

[1] Barr, M. :
**-autonomous categories*, Springer Lecture Notes 752.

[2] Danos, V. & Regnier, L. :
Multiplicatives bis, draft, université Paris VII, 1988.

[3] Dunn, J.M. :
Relevance logic and entailment, Handbook of philosphical logic, vol III, ed Gabbay & Guenthner, D.Reidel 1986.

[4] Freyd, P. & Girard, J.Y. & Scedrov, A. & Scott, P. :
Semantic parametricity in typed λ-calculus, proceedings of the Congress "Logic in Computer Science 1988", to be held in Edimburgh.

[5] Girard, J.Y. :
Une extension de l'interprétation fonctionnelle de Gödel à l'analyse et son application à l'élimination des coupures dans l'analyse et la théorie des types, Proc. Second Scand. Log Symp., ed. Fenstad, North Holland 1971.

[6] Girard, J.Y. :

Proof-theory and Logical complexity, Bibliopolis, Napoli, 1987, ISBN 88-7088-123-7.

[7] Girard, J.Y. :

Linear Logic, Theoretical Computer Science 50:1, 1987.

[8] Girard, J.Y. :

Multiplicatives, Rendiconti del seminario matematico dell'università e politecnico di Torino, special issue on logic and computer science, 1988

[9] Girard, J.Y. :

Quantifiers in linear logic, to appear in the Proceedings of the SILFS conference, held in Cesena, January 1987.

[10] Girard, J.Y. :

Typed λ-calculus, in preparation for Cambridge Tracts in Theoretical Computer Science.

[11] Girard, J.Y. & Lafont, Y. :

Linear logic and lazy computation, Proceedings of TAPSOFT '87, Pisa. SLNCS 250.

[12] Howard, W.A. :

The formulae-as-types notion of construction, in Curry Volume, eds Hindley & Seldin, Academic Press, London 1980

[13] Ketonen, J. & Weyhrauch, R. :

A decidable fragment of predicate calculus, Theoretical Computer Science 32:3, 1984.

[14] Lambek, J. :

The mathematics of sentence structure, Am. Math. Monthly 65, 1958.

[15] Lambek, J. & Scott, P. :

Introduction to higher order categorical logic, Cambridge University Press, Cambridge 1986

[16] Prawitz, D. :
Natural Deduction, Almqvist & Wiksell, Stockholm 1965.

[17] Scott, D. :
Domains for denotational semantics, Proceedings of ICALP '82, SLNCS 140.

[18] Statman, R. :
Structural complexity of proofs, Ph. D., Stanford, 1975.

Équipe de Logique, UA 753 du CNRS
Mathématiques, t 45-55, 5° Étage
2 Place Jussieu, 75251 Paris Cedex 05

Contemporary Mathematics
Volume **92**, 1989

The Category of Sketches as a Model for Algebraic Semantics[1]

John W. Gray

1. INTRODUCTION

There are two quite distinct ways to talk about type theory as it occurs in theoretical computer science; namely, logical type theory versus algebraic type theory. Many computer languages are based on some version of logical type theory; e.g., Pascal and ML [1]. Languages based on algebraic type theory are less common; e.g., OBJ2 [4], Act One [2], Scratchpad [8], etc.

Logical type theory is holistic. In it, the interesting thing is the collection of types as a whole. There are basic types which are not analysed further. They have no internal structure and are simply names; e.g., Int, Bool, Nat, Char, Real, etc. Other types are built up from them recursively using type constructors such as $\rightarrow$ (function space), $\times$ (product), and + (sum). For instance, a series of reals has type [Nat $\rightarrow$ Real], a series of series has type [Nat $\rightarrow$ [Nat $\rightarrow$ Real]] or [Nat $\times$ Nat $\rightarrow$ Real]. The sum of a series is a function of type [[Nat $\rightarrow$ Real] $\rightarrow$ Real]. The only structure that such a type has is the tree structure describing its construction from the type constructors. The usual theory for discussing logical types is the lambda calculus in one of its many guises: the pure lambda calculus, typed lambda calculus, polymorphic typed lambda calculus, Martin-Löf type theory, etc. However, from the point of view of computer science, logical types are just syntactic entities whose only purpose in programs is to make type checking possible.

Algebraic type theory is fragmentary. It looks at types like Int or Bool and observes that, of course, these types have relevant internal structure. They support operations (such as "addition" or "and") with equations (such as the commutative law or the idempotent law) between the operations. Thus a type is described by a "specification" which lists the operations and relevant equations. In this approach, types themselves are initial algebras for algebraic theories. Each one is an autonomous bundle of sorts, operations and equations. It is a small world in itself. An individual type contains all the information about some (many-sorted) algebraic situation. Everything that can be done with a given type is derivable from its specification. Little or nothing is said about relations between various types or about the collection of types as a whole. This

[1] This research was partially supported by the National Science Foundation.

fragmentation is somewhat meliorated by allowing one type to be a subtype of another (but normally this is just shorthand for explicitely including the sorts, operations and equations of one type in those of another) and by allowing for parametric types.

The aim of this work is to provide a more holistic setting for algebraic types; i.e., to provide a logical theory of algebraic types. In particular, the collection of all algebraic types will be something like a model of Martin-Löf type theory, using constructions derived from category theory. It is difficult to discuss the collection of all algebraic types if such types are presented by specifications for algebraic theories because there is really no good notion of a morphism between specifications. Instead, we shall present algebraic theories by "sketches", with an obvious notion of homomorphism between sketches. The category of sketches built in this way supports many constructions of interest in the semantics of programming languages. It is cartesian closed and hence a model of the typed lambda calculus, as well as being locally cartesian closed and hence a model of Martin-Löf type theory. Furthermore, it is complete and cocomplete so it contains recursive types. However, there is a richer notion of dependent type than that given by the local cartesian closed structure which precisely captures the kind of algebraic structure that dependent types should have. Thus, a subsidiary aim of this work is to provide an algebraic semantics for dependent types.

2. CATEGORIES OF SKETCHES

The basic structure underlying a sketch is a reflexive graph, so sketches will be reflexive graphs with extra structure. Recall that a reflexive graph, G, consists of a pair of sets, G_0 = set of *nodes* or *vertices* or *objects*, and G_1 = set of *directed edges* or *arrows*, together with two functions $d_0 : G_1 \to G_0$ the *domain* or *source* function, and $d_1 : G_1 \to G_0$ the *codomain* or *target* function, as well as a function $U : G_0 \to G_1$ the *identity* or *loop* function, satisfying $d_i \bullet U$ = identity, for i = 0, 1. If $\alpha \in G_1$ satisfies $d_i(\alpha) = a_i$, then we write $\alpha : a_0 \to a_1$. A homomorphism of reflexive graphs, $h : G \to H$ consists of a pair of functions $h_i : G_i \to H_i$ for i = 0, 1, such that $h_0 \bullet d_i = d_i \bullet h_1$ for i = 0, 1, and $h_1 \bullet U = U \bullet h_0$. Here d_0, d_1, and U refer to G on the left and to H on the right. The category of reflexive graphs and graph homomorphisms will be denoted by **GRAPH**. Equivalently, if **E** denotes the category with two objects 0 and 1 and three non-identity morphisms d_0, d_1, U as illustrated such that $d_i \bullet U = id_0$ for i = 0, 1, then **GRAPH** is isomorphic to the functor category **SET**$^{\mathbf{E}}$ of all **SET**-valued functors from **E** to the category **SET** of sets and functions. Morphisms in **SET**$^{\mathbf{E}}$ are natural transformations. It follows that **GRAPH** is a topos (cf., Johnstone [9]) which, as far as we are concerned here, means that **GRAPH** has very many desirable properties, some of which will be derived later in order to use their explicit forms in making constructions for sketches.

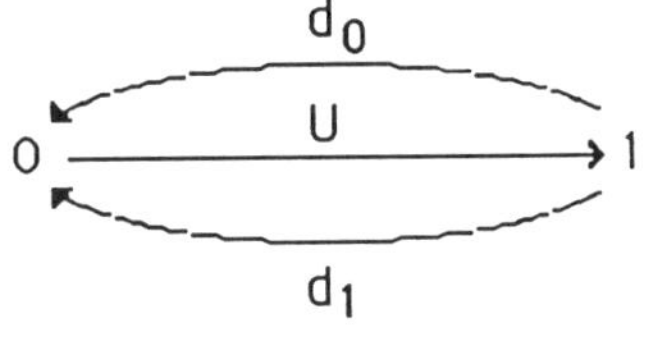

In everything that follows, the term "graph" always means "reflexive graph". In illustrating particular graphs we normally omit the values of U, which for each object give a loop at that object.

2.1 Definitions.

i) ***D*** denotes a fixed class of graphs, called *shapes of diagrams* . A *D–diagram in a graph* G, or a *diagram of shape* ***D*** in G, is a graph homomorphism $\delta : D \to G$ for some diagram shape $D \in \boldsymbol{D}$. In particular, $\boldsymbol{D}_0$ is the collection of all graphs of the upper shape for $0 \leq p, g < \infty$. We make the blanket assumption that ***D*** contains the lower two shapes called, respectively, the *loop* and the *rectangle*

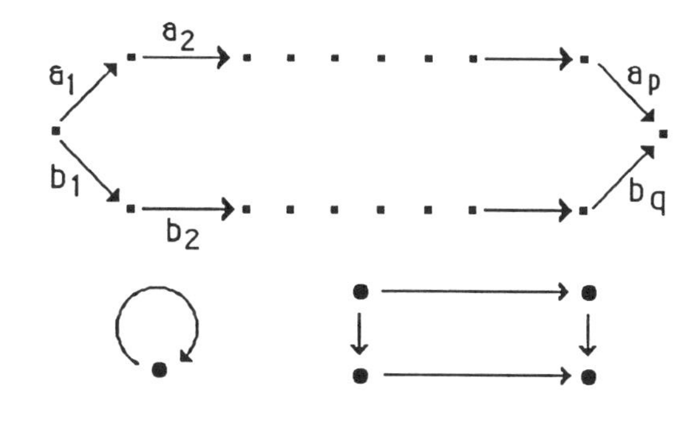

ii) Let (C, v) be a pointed graph; i.e., a graph C together with a chosen vertex v in C. (C, v) is called a *cone* if there is exactly one edge from v to every vertex of C. The graph C-v given by deleting v and all edges from v is called the *base* of (C, v). ***C*** denotes a fixed class of cones called *shapes of cones.* A ***C****-cone in a graph* G, or a *cone of shape* ***C*** in G, is a graph homomorphism $\gamma : C \to G$ for some cone shape $(C, v) \in \boldsymbol{C}$. $\gamma(v)$ is called the *vertex* of γ and $\gamma(C\text{-}v)$ is called the *base* of γ. In particular, ***FP*** is the class of all cone shapes of the adjoining form for $0 \leq p \leq \infty$. ***FP*** stands for "finite product".

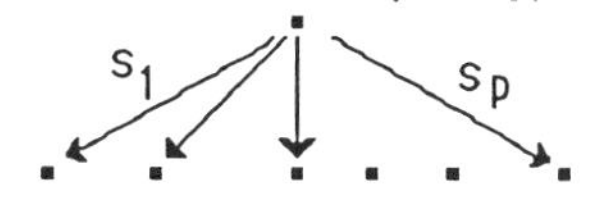

iii) (C, v) as in ii) is called a *cocone* if there is exactly one edge to v from every vertex of C. The graph C-v given by deleting v and all edges to v is called the *base* of (C, v). ***coC*** denotes a fixed class of cocones called *shapes of cocones.* A ***coC****-cocone in a graph* G, or a *cocone of shape* ***coC*** in G, is a graph homomorphism $\gamma' : C \to G$ for some cocone shape $(C, v) \in \boldsymbol{coC}$. $\gamma(v)$ is called the *vertex* of γ and $\gamma(C\text{–}v)$ is called the *base* of γ. In particular, ***FcoP*** is the class of all cocone shapes of the adjoining form for $0 \leq p < \infty$. ***FcoP*** stands for "finite coproduct".

2.2 Definition.

A ***D-C-coC sketch*** is a 4-tuple $\mathbf{A} = (|A|, \boldsymbol{D_A}, \boldsymbol{C_A}, \boldsymbol{coC_A})$ where $|A|$ is a graph, called the *underlying reflexive graph* of A, $\boldsymbol{D_A}$ is a collection of *diagrams in* $|A|$ *of shape* $\boldsymbol{D}$, $\boldsymbol{C_A}$ is a collection of *cones in* $|A|$ *of shape* $\boldsymbol{C}$, and $\boldsymbol{coC_A}$ is a collection of *cocones in* $|A|$ *of shape* $\boldsymbol{coC}$. The diagrams in $\boldsymbol{D_A}$ (resp., cones in $\boldsymbol{C_A}$, cocones in $\boldsymbol{coC_A}$) are called the ***admissable diagrams*** (resp., ***cones, cocones***) for **A.** The only conditions on the admissable diagrams are that for each object $a \in |A|_0$ and for each arrow $\alpha : a \to b$ the loop and the rectangle are admissable. These are called the *trivial* admissable diagrams.

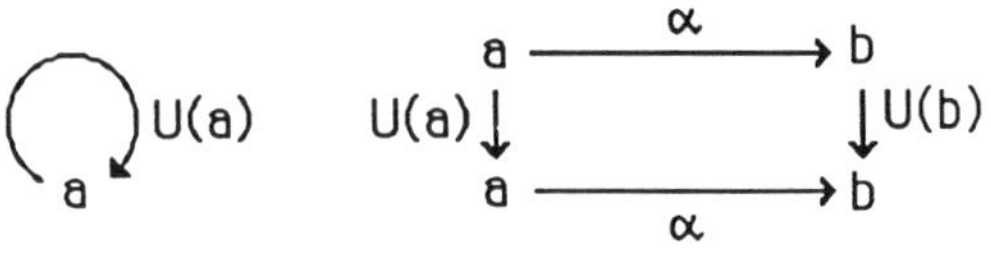

$$a \circlearrowleft U(a) \qquad \begin{array}{ccc} a & \xrightarrow{\ \alpha\ } & b \\ {\scriptstyle U(a)}\downarrow & & \downarrow{\scriptstyle U(b)} \\ a & \xrightarrow[\ \alpha\ ]{} & b \end{array}$$

There are no conditions on the admissable cones and cocones.

In nearly all circumstances, we will work with fixed collections $\boldsymbol{D}$, $\boldsymbol{C}$, and $\boldsymbol{coC}$ so they can be dropped from the notation. For instance, to describe FP sketches, we would take $\boldsymbol{D}$ to be either the collection of all graphs or the collection $\boldsymbol{D_0}$, $\boldsymbol{C}$ to be the collection $\boldsymbol{FP}$, and $\boldsymbol{coC}$ to be empty.

2.3 Definition.

i) Let $\mathbf{A} = (|A|, \boldsymbol{D_A}, \boldsymbol{C_A}, \boldsymbol{coC_A})$ and $\mathbf{B} = (|B|, \boldsymbol{D_B}, \boldsymbol{C_B}, \boldsymbol{coC_B})$ be ***D-C-coC*** sketches. A *homomorphism* h from **A** to **B** is a graph homomorphism $|h| : |A| \to |B|$ such that :

a) if $\delta : D \to |A|$ is admissable, then so is $|h| \cdot \delta : D \to |B|$.

b) if $\gamma : (C, v) \to |A|$ is admissable, then so is $|h| \cdot \gamma : (C, v) \to |B|$.

c) if $\gamma' : (C, v) \to |A|$ is admissable cocone, then so is $|h| \cdot \gamma' : (C, v) \to |B|$.

ii) ***D-C-coC*-SK**, denotes the category of ***D-C-coC***-sketches and homomorphisms of ***D-C-coC***-sketches. Usually we just write **SK**, meaning the category ***D-C-coC*- SK** for some fixed choices of $\boldsymbol{D}$, $\boldsymbol{C}$, and $\boldsymbol{coC}$.

2.4 Definition.

Let $\mathbf{A} = (|A|, \boldsymbol{D_A}, \boldsymbol{C_A}, \boldsymbol{coC_A})$ be a fixed ***D-C-coC*** sketch and let **C** be a category. Let

$$^\wedge\mathbf{C} = (|C|, \boldsymbol{D}_{^\wedge C}, \boldsymbol{C}_{^\wedge C}, \boldsymbol{coC}_{^\wedge C})$$

denote the underlying ***D-C-coC*** sketch of **C** in which $|C|$ is the underlying graph of **C**, $\boldsymbol{D}_{^\wedge C}$ is the collection of all commutative diagrams of shape $\boldsymbol{D}$ in **C**, $\boldsymbol{C}_{^\wedge C}$ is the collection of all limit cones of shape $\boldsymbol{C}$ in **C**, and $\boldsymbol{coC}_{^\wedge C}$ is the collection of all colimit cones of shape $\boldsymbol{coC}$ in **C**. Note that the condition that the loop U(a) be a commutative diagram in **C** implies that $U(a) = id_a$ for all objects a in **C**. $^\wedge\mathbf{C}$ should have ***D-C-coC*** as a subscript, but that will be omitted whenever possible.

i) A *model* of **A** in **C** is a sketch homomorphism $M : \mathbf{A} \to {}^\wedge\mathbf{C}$.

ii) If M and M' are models of **A** in **C**, then a *homomorphism* $t : M \to M'$ of models is a "natural transformation"; i.e., a family of morphisms $\{t_a : M(a) \to M'(a) \mid \text{for all } a \in |A|_0\}$ in **C**, such that for all arrows $\alpha : a \to a'$ in $|A|_1$, the adjoining diagram commutes in **C**; i.e., is an admissable diagram in ^**C**. (This makes sense since **C** is a category. Commutativity of diagrams in a sketch doesn't make sense but being admissable always does.)

$$\begin{array}{ccc} M(a) & \xrightarrow{M(\alpha)} & M(a') \\ \downarrow{\scriptstyle t_a} & & \downarrow{\scriptstyle t_{a'}} \\ M'(a) & \xrightarrow{M'(\alpha)} & M'(a') \end{array}$$

iii) $\mathbf{MOD_C(A)}$ denotes the category of models of **A** in **C** and homomorphisms of such models. As before, if **C** = **SET**, then it is omitted from the notation.

2.4 Remarks.

There are a number a ways that the basic definitions can be modified without affecting the collection of models of a sketch.

i) One could require that if $\delta : D \to |A|$ is admissable and if $\delta' : D' \to D$ is any homomorphism between graphs in $\boldsymbol{D}$, then $\delta \bullet \delta' : D' \to |A|$ is admissable. This would destroy any possibility of talking about finite structures unless $\boldsymbol{D}$ were itself finite, and hence one would have to discuss generating families of diagrams. However, this does suggest a different notion of sketch homomorphism; namely, a graph homomorphism $h : |A| \to |B|$ such that for each admissable $\delta : D \to |A|$, there is an admissable $\delta' : D' \to |B|$ and a graph homomorphism $\delta'' : D \to D'$ such that $h \bullet \delta = \delta' \bullet \delta''$. This might turn out to be a useful notion, but we shall make no use of it here.

ii) Let $\mathbf{A} = (|A|, \boldsymbol{D}_\mathbf{A}, \boldsymbol{C}_\mathbf{A}, \boldsymbol{coC}_\mathbf{A})$ be a fixed ***D-C-coC*** sketch and define

$\boldsymbol{D}_\mathbf{A}\# = \{\delta : D \to |A| \mid M \bullet \delta$ is a commutative diagram for all models M of **A**.$\}$

$\boldsymbol{C}_\mathbf{A}\# = \{\gamma : (C, v) \to |A| \mid M \bullet \gamma$ is a limit cone for all models M of $|A|$.$\}$

$\boldsymbol{coC}_\mathbf{A}\# = \{\gamma : (C, v) \to |A| \mid M \bullet \gamma$ is a colimit cocone for all models M of $|A|$.$\}$

Call $\mathbf{A}^\# = (|A|, \boldsymbol{D}_\mathbf{A}\#, \boldsymbol{C}_\mathbf{A}\#, \boldsymbol{coC}_\mathbf{A}\#)$ the *model completion* of **A**. $(-)^\#$ clearly determines an endofunctor on the category of sketches which underlies an idempotent triple. It makes sense to call a graph homomorphism $h : |A| \to |B|$ a *weak sketch homomorphism* if $h : \mathbf{A} \to \mathbf{B}^\#$ is a sketch homomorphism; i.e., if h is a Kleisli morphism for this triple. I thank the referee for pointing out that these are essentially the same as morphisms from **A** to the classifying category of **B.** Again, we shall make no use of this notion here.

iii) There is clearly a Galois connection between sketch structures on a given graph G and families of graph homorphisms from G to ^**SET** (or, in fact, to **A** for any sketch A) of which the construction in ii) is a special case.

3. CARTESIAN PROPERTIES OF SK.

For any fixed choice of $\boldsymbol{D}$, $\boldsymbol{C}$, and $\boldsymbol{coC}$, the category **SK** has many interesting properties which show how to carry out useful constructions on sketches. In particular, we want to show is that **SK** is cartesian closed, locally cartesian closed and complete and cocomplete. In order to do this, we will need the preliminary results that the category **GRAPH** of graphs has these properties. These are a well-known facts, but the required constructions here will guide our constructions in **SK.**

3.1 Proposition.

GRAPH is cartesian closed.

Proof. It must be shown that **GRAPH** has a terminal object, binary products and function space objects.

i) The terminal object is given by the graph 1 with one object, one arrow, and the only possible functions for d_0, d_1, and U.

ii) Binary products are constructed by forming the products independently of the sets of objects and arrows and taking the product functions for d_0, d_1 and U.

iii) Function space object are somewhat more complicated to construct but there is a general formula for **SET**-valued functor categories which guides the construction. (Cf., Mac Lane [12].)

Let G and H be graphs.

a) The set of objects $[G \to H]_0$ is the set of graph homomorphisms from G to H.

b) The set of arrows is

$$[G \to H]_1 = \{(h^0, t, h^1) \mid h^i \text{ is a graph homomorphism and } t : G_1 \to H_1 \text{ is a function such that } d_i \bullet t = (h^i)_0 \bullet d_i \text{ for } i = 0, 1\}.$$

These equations say that, for any arrow $\alpha : a \to b$ in G, there are arrows as illustrated in H:

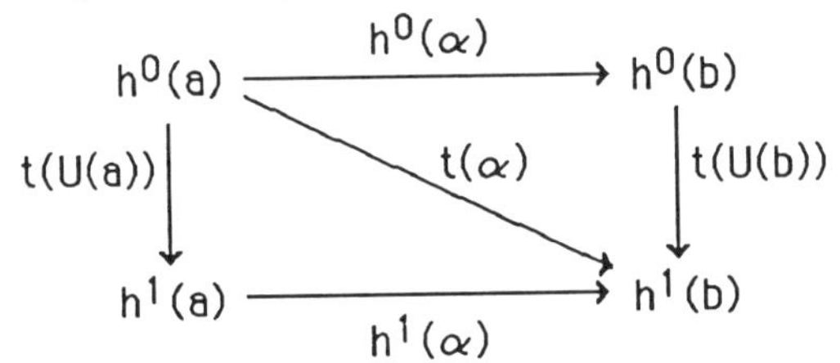

c) d_i and U for $[G \to H]$ are given by the formulas: $d_i(h^0, t, h^1) = h^i$, for $i = 0, 1$, and $U(h) = (h, h_1, h)$.

d) The graph homomorphism $app_{G,H} : [G \to H] \times G \to H$ has components given by the formulas: $(app_{G,H})_0(h, a) = h_0(a)$ and $(app_{G,H})_1((h^0, t, h^1), \alpha) = t(\alpha)$.

e) If $m : K \times G \to H$ is a graph homomorphism, then $m\# : K \to [G \to H]$ is the graph homomorphism with components given as follows:

$(m\#)_0 : K_0 \to [G \to H]_0$ is given by the formulas

$$((m\#)_0(a))_0(b) = m_0(a, b) \quad \text{where } a \in K_0 \text{ and } b \in G_0$$

$$((m\#)_0(a))_1(\beta) = m_1(U_K(a), \beta) \quad \text{where } a \in K_0 \text{ and } \beta \in G_1$$

$(m\#)_1 : K_1 \to [G \to H]_1$ is given by the formula: if $\alpha : a \to a'$ in K, then

$$(m\#)_1(\alpha) = ((m\#)_0(a)), t_\alpha, (m\#)_0(a')) \text{ where } t_\alpha(\beta) = m_1(\alpha, \beta)$$

It is easy to show that m# is the unique graph homomorphism such that $app_{G,H} \bullet (m\# \times id_G) = m$.

3.2 The model structures.

i) To each graph $D \in \boldsymbol{D}$ we associate a sketch, called a model diagram, $\mathbf{Md}(D) = (D, \{id_D\}, \emptyset, \emptyset)$; i.e., $|\mathbf{Md}(D)| = D$, the only diagram is the identity morphism from D to D, while the two $\emptyset_D$'s mean that there are no cones or cocones. $\mathbb{D}$ denotes the collection of all such model diagrams.

ii) To each pointed graph $(C, v) \in \boldsymbol{C}$ we associate a sketch, called a *model cone* , $\mathbf{Mc}(C,v) = (C, \emptyset, \{id_C\}, \emptyset)$; i.e., $|\mathbf{Mc}(C,v)| = C$, there are no diagrams, the only cone is the identity morphism from C to C, and there are no cocones. $\mathbb{C}$ denotes the collection of all such model cones.

iii) To each pointed graph $(C, v) \in \boldsymbol{coC}$, we associate a sketch, called a *model cocone* , $\mathbf{Mcc}(C,v) = (C, \emptyset, \emptyset, \{id_C\})$; i.e., $|\mathbf{Mcc}(C,v)| = C$, there are no diagrams or cones and the only cocone is the identity morphism from C to C. $\mathbb{coC}$ denotes the collection of all such model cocones.

We use the term *model* here in the sense of model-generated categories. For instance, manifolds are described in terms of certain model objects; namely, open balls in Euclidean space. A general manifold is a topological space equipped with an atlas of continuous maps from the models. In our case, the models are the structures described above. A general sketch is a graph equipped with an atlas of graph homomorphisms from the models; i.e., the admissable diagrams, cones, and cocones. Note that a graph homomorphism $\delta: D \to \mathbf{A}$ is admissable if and only if $\delta: \mathbf{Md}(D) \to \mathbf{A}$ is a sketch homomorphism. Similarly for cones and cocones. It is occasionally helpful to confuse the two notions and speak of $\delta: D \to \mathbf{A}$ being a sketch homomorphism.

3.3 Proposition.

SK is cartesian closed.

Proof. Fix a choice of $\boldsymbol{D}$, $\boldsymbol{C}$, and $\boldsymbol{coC}$. Then it must be shown that **SK** has a terminal object, binary products and function space objects.

i) **SK** has a terminal object given by $\mathbf{1} = (1, \mathbf{All}_1, \mathbf{All}_1, \mathbf{All}_1)$ where $\mathbf{All}_1$ means successively, all diagrams of shape $\boldsymbol{D}$, all cones of shape $\boldsymbol{C}$, and all cocones of shape $\boldsymbol{coC}$, there being exactly one for each such diagram shape, cone shape, or cocone shape.

ii) To show that **SK** has binary products, let $\mathbf{A} = (|\mathbf{A}|, \boldsymbol{D}_\mathbf{A}, \boldsymbol{C}_\mathbf{A}, \boldsymbol{oC}_\mathbf{A})$ and $\mathbf{B} = (|\mathbf{B}|, \boldsymbol{D}_\mathbf{B}, \boldsymbol{C}_\mathbf{B}, \boldsymbol{coC}_\mathbf{B})$ belong to **SK**. Then $\mathbf{A} \times \mathbf{B}$ is constructed as follows:

a) $|\mathbf{A} \times \mathbf{B}| = |\mathbf{A}| \times |\mathbf{B}|$. Let $pr_{|\mathbf{A}|}$ and $pr_{|\mathbf{B}|}$ denote the two projection graph homomorphisms.

b) If $\delta : D \to |\mathbf{A} \times \mathbf{B}|$ is a graph homomorphism with $D \in \boldsymbol{D}$, then δ is an admissable diagram for $\mathbf{A} \times \mathbf{B}$ if and only if $pr_{|\mathbf{A}|} \bullet \delta$ and $pr_{|\mathbf{B}|} \bullet \delta$ are admissable diagrams for $\mathbf{A}$ and $\mathbf{B}$ respectively.

c) Admissable cones and cocones for $\mathbf{A} \times \mathbf{B}$ are described analogously. It is immediate that if $pr_\mathbf{A}$ and $pr_\mathbf{B}$ are the morphisms with underlying graph homomorphisms given by $pr_{|\mathbf{A}|}$ and $pr_{|\mathbf{B}|}$ then $pr_{\mathbf{A}}$ and $pr_{\mathbf{B}}$ are sketch homomorphisms making $\mathbf{A} \times \mathbf{B}$ the product of $\mathbf{A}$ and $\mathbf{B}$.

ii) The construction of the function space object for sketches is based on that for graphs together with the model structures described in 3.2. Let $\mathbf{A}$ and $\mathbf{B}$ be sketches. Then $[\mathbf{A} \to \mathbf{B}]$ is constructed as follows:

a) The underlying graph of $[\mathbf{A} \to \mathbf{B}]_{\mathbf{SK}}$ is $[|\mathbf{A}| \to |\mathbf{B}|]_{\mathbf{GRAPH}}$, where the subscripts are added for clarity here, but will be omitted from now on.

b) If $\delta : D \to [|\mathbf{A}| \to |\mathbf{B}|]$ is a graph homomorphism with $D \in \boldsymbol{D}$, then δ is an admissable diagram for $[\mathbf{A} \to \mathbf{B}]$ if and only if $\delta^b : \mathbf{Md}(D) \times \mathbf{A} \to \mathbf{B}$ is a sketch homomorphism. Here δ^b is "uncurry(δ)"; i.e., $\delta^b = \mathrm{app} \bullet (\delta \times id_{\mathbf{A}})$.

c) Admissable cones and cocones for $[\mathbf{A} \to \mathbf{B}]$ are described analogously.

It is worth pointing out that this construction is not what it might seem to be. Consider $\delta^b : \mathbf{Md}(D) \times \mathbf{A} \to \mathbf{B}$. Since $\mathbf{Md}(D)$ has no cones or cocones, neither does $\mathbf{Md}(D) \times \mathbf{A}$ by the construction of binary products in ii). Furthermore, the only diagram in $\mathbf{Md}(D)$ is the identity graph homomorphism from D to D. Hence the only diagrams in $\mathbf{Md}(D) \times \mathbf{A}$ are those of the form $\langle id_D, \gamma \rangle : D \to |\mathbf{Md}(D) \times \mathbf{A}|$ where $\gamma : D \to |\mathbf{A}|$ is a diagram of shape D in $\mathbf{A}$. Note that $\langle id_D, \gamma \rangle$ factors as the composition

$$D \xrightarrow{\Delta} D \times D \xrightarrow{id_D \times \gamma} \mathbf{Md}(D) \times \mathbf{A}$$

where Δ is the diagonal map. Hence, $\delta : D \to [|\mathbf{A}| \to |\mathbf{B}|]$ is admissable if and only if $\delta^b : \mathbf{Md}(D) \times \mathbf{A} \to \mathbf{B}$ takes such "diagonal" diagrams of shape D in $\mathbf{Md}(D) \times \mathbf{A}$ to diagrams of shape D in $\mathbf{B}$. Similar comments apply to cones and cocones.

iii) We must now check that the sketch $[\mathbf{A} \to \mathbf{B}]$ is the function space object for $\mathbf{A}$ and $\mathbf{B}$. First of all, let $app_{\mathbf{A},\mathbf{B}} : [\mathbf{A} \to \mathbf{B}] \times \mathbf{A} \to \mathbf{B}$ be the morphism whose underlying graph homomorphism is

$$app_{|\mathbf{A}|,|\mathbf{B}|} : [|\mathbf{A}| \to |\mathbf{B}|] \times |\mathbf{A}| \to |\mathbf{B}|.$$

It must be shown that $app_{\mathbf{A},\mathbf{B}}$ is a sketch homomorphism. a) $App_{\mathbf{A},\mathbf{B}}$ preserves diagrams. Let $\delta = \langle \delta_1, \delta_2 \rangle : D \to [\mathbf{A} \to \mathbf{B}] \times \mathbf{A}$ be an admissable diagram in $[\mathbf{A} \to \mathbf{B}] \times \mathbf{A}$. Then $App_{\mathbf{A},\mathbf{B}} \bullet \delta : D \to \mathbf{B}$ has the following factorization:

$$\begin{array}{ccccc} D & \xrightarrow{\langle \delta_1, \delta_2 \rangle} & [\mathbf{A} \to \mathbf{B}] \times \mathbf{A} & & \\ & \searrow^{\langle id_D, \delta_2 \rangle} & \uparrow^{\langle \delta_1, id_{\mathbf{A}} \rangle} & \searrow^{app_{\mathbf{A},\mathbf{B}}} & \\ & & D \times \mathbf{A} & \xrightarrow{\delta_1^b} & \mathbf{B} \end{array}$$

Here, $\delta_1 : D \to [\mathbf{A} \to \mathbf{B}]$ is admissable, so $\delta_1{}^b : \mathbf{Md}(D) \times \mathbf{A} \to \mathbf{B}$ preserves diagrams. But $\delta_2 : D \to \mathbf{A}$ is an admissable diagram in $\mathbf{A}$, so $\langle id_D, \delta_2 \rangle : D \to \mathbf{Md}(D) \times \mathbf{A}$ is an admissable diagram in $\mathbf{Md}(D) \times \mathbf{A}$. Hence

$$App_{\mathbf{A},\mathbf{B}} \bullet \delta = \delta_1{}^b \bullet \langle id_D, \delta_2 \rangle : D \to \mathbf{B}$$

is an admissable diagram in **B**. b) Similar proofs show that $App_{\mathbf{A},\mathbf{B}}$ preserves cones and cocones.

iv) Finally, we have to check the universal mapping property of $[\mathbf{A} \to \mathbf{B}]$. Let $f : \mathbf{X} \times \mathbf{A} \to \mathbf{B}$ be a sketch homomorphism. Then there is a unique graph homomorphism $f^\# : |\mathbf{X}| \to |[\mathbf{A} \to \mathbf{B}]|$ such that $App_{\mathbf{A},\mathbf{B}} \bullet (f^\# \times id_{\mathbf{A}}) = f$. It is sufficient to show that $f^\#$ is a sketch homomorphism.

a) Suppose $\delta : D \to \mathbf{X}$ is an admissable diagram for **X**. We must show that $f^\# \bullet \delta : D \to [\mathbf{A} \to \mathbf{B}]$ is an admissable diagram for $[\mathbf{A} \to \mathbf{B}]$; i.e., that $(f^\# \bullet \delta)^b : \mathbf{Md}(D) \times \mathbf{A} \to \mathbf{B}$ preserves diagrams. Consider the factorization:

$$(f^\# \bullet \delta)^b = app_{\mathbf{A},\mathbf{B}} \bullet ((f^\# \bullet \delta) \times id_{\mathbf{A}}) = app_{\mathbf{A},\mathbf{B}} \bullet (f^\# \times id_{\mathbf{A}}) \bullet (\delta \times id_{\mathbf{A}}) = f \bullet (\delta \times id_{\mathbf{A}}).$$

But $\delta \times id_{\mathbf{A}}$ preserves diagrams trivially, and f preserves diagrams by hypothesis, so their composition $(f^\# \bullet \delta)^b$ preserves diagrams.

b) Similar proofs show that $f^\#$ preserves cones and cocones.

Next, we turn to the question of local cartesian closure.

3.4 Definition.

i) Let **C** be a category and let C be an object in **C**. The *slice* (or *comma*) category $\mathbf{C}\downarrow C$ is the category (cf. Mac Lane [12]) whose objects are pairs $(A, f : A \to C)$, abbreviated by (A,f). A morphism from (A, f) to (A', f') is a morphism $h : A \to A'$ in **C** such that $f' \bullet h = f$.

ii) If $g : C \to C'$, then $\Sigma g : \mathbf{C}\downarrow C \to \mathbf{C}\downarrow C'$ is the functor whose value on objects is given by $\Sigma g(f) = g \bullet f$ and whose value on morphisms is given by $\Sigma g(h) = h$.

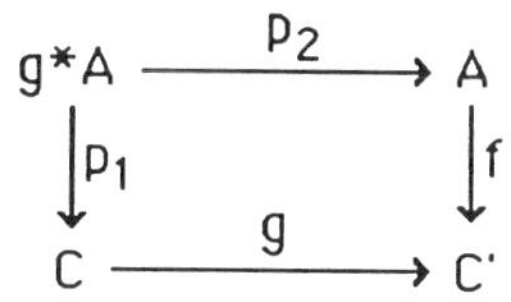

iii) If **C** has chosen pullbacks and if $g : C \to C'$, then $g^* : \mathbf{C}\downarrow C' \to \mathbf{C}\downarrow C$ denotes the functor given by *pulling back along* g; i.e., $g^*(A, f : A' \to C') = (g^*A, p_1 : g^*A \to C)$ where p_1 is the chosen morphism so that the adjoining diagram is a pullback diagram in **C**.

iv) Let **C** be a category with a terminal object 1. **C** is called *locally cartesian closed* if for all C in **C**, the slice category $\mathbf{C}\downarrow C$ is cartesian closed. It is easily checked that $\mathbf{C}\downarrow 1$ is isomorphic to **C**, so locally cartesian closed implies cartesian closed.

3.5 Proposition.

Let **C** be a category with chosen pullbacks and a terminal object.

i) For all $g : C \to C'$ in **C**, $\Sigma g : \mathbf{C}\downarrow C \to \mathbf{C}\downarrow C'$ is left adjoint to $g^* : \mathbf{C}\downarrow C' \to \mathbf{C}\downarrow C$.

ii) **C** is locally cartesian closed if and only if g^* has a right adjoint $\Pi g : \mathbf{C}\downarrow C \to \mathbf{C}\downarrow C'$.

Proof. i) This is just a simple restatement of the universal property of pullbacks.

ii) First of all, $C{\downarrow}C$ has a terminal object given by (C, id_C) and binary products given by pullbacks; i.e., the product of $(A, f : A \to C)$ and $(A', f' : A' \to C)$ is $(A \times_C A', f \cdot p_1 : A \times_C A' \to C)$ as illustrated

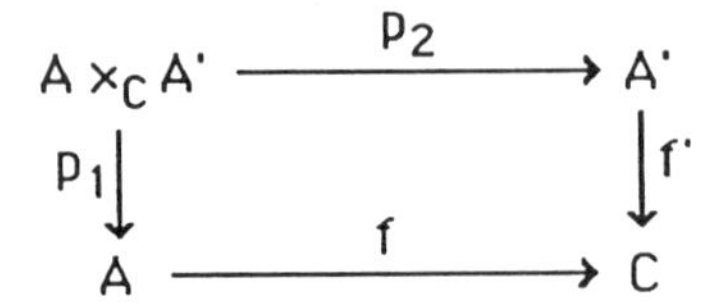

(Note that we use two different notations for pullbacks depending on whether we are thinking of the change of base functor g^* or of products in $C{\downarrow}C$.) Now, suppose a right adjoint, $\Pi g : C{\downarrow}C \to C{\downarrow}C'$ to g^* exists. Then it is easily checked that the function space object for $(A, f : A \to C)$ and $(A', f' : A' \to C)$ is given by the construction $([A \to A']_C, p : [A \to A']_C \to C)$ where

$$[A \to A']_C = (\Pi f)(f^*(A')) = (\Pi f)(A \times_C A').$$

In the second expression, $A \times_C A'$ is regarded as an object over A via the first projection, p_1 and p is the corresponding morphism to C. (See, [3] or [15].)

3.6 Proposition.

i) **SET** is locally cartesian closed.

ii) **GRAPH** is locally cartesian closed.

iii) **SK** is locally cartesian closed.

Proof. We use part two of the preceeding proposition.

i) **SET** is complete and cartesian closed. We assume products of sets are canonically chosen as sets of ordered pairs, and pullbacks are canonical chosen as subsets of products so the functors Σg and g^* exist for $g : C \to C'$. Now, consider an object $(A, f : A \to C)$ in **SET**$\downarrow$C. Given any $c \in C$, let $A_c = f^{-1}(c)$. A, as a set over C, is completely determined by the family of sets $\{A_c \mid c \in C\}$ indexed by the elements of C. A_c is called the *fibre* of $f: A \to C$ over c. Alternatively, it can be described as the adjoining pullback where 'c is the function such that 'c(1) = c. Thus, A_c = 'c*(A) = $1 \times_C A$.

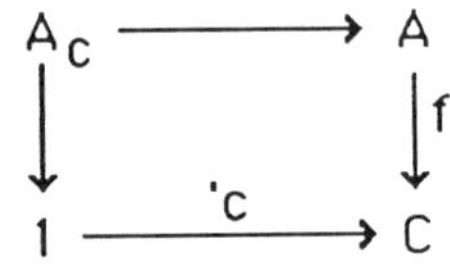

If X is a subset of C with inclusion function inc : $X \to C$, then

$$\Gamma(X, A) = \{s : X \to A \mid f \cdot s = inc\} \approx \Pi\{A_c \mid c \in X\}.$$

$\Gamma(X, A)$ is called the *set of sections* of A over X. To show that Πg exists, consider the adjoining situation in **SET**: We can describe $\Pi g(A)$ by specifying a family of sets indexed by the elements of C'; namely,

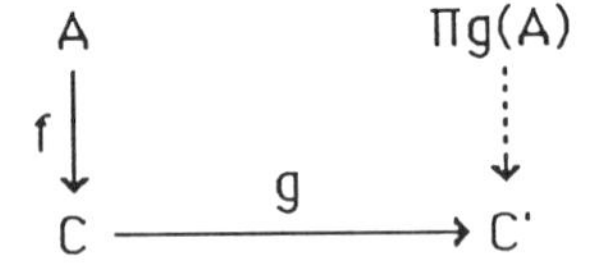

$$(\Pi g(A))_{c'} = \Pi\{A_c \mid f(c) = c'\} \approx \Gamma(f^{-1}(c'), A) = \Gamma(C_{c'}, A).$$

We leave it as an exercise to show that this is the construction of $\prod g$ in **SET** by considering functions from 1 to $\prod g(A)$. Note that Σg has a similar description in **SET**; namely, $(\Sigma g(A))_{c'} = \Sigma\{A_c \mid f(c) = c'\}$. Here the Σ on the right hand side means "coproduct".

Finally, using the construction in 3.5, the function space object in **SET**$\downarrow$C for a pair of objects $(A, f : A \to C)$ and $(A', f' : A' \to C)$ is given by the indexed family of sets:

$$([A \to A']_C)_c = (\prod f(A \times_C A'))_c = \Gamma(A_c, A \times_C A') = \{h : A_c \to A \times_C A' \mid p_1 \bullet h = inc\}$$
$$\approx \{k : A_c \to A' \mid f' \bullet k = f \bullet inc\} \approx \{k : A_c \to A' \mid f' \bullet k = c\}$$
$$\approx \mathbf{SET}(A_c, A'_c)$$

ii) **GRAPH** has a terminal object, the graph with one object and one arrow. Since it is a **SET**-valued functor category, it has arbitrary limits that are computed object-wise. In more detail, the product of two graphs is formed from the products of the sets of objects and sets of arrows in the obvious way. Finally, **GRAPH** has chosen pullbacks given by chosing pullbacks of sets of objects and sets of morphisms. Thus, the functors Σg and g^* exist. In **GRAPH**, we just carry out the construction in **SET** for the sets of objects and sets of morphisms independently to get the construction in **GRAPH**. The formula for $\prod g$ is somewhat more complicated. Some of the expressions for **SET** generalize to an arbitrary topos, providing one considers generalized elements $x : X \to C$ rather than just elements of the form $'c : 1 \to C$. In the case of a **SET**-valued functor category like **GRAPH** $\approx$ **SET**^{E}, it is sufficient to look at generalized elements where X is a representable functor. Since E has only two objects, there are only two such functors which, when viewed as graphs, look as follows:

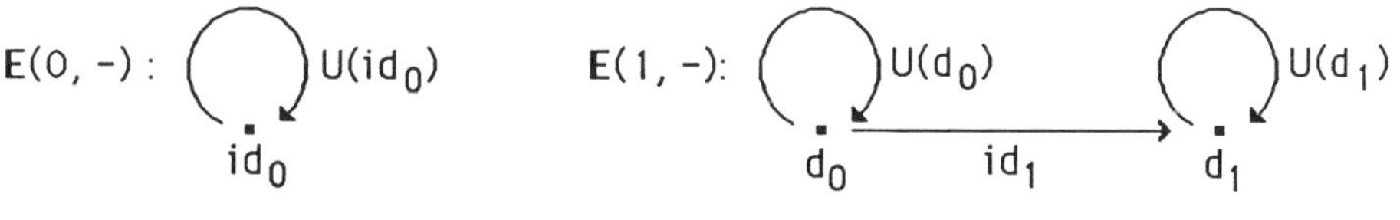

$E(0, -)$: loop $U(id_0)$ at id_0; $E(1, -)$: loop $U(d_0)$ at d_0, arrow $id_1 : d_0 \to d_1$, loop $U(d_1)$ at d_1

For the purposes of the discussion here, instead of talking about objects and arrows of a graph, we talk about 0-cells and 1-cells; i.e., i-cells for i = 0, 1. If C is a graph, then to any i-cell x_i of C there is a uniquely determined graph homomorphism $'x_i : \mathbf{E}(i, -) \to C$. If $(A, f : A \to C)$ is a graph over C and x_i is an i-cell of C, then the fibre of A over x_i is defined to be the adjoining pullback; i.e., it is $\mathbf{E}(i, -) \times_C A$, the product of $'x_i$ and f in **GRAPH**$\downarrow$C. It is easily checked that $A_{x0} = \mathbf{E}(0, -) \times_C A$ is iso-morphic to the subgraph of A consisting of 0-cells y_0 of A with $f_1(y_1) = U(x_0)$. Similarly, $A_{x1} = \mathbf{E}(1, -) \times_C A$ is isomorphic to the subgraph of A consisting of 0-cells y_0 of A with $f_0(y_0) = d_0(x_1)$ or $d_1(x_1)$ and 1-cells y_1 of A with $f_1(y_1) = x_1$ or $U(d_0(x_1))$ or $U(d_1(x_1))$. Finally, if X is any subgraph of C with inclusion homomorphism

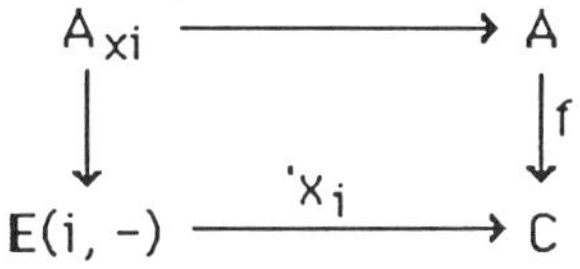

$$\begin{array}{ccc} A_{xi} & \longrightarrow & A \\ \downarrow & & \downarrow f \\ E(i, -) & \xrightarrow{'x_i} & C \end{array}$$

inc : X → C, then Γ(X, A) is defined to be the adjoining pullback (in **SET**).Thus, it is the set of graph homomorphism sections of A over X.

$$\begin{array}{ccc} \Gamma(X, A) & \longrightarrow & \mathbf{GRAPH}(X, A) \\ \downarrow & & \downarrow \mathbf{GRAPH}(X, f) \\ 1 & \xrightarrow{\ 'inc\ } & \mathbf{GRAPH}(X, C) \end{array}$$

Let g : C → C' and let (A, g : A → C) ∈ **GRAPH↓C**. To calculate (Πg(A), q : Πg(A) → C'), note that an i-cell, w_i, of Πg(A) determines a graph homomorphism $'w_i : \mathbf{E}(i, -) \to \Pi g(A)$ and hence a graph homomorphism $'x_i = q \bullet 'w_i : \mathbf{E}(i, -) \to C'$. By adjointness, w_i corresponds to a graph homomorphism $'w_i{}^b : \mathbf{E}(i, -) \times_{C'} C \to A$ over C as illustrated;

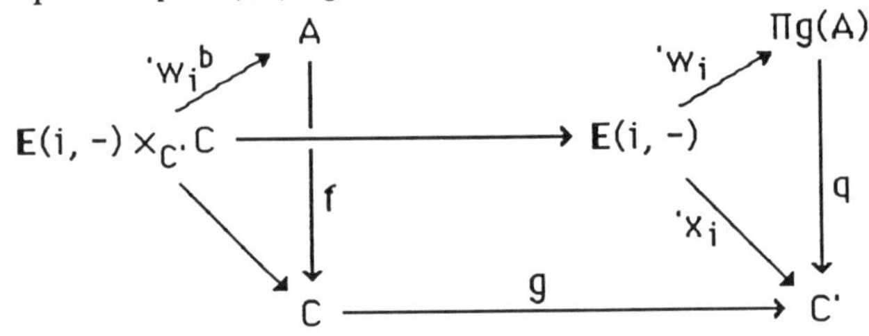

i.e., $'w_i{}^b : C_{xi} \to A$ is a section of A over C_{xi}. Thus, the i-cells of Πg(A) are given by the formula:

$$\Pi g(A)_i = \sum_{x_i \in C'} \Gamma(C_{xi}, A)$$

(Σ means coproduct here.) The source and target functions d_0 and d_1 are given by the obvious restriction functions and $U(s_0) = s_0$, which makes sense since $\Gamma(C_{U(x0)}, A) = \Gamma(C_{x0}, A)$.

Finally, given (A, f : A → C) and (A', f' : A' → C), the function space object $[A \to A']_C$ in **GRAPH↓C** is given by the formulas

$$([A \to A']_C)_i = \sum_{x_i \in C} \Gamma(A_{xi}, A \times_C A') = \sum_{x_i \in C} \mathbf{GRAPH{\downarrow}C}(A_{xi}, A'_{xi})$$

for i = 0, 1, where A_{xi} and A'_{xi} are graphs over C via the composed graph homomorphisms

$$A_{xi} \xrightarrow{\ inc\ } A \xrightarrow{\ f\ } C \qquad\qquad A'_{xi} \xrightarrow{\ inc'\ } A' \xrightarrow{\ f'\ } C$$

The structure functions d_0, d_1, and U are given as above.

iii) Now, for the category **SK↓C** and g : **C** → **C'**, consider the adjoining situation in **SK**. Here Πg(**A**) is the sketch in which |Πg(**A**)| = Πg(|**A**|). For D ∈ *D* and δ : D → Πg(|**A**|) the conditions for δ to be an admissable diagram for Πg(**A**) are the following:

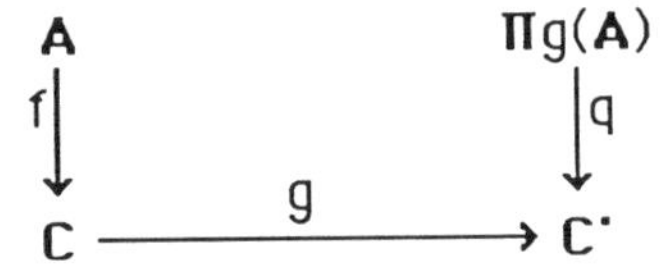

a) q • δ : D → |**C'**| is an admissable diagram for **C'**. Note that then q • δ : **Md**(D) → **C'** is a sketch homomorphism.

b) $\delta : (D, q \cdot \delta : D \to |C'|) \to (\Pi g(|A|), q : \Pi g(|A|) \to |C'|)$ is a morphism in **GRAPH**$\downarrow|C'|$ so it corresponds by adjointness to a graph homomorphism $\delta^{\#} : g^*(D) \to |A|$ over **C**, in **GRAPH**$\downarrow|C|$. Then, δ is defined to be admissable for $\Pi g(\mathbf{A})$ iff $\delta^{\#} : g^*(\mathbf{Md}(D) \to \mathbf{A}$ is a sketch homomorphism, as illustrated.

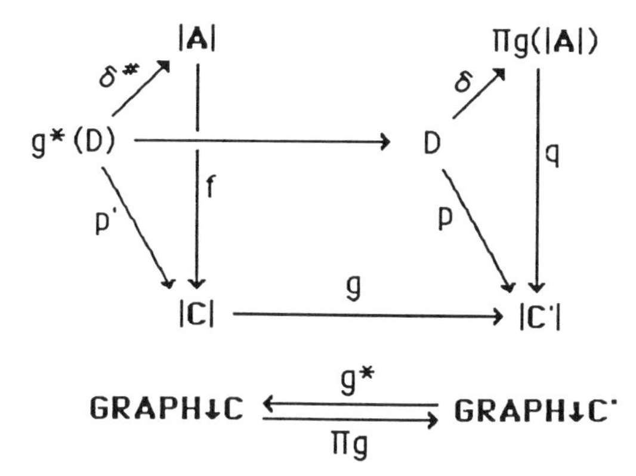

Exactly similar constructions determine the admissable cones and cocones for $\Pi g(\mathbf{A})$. To show that this construction works, consider the following situation. Here, the right hand triangle is over **C'** and the left hand triangle is over **C**. $k : \mathbf{B} \to \Pi g(\mathbf{A})$ is a graph homomorphism over **C'** and $k^{\#} : g^*(\mathbf{B}) \to \mathbf{A}$ is the transpose graph homomorphism over **C**. It must be shown that k is a sketch homomorphism if and only if $k^{\#}$ is.

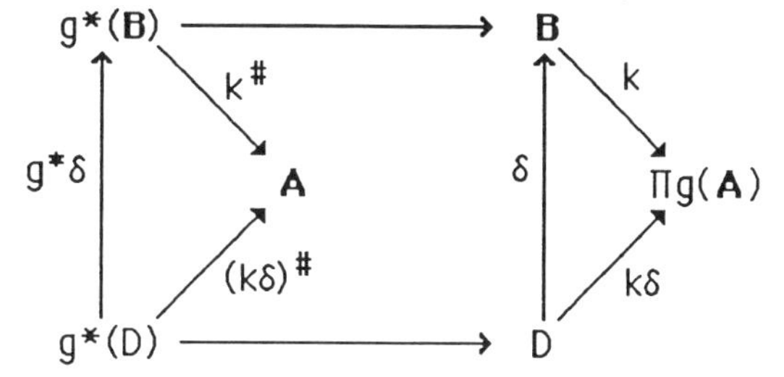

i) Suppose $k^{\#}$ is a sketch homomorphism. Consider an admissable diagram $\delta : D \to \mathbf{B}$. Put D over **C'** by $p\ \delta$ where $p : \mathbf{B} \to \mathbf{C'}$ so $p\ \delta$ is admissable for **C'**. Look at $k\ \delta$ and its transpose $(k\ \delta)^{\#}$. Since $\delta : D \to \mathbf{B}$ is a sketch homomorphism so is $g^*(\delta) : g^*(D) \to g^*(\mathbf{B})$. Hence $(k\ \delta)^{\#} = k^{\#}\ g^*(\delta)$ is a sketch homomorphism, so $k\ \delta$ is admissable for $\Pi g(\mathbf{A})$; i.e., k preserves admissable diagrams. Exactly similar arguments show that it also preserves admissable cones and cocones. Therefore, k is a sketch homomorphism.

ii) Suppose k is a sketch homomorphism. Then we need the following picture:

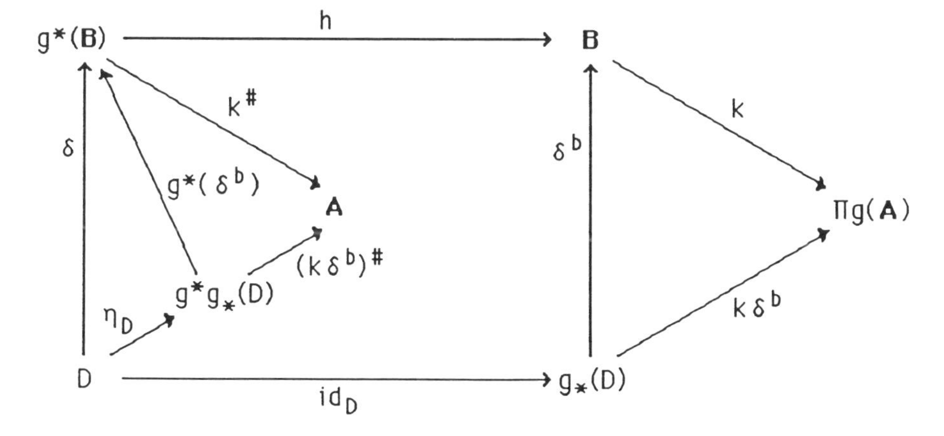

Again the left hand part is over **C** and the right hand part over **C'**. This time $\delta: D \to g^*(\mathbf{B})$ is an admissable diagram for $g^*(\mathbf{B})$. Put D over **C** by $p'\,\delta$ where $p' : g^*(\mathbf{B}) \to \mathbf{C}$ so $p'\,\delta$ is admissable for **C**. The left adjoint to g^* is the functor g_* given by composition with g, so $g_*(D)$ is just D equipped with the structure map $g\,p'\,\delta : D \to \mathbf{C'}$. The transpose of δ is $\delta^b = h\,\delta : g_*(D) \to \mathbf{B}$. Since h is a sketch homomorphism (see 3.8 for limits in the category of sketches), $\delta^b = h\,\delta$ is admissable for **B** and hence $k\,\delta^b$ is admissable for $\prod g(\mathbf{A})$ since k is a sketch homomorphism. By the definition of admissable diagrams for $\prod g(\mathbf{A})$, it follows that $(k\,\delta^b)^\# : g^*\,g_*(D) \to \mathbf{A}$ is a sketch homomorphism. But the adjunction morphism $\eta_D : D \to g^*\,g_*(D)$ is a sketch homomorphism by the construction of pullbacks (see 3.8) since $\eta_D = \langle id_D, p'\,\delta \rangle$. Finally, note that $k^\#\,\delta = k^\#\,g^*(\delta^b)\,\eta_D = (k\,\delta^b)^\#\,\eta_D$ is a sketch homomorphism; i.e., $k^\#\,\delta$ is admissable for **A**. Thus $k^\#$ preserves admissable diagrams. Exactly similar arguments show that it also preserves admissable cones and cocones. Therefore, $k^\#$ is a sketch homomorphism.

We note in passing that locally cartesian closed categories are the appropriate categories to be models of Martin-Löf type theory. (See, Seely, [14].) In particular, Σg and $\prod g$ provide constructions for "dependent types".

Next we want to show that the category **SK** is complete and cocomplete. This involves techniques that were developed to generalize the properties of the category **TOP** of topological spaces, regarded as a category equipped with an underlying functor to **SET**. In our case we have an underlying functor from **SK** to **GRAPH**, and **GRAPH** is almost as nice a category as **SET**. In particular, it is complete and cocomplete, all limits and colimits being given by their separate constructions for objects and arrows. This is enough for the following theorem to imply the completeness and cocompleteness of **SK**.

3.7 Proposition.

SK is topological over **GRAPH**.

Proof. The term "topological" means that the following properties are satisfied:

i) For each $G \in$ **Graph**, the subcategory $\mathbf{SK}_G$ of **SK** determined by sketches whose underlying graph equals G and sketch homomorphisms which are the identity on G, is a complete lattice. To see this, let $\{\mathbf{A}_i = (G, \boldsymbol{D}_{Ai}, \boldsymbol{C}_{Ai}, \boldsymbol{coC}_{Ai}) \mid i \in I\}$ be a family of sketches in $\mathbf{SK}_G$. Then it is easily checked that

$$\vee\{\mathbf{A}_i \mid i \in I\} = (G, \cup\{\boldsymbol{D}_{Ai} \mid i \in I\}, \cup\{\boldsymbol{C}_{Ai} \mid i \in I\}, \cup\{\boldsymbol{coC}_{Ai} \mid i \in I\})$$

$$\wedge\{\mathbf{A}_i \mid i \in I\} = (G, \cap\{\boldsymbol{D}_{Ai} \mid i \in I\}, \cap\{\boldsymbol{C}_{Ai} \mid i \in I\}, \cap\{\boldsymbol{coC}_{Ai} \mid i \in I\}).$$

The bottom object in $\mathbf{SK}_G$ is (G, Triv, Ø, Ø) where Triv consists of the trivial diagrams that are required to be present in any sketch (cf., 2.2), and the top object is (G, All$\boldsymbol{D}$, All$\boldsymbol{C}$, All$\boldsymbol{coC}$), where **SK** means $\boldsymbol{D}$-$\boldsymbol{C}$-$\boldsymbol{coC}$-**SK**, and All$\boldsymbol{D}$, for instance, means all diagrams of type $\boldsymbol{D}$, etc.

ii) If $f : G \to G'$ is a graph homomorphism, then there are functors $f^* : \mathbf{SK}_{G'} \to \mathbf{SK}_G$ and $f_* : \mathbf{SK}_G \to \mathbf{SK}_{G'}$ and natural isomorphisms:

$$\mathbf{SK}_{G'}(f_*(\mathbf{A}), \mathbf{A'}) \approx \mathbf{SK}(\mathbf{A}, \mathbf{A'}) \approx \mathbf{SK}_G(\mathbf{A}, f^*(\mathbf{A'})).$$

Here, $\mathbf{A} \in \mathbf{SK}_G$ and $\mathbf{A}' \in \mathbf{SK}_{G'}$, and

$$\mathcal{D}_{f_* \mathbf{A}} = \{f \circ \delta \mid \delta \in \mathcal{D}_{\mathbf{A}}\}, \qquad \mathcal{D}_{f^* \mathbf{A}'} = \{\delta : D \to G \mid f \circ \delta \in \mathcal{D}_{\mathbf{A}'}\}$$

with similar formulas for cones and cocones. $f_*\mathbf{A}$ is called the *final* structure on G' determined by **A**, and $f^*\mathbf{A}'$ is called the *initial* structure on G determined by **A'**.

3.8 Proposition.

SK is complete and cocomplete.

Proof. This follows immediately from 3.7. For instance, completeness follows from the existence of products and equalizers which are constructed as follows: The product of a family of sketches in given by the product of their underlying graphs equipped with the infimum of the initial sketch structures determined by the projections onto the factors. Similarly, the equalizer of a pair of sketch homomorphisms is given by the equalizer of the underlying graph homomorphisms equipped with the initial sketch structure from its graph homomorphism into the domain sketch.

Once we know that **SK** is cocomplete, it follows that we can solve domain equations in order to describe sketches recursively as least fixed objects for suitable endofunctors. Such least fixed objects certainly exist for endofunctors $F : \mathbf{SK} \to \mathbf{SK}$ that have right adjoints. Since **SK** is cartesian closed, we already have a whole class of functors with right adjoints; namely, the functors $- \times \mathbf{A} : \mathbf{SK} \to \mathbf{SK}$ which have the functors $[\mathbf{A} \to -]$ as right adjoints. E.g., if we define $F(\mathbf{X}) = \mathbf{1} + \mathbf{X} \times \mathbf{A}$, then F(-) has a least fixed object which is a sketch $\mathbf{A}^*$ satisfying the condition that $\mathbf{A}^* \approx \mathbf{1} + \mathbf{A}^* \times \mathbf{A}$.

4. A SECOND CLOSED CATEGORY STRUCTURE.

There is another closed category structure on **SK** consisting of a tensor product functor $\mathbf{A} \otimes \mathbf{B}$ and a right adjoint internal hom functor $\mathbf{Hom}(\mathbf{A}, \mathbf{B})$. In many ways, this structure is more useful than the cartesian closed structure. Models of $\mathbf{A} \otimes \mathbf{B}$ are more interesting than models of $\mathbf{A} \times \mathbf{B}$, objects of $\mathbf{Hom}(\mathbf{A}, \mathbf{B})$ are actual sketch homomorphisms, it leads to further interesting structures (as in Gray [7]), etc. The constructions are quite simple. (Cf., Lair [11], Gray [7])

i) The tensor product. We begin by defining a tensor product for graphs. Let G and H be graphs. Then $G \otimes H$ is the graph such that $(G \otimes H)_0 = G_0 \times H_0$ and

$(G \otimes H)_1 = \{(\alpha, U(b)) \mid \alpha \in G_1 \text{ and } b \in H_0\} \cup \{(U(a), \beta) \mid a \in G_0 \text{ and } \beta \in H_1\}$.

with obvious component-wise definitions for d_0, d_1, and U. For each $a \in G_0$ and $b \in H_0$, there are graph homomorphisms $j_a : H \to G \otimes H$ and $j_b : G \to G \otimes H$ given by $(j_a)_0(b) = (a, b)$ and $(j_a)_1(\beta) = (U(a), \beta)$, for instance. Then $\mathbf{A} \otimes \mathbf{B}$ is the sketch whose underlying graph is $|\mathbf{A}| \otimes |\mathbf{B}|$ with

$$\mathbf{A} \otimes \mathbf{B} = \underbrace{\bigvee\{(j_b)_*(\mathbf{A}) \mid b \in |\mathbf{B}|_0\}}_{\text{horizontal structure from } \mathbf{A}} \vee \underbrace{\bigvee\{(j_a)_*(\mathbf{B}) \mid a \in |\mathbf{A}|_0\}}_{\text{vertical structure from } \mathbf{B}} \vee \mathbf{R}$$

where **R** is the "rectangular" structure on $|\mathbf{A} \otimes \mathbf{B}|$ which has no cones or cocones, but all diagrams of the adjoining form for all $\alpha : a \to a'$ in $|\mathbf{A}|_1$ and all $\beta : b \to b'$ in $|\mathbf{B}|_1$. $\otimes$ is made into a functor by defining

$$\begin{array}{ccc} (a, b) & \xrightarrow{(\alpha, U(b))} & (a', b) \\ \downarrow{\scriptstyle (U(a), \beta)} & & \downarrow{\scriptstyle (U(a'), \beta)} \\ (a, b') & \xrightarrow{(\alpha, U(b'))} & (a', b') \end{array}$$

$f \otimes g : \mathbf{A} \otimes \mathbf{B} \to \mathbf{A'} \otimes \mathbf{B'}$

by the formulas:

$(f \otimes g)_0 = f_0 \times g_0$

$(f \otimes g)_1 = (f_1 \times g_1) \,|\, (|\mathbf{A} \otimes \mathbf{B}|)_1$.

Note that $\otimes$ is commutative, associative, and has the sketch **obj** = (1, Min, Ø, Ø) = (**E**(0, -), Min, Ø, Ø) as left and right unit (all up to coherent natural isomorphisms.)

ii) The internal hom. Define $|\mathbf{Hom}(\mathbf{A}, \mathbf{B})|_0 = \mathbf{SK}(\mathbf{A}, \mathbf{B})$ and $|\mathbf{Hom}(\mathbf{A}, \mathbf{B})|_1 = \mathbf{SK}(\mathbf{2} \otimes \mathbf{A}, \mathbf{B})$ where $\mathbf{2}$ = (**E**(1, -), Min, Ø, Ø). The graph homomorphisms induce the graph structure functions

$$\mathbf{E}(0, -) \underset{\mathbf{E}(U, -)}{\overset{\mathbf{E}(d_i, -)}{\rightleftarrows}} \mathbf{E}(1, -)$$

$$\mathbf{SK}(\mathbf{A}, \mathbf{B}) \approx \mathbf{SK}(\mathbf{obj} \otimes \mathbf{A}, \mathbf{B}) \underset{U}{\overset{d_i}{\leftrightarrows}} \mathbf{SK}(\mathbf{2} \otimes \mathbf{A}, \mathbf{B})$$

for **Hom(A, B)**. An arrow in **Hom(A, B)** can be described as a triple (f, τ, g) where f, g : **A** → **B** are sketch homomorphisms and τ is a family of arrows $\{(\tau_a : f(a) \to g(a)) \in |\mathbf{B}|_1 \mid a \in |\mathbf{A}|_0\}$ such that for all $\alpha : a \to a'$ in $|\mathbf{A}|_1$, the adjoining diagram is admissable in **B**. In particular, if **C** is a category, then arrows in **Hom(A, ^C)** look like natural transformations. (Note that there is an analogous **Hom** for graphs.)

$$\begin{array}{ccc} f(a) & \xrightarrow{f(\alpha)} & f(a') \\ \downarrow{\scriptstyle \tau_a} & & \downarrow{\scriptstyle \tau_b} \\ g(a) & \xrightarrow{g(\alpha)} & g(a') \end{array}$$

4.1 **Theorem.** There are natural isomorphisms

i) **SK(A ⊗ B, C) ≈ SK(A, Hom(B, C))**,

ii) **Hom(A ⊗ B, C) ≈ Hom(A, Hom(B, C))**.

Note that the first statement says that for all **B**, - ⊗ **B** is left adjoint to **Hom(B**, -) and the second says that this adjunction is enriched in the closed category structure on **SK** given by this adjoint pair of functors.

Proof. There is an "evaluation" sketch homomorphism $ev_{\mathbf{B},\mathbf{C}} : \mathbf{Hom}(\mathbf{B}, \mathbf{C}) \otimes \mathbf{B} \to \mathbf{C}$ given by the formulas: $(ev_{\mathbf{B},\mathbf{C}})_0(f, b) = f_0(b)$, $(ev_{\mathbf{B},\mathbf{C}})_1((f, \tau, g), U(b)) = \tau_b$, and $(ev_{\mathbf{B},\mathbf{C}})_1(U(f), b) = f_1(b)$. It satisfies the universal mapping property that, given any sketch homomorphism $h : \mathbf{A} \otimes \mathbf{B} \to \mathbf{C}$, there is a unique sketch homomorphism $h\# : \mathbf{A} \to \mathbf{Hom}(\mathbf{B}, \mathbf{C})$ such that $ev_{\mathbf{B},\mathbf{C}} \bullet (h\# \otimes id_{\mathbf{B}}) = h$. h# is given by the following formulas:

$$(h\#_0(a))_0(b) = h_0(a, b),\ (h\#_0(a))_1(\beta) = h_1(U(a), \beta), \text{ and}$$
$$h\#_1(\alpha : a \to a') = (h\#_0(a), \tau_a, h\#_0(a')),$$

where $\tau_a = \{(h_1(\alpha), U(b)) \mid b \in |\mathbf{B}|_0\}$. This describes the bijection in part i) of the theorem. Naturality is proved in the usual way. Part ii) follows by applying part i) with **A** replaced by $\mathbf{2} \otimes \mathbf{A}$.

4.2 **Proposition.**

For any category **C** and any sketch **A**, $|\mathbf{MOD_C(A)}| \approx |\mathbf{Hom(A, {}^\wedge C)}|$. In particular,

$$\mathbf{MOD(A \otimes B) \approx MOD_{MOD(B)}(A)}.$$

Proof. The first isomorphism follows immediately from the definitions. For the second, one has

$$\begin{aligned}\mathbf{MOD(A \otimes B)} &= \mathbf{MOD_{SET}(A \otimes B) \approx Hom(A \otimes B, {}^\wedge SET)}\\ &\approx \mathbf{Hom(A, Hom(B, {}^\wedge SET)) \approx Hom(A, MOD_{SET}(B))}\\ &\approx \mathbf{MOD_{MOD(B)}(A)}\end{aligned}$$

5. LAX COLIMITS AND DEPENDENT TYPES.

The category $\mathbf{SK}\!\downarrow\!\mathbf{A}$ provides one notion of a family of sketches indexed by **A** and the constructions Σg and Πg provide one notion of the associated dependent types for such a family. However, there is a more refined notion of a family of sketches indexed by a sketch with constructions of associated dependent types which seems to be more satisfactory in the light of examples. These constructions are closely related to the notion of lax colimits which arise in the theory of 2-categories. For a discussion of them see Gray [5], [6], or Kelly and Street [10]. We want to develop an analogous notion here that makes sense for categories enriched in **GRAPH**, where **GRAPH** is regarded as a monoidal category with respect to the tensor product which is described in the preceeding section in the construction of the tensor product of sketches.

5.1. **Definition.** Let **A** and **B** be sketches.

i) A *"functor"* (or *1-cell*) $f : \mathbf{A} \to \mathbf{B}$ is a graph homomorphism $f : |\mathbf{A}| \to |\mathbf{B}|$ which preserves diagrams; i.e., f satisfies condition 2.3, a), i) in the definition of a sketch homomorphism. The category of sketches (for a fixed collection of diagrams, cones, and cocones) and "functors" will be denoted by $\mathbf{SK_f}$.

ii) Let f and $g : \mathbf{A} \to \mathbf{B}$ be "functors". A *"natural transformation"* (or *2-cell*) $t : f \Rightarrow g$ is a family $\{t_a : f(a) \to g(a)\}$ of arrows in **B** such that for all arrows $\alpha : a \to a'$ in **A**, the adjoining diagram is admissable in **B**

$$\begin{array}{ccc} f(a) & \xrightarrow{f(\alpha)} & f(a') \\ \downarrow{\tau_a} & & \downarrow{\tau_b} \\ g(a) & \xrightarrow{g(\alpha)} & g(a') \end{array}$$

$\mathbb{SK}_f(\mathbf{A}, \mathbf{B})$ denotes the graph whose objects are "functors" from **A** to **B** and whose arrows are "natural transformations" between such "functors".

iii) Given "functors" and a "natural transformation" as illustrated,

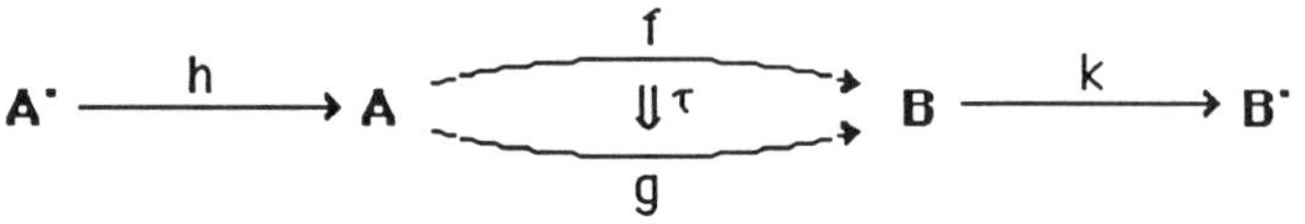

$$A' \xrightarrow{h} A \underset{g}{\overset{f}{\underset{\Downarrow \tau}{\rightrightarrows}}} B \xrightarrow{k} B'$$

then compositions $\tau * h : f \bullet h \Rightarrow g \bullet h$ and $k * \tau : k \bullet f \Rightarrow k \bullet g$ are defined by the formulas $(\tau * h)_{a'} = \tau_{h(a')}$ and $(k * \tau)_a = k(\tau_a)$. Let $* : \mathbb{SK}_f(\mathbf{A}, \mathbf{B}) \otimes \mathbb{SK}_f(\mathbf{B}, \mathbf{C}) \to \mathbb{SK}_f(\mathbf{A}, \mathbf{C})$ be given by $*(f, h) = h \bullet f$, $*(U(f), \beta) = \beta * f$, and $*(\alpha, U(h)) = h * \alpha$.

5.2 **Proposition.** The graphs $\mathbb{SK}_f(\mathbf{A}, \mathbf{B})$ and the operations $*$ describe an enrichment of the category $\mathbf{SK_f}$ in the category **GRAPH.**

Proof. This just means that $*(id_\mathbf{A}, f) = f = *(f, id_\mathbf{B})$ and that $*$ is associative up to appropriate coherence isomorphisms. The details are easily supplied. Note that in distinction to the 2-category case, there is no "vertical" composition of 2-cells and no "horizontal" composition unless one of the factors is an "identity" 2-cell.

We can now describe what will be meant by a family of sketches indexed by a sketch.

5.3 **Definition.** Let **A** be a sketch.

i) A *co* (or *contra*) *variant* **A**-*indexed family of sketches* is a "functor" $\mathbf{B}(-) : \mathbf{A} \to {}^\wedge\mathbf{SK_f}$ (or $\mathbf{B}(-) : \mathbf{A} \to {}^\wedge(\mathbf{SK_f}^{op})$). Thus, for each object a in **A**, **B**(a) is a sketch and for each arrow $\alpha : a \to a'$ in **A**, $\mathbf{B}(\alpha) : \mathbf{B}(a) \to \mathbf{B}(a')$ (or $\mathbf{B}(\alpha) : \mathbf{B}(a') \to \mathbf{B}(a)$) is a "functor". These data satisfy $\mathbf{B}(U(a)) = id_{\mathbf{B}(a)}$ and **B** applied to any diagram in **A** is a commutative diagram in $\mathbf{SK_f}$ (or $\mathbf{SK_f}^{op}$). If each $\mathbf{B}(\alpha)$ is a sketch homomorphism, then **B** is called a *homomorphic* family. In this case it takes values in the subsketch $^\wedge\mathbf{SK}$.

ii) Let **B**(-) and **B'**(-) be **A**-indexed families of sketches. An **A**-*indexed family of "functors"* is a "natural transformation" $\tau : \mathbf{B}(-) \Rightarrow \mathbf{B'}(-)$. If all components $\tau_a : \mathbf{B}(a) \to \mathbf{B'}(a)$ are sketch homomorphisms, then τ is called an **A**-*indexed family of sketch homomorphisms*. Note that "natural transformation" here means that for all arrows $\alpha : a \to a'$ in **A**, the diagram of "functors",

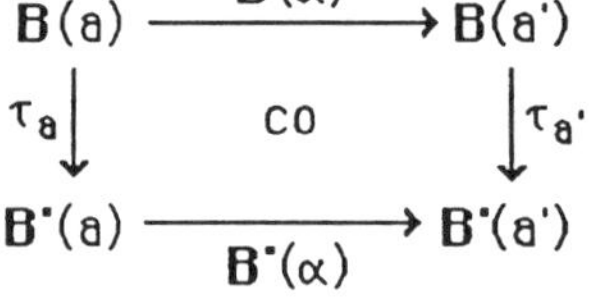

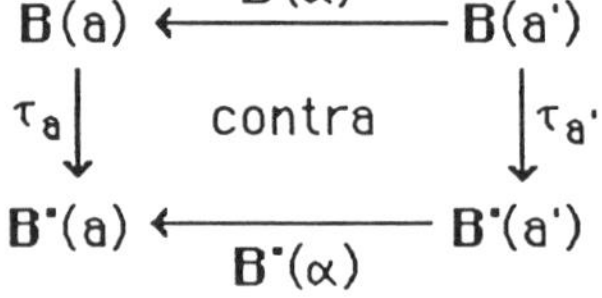

$$\begin{array}{ccc} B(a) & \xrightarrow{B(\alpha)} & B(a') \\ {\scriptstyle\tau_a}\downarrow & co & \downarrow{\scriptstyle\tau_{a'}} \\ B'(a) & \xrightarrow[B'(\alpha)]{} & B'(a') \end{array} \qquad \begin{array}{ccc} B(a) & \xleftarrow{B(\alpha)} & B(a') \\ {\scriptstyle\tau_a}\downarrow & contra & \downarrow{\scriptstyle\tau_{a'}} \\ B'(a) & \xleftarrow[B'(\alpha)]{} & B'(a') \end{array}$$

commutes, since $^\wedge\mathbf{SK}$ is the underlying sketch of a category.

iii) Let **B**(-) and **B'**(-) be **A**-indexed families of sketches. A *lax* **A**-*indexed family of sketch homomorphisms* $\tau : \mathbf{B}(-) \Rightarrow \mathbf{B'}(-)$ is a family of sketch homomorphisms

$$\{\tau_a : \mathbf{B}(a) \to \mathbf{B'}(a) \mid a \in |\mathbf{A}|_0\}$$

together with a family of "natural transformations"

$\{\tau_\alpha : \mathbf{B'}(\alpha) \bullet \tau_a \Rightarrow \tau_{a'} \bullet \mathbf{B}(\alpha) \mid (\alpha : a \to a') \in |\mathbf{A}|_1\}$
(or $\{\tau_\alpha : \tau_a \bullet \mathbf{B}(\alpha) \Rightarrow \mathbf{B'}(\alpha) \bullet \tau_{a'} \mid (\alpha : a \to a') \in |\mathbf{A}|_1\}$)

as illustrated:

iv) Let $\mathbf{B}(\text{-})$ be an $\mathbf{A}$-indexed family of sketches and let $\mathbf{C}$ be a sketch. $K_\mathbf{A}\mathbf{C}$ denotes the constant $\mathbf{A}$-indexed family equal to $\mathbf{C}$. A *lax cocone* on $\mathbf{B}(\text{-})$ with vertex $\mathbf{C}$ is a lax $\mathbf{A}$ indexed family of sketch homomorphisms $\tau : \mathbf{B}(\text{-}) \Rightarrow K_\mathbf{A}\mathbf{C}$; i.e., for each object a in $\mathbf{A}$, $\tau_a : \mathbf{B}(a) \to \mathbf{C}$ is a sketch homomorphism and for each arrow $\alpha : a \to a'$ in $\mathbf{A}$, $\tau_\alpha : \tau_a \Rightarrow \tau_{a'} \bullet \mathbf{B}(\alpha)$
(or $\tau_\alpha : \tau_a \bullet \mathbf{B}(\alpha) \Rightarrow \tau_{a'}$) as illustrated:

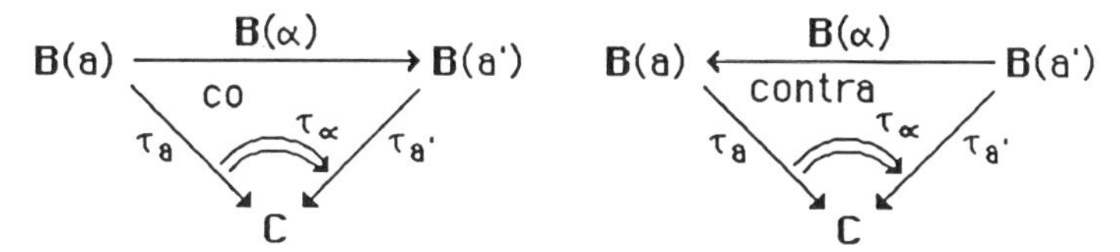

v) Let $\tau : \mathbf{B}(\text{-}) \Rightarrow K_\mathbf{A}\mathbf{C}$ and $\tau' : \mathbf{B}(\text{-}) \Rightarrow K_\mathbf{A}\mathbf{C'}$ be lax cocones on $\mathbf{B}(\text{-})$. A *morphism of lax cocones* from τ to τ' is a sketch homomorphism $k : \mathbf{C} \to \mathbf{C'}$ such that $k \bullet \tau = \tau'$; i.e., $k \bullet \tau_a = \tau'_a$ and $k * \tau_\alpha = \tau'_\alpha$. The category of lax cocones on $\mathbf{B}(\text{-})$ and morphisms of such is denoted by **LAX_COCONE(B(-))**.

vi) An initial object of **LAX_COCONE(B(-))** is called a *universal lax cocone* for $\mathbf{B}(\text{-})$, or a *lax colimit* of $\mathbf{B}(\text{-})$.

Unfortunately, the lax colimit of $\mathbf{B}(\text{-})$ is not exactly the right construction for the Σ-dependent type determined by the $\mathbf{A}$-indexed family $\mathbf{B}(\text{-})$ because it does not depend tightly enough on the sketch structure of $\mathbf{A}$. Instead, we have to consider a still more precise concept - the "admissable" lax colimit, based on the notion of an "admissable" lax cocone. We require some more machinery to describe this concept. As with comma categories, one can construct comma graphs, comma sketches, lax comma graphs and lax comma sketches. We just describe the particular cases that concern us here.

5.4 **Definition.** Let $\mathbf{B}(\text{-})$ be an $\mathbf{A}$-indexed family of sketches.

i) Let $\mathbf{obj} = (1, \text{Min}, \emptyset, \emptyset)$ and let $|\mathbf{obj}\downarrow\mathbf{B}(\text{-})|$ denote the graph whose objects are pairs $(b : \mathbf{obj} \to \mathbf{B}(a), a)$ where $a \in |\mathbf{A}|_0$ and b is a sketch homomorphism. An arrow from $(b : \mathbf{obj} \to \mathbf{B}(a), a)$ to $(b' : \mathbf{obj} \to \mathbf{B}(a'), a')$ is an arrow $\alpha : a \to a'$ in $\mathbf{A}$ such that $\mathbf{B}(\alpha) \bullet b = b'$ (or $\mathbf{B}(\alpha) \bullet b' = b$), as illustrated:

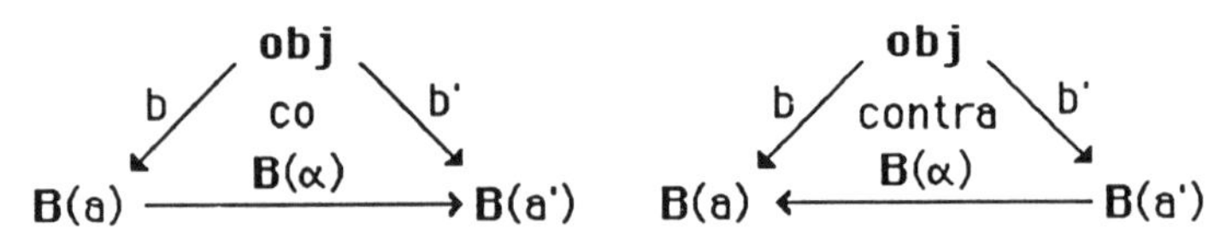

Let $p : |\mathbf{obj}{\downarrow}\mathbf{B}(-)| \to |\mathbf{A}|$ denote the graph homomorphism given by $p(b : \mathbf{obj} \to \mathbf{B}(a), a) = a$ and $p(\alpha) = \alpha$. Then $\mathbf{obj}{\downarrow}\mathbf{B}(-)$ is the sketch on the graph $|\mathbf{obj}{\downarrow}\mathbf{B}(-)|$ with the initial sketch structure, $p^*(\mathbf{A})$, from the graph homomorphism p; i.e., a diagram δ (resp., a cone or cocone γ) is admissable if and only if $p \bullet \delta$ (resp., $p \bullet \gamma$) is admissable in $\mathbf{A}$.

ii) Let $\mathbf{C}$ be a sketch. $|\mathbf{B}(-){\downarrow\downarrow}\mathbf{C}|$ denotes the graph whose objects are pairs $(a, f : \mathbf{B}(a) \to \mathbf{C})$ where $a \in |\mathbf{A}|_0$ and f is a sketch homomorphism. An arrow from $(a, f : \mathbf{B}(a) \to \mathbf{C})$ to $(a', f' : \mathbf{B}(a') \to \mathbf{C})$ is a pair (α, τ) where $\alpha : a \to a'$ is an arrow in $\mathbf{A}$ and $\tau : f \Rightarrow f' \bullet \mathbf{B}(\alpha)$ (or $\tau : f \bullet \mathbf{B}(\alpha) \Rightarrow f$) is a "natural transformation", as illustrated:

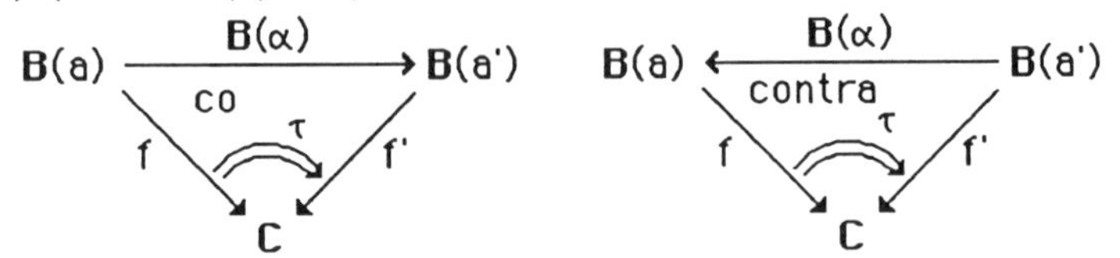

Let $q : |\mathbf{B}(-){\downarrow\downarrow}\mathbf{C}| \to |\mathbf{A}|$ denote the graph homomorphism given by $q(a, f : \mathbf{B}(a) \to \mathbf{C}) = a$ and $q(\alpha, \tau) = \alpha$. Now, a lax cocone on $\mathbf{B}(-)$ with vertex $\mathbf{C}$ given by components

$$\{\tau_a : \mathbf{B}(a) \to \mathbf{C} \mid a \in |\mathbf{A}|_0\} \text{ and } \{\tau_\alpha : \tau_a \Rightarrow \tau_{a'} \bullet \mathbf{B}(\alpha) \mid (\alpha : a \to a') \in |\mathbf{A}|_1\}$$

(or, $\{\tau_\alpha : \tau_a \bullet \mathbf{B}(\alpha) \Rightarrow \tau_{a'} \mid (\alpha : a \to a') \in |\mathbf{A}|_1\}$) corresponds to a section (i.e., a graph homomorphism whose composition with q is the identity) $t : |\mathbf{A}| \to |\mathbf{B}(-){\downarrow\downarrow}\mathbf{C}|$ of q in the obvious way; namely., $t(a) = t_a$ and $t(\alpha) = (\alpha, \tau_\alpha)$. We choose a fixed section t (i.e., a fixed lax cocone on $\mathbf{B}(-)$), and then $\mathbf{B}(-){\downarrow\downarrow_t}\mathbf{C}$ denotes the sketch on $|\mathbf{B}(-){\downarrow\downarrow}\mathbf{C}|$ with the final sketch structure from the graph homomorphism t. Thus, if δ is an admissable diagram (resp., γ is an admissable cone or cocone) of $\mathbf{A}$, then $t \bullet \delta$ (resp., $t \bullet \gamma$) is admissable for $\mathbf{B}(-){\downarrow\downarrow_t}\mathbf{C}$.

iii) Consider the pullback graph

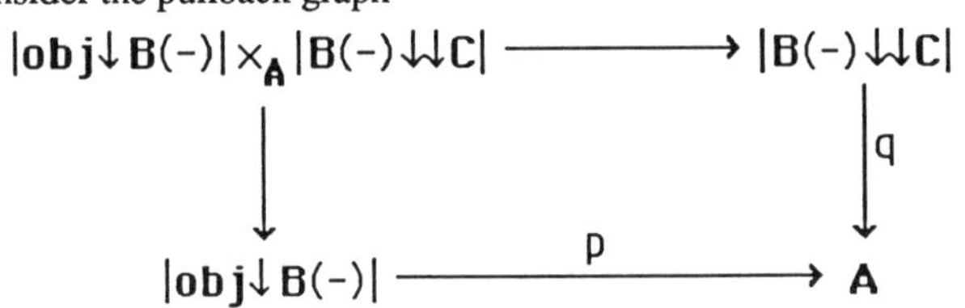

There is a composition operation $\mathrm{comp} : |\mathbf{obj}{\downarrow}\mathbf{B}(-)| \times_{\mathbf{A}} |\mathbf{B}(-){\downarrow\downarrow}\mathbf{C}| \to |\mathbf{obj}{\downarrow\downarrow}\mathbf{C}| \approx |\mathbf{C}|$ given by $\mathrm{comp}((b : \mathbf{obj} \to \mathbf{B}(a), a), (a, f : \mathbf{B}(a) \to \mathbf{C})) = f \bullet b = f(b)$, and $\mathrm{comp}(\alpha, (\alpha, \tau)) = \tau_\alpha$, where $\alpha : a \to a'$ and $\tau : f \Rightarrow f' \bullet \mathbf{B}(\alpha)$ (or $\tau : f \bullet \mathbf{B}(\alpha) \Rightarrow f'$). This makes sense since $f'(\mathbf{B}(\alpha))(b) = f'(b')$ (or $f(\mathbf{B}(\alpha))(b') = f(b)$). (I.e., paste the diagram for α in i) on top of the diagram for (α, τ) in ii) and compose everything.)

iv) A lax cocone with corresponding section t of $|\mathbf{B}(-){\downarrow\downarrow}\mathbf{C}|$ is called *admissable* if

$$\mathrm{comp} : |\mathbf{obj}{\downarrow}\mathbf{B}(-)| \times_{\mathbf{A}} |\mathbf{B}(-){\downarrow\downarrow_t}\mathbf{C}| \to |\mathbf{obj}{\downarrow\downarrow}\mathbf{C}| \approx |\mathbf{C}|$$

is a sketch homomorphism. The full subcategory of **LAX_COCONE(B(-))** determined by the admissable lax cocones is denoted by **ADM_LAX_COCONE(B(-))**.

v) An initial object of **ADM_LAX_COCONE(B(-))** is called a *universal admissable lax cocone* for **B**(-), or an *admissable lax colimit* of **B**(-).

A typical diagram of $\mathbf{obj}{\downarrow}\mathbf{B}(\text{-}) \times_{\mathbf{A}} \mathbf{B}(\text{-}){\downarrow\downarrow_t}\mathbf{C}$ is of the form $<\delta, t \bullet p \bullet \delta>$ where $p \bullet \delta$ is an admissable diagram in **A**. Similar descriptions hold for cones and cocones. These diagrams, cones and cocones reflect the structure of **A** so that requiring comp to be a sketch homomorphism says that **C** not only has the "vertical" sketch structures coming from the fibres **B**(a) but it also has the "horizontal" sketch structure coming from **A**.

We adopt the notation $\sum_{a:\mathbf{A}} \mathbf{B}(a)$ for the admissable lax colimit of **B**(-) to suggest the similar notation in Martin-Löf type theory for the dependent sum. The following construction shows that it is a reasonable candidate for such a dependent sum in our context.

5.6 **Theorem.**

The sketch $\sum_{a:\mathbf{A}} \mathbf{B}(a)$ exists for an **A**-indexed family of sketches.

Proof. The construction is best conveyed by a picture together with a brief description of the contents of the picture. Essentially, $\sum_{a:\mathbf{A}} \mathbf{B}(a)$ is built from the disjoint union of the sketches **B**(a) for $a \in |\mathbf{A}|_0$ by adding arrows $(b, \alpha) : b \to \mathbf{B}(\alpha)(b)$ (or $(b', \alpha) : \mathbf{B}(\alpha)(b') \to b$) for each $b \in |\mathbf{B}(a)|_0$ and each $\alpha : a \to a'$ in $|\mathbf{A}|_1$. This has to be augmented by the appropriate "horizontal" sketch structure coming from **A**. The picture for the co case is as follows:

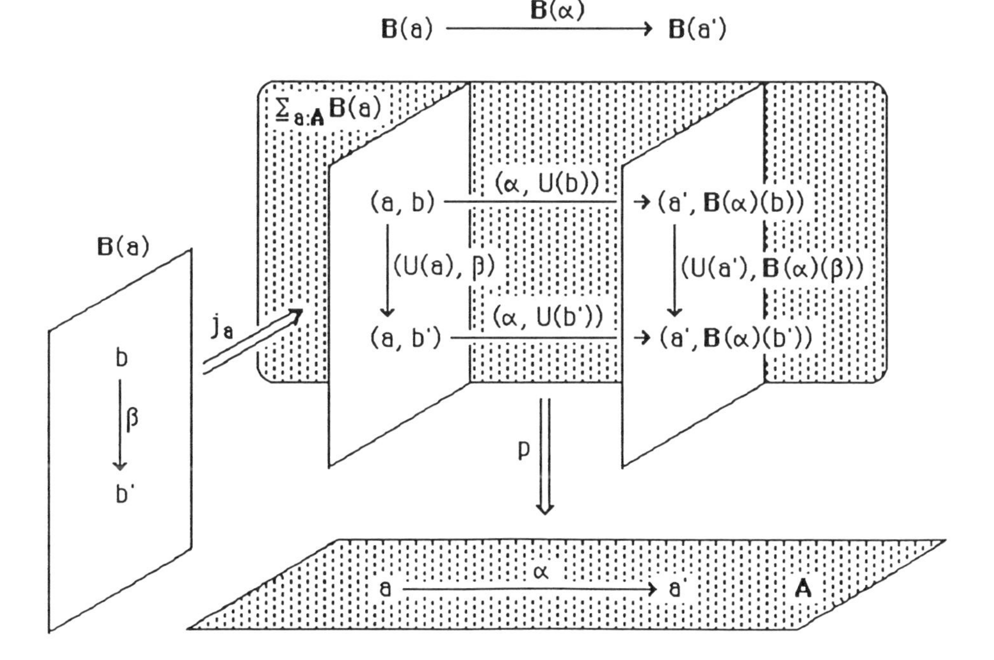

As shown,

$$|\Sigma_{a:\mathbf{A}}\mathbf{B}(a)|_0 = \Sigma_{a\in|\mathbf{A}|}|\mathbf{B}(a)|_0 = \{(a, b) \mid a \in |\mathbf{A}|_0 \text{ and } b \in |\mathbf{B}(a)|_0\}$$
$$|\Sigma_{a:\mathbf{A}}\mathbf{B}(a)|_1 = \{(U(a), \beta) \mid a \in |\mathbf{A}|_0 \text{ and } \beta \in |\mathbf{B}(a)|_1\}$$
$$\cup \{(\alpha, U(b)) \mid \alpha \in |\mathbf{A}|_1 \text{ and } b \in \mathbf{B}(d_0(a))\}.$$
$$d_i(U(a), \beta) = (a, d_i(\beta)) \text{ for } i = 0, 1$$
$$d_0(\alpha, U(b)) = (d_0(\alpha), b) \text{ and } d_1(\alpha, b) = (d_1(\alpha), \mathbf{B}(\alpha)(b))$$
$$U(a, b) = (U(a), U(b)).$$

There are graph homomorphisms $j_a : \mathbf{B}(a) \to |\Sigma_{a:\mathbf{A}}\mathbf{B}(a)|$ given by $j_a(b) = (a, b)$ and $j_a(\beta) = (U(a), \beta)$ for all $a \in |\mathbf{A}|_0$. Also there is a graph homomorphism $p : |\Sigma_{a:\mathbf{A}}\mathbf{B}(a)| \to |\mathbf{A}|$ given by $p(a, b) = a$, $p(U(a), \beta) = U(a)$, and $p(\alpha, U(b)) = \alpha$. The sketch structure is given by

$$\vee\{(j_a)_*(\mathbf{B}(a)) \mid a \in |\mathbf{A}|_0\} \vee \mathbf{R} \vee \mathbf{H};$$

i.e., the sup of the final structures from the inclusions j_a of all of the fibres $\mathbf{B}(a)$ for $a \in |\mathbf{A}|_0$ together with two other structures, $\mathbf{R}$ and $\mathbf{H}$, to be described. The sketch structure $\mathbf{R}$ consists of no cones, no cocones, and all diagrams of the form

$$\begin{array}{ccc}
(a, b) & \xrightarrow{(\alpha,\, U(b))} & (a', \mathbf{B}(\alpha)(b)) \\
\Big\downarrow{\scriptstyle (U(a),\, \beta)} & & \Big\downarrow{\scriptstyle (U(a'),\, \mathbf{B}(\alpha)(\beta))} \\
(a, b') & \xrightarrow{(\alpha,\, U(b'))} & (a', \mathbf{B}(\alpha)(b'))
\end{array}$$

for $(\alpha : a \to a') \in |\mathbf{A}|_1$ and $(\beta : b \to b') \in |\mathbf{B}(a)|_1$. To explain the last term, $\mathbf{H}$, observe that the graph homomorphisms j_a are now sketch homomorphisms and given $\alpha : a \to a'$, there is a "natural transformation" $j_\alpha : j_a \Rightarrow j_{a'} \cdot \mathbf{B}(\alpha)$. Thus, these determine a lax cocone $j : \mathbf{B}(\text{-}) \Rightarrow K_{\mathbf{A}}\,\Sigma_{a:\mathbf{A}}\mathbf{B}(a)$ with a corresponding section $t : |\mathbf{A}| \to |\mathbf{B}(\text{-})\!\downarrow\!\downarrow\!(\Sigma_{a:\mathbf{A}}\mathbf{B}(a))|$. As before,

$$\text{comp} : |\mathbf{obj}\!\downarrow\!\mathbf{B}(\text{-})| \times_{\mathbf{A}} |\mathbf{B}\!\downarrow\!\downarrow_t\!(\Sigma_{a:\mathbf{A}}\mathbf{B}(a))| \to |\Sigma_{a:\mathbf{A}}\mathbf{B}(a)|$$

and we set $\mathbf{H} = (\text{comp})_*(\mathbf{obj}\!\downarrow\!\mathbf{B}(\text{-}) \times_{\mathbf{A}} \mathbf{B}(\text{-})\!\downarrow\!\downarrow_t\!(\Sigma_{a:\mathbf{A}}\mathbf{B}(a)))$; i.e., $\mathbf{H}$ is the final structure from the graph homomorphism comp. Hence $\mathbf{H}$ provides the required "horizontal" structure and j becomes an admissable lax cocone. Finally, let $\tau : \mathbf{B}(\text{-}) \Rightarrow K_{\mathbf{A}}\mathbf{C}$ be an admissable lax cocone on $\mathbf{B}(\text{-})$ with vertex $\mathbf{C}$. Then there is a unique sketch homomorphism $f : \Sigma_{a:\mathbf{A}}\mathbf{B}(a) \to \mathbf{C}$ such that $f \cdot j = \tau$; namely, $f(a, b) = \tau_a(b)$, $f(U(a), \beta) = \tau_a(\beta)$, and $f(\alpha, U(b)) = (\tau_\alpha)_b$. Hence, $\Sigma_{a:\mathbf{A}}\mathbf{B}(a)$ is the admissable lax colimit of $\mathbf{B}(\text{-})$. In the contra case, the picture is altered by replacing the top line by

$$\mathbf{B}(a) \xleftarrow{\mathbf{B}(\alpha)} \mathbf{B}(a')$$

and replacing the rectangles that make up the structure **R** by rectangles of the form

$$\begin{array}{ccc} (a, \mathbf{B}(\alpha)(b)) & \xrightarrow{(\alpha,\, U(b))} & (a', b) \\ \Big\downarrow{\scriptstyle (U(a),\, \mathbf{B}(\alpha)(\beta))} & & \Big\downarrow{\scriptstyle (U(a'),\, \beta)} \\ (a, \mathbf{B}(\alpha)(b')) & \xrightarrow{(\alpha,\, U(b'))} & (a', b') \end{array}$$

for $\alpha : a \to a'$ in **A** and $\beta : b \to b'$ in $\mathbf{B}(a')$. Similar alterations in the rest of the description complete the construction in this case.

5.7 Examples.

1. As before, let $K_\mathbf{A}\mathbf{B} : \mathbf{A} \to$ **^SK** denote the constant **A**-indexed family equal to **B**. Then $\sum_{a:\mathbf{A}} K_\mathbf{A}\mathbf{B} \approx \mathbf{A} \otimes \mathbf{B}$. In particular, $\sum_{a:\mathbf{A}} K_\mathbf{A}\mathbf{obj} \approx \mathbf{A}$. (This is why the rectangular trivial diagrams are required.) The proof is immediate by inspection of the two constructions. We leave as an exercise for the interested reader the construction of an appropriate notion of dependent sum which specializes for a constant family to $\mathbf{A} \times \mathbf{B}$.

2. Let $h : \mathbf{B_1} \to \mathbf{B_2}$ be a "functor" between sketches $\mathbf{B_1}$ and $\mathbf{B_2}$. Then h can be viewed as the value of a "functor" from the sketch **2** to **^SK** taking 0 to $\mathbf{B_1}$, 1 to $\mathbf{B_2}$ and the arrow α from 0 to 1 to h. Denote the sketch $\sum_{i:\mathbf{2}} \mathbf{B_i}$ by $[h : \mathbf{B_1} \to \mathbf{B_2}]$. It can be regarded as the disjoint union of $\mathbf{B_1}$ and $\mathbf{B_2}$ together with extra arrows from $b \in \mathbf{B_1}$ to $h(b) \in \mathbf{B_2}$ for each object b in $\mathbf{B_1}$ such that for all arrows $\beta : b \to b'$ in $\mathbf{B_1}$ the adjoining diagram is admissable.

$$\begin{array}{ccc} b & \longrightarrow & h(b) \\ \Big\downarrow{\scriptstyle \beta} & & \Big\downarrow{\scriptstyle h(\beta)} \\ b' & \longrightarrow & h(b') \end{array}$$

A model of $[h : \mathbf{B_1} \to \mathbf{B_2}]$ therefore consists of a pair of models M_i of $\mathbf{B_i}$ for i = 1, 2, together with extra operations $t_b : M_1(b) \to M_2(h(b))$ such that for all arrows $\beta : b \to b'$ the adjoining diagram of functions commutes

$$\begin{array}{ccc} M_1(b) & \longrightarrow & M_2(h(b)) \\ \Big\downarrow{\scriptstyle M_1(\beta)} & & \Big\downarrow{\scriptstyle M_2(h(\beta))} \\ M_1(b') & \longrightarrow & M_2(h(b')) \end{array}$$

If h is a sketch homomorphism, then $M_2 \bullet h$ is a model of $\mathbf{B_1}$ and $t : M_1 \Rightarrow M_2 \bullet h$ is a natural transformation; i.e., a homomorphism of models of **A**. In this case, $\mathbf{MOD}([h : \mathbf{B_1} \to \mathbf{B_2}])$ can be described as the category of models of the sketch homomorphism h.

3. In [7], we discussed a construction called $\mathbf{A}^\dagger\mathbf{of}\mathbf{B}$. This is almost a special case of the dependent sum construction as follows: $\mathbf{A}^\dagger$ denotes a triple (**A**, **A'**, **A"**) where **A'** and **A"** are full (cf., [7], 3.10) subsketches of **A** whose intersection is an object 1 which is the vertex of the empty cone. **B** is a given sketch and **B'** denotes the same sketch as **B** except that **B'** has no cones or

cocones. Define $\mathbf{B}(\text{-}) : \mathbf{A} \to \,^\wedge\mathbf{SK}$ by the rules: $\mathbf{B}(a') = \mathbf{B}$ for $a' \in \mathbf{A'}$, $\mathbf{B}(a'') = \mathbf{obj}$ for $a'' \in \mathbf{A''}$, and $\mathbf{B}(a) = \mathbf{B'}$ otherwise. If α is an arrow whose domain and codomain do not belong to $\mathbf{A''}$, then $\mathbf{B}(\alpha)$ is the identity sketch homomorphism. If the codomain of α belongs to $\mathbf{A''}$, then $\mathbf{B}(\alpha)$ is the unique "functor" with codomain $\mathbf{obj}$. If the domain of α belongs to $\mathbf{A''}$, then $\mathbf{B}(\alpha)$ is some chosen sketch homomorphism with domain $\mathbf{obj}$. Then $\Sigma_{a:\mathbf{A}}\mathbf{B}(\text{-})$ is almost the same as $\mathbf{A}^{\dagger}\mathrm{of}\mathbf{B}$. The only difference is that if there is an arrow α whose domain belongs to $\mathbf{A''}$, then in $\mathbf{A}^{\dagger}\mathrm{of}\mathbf{B}$ there is an arrow over it from $\mathbf{obj}$ to every object in $\mathbf{B}(\text{-})$ applied to the codomain of α.

6. DEPENDENT PRODUCTS.

Dependent sums make sense for sketches while dependent products make sense for categories of models of sketches. Given an $\mathbf{A}$-indexed family of sketches $\mathbf{B}(\text{-})$, our first definition of a dependent product is

$$\Pi_{a:\mathbf{A}}\mathbf{MOD}(\mathbf{B}(a)) = \mathbf{MOD}(\Sigma_{a:\mathbf{A}}\mathbf{B}(a)).$$

This is appealing since $\mathbf{MOD}(\text{-}) \approx \mathbf{Hom}(\text{-}, \,^\wedge\mathbf{SET})$ and hence it should turn colimit constructions into the analogous limit constructions. Furthermore, models of $\Sigma_{a:\mathbf{A}}\mathbf{B}(a)$ have an appropriate structure; namely, a model M of $\Sigma_{a:\mathbf{A}}\mathbf{B}(a)$ consists of a family of models $\{M_a \in \mathbf{MOD}(\mathbf{B}(a)) \mid a \in \mathbf{A}\}$ together with extra functions

$$\{M_\alpha(b) : M_a(b) \to M_{a'}((\mathbf{B}(\alpha))(b)) \mid \alpha : a \to a' \text{ in } \mathbf{A} \text{ and } b \in \mathbf{B}\}$$

$$(\text{or } \{M_\alpha(b) : M_a((\mathbf{B}(\alpha))(b')) \to M_{a'}(b') \mid \alpha : a \to a' \text{ in } \mathbf{A} \text{ and } b' \in \mathbf{B'}\})$$

such that if $\beta : b_1 \to b_2$ is an arrow in the appropriate value of $\mathbf{B}(\text{-})$, then the diagram

$$\begin{array}{ccc} M_a(b_1) & \xrightarrow[\text{co}]{M_\alpha(b_1)} & M_{a'}(\mathbf{B}(\alpha)(b_1)) \\ {\scriptstyle M_a(\beta)}\downarrow & & \downarrow{\scriptstyle M_{a'}(\mathbf{B}(\alpha)(\beta))} \\ M_a(b_2) & \xrightarrow[M_\alpha(b_2)]{} & M_{a'}(\mathbf{B}(\alpha)(b_2)) \end{array} \qquad \begin{array}{ccc} M_a(\mathbf{B}(\alpha)(b_1)) & \xrightarrow[\text{contra}]{M_\alpha(b_1)} & M_{a'}(b_1) \\ {\scriptstyle M_a(\mathbf{B}(\alpha)(\beta))}\downarrow & & \downarrow{\scriptstyle M_{a'}(\beta)} \\ M_a(\mathbf{B}(\alpha)(b_2)) & \xrightarrow[M_\alpha(b_2)]{} & M_{a'}(b_2) \end{array}$$

commutes. There is a further condition that "horizontal" cones and cocones in $\Sigma_{a:\mathbf{A}}\mathbf{B}(a)$, which are taken to cones and cocones whose arrows are all of the form $M_\alpha(b)$ for suitable α's and b's, become limit cones and colimit cocones respectively.

However, again, examples suggest that this construction is not satisfactory because the structure of $\mathbf{A}$ is not tightly enough involved in it. The dependent product should somehow involve sections of the dependent sum, but there are no sections of $\Sigma_{a:\mathbf{A}}\mathbf{B}(a)$ here. This is remedied in our official definition.

6.1 **Definition.** Let $s : \mathbf{A} \to \Sigma_{a:\mathbf{A}}\mathbf{B}(a)$ be a section. (I.e., a sketch homomorphism such that $p \cdot s = \mathrm{id}_{\mathbf{A}}$.) Then ${}_s\Pi_{a:\mathbf{A}}\mathbf{MOD}(\mathbf{B}(a)) = \mathbf{MOD}([s : \mathbf{A} \to \Sigma_{a:\mathbf{A}}\mathbf{B}(a)])$.

Thus, an object of ${}_s\Pi_{a:\mathbf{A}}\mathbf{MOD}(\mathbf{B}(a))$ is a model of $[s : \mathbf{A} \to \Sigma_{a:\mathbf{A}}\mathbf{B}(a)]$ which, by example 2 of 5.7 consists of a model M of $\mathbf{A}$, a model $\{M_a \in \mathbf{MOD}(\mathbf{B}(a)) \mid a \in \mathbf{A}\}$ of $\Sigma_{a:\mathbf{A}}\mathbf{B}(a)$ and a family of functions $\{\mathrm{select}_a : M(a) \to M_a(s(a)) \mid a \in \mathbf{A}\}$ such that if $\alpha : a \to a'$ is an arrow in $\mathbf{A}$ then the adjoining diagram commutes. In the contra case, $M_\alpha(s(a))$ is replaced by $M_\alpha(s(a'))$.

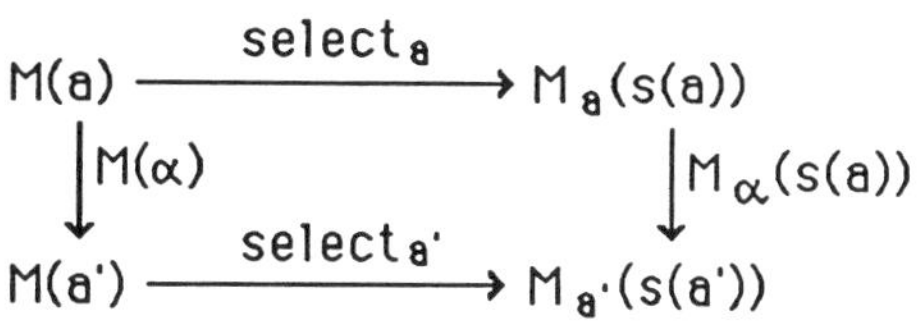

$$\begin{array}{ccc} M(a) & \xrightarrow{\mathrm{select}_a} & M_a(s(a)) \\ \downarrow{\scriptstyle M(\alpha)} & & \downarrow{\scriptstyle M_\alpha(s(a))} \\ M(a') & \xrightarrow{\mathrm{select}_{a'}} & M_{a'}(s(a')) \end{array}$$

6.2 **Examples.**

i) As before, let $K_{\mathbf{A}}\mathbf{B} : \mathbf{A} \to {}^\wedge\mathbf{SK}$ denote the constant $\mathbf{A}$-indexed family equal to $\mathbf{B}$, so $\Sigma_{a:\mathbf{A}}\, K_{\mathbf{A}}\mathbf{B} \approx \mathbf{A} \otimes \mathbf{B}$. Then a section of $\Sigma_{a:\mathbf{A}}\, K_{\mathbf{A}}\mathbf{B}$ reduces to a sketch homomorphism $s : \mathbf{A} \to \mathbf{B}$. Now ${}_s\Pi_{a:\mathbf{A}}\mathbf{MOD}(K_{\mathbf{A}}\mathbf{B})$ contains as a subcategory $\mathbf{MOD}([s : \mathbf{A} \to \mathbf{B}])$. They are not equal, but $\mathbf{MOD}([s : \mathbf{A} \to \mathbf{B}])$ corresponds to the constant models in ${}_s\Pi_{a:\mathbf{A}}\mathbf{MOD}(K_{\mathbf{A}}\mathbf{B})$; i.e., the models such that M_a is independent of a. Thus, ${}_s\Pi_{a:\mathbf{A}}\mathbf{MOD}(K_{\mathbf{A}}\mathbf{B})$ contains the category of models of the sketch homomorphism s.

ii) We illustrate the general construction with an example taken from Meyer [13]. Let ZeroVector[-] be the function f whose value for $n \geq 0$ is the zero vector of length n. We seek a type for f. The problem is that the type of the value f(n) is NatVect(n), the space of vectors of length n of natural numbers, which depends on n, so the type of f cannot be a simple function space type. Instead, it should be of type $\Pi_{n:\mathrm{Nat}}\mathrm{NatVect}(n)$. We will describe a sketch **Nat** for Nat, a contra **Nat**-indexed family of sketches **NatVect**(-) and a section s of $\Sigma_{n:\mathbf{Nat}}\mathbf{NatVect}(\text{-})$ over **Nat** such that ${}_s\Pi_{n:\mathbf{Nat}}\mathbf{MOD}(\mathbf{NatVect}(n))$ contains the function f.

a) **Nat** is the sketch illustrated by:

$$0 \to 1 \to 2 \to \dots \to n \to (n+1) \to \dots$$

It has no diagrams or cocones. The only cone is the empty cone with vertex 0. Thus a model looks like a sequence of sets beginning with the one element set 1:

$$1 \to Y_1 \to Y_2 \to \dots \to Y_n \to Y_{(n+1)} \to \dots$$

The initial model therefore is just the sequence of one element sets:

$$1 \to 1 \to 1 \to \dots \to 1 \to 1 \to \dots$$

(We use this sketch for **Nat** rather than the one in [7] since the only structure of **Nat** that is relevant here is the order structure.)

b) Next we construct a **Nat**-indexed family of sketches $\mathbf{NatVect}(\text{-}) : \mathbf{Nat} \to {}^\wedge\mathbf{SK}^{op}$ as follows: **NatVect**(n) is the sketch illustrated to the right. It has no diagrams or cocones.

$$x^n \underset{pr_n}{\overset{pr_1}{\underset{\vdots}{\rightrightarrows}}} x \circlearrowleft \mathrm{succ}$$

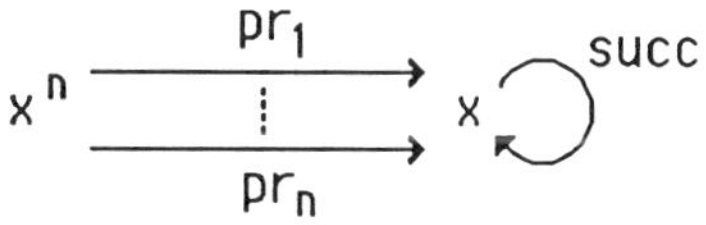

The only cone in the n-fold product cone with vertex x^n and projection arrows pr_i, $i = 1, \dots, n$. Thus a model looks like the picture at the right; i.e., a set X, its n-fold product X^n with projection functions pr_i, $i = 1, \dots, n$, and an endomorphism f on X.

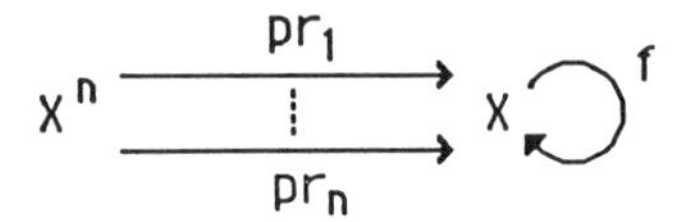

i.e., a set X, its n-fold product X^n with projection functions pr_i, $i = 1, \dots, n$, and an endomorphism f on X. The initial model is the one with $X = \emptyset$ except for $n = 0$ where it reduces to the natural numbers with the usual Peano structure of a zero element and a successor function. Now, for each n we have to describe a "functor" h_n : **NatVect**(n+1) $\to$ **NatVect**(n) to be the value of **NatVect**(-) for the arrow $n \to n+1$ in **Nat**. h_n is given by the values:

$$h_n(x) = x, \qquad h_n(\text{succ}) = \text{succ},$$
$$h_n(x^{n+1}) = x^n,\ h_n(pr_i) = pr_i \text{ for } i = 1, \dots, n, \text{ and } h_n(pr_{n+1}) = pr_n.$$

With this structure, the dependent sum $\Sigma_{n:\mathbf{Nat}}$**NatVect**(n) looks like:

$$\begin{array}{ccccc}
\cdots \longrightarrow (n, x^n) & \xrightarrow{(\to,\, x^{n+1})} & (n+1, x^{n+1}) \longrightarrow \cdots \\
(n, p_1) \downarrow \cdots \downarrow (n, p_n) & & (n+1, p_1) \downarrow \cdots \downarrow\downarrow (n+1, p_{n+1}) \\
\cdots \longrightarrow (n, x) & \xrightarrow{(\to,\, x)} & (n+1, x) \longrightarrow \cdots \\
\circlearrowleft (n, \text{succ}) & & \circlearrowleft (n+1, \text{succ})
\end{array}$$

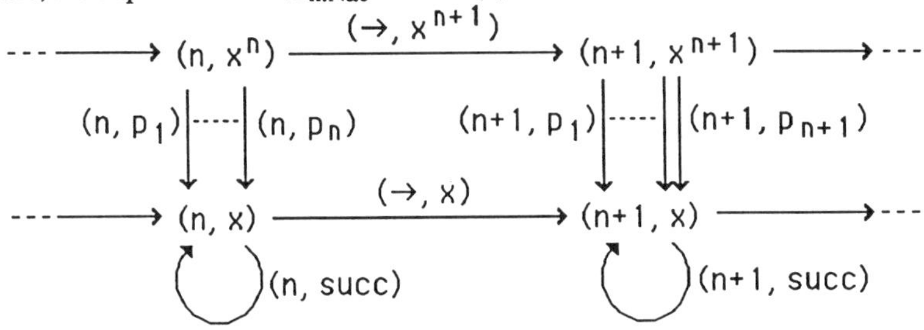

Here, all rectangles with vertical arrows (n, p_i) and $(n+1, p_i)$ as well as the rectangle with vertical arrows (n, p_n) and $(n+1, p_{n+1})$ belong to the admissable diagrams. There are no vertical diagrams or cocones and the only vertical cones are those that say that (n, x^n) is the n-fold product of (n, x). Similarly, there are no horizontal diagrams or cocones and the only horizontal cones are the ones that say that the objects $(0, x)$ and $(0, x^0)$ are vertices of empty cones.

Finally, we choose the sketch homomorphism s : **Nat** $\to \Sigma_{n:\mathbf{Nat}}$**NatVect**(n) given by $s(n) = (n, x^n)$ and $s(\to) = (\to, x^{n+1})$. A typical model of [s : **Nat** $\to \Sigma_{n:\mathbf{Nat}}$**NatVect**(n)] looks like:

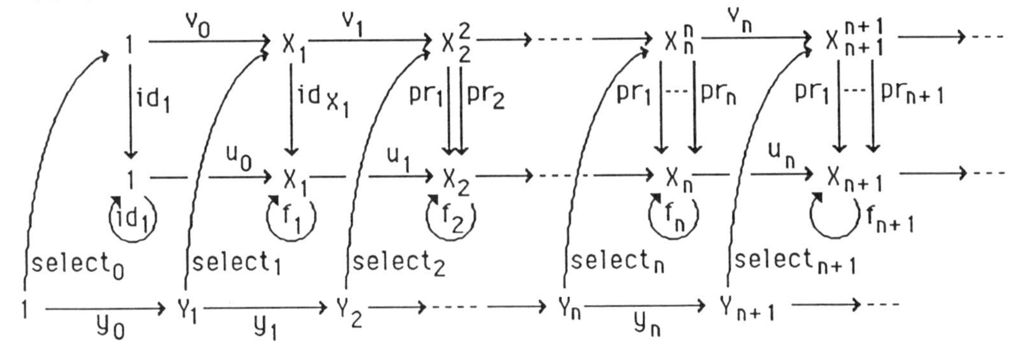

where

$$u_n \bullet f_n = f_{n+1} \bullet u_n, \quad u_n \bullet pr_j = pr_j \bullet v_n \text{ for } j \leq n,$$

$$u_n \bullet pr_n = pr_n \bullet v_{n+1} \text{ and } v_n \bullet select_n = select_{n+1} \bullet y_n \text{ for all } n \geq 0.$$

In terms of coordinates, this says that $v_n(x_1, \ldots x_n) = (u_n(x_1), \ldots, u_n(x_n), u_n(x_n))$. The initial model consists of $X_n = N$ and $Y_n = 1$ for all n, where N denotes the set of natural numbers. In this model, $v_0 = u_0 = '0$, $u_n = id_N$, $v_1 = \Delta$ (the diagonal map) $v_n = id^{n-1} \times \Delta$, f_n = succ, and $select_n = '(0, \ldots, 0)$ for $n \geq 1$. Thus, the initial model of $_s\Pi_{n:\mathbf{Nat}}\mathbf{MOD}(\mathbf{NatVect}(n))$ = $\mathbf{MOD}([s : \mathbf{Nat} \to \Sigma_{n:\mathbf{Nat}}\mathbf{NatVect}(n)]$ represents the function whose value at n is the zero vector of length n.

References.

[1] M. Barr and C. F. Wells, *Category Theory for Computer Science*, to appear.

[1] R. Burstall and D. Rydeheard, *Computational Category Theory*, to appear.

[2] H. Ehrig and B. Mahr, *Fundamentals of Algebraic Specification* 1, Springer-Verlag, New York, 1985.

[3] P. J. Freyd, Algebra valued functors in general and tensor products in particular, Colloq. Math. 14 (1966), 89 - 106

[4] J. Goguen and J. Meseguer, Initiality, induction, and computability, in *Algebraic Mathods in Semantics*, M. Nivat and J. C. Reynolds (eds.), Cambridge University Press, 1985.

[5] J. W. Gray, The 2-adjointness of the fibred-category construction, in *Symposia Mathematica* IV, Istituto Nazionale di Alta Matematica, Academic Press, New York, 1970, 457-492.

[6] –, *Formal Category theory: Adjointness for 2-Categories*, Lecture Notes in Mathematics 391, Springer -Verlag, New York, 1974.

[7] –, Categorical aspects of data type constructors, Theoretical Computer Science 50(1987), 103-135.

[8] R. D. Jenks, A primer: 11 Keys to New SCRATCHPAD, Lecture Notes in Computer Science 174 (1984), Springer-Verlag, New York, 123-147.

[9] P. Johnstone, *Topos Theory*, Academic Press, New York, 1977.

[10] G. M. Kelly and R. Street, Review of the elements of 2-categories, in *Category Seminar, Sydney , 1972/73*, Lecture Notes in Mathematics 420, Springer-Verlag, New York, 1974.

[11] C. Lair, Etude Générale de la Catégorie des Esquisses, Esquisses Mathématiques 23, Paris 1975, 1 - 62.

[12] S. Mac Lane, *Categories for the Working Mathematician*, Springer Verlag, New York, 1972.

[13] A. R. Meyer and M. B. Reinhold, Type is not a type: Preliminary Report, Proceedings, 13th Annual ACM Symposium on Principles of Programming Languages, 1986, 187 - 295.

[14] R. Seely, Locally cartesian closed categories and type theory, Proc. Camb. Philos. Soc. 95 (1984), 33 - 48.

[15] P. Taylor, Recursive Domains, Indexed Category Theory, and Polymorphism, dissertation University of Cambridge, Cambridge, England, 1986.

University of Illinois at Urbana-Champaign

Urbana, Ill.

Contemporary Mathematics
Volume **92**, 1989

THE THEORY OF CONSTRUCTIONS: CATEGORICAL SEMANTICS AND TOPOS-THEORETIC MODELS

J. Martin E. Hyland and Andrew M. Pitts

ABSTRACT. A syntactically rich version of the Coquand-Huet *theory of constructions* is described as a theory of dependent types involving expressions at three different levels (Terms, Types and Orders) together with indexed sums and products of various kinds. Two extensions of the theory involving universal types are also discussed. A complete category-theoretic explanation of the meaning of the theory is built up, based upon a careful analysis of the categorical semantics of Martin-Lōf's theory of dependent types. Finally, two particular models of the theory of constructions are described (modelling the two extensions of the theory mentioned above). In these models, the Orders and Types are denoted by particular kinds of Grothendieck topos, namely *algebraic toposes* (toposes of presheaves on small categories with finite limits) and *algebraic-localic toposes* (toposes of presheaves on meet semi-lattices).

CONTENTS

Introduction

The *Theory of Constructions* is the very high level functional programming language due to Coquand and Huet [CH]; it contains Girard's higher order lambda calculus [G1] and the core of Martin-Lōf's theory of dependent types [M-L2] as subsystems. In this paper we describe a syntactically rich version of this theory, introduce a general class of models for it and give two particular models. Category theory comes into our work in two distinct ways which it is as well to distinguish here.

In the first place, in obtaining our general notion of model we give an explanation in terms of category theory of what constitutes a semantics for the various parts of the

1980 *Mathematics Subject Classification* (1985 *Revision*). 03B15, 18B25, 68Q55.
The first author's research was supported by grants from the SERC and the Alvey Project.
The second author is a Royal Society University Research Fellow and gratefully acknowledges the financial support of that institution.

theory of constructions. We certainly believe this is the right approach. For one thing, the descriptions of semantics (even for a simple subsystem like the second order lambda calculus) appear mathematically uncivilized when couched in non-categorical terms. What is more, the categorical concepts we use in defining the semantics have (we think) a remarkable degree of simplicity and elegance—especially in comparison with the syntactic complexities of the theory of constructions itself. For another thing, the category-theoretic perspective provides all kinds of valuable insights into the underlying *meaning* of the formal constructs of the theory. The main mathematical ideas which underlie our explanation of the semantics are those of (*locally*) *cartesian closed category*, Grothendieck's *fibrations* and Lawvere's *hyperdoctrines*. (These are part of the category theorist's inheritance from the 1960s.)

Secondly, the objects that we use to model the types (and "orders") of the language are not simply sets, or domains, or objects *in* some category—as in earlier work. They are *themselves* categories. (Hence they are objects in a 2-category, although we do not exploit this extra level of structure in this paper.) This is in line with an old idea of Lawvere that the structures of mathematical interest can not only be organized into categories, but also often *are* (usefully seen as) categories. The categories we use are certain kinds of *Grothendieck topos*. So the main mathematical ideas we use to develop our models are those of *classifying topos*, of the *internal logic* of toposes and of *relativization* to an arbitrary base topos of mathematical constructions and proofs. (These are for the most part the category theorist's inheritance from the 1970s.)

We would be the first to admit that this puts rather large demands on the understanding of those readers who are not topos-theorists. (We hope there are many.) Since there are equivalent, more concrete descriptions of our models (which we give briefly in 4.20 and 5.11), it seems worth saying something in justification of our chosen treatment. Certainly some aspects of the models can be developed in a way which is a straight forward imitation and generalization (from the order-theoretic to the fully category-theoretic) of ideas from domain theory. However, other aspects are really very much simpler from the topos-theoretic viewpoint. To give just one example, the existence of an Order of Types (see 2.11) in the models is an easy consequence of the notion of *classifying topos*. Indeed, there are delicate points involving variations on the notion of "internal category with finite limits" in a topos which it would be hard to steer through without some understanding of the internal logic. So our advice to computer scientists has to be not to try to avoid topos theory, but to try to learn it!

Ours are not the first models for the theory of constructions. First of course there is the term model, features of which we understand well in view of the strong normalization theorem [Coq1] which holds for the system. Then there is a whole class of models which are based on sufficiently complete, internal categories in realizability toposes (see [HRR, section 8). Here incidentally category theory provided the kind of insight we referred to above in showing how to extend what were known models of higher order polymorphic lambda calculus to models for the theory of constructions. Finally, there are models based on models of a system of types with "Type:Type" and fixpoint recursion (see [Cd] and the references therein); these are derived from variants of the closure operator model described by Scott in [Sc]. It is these models which we feel that Girard was most justified in criticising in [G2] for their *ad hoc* nature. We claim that our models can be seen as "natural" substitutes (i.e. relatively free of coding) for the *ad hoc* models. (In particular, it is the case that our models satisfy a version of "Type:Type" and support initial fixpoint recursion both for Operators (including Types) and for Terms; we will describe the modelling of "Type:Type" quite fully, but leave the discussion of recursion to another occasion.)

The structure of the paper is as follows:

In section 1 we give a detailed description of the theory of constructions. The system is presented in two tiers, each of which is a theory in the style of Martin-Löf, with rules for sums and products of dependent types and also with unit type (which plays a crucial role in the theory as we formulate it and is not there merely for aesthetic reasons). The types of the top layer are called "Orders" and their elements (terms) are called "Operators", whilst those of the lower layer are "Types" and their elements "Terms". If you will, the top layer consists of the "large" types and the bottom layer of the "small" ones—although we will see through the models that the theory actually does not carry any import of size. The connection between the two layers is provided by a particular Order *Type* whose Operators are precisely the Types: in other words *Type* is the "Order of all Types". So far this amounts to a restricted version of Martin-Löf type theory (no equality types, no finite sum types, no natural number type) with a single Martin-Löf universe. On top of this we require not only the closure of Orders under Type-indexed products and sums, but also the closure of Types under Order-indexed products and sums. The latter property requires careful formulation (for the sum clauses) and is the aspect which makes finding models rather hard. We enrich the basic theory of constructions by allowing constants of various kinds and consider equational theories extending the basic theory. In particular, at the end of the section we describe two extensions of the basic theory, one with an "Order of all Orders" (1.11) and one where the Types and Orders are essentially the same (1.12).

In section 2 we build up a description of the categorical structures needed to model the theory of constructions: see 2.13 for a summary of what we require. It should be emphasised that the aim (which we achieve) is to give a notion of model which is truly *general*: the criterion for success is the familiar one in categorical logic—to have an equivalence between a 2-category of theories (over the basic theory of constructions) and a 2-category of suitable categorical structures. A fundamental part of this work is an account (2.2) of a general category-theoretic semantics for the pure notion of "dependent type" with no closure assumptions other than that of substitutivity. The semantics of sums and products is then considered on top of this (2.4), leading to the notion of *relatively cartesian closed category* (2.7 *et seq.*), taken from [Ta] and which we see (in Proposition 2.6) generalizes that of locally cartesian closed category. We believe that this material is of interest in its own right. It unifies previous work in this area by Seely [Se1], by

Cartmell [Ca] and more recently by Taylor [Ta]. (We believe also that it encompasses Obtułowicz's recent work [Ob], by applying an iterated version of the Grothendieck construction to one of his hierarchies of indexed categories to obtain one of our relatively cartesian closed categories. It is important for Obtułowicz that his structures are (essentially) algebraic; however, ours are models of a particular lim theory and the latter are in many respects just as well behaved as algebraic theories.)

The rest of the paper is devoted to describing our particular examples of the general structure which emerged in section 2. These examples involve properties of internal categories with finite limits in toposes (and the special case of internal meet semilattices). In order to develop some of these properties we employ the "lim theories" of M.Coste [Co1]; the necessary material is reviewed in section 3. Then in section 4 we present the model of the theory of constructions in which both Orders and Types are denoted by *algebraic toposes*—by definition these are the Grothendieck toposes which are equivalent to a category of presheaves on a small category with finite limits. The topos theory needed to verify that we have a model is nearly all standard; the exception to this comes when we prove (in 4.16) that the geometric morphisms which determine relative algebraic toposes are closed under exponentiation by algebraic toposes—our proof requires some work on "strictifying" *lex fibrations* (see 4.11 *et seq.*). Finally in section 5 we describe a model in which the Orders are still denoted by algebraic toposes, but the Types are now denoted by algebraic toposes which are also *localic*. With one exception, the material we need for this model is part of the well-developed theory of localic toposes. The exception has once again to do with closure under exponentiation. We prove a purely topos-theoretic result (Proposition 5.6) which as far as we know is new—namely that *for each exponentiable topos, exponentiation preserves geometric inclusions and localic geometric morphisms.*

Clearly this paper is only a beginning. Some extensions of the models, for example to injective toposes and continuous lattices, appear quite straightforward. Others, such as extensions to more general classes of toposes and domains, seem more problematic. Frustratingly, we are still unable to use topos-theoretic machinery (such as the theory of *atomic* toposes) to extend the Girard style models of polymorphism [G2]. There are also problems with understanding the proper rules for weak equality types (it seems that there is more than one notion), some of which are certainly interpretable in our models. Having only *weak* equality types (and *weak* finite sum types) is an inevitable consequence of a facet of our models which we have not addressed at all here, namely that they support the interpretation of recursive Terms and Types. We believe that the proper handling of these aspects of the models is part of the bigger problem, which we intend to address in further work, of understanding how the 2-categorical (in fact, bicategorical) structure of the models should be reflected in a richer syntax than the present one.

This paper emerged from the authors' joint interest in the "natural" model for polymorphism given by Girard [G2]. We owe an important stimulus to Thierry Coquand who, in the course of his work with Gunter and Winskel [CGW] on extensions of Girard's idea to more general "Berry domains", noticed that by relaxing the definitions slightly, the same idea worked for Scott domains. Coquand has since pushed these ideas further in a concrete fashion—as has Lamarche [La], independently. A general notion of indexed products for Scott domains had been considered by Paul Taylor in his thesis [Ta], but he did not explicitly extend the notion to products over a base *category*. Thus in the air was a model for the theory of Types (but not Orders) corresponding to the localic component of the model described in section 5. Initially, the first author developed this part of the model via the notion of internal algebraic lattices in toposes. The second author took the step of formulating things entirely in terms of the relevant toposes, from which the full model (of Orders as well as of Types) emerged quite rapidly.

Finally we would like to pay tribute to the open atmosphere and spirit of co-operation created by the mathematicians (Johnstone, Moerdijk, dePaiva, Robinson, Rosolini) and computer scientists (Coquand, Gunter, Winskel) in Cambridge during 1986-87. All of them commented in various ways on early versions of the ideas presented here; and they contributed crucially to our understanding of what we were trying to model and how.

1 The theory of constructions

1.1. Introduction. In this section we are going to describe a syntactically rich version of the *theory of constructions* of Coquand and Huet [CH]. This language is a natural amalgam of Martin-Lōf's predicative theory of dependent types [M-L2] and Girard's impredicative higher order lambda calculus [G1]. We do not assume familiarity with these languages. Many readers, whom we hope to interest in our semantics, will be familiar only with the syntax of the predicate calculus and simple type theory. Hence we give a fairly leisurely account of the formal syntax. We also refer the reader to Troelstra's careful account of the syntax of Martin-Lōf type theories [Tr].

In the interests of clarity, the well-formed expressions of the language are divided into three levels: *Terms, Operators* and *Orders*. This is the practice in Edinburgh LF [HHP], which is a predicative fragment of the theory of constructions: there the corresponding terminology is *Terms, Types* (misleadingly, as most expressions of the middle level are not of kind *"Type"*) and *Kinds*. Coquand and Huet see their theory as (an extension of) the proof theory of Church's theory of types [C]. So their terminology is *Proofs, Terms* (including *Propositions*) and *Types*. This interpretation is based on the idea that *Propositions* are some special *Types* (but not all *Types* are *Propositions*).

To describe the syntax, we use metavariables $K,L,M,\ldots$ for Orders, metavariables $S,T,U,\ldots$ for Operators and metavariables $s,t,u,\ldots$ for Terms. *Types* are Operators of Order "*Type*" and we use metavariables $A,B,C,\ldots$ for them. In the Coquand-Huet interpretation, their *Propositions* (≡ our *Types*) are explicitly a special case of their *Types* (≡ our *Orders*). We find it conceptually helpful to keep the syntax for Orders and Operators distinct from the syntax for Types and Terms. The similarities are such however as to make it useful to have a generic notation. We will use $P,Q,R,\ldots$ to denote things which are either Orders or Types and correspondingly $p,q,r,...$ to denote things which are either Operators or Terms.

The system we describe has Operator variables $X,Y,Z,...$ (which run over suitable Orders) and Term variables $x,y,z,...$ (which run over suitable Types). The notion of free variable is standard and we assume the reader can supply the definitions as the syntax is presented. We write $FV(K)$, $FV(S)$ and $FV(s)$ for the finite sets of free variables in an Order K, an Operator S and a Term s respectively. As a generic notation for variables we use $\xi,\eta,\zeta,...$.

We also assume we have a suitable notion of substitution of Operators and Terms for free variables of given Orders and Types. We use the notation $E(p/\xi)$ for the result of substituting p for ξ throughout the expression E.

The basic theory and extensions of it by axioms consist of sets of *judgements* J, made in *contexts* $\mathbf{\Gamma}$. We write these as

$$J\ [\mathbf{\Gamma}].$$

All the variables free in a given judgement must be "declared" in the context $\mathbf{\Gamma}$ in which the judgement is made. The well-formed expressions of the three levels, the judgements involving them and the contexts (containing the variable declarations) in which the judgements are made, are all defined together by a mammoth simultaneous recursive definition. Such a definition presents problems of exposition. We explain the forms of judgement and the notion of a context outside the recursive definition, and then give the basic clauses of the definition in the style used by Martin-Lōf [M-L2]. So the general form of a clause is:

$$\frac{J_1\ [\mathbf{\Gamma}_1]\ \ldots .\ J_n\ [\mathbf{\Gamma}_n]}{J\ [\mathbf{\Gamma}]}\ ;$$

and its meaning is that if judgements of the form $J_1\ [\mathbf{\Gamma}_1],...,J_n\ [\mathbf{\Gamma}_n]$ are in our set, then so is $J\ [\mathbf{\Gamma}]$. (Usually the reference to the contexts will be supressed.)

1.2. Forms of judgement. We adopt the conceptual framework of Martin-Lōf type theory [M-L2]. Thus a theory will consist of judgements (*structural judgements*) of the form

$$K \in ORDER, \quad S \in K, \quad \text{or} \quad s \in A$$

and judgements (*equality judgements*) of the form

$$K=L \in ORDER, \quad S=T \in K, \quad \text{or} \quad s=t \in A,$$

all made in suitable contexts. Part of the force of the structural judgements is that the expression to the left of "$\in$" is well-formed. Indeed an expression is well-formed just by virtue of appearing to the left of "$\in$" in a structural judgement. Of course $s \in A$ presupposes that A is well-formed, that is, that we already have the judgement $A \in Type$. Similarly, $S \in K$ presupposes that K is well-formed, that is, that we already have the judgement $K \in ORDER$.

A judgement of form $K \in ORDER$ has no further force than that K is well-formed.

Similarly a judgement of form $K=L\in ORDER$ plays a relatively weak role in the theory. There are no $ORDER$ variables (as there is no level of syntax above that of $ORDER$) and hence no question of "the substitution of equals for equals". In giving the formal rules of the language we will write "K" for "$K\in ORDER$" and "$K=L$" for "$K=L\in ORDER$".

However the "$\in$" has greater force in the other judgements—generically of the form $p\in P$ and $p=q\in P$. The judgement $p\in P$ states not only that p is well-formed, but also that P has a member (namely p). Thus in the Coquand-Huet interpretation, where A is a proposition, the judgement $s\in A$ presents a proof of A. It is worth stressing that while equality judgements *seem* familiar, $p=q\in P$ is not an *atomic* proposition appearing in more complicated ones in the theory of constructions. Equality judgements play a lively role in the production of new judgements by "the substitution of equals for equals" and have the force of "definitional" equality—for which see [M-L2]. In extensions of the basic system we may have the quite distinct notion of *equality* Types or Propositions: these will appear in complex expressions. In the pure theory, equality judgements are generated by "reductions". That is, there is a direction to them and they form a rewrite system. A strong normalization theorem can be proved (compare [Coq1]).

1.3. Variable declarations. All judgements are made in a context which, as we will shortly describe, is a structure on judgements of the form $X\in K$ or $x\in A$, where X and x are variables. Judgements of these forms, that is, judgements of the general form

$$\xi\in P$$

are called *variable declarations*. If the judgement J is a variable declaration, we write ξ_J for the variable to the left of "$\in$"—the *variable declared* by J—and P_J for the Order or Type to the right of "$\in$"—the *kind* of the declaration.

The need for variable declarations as an integral part of the language comes about as follows. In a calculus of dependent types we have to consider judgements of the form

$$y\in B(x) \quad [x\in A],$$

that is, y a free variable of type $B(x)$ varying as x varies over A. Suppose now we wish to substitute $a\in A$ for x. We should obtain a free variable of type $B(a)$. It is pointless to attempt to mangle or decorate "y". After substituting, we must have $y\in B(a)$. Thus type expressions can not be provided with their own inviolate collection of variables. An expression must involve variables which have been declared to be of given type.

In our version of the theory of constructions we are taking variables $X,Y,Z,\ldots$ which may be declared to be of given Order, and variables $x,y,z,\ldots$ which may be declared to be of given Type. From the clauses of the recursive definition, the reader will see that the variables of either level may depend on (that is, be declared in the context of) variables of either level. (As mentioned in 1.1, Coquand and Huet regard their Propositions as special cases of their Types and so do without this distinction between the variables at the two levels.)

Once a variable has been declared it can be used to build up complex expressions which appear in further judgements. The variable declaration forms part of the context for these further judgements.

1.4. Contexts. We describe here a notion of *context* slightly more general than that used by Martin-Löf [M-L2] or Coquand and Huet [CH]. For us contexts are certain partial orders $<$ on finite sets $\mathbf{\Gamma}$ of variable declarations. We read "$J_1 < J_2$" as "J_1 is a prerequisite for J_2", or "J_1 is presupposed by J_2", or "J_1 precedes J_2".

If $\Gamma = (\mathbf{\Gamma}, <)$ is a poset and $J \in \mathbf{\Gamma}$, write $\Gamma|_J = (\mathbf{\Gamma}|_J, <)$ for the restriction of the partial order $<$ to

$$\mathbf{\Gamma}|_J = \downarrow(J) = \{ J' \mid J' \leq J \}.$$

If $J \notin \mathbf{\Gamma}$ write $(J{>}\Gamma) = ((J{>}\mathbf{\Gamma}), <)$ for the poset obtained by adding J as a greatest element above all of Γ; more generally, if Γ and Γ' are disjoint posets $(\Gamma' > \Gamma)$ will denote their union, ordered so that everything in Γ' is greater than anything in Γ. Finally we write (J, Γ) for a poset with J as *a* maximal element whose removal leaves the poset Γ.

Define $\Gamma = (\mathbf{\Gamma}, <)$ to be a *context* if and only if:

(i) the elements of $\mathbf{\Gamma}$ are all variable declarations and distinct variables are declared by distinct declarations;

(ii) if $J \in \mathbf{\Gamma}$, then $J\ [\Gamma|_J]$ (the judgement J in the context $[\Gamma|_J]$) is a judgement of the theory.

(It will be a consequence of the definitions that—in line with the discussion in 1.1 and 1.2—if Γ is a context and $J \in \mathbf{\Gamma}$, then $FV(P_J) \subseteq \{ \xi_{J'} \mid J' < J \}$.)

A context in Martin-Löf's sense is a linearly ordered context in ours. Given one of our contexts $(\mathbf{\Gamma}, <)$, any extension of $<$ to a linear order on $\mathbf{\Gamma}$ will provide a more or less equivalent context in Martin-Löf's sense. Martin-Löf's procedure is in harmony with the way in which the lambda calculus reduces functions of many variables to functions of one. As such it seems well-adapted to implementation. However, we have reasons for preferring a more liberal notion, as we now explain.

Firstly, if one imagines the clauses of the recursive definition as providing natural deduction trees leading to judgements, then the relevant variable declarations will sit at nodes of the tree and a partial order will be induced in a natural way. This suggests adopting the idea familiar from the predicate calculus that we can always regard an expression as involving extra free variables. If a judgement can be made in a context, it can always be made in a wider context. (So the notation $J\ [\Gamma]$ corresponds here to the predicate calculus notation $\phi(\bar{x})$ for a formula whose free variables appear in the list $\bar{x}$.) Secondly, we wish to allow for the introduction of constant Orders (Order constructor symbols) with free variables. (Also we have constant Types with free variables, but as we could in principle do without them, they are not so pressing.) It is natural then to have constant Orders

$$K(\xi_1, \ldots, \xi_n)$$

which depend on *discrete* contexts

$[\xi_1 \in P_1, \ldots, \xi_n \in P_n]$.

(Compare with the introduction of predicate symbols in the predicate calculus.) In particular there is no natural choice of total order on such a set of variable declarations.

However one regards contexts, one should think of the clauses in the recursive definition as defining the central notion—namely the set of (correct) judgements-in-contexts. When giving the clauses however, we will supress mention of the contexts as far as possible.

1.5. General rules. In this section we display those clauses of the recursive definition which do not involve particular operators on Orders and Types. We use notation from 1.1 and 1.2.

- *Reflexivity* $\quad \dfrac{s \in A}{s = s \in A} \qquad \dfrac{S \in K}{S = S \in K} \qquad \dfrac{K}{K = K}$

- *Symmetry* $\quad \dfrac{s = t \in A}{t = s \in A} \qquad \dfrac{S = T \in K}{T = S \in K} \qquad \dfrac{K = L}{L = K}$

- *Transitivity* $\quad \dfrac{s = t \in A \quad t = u \in A}{s = u \in A} \qquad \dfrac{S = T \in K \quad T = U \in K}{S = U \in K} \qquad \dfrac{K = L \quad L = M}{K = M}$

- *Equality* $\quad \dfrac{s \in A \quad A = B \in Type}{s \in B} \qquad \dfrac{S \in K \quad K = L}{S \in L}$

  $\dfrac{s = t \in A \quad A = B \in Type}{s = t \in B} \qquad \dfrac{S = T \in K \quad K = L}{S = T \in L}$

We condense the next collection of rules by using our notation $p \in P$, $p = q \in P$ for judgements where P may be either an Order or a Type.

- *Substitution* $\quad \dfrac{p \in P \ [\Gamma] \quad q \in Q \ [\Gamma', \xi \in P \rangle \Gamma]}{q(p/\xi) \in Q(p/\xi) \ [\Gamma'(p/\xi) \rangle \Gamma]} \qquad \dfrac{p = p' \in P \ [\Gamma] \quad q = q' \in Q \ [\Gamma', \xi \in P \rangle \Gamma]}{q(p/\xi) = q'(p'/\xi) \in Q(p/\xi) \ [\Gamma'(p/\xi) \rangle \Gamma]}$

  $\dfrac{p \in P \ [\Gamma] \quad K \ [\Gamma', \xi \in P \rangle \Gamma]}{K(p/\xi) \ [\Gamma'(p/\xi) \rangle \Gamma]} \qquad \dfrac{p = p' \in P \ [\Gamma] \quad K = K' \ [\Gamma', \xi \in P \rangle \Gamma]}{K(p/\xi) = K'(p'/\xi) \ [\Gamma'(p/\xi) \rangle \Gamma]}$

Finally we give the rules for the declaration of variables and extension of contexts.

- *Assumption* $\quad \dfrac{A \in Type \ [\Gamma]}{x \in A \ [x \in A \rangle \Gamma]} \qquad \dfrac{K \ [\Gamma]}{X \in K \ [X \in K \rangle \Gamma]}$

  (under the assumption x and X respectively do not appear in Γ)

- *Weakening* $\quad \dfrac{A \in Type \ [\Gamma] \quad J \ [\Gamma' \rangle \Gamma]}{J \ [\Gamma', x \in A \rangle \Gamma]} \qquad \dfrac{K \ [\Gamma] \quad J \ [\Gamma' \rangle \Gamma]}{J \ [\Gamma', X \in K \rangle \Gamma]}$

  (under the assumption x and X respectively do not appear in Γ or Γ')

Remark. The *Substitution, Assumption* and *Weakening clauses* give rise to a general principle of substitution which we now describe.

An *interpretation* of a context Γ in a context Δ consists of a function p associating to each judgement J in Γ an Operator or Term p_J (as appropriate) such that for each $J \in \Gamma$

$$p_J \in P_J(p_{J'}/\xi_{J'} \mid J' < J) \; [\Delta]$$

is a judgement of our theory.

General principle of substitution. *Whenever*

$$J \; [\Gamma]$$

is a judgement of our theory and p *is an interpretation of* Γ *in* Δ, *then*

$$J(p_{J'}/\xi_{J'} \mid J' \in \Gamma) \; [\Delta]$$

is also a judgement of our theory.

1.6. Constant Orders. There is a distinguished constant Order, *Type*, the Order of Types, and any number of other constant Orders which may be introduced at various stages of the recursive definition. These latter partly determine the signature of the theory. A constant Order has an "arity" given by Orders and Types for its (free) variables. The collection of *atomic* Orders is formed from the constant Orders by general substitution.

· *Type is an Order* $\dfrac{}{Type}$

(That is, we make the judgement $Type \in ORDER$ outright.)

· *Constant Orders* $\dfrac{}{L(\xi_1,\ldots,\xi_n) \; [\Gamma]}$

(where all the variables ξ_i are declared in Γ)

Remarks.

(i) It is possible that in the *Constant Orders* clause, Γ declares more variables than appear in the list $\xi_1,\ldots,\xi_n$.

(ii) Serious substitutions may make no visible difference to the main judgement. For example, if we introduce

$$L(y) \; [y \in B(x) > x \in A]$$

and a is a closed term of Type A, then by substitution we get

$$L(y) \; [y \in B(a)].$$

1.7. The structure of Orders and Operators. Clauses giving the closure properties of Orders (*formation clauses*) are naturally associated with clauses which give Operators or Terms of respectively the Orders or Types involved (*introduction and elimination clauses*); and these are naturally associated with clauses giving the fundamental equality judgements associated with the Operators or Terms (*equality clauses*). So we simultaneously give closure properties of Orders, of Operators and of Terms.

Orders are closed under "quantification" (that is, indexed sums and products) over both Types and Orders. We give first the clauses relating to "quantification" over Types.

- *Unit clauses*:
  - *formation* $\dfrac{}{1_O}$
  - *introduction* $\dfrac{}{\langle\rangle_O \in 1_O}$
  - *equality* $\dfrac{T \in 1_O}{T = \langle\rangle_O \in 1_O}$
- *Sum clauses*:
  - *formation* $\dfrac{K\ [x \in A, \Gamma]}{\sum x \in A.K\ [\Gamma]}$ $\qquad \dfrac{A = A' \in Type\ [\Gamma] \quad K = K'\ [x \in A, \Gamma]}{\sum x \in A.K = \sum x \in A'.K'\ [\Gamma]}$
  - *introduction* $\dfrac{s \in A \quad S \in K(s/x)}{\langle s, S\rangle \in \sum x \in A.K}$ $\qquad \dfrac{s = s' \in A \quad S = S' \in K(s/x)}{\langle s, S\rangle = \langle s', S'\rangle \in \sum x \in A.K}$
  - *elimination* $\dfrac{T \in \sum x \in A.K}{\mathrm{fst}(T) \in A}$ $\qquad \dfrac{T = T' \in \sum x \in A.K}{\mathrm{fst}(T) = \mathrm{fst}(T') \in A}$

    $\dfrac{T \in \sum x \in A.K}{\mathrm{Snd}(T) \in K(\mathrm{fst}(T)/x)}$ $\qquad \dfrac{T = T' \in \sum x \in A.K}{\mathrm{Snd}(T) = \mathrm{Snd}(T') \in K(\mathrm{fst}(T)/x)}$
  - *equality* $\dfrac{s \in A \quad S \in K(s/x)}{\mathrm{fst}(\langle s, S\rangle) = s \in A}$ $\qquad \dfrac{s \in A \quad S \in K(s/x)}{\mathrm{Snd}(\langle s, S\rangle) = S \in K(s/x)}$

    $\dfrac{T \in \sum x \in A.K}{\langle \mathrm{fst}(T), \mathrm{Snd}(T)\rangle = T \in \sum x \in A.K}$
- *Product clauses*:
  - *formation* $\dfrac{K\ [x \in A, \Gamma]}{\prod x \in A.K\ [\Gamma]}$ $\qquad \dfrac{A = A' \in Type\ [\Gamma] \quad K = K'\ [x \in A, \Gamma]}{\prod x \in A.K = \prod x \in A'.K'\ [\Gamma]}$
  - *introduction* $\dfrac{S \in K\ [x \in A, \Gamma]}{\lambda x \in A.S \in \prod x \in A.K\ [\Gamma]}$ $\qquad \dfrac{A = A' \in Type\ [\Gamma] \quad S = S' \in K\ [x \in A, \Gamma]}{\lambda x \in A.S = \lambda x \in A'.S' \in \prod x \in A.K\ [\Gamma]}$
  - *elimination* $\dfrac{T \in \prod x \in A.K \quad s \in A}{Ts \in K(s/x)}$ $\qquad \dfrac{T = T' \in \prod x \in A.K \quad s = s' \in A}{Ts = T's' \in K(s/x)}$
  - *equality* $\dfrac{s \in A\ [\Gamma] \quad S \in K\ [x \in A, \Gamma]}{(\lambda x \in A.S)s = S(s/x) \in K(s/x)}$ $\qquad \dfrac{T \in \prod x \in A.K}{\lambda x \in A.Tx = T \in \prod x \in A.K}$

Now we consider the analogous clauses giving the closure of Orders under "quantification" over Orders.

- *Sum clauses*:
  - *formation* $\dfrac{L\ [X \in K, \Gamma]}{\sum X \in K.L\ [\Gamma]}$ $\qquad \dfrac{L = L'\ [\Gamma] \quad K = K'\ [X \in K, \Gamma]}{\sum X \in K.L = \sum X \in K'.L'\ [\Gamma]}$
  - *introduction* $\dfrac{S \in K \quad T \in L(S/X)}{\langle S, T\rangle \in \sum X \in K.L}$ $\qquad \dfrac{S = S' \in K \quad T = T' \in L(S/X)}{\langle S, T\rangle = \langle S', T'\rangle \in \sum X \in K.L}$

· *elimination*

$$\frac{U \in \sum X \in K . L}{\mathrm{Fst}(U) \in K} \qquad \frac{U = U' \in \sum X \in K . L}{\mathrm{Fst}(U) = \mathrm{Fst}(U') \in K}$$

$$\frac{U \in \sum X \in K . L}{\mathrm{Snd}(U) \in K(\mathrm{Fst}(U)/X)} \qquad \frac{U = U' \in \sum X \in K . L}{\mathrm{Snd}(U) = \mathrm{Snd}(U') \in K(\mathrm{Fst}(U)/X)}$$

· *equality*

$$\frac{S \in K \quad T \in L(S/X)}{\mathrm{Fst}(\langle S,T \rangle) = S \in K} \qquad \frac{S \in K \quad T \in L(S/X)}{\mathrm{Snd}(\langle S,T \rangle) = T \in L(S/X)}$$

$$\frac{U \in \sum X \in K . L}{\langle \mathrm{Fst}(U), \mathrm{Snd}(U) \rangle = U \in \sum X \in K . L}$$

· *Product clauses*:

· *formation*

$$\frac{L \ [X \in K, \Gamma]}{\prod X \in K . L \ [\Gamma]} \qquad \frac{K = K' \ [\Gamma] \quad L = L' \ [X \in K, \Gamma]}{\prod X \in K . L = \prod X \in K' . L' \ [\Gamma]}$$

· *introduction*

$$\frac{T \in L \ [X \in K, \Gamma]}{\lambda X \in K . T \in \prod X \in K . L \ [\Gamma]} \qquad \frac{K = K' \ [\Gamma] \quad T = T' \in L \ [X \in K, \Gamma]}{\lambda X \in K . T = \lambda X \in K' . T' \in \prod X \in K . L \ [\Gamma]}$$

· *elimination*

$$\frac{U \in \prod X \in K . L \quad S \in K}{US \in L(S/X)} \qquad \frac{U = U' \in \prod X \in K . L \quad S = S' \in K}{US = U'S' \in L(S/X)}$$

· *equality*

$$\frac{S \in K \ [\Gamma] \quad T \in L \ [X \in K, \Gamma]}{(\lambda X \in K . T) S = T(S/X) \in L(S/X)} \qquad \frac{U \in \prod X \in K . L}{\lambda X \in K . UX = U \in \prod X \in K . L}$$

Remarks.

(i) The only way in which we have varied the presentation from that of Martin-Löf [M-L2] is in the elimination and equality rules for sum types. There we have followed Seely [Se1] and used constants for first and second projection rather than "elimination" constants. The latter would involve using the following rules:

$$\frac{S \in \sum x \in A . K \ [\Gamma] \quad p \in P(\langle x,X \rangle / Z) \ [X \in K, x \in A, \Gamma]}{\mathrm{E}(S, (x,X).p) \in P(S/Z) \ [\Gamma]}$$

$$\frac{S = S' \in \sum x \in A . K \ [\Gamma] \quad p = p' \in P(\langle x,X \rangle / Z) \ [X \in K, x \in A, \Gamma]}{\mathrm{E}(S, (x,X).p) = \mathrm{E}(S', (x,X).p') \in P(S/Z) \ [\Gamma]}$$

$$\frac{s \in A \ [\Gamma] \quad S \in K(s/x) \ [\Gamma] \quad p \in P(\langle x,X \rangle / Z) \ [X \in K, x \in A, \Gamma]}{\mathrm{E}(\langle s,S \rangle, (x,X).p) = p(s/x, S/X) \in P(\langle s,S \rangle / Z) \ [\Gamma]}$$

$$\frac{S \in \sum x \in A . K \ [\Gamma] \quad p \in P \ [Z \in \sum x \in A . K, \Gamma]}{\mathrm{E}(S, (x,X).p(\langle x,X \rangle / Z)) = p(S/Z) \in P(S/Z) \ [\Gamma]}$$

for sums of Orders over Types and similar rules for sums of Orders over Orders. This formulation is equivalent to the one we have given as regards equalities (though not as regards reductions): in one direction we can define $\mathrm{E}(S,(x,X).p)$ to be $p(\mathrm{fst}(S)/x, \mathrm{Snd}(S)/X)$ and derive the above rules from the *Sum clauses*; and in the other direction we can define $\mathrm{fst}(S)$ to be $\mathrm{E}(S,(x,X).x)$, $\mathrm{Snd}(S)$ to be $\mathrm{E}(S,(x,X).X)$ and derive the *Sum clauses* from the above rules. (The first three of the above rules appear in [M-L2], where they are used to derive the rules involving "fst" and "Snd" *in the presence of rules for equality types.* We do not introduce equality types here because the models we are going to consider in sections 4 and 5 do not support them—or at least do so only with very weak rules.)

(*ii*) As in [M-L2], we can have notations $A\times K$ and $A\to K$ for $\sum x\in A.K$ and $\prod x\in A.K$ respectively when x is not free in K; and notations $K\times L$ and $K\to L$ for $\sum X\in K.L$ and $\prod X\in K.L$ respectively when X is not free in K. Note that for any Type A, the Order $A\times 1_O$ is "essentially equivalent" to A: up to provable equality, there is a bijective correspondence between Terms of Type A and Operators of Order $A\times 1_O$.

(*iii*) Strictly speaking, the notation $\langle s,S\rangle, \langle S,T\rangle$ for members of sum Orders, is not satisfactory because of the ambiguity involved in $K(s/x), K(S/X)$. An unambiguous notation is preferred by computer scientists (see [MiPl] for example).

1.8. Constant Operators. We can introduce any number of constant Operators of various Orders at appropriate stages of the recursive definition. These partly determine the signature. A constant Operator has "arity" given by Orders and Types for its (free) variables. The collection of *atomic* Operators is formed from the constant Operators by general substitution.

· *Constant Operators*
$$\frac{K\ [\Gamma]}{T(\xi_1,\ldots,\xi_n)\in K\ [\Gamma]}$$
(where all the variables ξ_i are declared in Γ)

Remark. It is worth noting however that there is no real call for constant Operators *with arities*. For example, instead of introducing

$$T(x)\in K(x)\ [x\in A]$$

we can introduce

$$T\in\prod x\in A.K\ .$$

Then the elimination clause for products will enable us to recapture $T(x)$ as Tx. (Note however, that there is one problem with this procedure. If Γ declares more variables than appear in $T(\xi_1,\ldots,\xi_n)$, we will want to declare explicitly that T is constant in some arguments. We have to do this by adding appropriate equality axioms—see 1.10.)

1.9. The structure of Types and Terms. As for Orders and Operators in 1.7, so for Types and terms here we have *formation, introduction, elimination* and *equality* clauses.

Just as for Orders, Types are closed under "quantification" (indexed sums and products) over both Types and Orders: there is an important difference however in the treatment of sums indexed over Orders. We treat first the clauses relating to "quantification" over Types.

- *Unit clauses*:

  - *formation* $\dfrac{}{1_T \in Type}$

  - *introduction* $\dfrac{}{\langle\rangle_T \in 1_T}$

  - *equality* $\dfrac{t \in 1_T}{t = \langle\rangle_T \in 1_T}$

- *Sum clauses*:

  - *formation* $\dfrac{B \in Type\ [x \in A, \Gamma]}{\sum x \in A.B \in Type\ [\Gamma]}$ $\qquad \dfrac{A = A' \in Type\ [\Gamma] \quad B = B' \in Type\ [x \in A, \Gamma]}{\sum x \in A.B = \sum x \in A'.B' \in Type\ [\Gamma]}$

  - *introduction* $\dfrac{s \in A \quad t \in B(s/x)}{\langle s,t\rangle \in \sum x \in A.B}$ $\qquad \dfrac{s = s' \in A \quad t = t' \in B(s/x)}{\langle s,t\rangle = \langle s',t'\rangle \in \sum x \in A.B}$

  - *elimination* $\dfrac{u \in \sum x \in A.B}{\mathsf{fst}(u) \in A}$ $\qquad \dfrac{u = u' \in \sum x \in A.B}{\mathsf{fst}(u) = \mathsf{fst}(u') \in A}$

    $\dfrac{u \in \sum x \in A.B}{\mathsf{snd}(u) \in B(\mathsf{fst}(u)/x)}$ $\qquad \dfrac{u = u' \in \sum x \in A.B}{\mathsf{snd}(u) = \mathsf{snd}(u') \in B(\mathsf{fst}(u)/x)}$

  - *equality* $\dfrac{s \in A \quad t \in B(s/x)}{\mathsf{fst}(\langle s,t\rangle) = s \in A}$ $\qquad \dfrac{s \in A \quad t \in B(s/x)}{\mathsf{snd}(\langle s,t\rangle) = t \in B(s/x)}$

    $$\dfrac{u \in \sum x \in A.B}{\langle \mathsf{fst}(u), \mathsf{snd}(u)\rangle = u \in \sum x \in A.B}$$

- *Product clauses*:

  - *formation* $\dfrac{B \in Type\ [x \in A, \Gamma]}{\prod x \in A.B \in Type\ [\Gamma]}$ $\qquad \dfrac{A = A' \in Type\ [\Gamma] \quad B = B' \in Type\ [x \in A, \Gamma]}{\prod x \in A.B = \prod x \in A'.B' \in Type\ [\Gamma]}$

  - *introduction* $\dfrac{t \in B\ [x \in A, \Gamma]}{\lambda x \in A.t \in \prod x \in A.B\ [\Gamma]}$ $\qquad \dfrac{A = A' \in Type \quad t = t' \in B\ [x \in A, \Gamma]}{\lambda x \in A.t = \lambda x \in A'.t' \in \prod x \in A.B\ [\Gamma]}$

  - *elimination* $\dfrac{u \in \prod x \in A.B \quad s \in A}{us \in B(s/x)}$ $\qquad \dfrac{u = u' \in \prod x \in A.B \quad s = s' \in A}{us = u's' \in B(s/x)}$

  - *equality* $\dfrac{s \in A\ [\Gamma] \quad t \in B\ [x \in A, \Gamma]}{(\lambda x \in A.t)s = t(s/x) \in B(s/x)}$ $\qquad \dfrac{u \in \prod x \in A.B}{\lambda x \in A.ux = u \in \prod x \in A.B}$

Now we treat the clauses giving the closure of Types under "quantification" over Orders.

- *Sum clauses*:

  - *formation* $\dfrac{A \in Type\ [X \in K, \Gamma]}{\sum X \in K.A \in Type\ [\Gamma]}$ $\qquad \dfrac{K = K'\ [\Gamma] \quad A = A' \in Type\ [X \in K, \Gamma]}{\sum X \in K.A = \sum X \in K'.A'\ [\Gamma]}$

· *introduction* $\dfrac{S \in K \quad s \in A(S/X)}{\langle S,s\rangle \in \sum X{\in}K\,.\,A}$ $\qquad \dfrac{S=S' \in K \quad s=s' \in A(S/X)}{\langle S,s\rangle = \langle S',s'\rangle \in \sum X{\in}K\,.\,A}$

· *elimination* $\dfrac{s \in \sum X{\in}K\,.\,A \ [\Gamma] \quad t \in B(\langle X,x\rangle/z) \ [x \in A, X \in K,\Gamma]}{\mathrm{E}(s,(X,x).t) \in B(s/z) \ [\Gamma]}$

$$\frac{s=s' \in \sum X{\in}K\,.\,A \ [\Gamma] \quad t=t' \in B(\langle X,x\rangle/z) \ [x \in A, X \in K,\Gamma]}{\mathrm{E}(s,(X,x).t) = \mathrm{E}(s',(X,x).t') \in B(s/z) \ [\Gamma]}$$

· *equality* $\dfrac{S \in K \ [\Gamma] \quad s \in A(S/X) \ [\Gamma] \quad t \in B(\langle X,x\rangle/z) \ [x \in A, X \in K,\Gamma]}{\mathrm{E}(\langle S,s\rangle,(X,x).t) = t(S/X,s/x) \in B(\langle S,s\rangle/z) \ [\Gamma]}$

$$\frac{s \in \sum X{\in}K\,.\,A \ [\Gamma] \quad t \in B \ [z \in \sum X{\in}K\,.\,A,\Gamma]}{\mathrm{E}(s,(X,x).t(\langle X,x\rangle/z)) = t(s/z) \in B(s/z) \ [\Gamma]}$$

· *Product clauses*:

· *formation* $\dfrac{A \in Type \ [X \in K,\Gamma]}{\prod X{\in}K\,.\,A \in Type \ [\Gamma]}$ $\qquad \dfrac{K=K' \ [\Gamma] \quad A=A' \in Type \ [X \in K,\Gamma]}{\prod X{\in}K\,.\,A = \prod X{\in}K'\,.\,A' \in Type \ [\Gamma]}$

· *introduction* $\dfrac{s \in A \ [X \in K,\Gamma]}{\lambda X{\in}K\,.\,s \in \prod X{\in}K\,.\,A \ [\Gamma]}$ $\qquad \dfrac{K=K' \ [\Gamma] \quad s=s' \in A \ [X \in K,\Gamma]}{\lambda X{\in}K\,.\,s = \lambda X{\in}K'\,.\,s' \in \prod X{\in}K\,.\,A \ [\Gamma]}$

· *elimination* $\dfrac{t \in \prod X{\in}K\,.\,A \quad S \in K}{tS \in A(S/X)}$ $\qquad \dfrac{t=t' \in \prod X{\in}K\,.\,A \quad S=S' \in K}{tS=t'S' \in A(S/X)}$

· *equality* $\dfrac{s \in A \ [X \in K,\Gamma] \quad S \in K \ [\Gamma]}{(\lambda X{\in}K\,.\,s)S = s(S/X) \in A(S/X) \ [\Gamma]}$ $\qquad \dfrac{t \in \prod X{\in}K\,.\,A}{\lambda X{\in}K\,.\,tX = t \in \prod X{\in}K\,.\,A}$

Finally, we can introduce any number of constant Terms at appropriate stages of the recursive definition. These partly determine the signature. A constant Term has "arity" given by Orders and Types for its (free) variables. (The remark we made in 1.8 about constant Operators *with arities* applies equally well here to constant Terms.) The collection of *atomic* Terms is formed from the constant Terms by general substitution.

· *Constant Terms* $\dfrac{A \in Type \ [\Gamma]}{f(\xi_1,\ldots,\xi_n) \in A \ [\Gamma]}$

Remarks.

(*i*) In the formulation of "quantification" over Orders, the clauses for sum elimination and equality resemble those familiar from Martin-Löf's presentations (with the exception of the last equality rule—*cf*. Remark (*i*) in 1.7). But the two different levels of "types" (Types and Orders) introduce a subtle distinction at this point. We are referring to the fact that in the *elimination* and *equality clauses* for sums of Types over Orders, B is only a Type and not an Order. Consequently it is not even possible to define a first projection "Fst" as we did in Remark (*i*) of 1.7, so that the rules given are no longer equivalent to a formulation involving projections. (We will discuss in 1.11 what happens if one strengthens the rules by replacing B by an Order K.) Nevertheless, the sum rules we have given are still strictly stronger than those in

Girard's original formulation of higher order lambda calculus (and so also stronger than the equivalent formulation in [MiPl]): we will explain why when we discuss their category theoretic interpretation in 2.12.

(*ii*) Just as in the case of quantification of Orders, we write $A \times B$ and $A \to B$ for $\sum x \in A.B$ and $\prod x \in A.B$ respectively when x is not free in B; and similarly write $K \times A$ and $K \to A$ for $\sum X \in K.A$ and $\prod X \in K.A$ respectively when X is not free in B. In particular, we can associate to any Order K a Type $K \times 1_T$. Unlike the situation in 1.7 for Types (where the stronger *Sum clauses* imply that the Type A is essentially equivalent to the Order $A \times 1_O$), it is *not* the case that K is essentially equivalent to $K \times 1_T$. Rather, the process of sending K to $K \times 1_T$ sets up a reflection of Orders into Types: this will be discussed further in 2.12.

(*iii*) Occurences of the variables X, x in t are *bound* in the elimination term $\mathrm{E}(s,(X,x).t)$—that is, $FV(\mathrm{E}(s,(X,x).t)) = FV(t) \setminus \{X,x\} \cup FV(s)$. The notation we use for this elimination term is Martin-Löf's; it is sometimes written as

$$\textbf{let } (X,x) \textbf{ be } s \textbf{ in } t$$

which conveys its intended meaning better, but is a less convenient notation in compound expressions.

1.10. Theories. We have now described the *basic theory* of constructions. A *theory* over this basic one is obtained essentially by adding as axioms equality judgements of the kind

$$S = T \in K \quad \text{or} \quad s = t \in A$$

(but *not* of the kind "$K=L$"—see the Remark below). Thus if we have for example $s \in A\ [\Gamma]$ and $t \in A\ [\Gamma]$ in the basic theory, then

$$s = t \in A\ [\Gamma]$$

is a possible axiom. However, the recursive definition of judgements means that the situation is more complicated. By adding equality judgements as axioms, one generates new structural judgements which lead to the possibility of new equality judgements as axioms. So a theory **T** should consist of an ordinal indexed family τ_α of sets of judgements such that for each

$$p = q \in P\ [\Gamma]$$

in τ_α, the judgements

$$p \in P\ [\Gamma] \quad \text{and} \quad q \in P\ [\Gamma]$$

are derivable from the basic theory of constructions plus the axioms in $\bigcup\{\tau_\beta \mid \beta < \alpha\}$.

Remark. Note that we are specifically excluding the possibility of having equality judgements of the form "$K = L \in ORDER$" in our theories. This means that the weak, definitional role played by this form of judgement in the basic theory is carried over to equational extensions. What we gain by this restriction is the possibility (outlined in 2.7 and 2.13) of a perfect correspondence between theories and instances of the kind of categorical structure to be described in the next section. Moreover, there is little loss of expressive power since the equality "$K=L$" can be simulated by an *isomorphism* $K\cong L$—by introducing Operators $S\in(K\to L)$, $T\in(L\to K)$ together with axioms saying that S and T are mutually inverse. This technique is used in the following extension of the basic theory:

1.11. Theory of constructions with "*Order*∈*ORDER*". We introduce a constant Order *Order* together with rules that make it a *universal* Order, or "Order of all Orders". Specifically we introduce constant Orders and constant Operators as follows:

$$\frac{}{Order} \qquad \frac{}{O(X)\ [X\in Order]} \qquad \frac{K\ [\Gamma]}{T_K(\xi)\in Order\ [\Gamma]}$$

$$\frac{K\ [\Gamma]}{I_K(\xi)\in O(T_K(\xi))\to K\ [\Gamma]} \qquad \frac{K\ [\Gamma]}{J_K(\xi)\in K\to O(T_K(\xi))\ [\Gamma]}$$

where $\xi=\xi_1,\ldots,\xi_n$ are the variables in the maximal declarations of Γ. Then whenever $K\ [\Gamma]$ is derivable, we introduce the axioms:

$$J_K(\xi)(I_K(\xi)X) = X\in O(T_K(\xi)) \qquad [X\in O(T_K(\xi)),\Gamma]$$

and

$$Y = I_K(\xi)(J_K(\xi)Y)\in K\ [Y\in K,\Gamma].$$

Thus *Order* is an Order of "names" of Orders: $O(X)$ is the Order named by $X\in Order$ and for each Order K there is a name $T_K\in Order$ whose corresponding Order $O(T_K)$ is isomorphic to K via I_K and J_K.

Of course this extension to the theory of constructions has some very odd consequences. For one thing we can carry out Girard's Paradox in it (see [Coq2]) and hence in particular every Order possesses a closed Operator. However, the system is very far from being contradictory (in the sense of all Operators of any particular Order being provably equal). Indeed the topos-theoretic models which we present in sections 4 and 5 are both very rich models of the theory of constructions with "*Order*∈*ORDER*". Moreover it is possible to make an even more radical extension of the theory without entailing contradiction:

1.12. Theory of constructions with "*Type*≃*ORDER*". We now consider what happens if we strengthen the basic theory of constructions by replacing the Type B by an Order K in the *elimination* and *equality clauses* for sums of Types indexed over Orders in 1.9. The resulting system does extend the original one because of the correspondence of Terms of Type B with Operators of Order $B\times 1_O$ noted in Remark (*ii*) of 1.7. We call the strengthened system the *theory of constructions with "Type≃ORDER"* for reasons which we now explain:

First note that Remark (*ii*) of 1.9 no longer applies and we can now define

$$\mathrm{Fst}(s) = \mathrm{E}(s,(X,x).X) \text{ and } \mathrm{snd}(s) = \mathrm{E}(s,(X,x).x)$$

and derive elimination and equality rules for $\sum X{\in}K.A$ entirely analogous to those for $\sum x{\in}A.K$ in 1.7. As a consequence we get a bijective correspondence up to provable equality between Operators of Order K and Terms of Type $K\times 1_T$. In fact the collections of Types and of Orders can essentially be identified, since the operations

$$A\in Type \longmapsto A\times 1_O \in ORDER \qquad \text{and} \qquad K\in ORDER \longmapsto K\times 1_T\in Type$$

establish an equivalence between Types and Orders: A is naturally isomorphic to $(A\times 1_O)\times 1_T$ and K is naturally isomorphic to $(K\times 1_T)\times 1_O$. One consequence of this identification of Types and Orders is that there is a universal Order, namely $Order = Type$. In other words the theory of constructions with "$Type\simeq ORDER$" manages to model (in a very strong way) the theory of constructions with "$Order\in ORDER$".

Note also that the Type $(Type\times 1_T)\in Type$ acts as a "type of all types" since the Terms of Type $Type\times 1_T$ correspond bijectively up to equality to Operators of Order $Type$, that is, to Types. So the theory of constructions with "$Type\simeq ORDER$" also models a theory "$type\in Type$" of a universal Type which is like 1.11 "moved one level down" and which we do not give explicitly here.

Despite its strength, the theory of constructions with "$Type\simeq ORDER$" is not contradictory. We will see in section 4 that the algebraic toposes provide a highly non-trivial model of this theory.

2 Categorical interpretations of type theories

2.1. Introduction. Our aim in this section is to give a plausible explanation of what categorical models of the theory of constructions look like. The fundamental idea behind categorical models of simple type theories is that the objects of a category will model the types of a theory, while the morphisms will model the terms. (One of the simplest significant cases is the well-known connection between the typed lambda calculus and cartesian closed categories; this is explained very fully in the book of Lambek and Scott [LS].) When one attempts to model more complicated type theories, especially those involving *dependent types*, then this fundamental idea of objects for types and morphisms for terms needs some elaboration. Therefore we first give a description of categorical models for calculi of dependent types.

This serves to describe models for the pure theory of Orders and Operators and for the pure theory of Types and Terms. However, Orders and Types are not independent of each other. In the first place we can take sums and products of each indexed over the other; and secondly, Type is an Order. Therefore we successively add on features required to model these ways in which Orders and Types interact, finally obtaining a complete description of categorical models for the theory of constructions.

Whilst our explanation of categorical models for the theory of constructions is based upon a categorical explanation of Martin-Löf type theory, there are many other interesting fragments of the theory of constructions whose categorical models we do not consider separately. (Explaining the models for a stronger system does not entail explaining the models for a weaker one.) In particular we regret not being able to comment usefully on models for the second and higher order lambda calculus. The reader will find accounts of categorical models for these calculi in [Se2] and [Pi2].

In the final part of this section we indicate what is required to model the extensions of the basic theory of constructions considered in 1.11 and 1.12.

2.2. Dependent types. Martin-Löf type theory is a calculus of *dependent types*, so before we can consider the rules for unit, sums and products we must explain how that notion is to be modelled categorically. So let us restrict our attention to that part of the theory presented in section 1 which is concerned with the pure theory of Types and Terms, without unit, sums or products. This leaves the general rules for Types and Terms (part of 1.5), together with rules for introducing constant Types (special case of 1.6) and constant Terms (last part of 1.9). We could just as well restrict to the pure theory of Orders and Operators. To use Cartmell's terminology [Ca], this is the "generalized algebraic" fragment of Martin-Löf's theory of dependent types.

Cartmell [Ca], Obtułowicz [Ob] and Seely [Se1] have given accounts of categorical models for calculi of dependent types. As we will explain shortly, Cartmell's notion of "contextual category" is too rigid for our purposes; and Obtułowicz's notion (involving a hierarchy of indexed categories) is for us unnecessarily general. The notion we need can be obtained by adapting the analysis in [Se1]. We consider the following category-theoretic structure:

- A category **B** with finite products. (The product of objects I and J in **B** will be denoted $I \times J$ with projection morphisms $\pi_1 : I \times J \longrightarrow I$ and $\pi_2 : I \times J \longrightarrow J$; the terminal object in **B** will be denoted 1.)
- A collection **A** of morphisms of **B** satisfying the following condition:

  Stability. *If $a : A \longrightarrow I$ is in* **A**, *$f : J \longrightarrow I$ is in* **B** *and*

  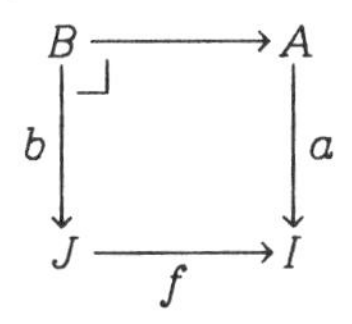

  is a pullback square in **B**, *then $b : B \longrightarrow J$ is in* **A**; *moreover there is a pullback square for each such a and f.*

Note that we are not assuming that **B** has *all* pullbacks. **A** should be regarded as the class of objects of a full subcategory, $\mathbf{A} \hookrightarrow \mathbf{B}^2$, of the arrow category $\mathbf{B}^2$. Composing with the codomain functor $cod : \mathbf{B}^2 \longrightarrow \mathbf{B}$ we obtain a functor $\mathbf{A} \longrightarrow \mathbf{B}$ which is a categorical *fibration* (*cf.* [B2]). If **B** does in fact have all pullbacks then $cod : \mathbf{B}^2 \longrightarrow \mathbf{B}$ is itself a fibration and we have a full and faithful cartesian functor over **B**:

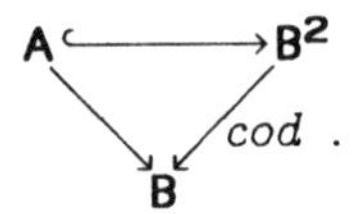

From this point of view **A** determines a "full subcategory of **B** as seen from **B**"—although not necessarily a "definable" full subcategory in the sense of [B2]. Even if $cod:\mathbf{B^2}\longrightarrow\mathbf{B}$ is not a fibration, it is convenient to say when **A** satisfies the **Stability** assumption above that

$\mathbf{A}$ is a full subcategory of $\mathbf{B^2}$ over $\mathbf{B}$.

We need some notation related to the category-theoretic structure just described. Recall that the *slice* of **B** by an object I, denoted $\mathbf{B}/I$, is the category whose objects are morphisms in **B** with codomain I and whose morphisms are commutative triangles in **B** with vertex I. For each object I in **B**, we write

$$\mathbf{A}(I)$$

for the full subcategory of $\mathbf{B}/I$ whose objects are the morphisms with codomain I which lie in **A**; this is precisely the fibre over I of the fibration $\mathbf{A}\longrightarrow\mathbf{B}$. Note in particular that we can identify $\mathbf{B}/1$ with **B**, in which case $\mathbf{A}(1)$ becomes a full subcategory of **B** in the usual sense: its objects are those $I\in\mathbf{B}$ for which the unique morphism $I\longrightarrow 1$ is in **A**. Given $f:J\longrightarrow I$ in **B**, we write

$$f^{\#}:\mathbf{A}(I)\longrightarrow\mathbf{A}(J)$$

for the pullback functor. We are tacitly assuming that a choice of such pullback functors is given along with the information specifying **A** (which is to say, in the language of fibrations, that $\mathbf{A}\longrightarrow\mathbf{B}$ is a "cloven" fibration); but note that the way we have phrased the **Stability** condition implies that *we can compose a morphism in* **A** *with an isomorphism on either side and still remain in* **A**. We use the above notation for pullback functors rather than the more usual " f^* " only because in sections 4 and 5, f will be a geometric morphism between toposes, in which case " f^* " conventionally denotes the inverse image part of f. However, applying the functor $f^{\#}$ to an object $A\longrightarrow I$ of $\mathbf{A}(I)$, the *domain* of the resulting object of $\mathbf{A}(J)$ will as usual be denoted by $J\times_I A$; in other words our standard notation for a pullback square in **B** is:

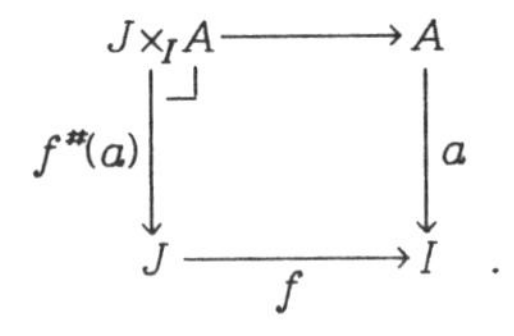

We now indicate in general terms how a category **B** with finite products equipped with a full subcategory $\mathbf{A}\hookrightarrow\mathbf{B^2}$ over **B** in the above sense, serves to interpret the calculus of dependent types. The interpretation is based upon having a "structure" in **B** for the language—in other words, having a particular choice of interpretation for the constant Types and Terms: a constant Type of arity n is interpreted by a morphism in the class **A**

of the form $J \longrightarrow I_1 \times \ldots \times I_n$; and a constant Term of arity n is interpreted by a morphism in **B** of the form $I_1 \times \ldots \times I_n \longrightarrow J$. (In case $n=0$, this means that a constant Type which is dependent on no variables is essentially interpreted as an object J of **B**, but one for which the unique morphism $J \longrightarrow 1$ is in **A**; and a constant Term depending upon no variables is interpreted as a global section $1 \longrightarrow J$ of an object in **B**.)

As formulated in section 1, the calculus consists of a collection of rules for deriving judgements (in contexts) about certain expressions. These judgements-in-context not only assert the typing and the equality of expressions, but also tell us which expressions are well-formed; moreover, these three kinds of assertion are built up simultaneously by mutual recursion. Therefore, it is not possible *first* to give a recursive definition of the denotations of well-formed expressions as objects and morphisms of **B** and *then* to give a recursive definition of when the various kinds of judgement are satisfied by the categorical structure. Instead one has to give a single definition by recursion on the derivation of a judgement of its satisfaction and (depending on the form of judgement) of the denotations of its constituent expressions which are used to express this satisfaction. We consider what is involved for each form of judgement in turn:

A judgement of the form $A \in Type\ [\Gamma]$ has the force that, in the context Γ, A is a well-formed Type. Its satisfaction then amounts to having built up a morphism in **A** of the form

$$[\![A]\!] \longrightarrow [\![\Gamma]\!] =_{def} \prod_J [\![A_J]\!]$$

where the codomain is a (finite) product indexed by the *maximal* elements J of the context Γ and A_J is the Type of the variable declared in J.

A judgement of the form $A=B \in Type\ [\Gamma]$ has the force that, in the context Γ, A and B are equal (well-formed) Types. It is satisfied if $[\![A]\!] \longrightarrow [\![\Gamma]\!]$ and $[\![B]\!] \longrightarrow [\![\Gamma]\!]$ are equal morphisms in **A**.

A judgement of the form $s \in A\ [\Gamma]$ has the force that s is a (well-formed) Term of Type A. As all relevant variables appear in the context Γ, the satisfaction of the judgement amounts to having built up a morphism

$$[\![s]\!] : [\![\Gamma]\!] \longrightarrow [\![A]\!]$$

in **B** which is a section of the **A**-morphism $[\![A]\!] \longrightarrow [\![\Gamma]\!]$ (that is, whose composition with this morphism is the identity on $[\![\Gamma]\!]$.)

Finally, a judgement of the form $s=t \in A\ [\Gamma]$ has the force that s and t are equal Terms of Type A. It is satisfied if $[\![s]\!] : [\![\Gamma]\!] \longrightarrow [\![A]\!]$ and $[\![t]\!] : [\![\Gamma]\!] \longrightarrow [\![A]\!]$ are equal morphisms in **B**.

The actual clauses of the recursive definition of satisfaction of judgements are rather straightforward for the pure theory of dependent types which we are considering at the moment: the only significant rule is that for *Substitution* and this is handled using the pullback functors $f^{\#}$. (Contexts and variable declarations are handled by finite products and projections, as above.)

2.3. Remarks.

(*i*) The idea of using a special class of morphisms closed under (at least) pullback is not new. For example Bénabou's notion of a "catégorie calibrée" in [B1] is the above notion with the addition of the condition for **Sums** from 2.4 and a further condition (his (P3)) which makes the special class of morphisms like a class of local homeomorphisms. Our approach was inspired by the notion of "display map" in Taylor's thesis [Ta], which is essentially our notion with the addition of the **Unit** and **Sums** conditions from 2.4 and the **Display** condition from 2.7.

(*ii*) The idea of a special class of maps is also the basis for Cartmell's *contextual categories* in [Ca]. Cartmell has a canonical choice of morphisms in **A** to represent the dependent types. Up to equivalence of categories, there is not much to choose between the two approaches. A contextual category gives rise to one of our subcategories (also satisfying the **Unit**, **Sums** and **Display** conditions below) just by closing the set of canonical morphisms under isomorphism. On the other hand, one of our subcategories (satisfying **Unit**, **Sums** and **Display**) can be "unravelled" to give an equivalent contextual category with the necessary canonical choices.

(*iii*) The version of Martin-Lōf type theory which Seely models has equality types and strong rules for equality as in [M-L1], as well as strong rules for sums. Equality is interpreted via equalizers in the category. It follows that *every* morphism

$$[\![s]\!]:[\![C]\!]\longrightarrow[\![A]\!]$$

in **B** interpreting a term s can be thought of as also representing an indexed family

$$B(x)\ [x\in A]$$

of types over A: one simply takes $B(x)$ to be

$$\Sigma y\in C.\,I_A(s(y),x)\ .$$

where I_A is the equality type for A. (See Sublemma 3.2.3.2 of [Se1].) So in this case *every* morphism of **B** should be in **A**. Without strong equality types—which we emphatically do not have in the topos-theoretic models presented in sections 4 and 5—one does not have this phenonomen. On the other hand, as we explain below, we are still able to model indexed families

$$B(x)\ [x\in A]$$

by morphisms $[\![B(x)]\!]\longrightarrow[\![A]\!]$ (in **A**) which also represent terms ("first projection" in this case).

2.4. Martin-Lōf type theory. We take the basic rules of Martin-Lōf type theory to be the *Unit, Sum* and *Product clauses* for "quantification" of Types over Types as given in 1.9. (In his usual presentation, Martin-Lōf has the unit rules appearing amongst the rules for *finite types*. We regard them as an essential part of the system: they denote an essential part of the semantics—*cf*. 2.8.)

The theory of constructions contains two instances of these basic rules: one in 1.7 for the pure theory of Orders and one in 1.9 for the pure theory of Types. We now consider what properties of a full subcategory **A** of $\mathbf{B}^2$ over **B** (as introduced in 2.2) are needed to model the basic rules.

For the *Unit clauses*, firstly *formation* gives us an object $[\![1_T]\!]$ in **B** for which the unique morphism $[\![1_T]\!] \longrightarrow 1$ is in **A**; and *introduction* gives us a morphism $[\![\langle\rangle_T]\!] : 1 \longrightarrow [\![1_T]\!]$ in **B** which is a section of $[\![1_T]\!] \longrightarrow 1$, i.e. is a right-sided inverse for it; but finally, *equality* implies that it is actually a two-sided inverse—since we can apply the clause in the context $\Gamma = [x \in 1_T]$ (for which $[\![\Gamma]\!] = [\![1_T]\!]$), to conclude that $\pi_1 : [\![\Gamma]\!] \times [\![1_T]\!] \longrightarrow [\![\Gamma]\!]$ has a *unique* section and hence that $[\![1_T]\!] \longrightarrow 1 \longrightarrow [\![1_T]\!]$ is the identity. So we conclude not only that $[\![1_T]\!]$ is isomorphic to the terminal object of **B**, but also that **A** contains an isomorphism with codomain *1*. Because of the **Stability** condition in 2.2, this is equivalent to requiring:

Unit. **A** *contains all the isomorphisms of* **B**.

Turning to the *Sum clauses*, categorically, indexed sums provide (stable) left adjoints to pullback functors. Suppose that $f : [\![B(x)]\!] \longrightarrow [\![A]\!]$ is the morphism in **A** interpreting $B(x)\ [x \in A]$. Then the left adjoint $f_!$ to $f^\#$ corresponds to taking

$$C(x,y)\ \ [y \in B(x) > x \in A]$$

to

$$\Sigma y{\in}B(x)\,.\,C(x,y)\ \ [x \in A]\ .$$

So we need left adjoints

$$f_! : \mathbf{A}(J) \longrightarrow \mathbf{A}(I)$$

to the pullback functor $f^\# : \mathbf{A}(I) \longrightarrow \mathbf{A}(J)$ for all morphisms $f : J \longrightarrow I$ *which are in* **A**. However, the rules for sums give something more. We have of course the unit of the adjunction, which syntactically is essentially

$$z \in C(x,y) \longmapsto \langle y,z\rangle \in \Sigma y{\in}B(x)\,.\,C(x,y)$$

and makes the diagram

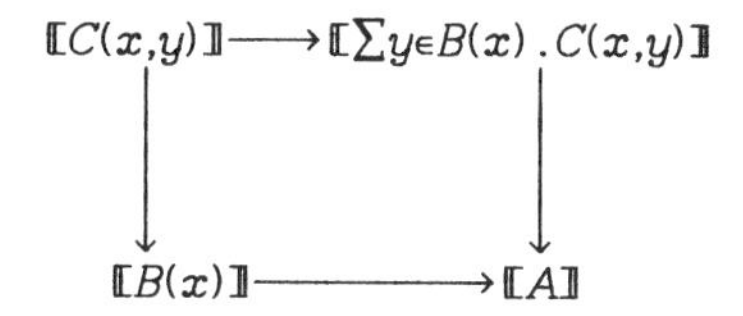

commute. "Snd" provides an inverse for this map, so it is an isomorphism. Consequently the indexed sums in **A** should be the standard ones given by *composition.* (If the pullback functor $f^\# : \mathbf{B}/I \longrightarrow \mathbf{B}/J$ exists, then it automatically has a left adjoint given by composing with $f : J \longrightarrow I$.) Then the expected "Beck-Chevalley" condition comes for free—namely that if

(2.1)
$$\begin{array}{ccc} L & \xrightarrow{k} & K \\ {\scriptstyle h}\downarrow & & \downarrow{\scriptstyle g} \\ J & \xrightarrow[f]{} & I \end{array}$$

is a pullback square in **B** with f (and hence also k) in **A**, then the canonical natural transformation $k_! \circ h^\# \longrightarrow g^\# \circ f_!$ is an isomorphism. So the *Sum clauses* are covered by the assumption:

> **Sums.** **A** *is closed under composition.*

Categorically, indexed products provide (stable) right adjoints to pullback functors. Thus as we argued above, we expect to have right adjoints

$$f_\# : \mathbf{A}(J) \longrightarrow \mathbf{A}(I)$$

to $f^\#$ for all morphisms $f : J \longrightarrow I$ which are in **A**. We will write

$$\Pi_f(B)$$

for the domain of the object of $\mathbf{A}(I)$ resulting from the application of $f_\#$ to an object $B \longrightarrow J$ of $\mathbf{A}(J)$. In contrast to the case for sums, we have to give an explicit condition for stability (which is needed to model the behaviour of products under substitution):

> **Beck-Chevalley condition.** *If (2.1) is a pullback square in* **B** *with f (and hence also k) in* **A**, *then the canonical natural transformation $g^\# \circ f_\# \longrightarrow k_\# \circ h^\#$ is an isomorphism.*

Thus the assumption needed to model indexed products is as follows:

> **Products.** *For all f in* **A** *, we have a right adjoint $f_\#$ to $f^\#$ which satisfies the Beck-Chevalley condition for pullbacks in* **B** *(along arbitrary morphisms).*

It is useful both conceptually and technically to have an equivalent formulation of this condition. First we give a preliminary result:

2.5. Lemma. *Suppose that* **A** *is a class of maps in a category* **B** *satisfying the above conditions of* **Stability, Unit, Sums** *and* **Products**. *Then the functor $f_\#$ preseves morphisms which are in* **A**.

Proof. Suppose we are given the following commutative diagram of morphisms in **A**:

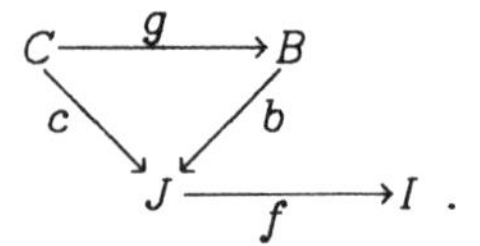

We have to show that $\Pi_f(g) : \Pi_f(C) \longrightarrow \Pi_f(B)$ is in **A**. Form the pullback squares

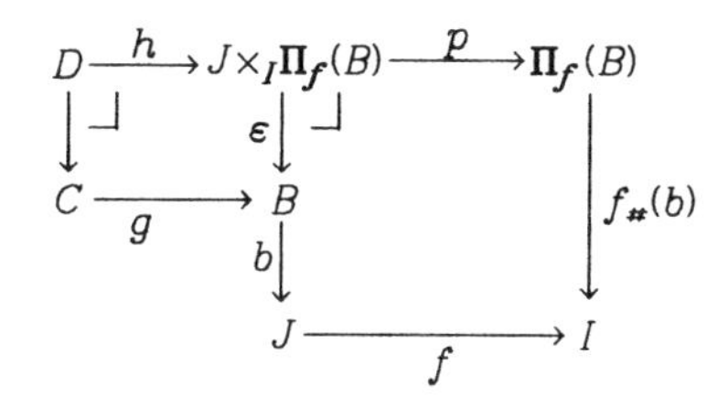

where $\varepsilon : f^{\#}f_{\#}(b) \longrightarrow b$ is the counit of the adjunction $f^{\#} \dashv f_{\#}$ at b. Note that p and h are both in **A** (because f and g are): hence we can form $p_{\#}(h) : \Pi_p(D) \longrightarrow \Pi_f(B)$ in $\mathbf{A}(\Pi_f(B))$. We claim that $p_{\#}(h)$ is isomorphic to $\Pi_f(g)$ in $\mathbf{B}/\Pi_f(B)$, which is sufficient to show that the latter is in **A**, bearing in mind our remark in 2.2 about the closure of **A** under composition with isomorphisms.

Whilst it is possible to prove this claim purely categorically, it is both easier and more illuminating to argue type-theoretically. So suppose that

$f : J \longrightarrow I$ denotes the type $J(i)\ [i \in I]$

$b : B \longrightarrow J$ denotes the type $B(i,j)\ [j \in J(i) > i \in I]$

and $g : C \longrightarrow B$ denotes the type $C(i,j,b)\ [b \in B(i,j) > j \in J(i) > i \in I]$.

Then $f_{\#}(b) : \Pi_f(B) \longrightarrow I$ denotes $\prod j \in J(i) . B(i,j)\ [i \in I]$,

and $f_{\#}(c) : \Pi_f(C) \longrightarrow I$ denotes

(2.2) $\prod j \in J(i) . \sum b \in B(i,j) . C(i,j,b)\ [i \in I]$.

Moreover, $\Pi_f(g)$ is the morphism corresponding to

(2.3) $z \in \prod j \in J(i) . \sum b \in B(i,j) . C(i,j,b) \longmapsto \lambda j \in J(i) . \mathsf{Fst}(z(j)) \in \prod j \in J(i) . B(i,j)$.

By construction $h : D \longrightarrow J \times_I \Pi_f(B)$ denotes $C(i,j',t(j'))\ [j' \in J(i), t \in \prod j \in J(i) . B(i,j) > i \in I]$, so that $p_{\#}(h) : \Pi_p(D) \longrightarrow \Pi_f(B)$ denotes $\prod j' \in J(i) . C(i,j',t(j'))\ [t \in \prod j \in J(i) . B(i,j) > i \in I]$ and hence the composition $f_{\#}(b) \circ p_{\#}(h)$ denotes

(2.4) $\sum t \in (\prod j \in J(i) . B(i,j)) . \prod j' \in J(i) . C(i,j',t(j'))\ [i \in I]$

But by Martin-Löf's form of the Axiom of Choice, the types in (2.2) and (2.4) are isomorphic over I; indeed the isomorphism is given in one direction by

(2.5) $z \in \prod j \in J(i) . \sum b \in B(i,j) . C(i,j,b) \longmapsto \langle \lambda j \in J(i) . \mathsf{Fst}(z(j)), \lambda j' \in J(i) . \mathsf{Snd}(z(j')) \rangle$

(*cf.* [M-L1], page 173) and in the other by

(2.6) $w \in \sum t \in (\prod j \in J(i) . B(i,j)) . \prod j' \in J(i) . C(i,j',t(j')) \longmapsto \lambda j \in J(i) . \langle (\mathsf{Fst} w) j, (\mathsf{Snd} w) j \rangle$.

Since the composition of (2.5) with the first projection is the map in (2.3), the interpretation of (2.5) and (2.6) give the required isomorphism between $\Pi_f(g)$ and $p_{\#}(h)$ over $\Pi_f(B)$.

□

2.6. Proposition. *Suppose that* **A** *is a class of maps in* **B** *satisfying* **Stability**, **Unit** *and* **Sums**. *Then the condition* **Products** *is equivalent to the combination of the following* *conditions*:

(*i*) *each* **A**(I) *is cartesian closed*;

(*ii*) *for each* f *in* **B**, $f^{\#}$ *preserves the cartesian closed structure;*

(*iii*) *for any* $a:A \longrightarrow I$ *in* **A**, *the exponential functor* $(a \to_I -):\mathbf{A}(I) \longrightarrow \mathbf{A}(I)$ *preserves morphisms which are in* **A**.

Proof. First note that if **A** satisfies the **Stability** and **Sums** conditions, then each **A**(I) has binary products given by pullbacks over I; and if **A** satisfies the **Unit** condition, then $id_I : I \longrightarrow I$ is a terminal object in **A**(I). If **A** also satisfies the **Products** condition, then we can calculate the exponential $(a \to_I b):(A \to_I B) \longrightarrow I$ of $a:A \longrightarrow I$ and $b:B \longrightarrow I$ in **A**(I) as $a_{\#}(a^{\#}(b)) : \Pi_a(A \times_I B) \longrightarrow I$, so we have (*i*). Moreover, the Beck-Chevalley condition ensures that these exponentials are preserved under pullback, so we have (*ii*). Finally, since **A** satisfies the **Sums** condition, then by Lemma 2.5 $(a \to_I -) = a_{\#}(a^{\#}(-))$ preserves morphisms in **A** (since the pullback functor $a^{\#}$ always does).

Conversely, suppose that **A** satisfies **Stability**, **Unit**, **Sums** and conditions (*i*) to (*iii*). Given $f:J \longrightarrow I$ and $b:B \longrightarrow J$ in **A**, then by the **Sums** condition fb is also in **A** and we can regard b as a morphism $fb \longrightarrow f$ in **A**(I). Then by (*iii*), $(f \to_I b):(f \to_I fb) \longrightarrow (f \to_I f)$ is given by a morphism which is also in **A**. Hence we can form the following pullback square in **B**:

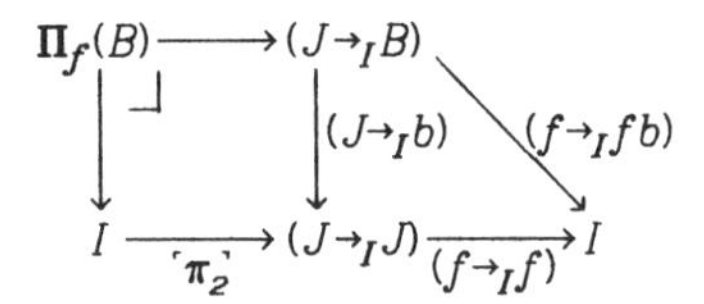

where $'\pi_2' : id_I \longrightarrow (f \to_I f)$ in **A**(I) is the exponential transpose of the isomorphism $\pi_2 : id_I \times_I f \cong f$. Thus the morphism $\Pi_f(B) \longrightarrow I$ is in **A** and a simple calculation shows that it has the correct universal property to be $f_{\#}(b)$. The fact that these right adjoints to pulling back satisfy the Beck-Chevalley condition follows from this recipe for their construction together with condition (*ii*).

□

When **A** consists of all the morphisms in **B**, then the above proposition reduces to the equivalence of two well known characterizations of locally cartesian closed categories. This justifies the following terminology, which we have borrowed from [Ta]:

2.7. Definition. A *relatively cartesian closed category* (or *rccc*, for short) is a category **B** with finite products equipped with a distinguished class of morphisms (called the *display morphisms* of **B**) satisfying the conditions **Stability** of 2.2 and **Unit**, **Sums** and **Products** of 2.4. A *morphism* of rccc's is a finite product preserving functor which sends display morphisms to display morphisms, preserves pullbacks of display morphisms along arbitrary morphisms and preserves the right adjoints to pulling back display morphisms along display morphisms. We will let **RCCC** denote the 2-category whose objects are rccc's, whose morphisms are rccc morphisms and whose 2-cells are natural transformations.

The situation for rccc's is almost analogous to that in [Se1], where an equivalence is established between locally cartesian closed categories and theories (in the sense of 1.10) over Martin-Löf type theory with unit, products and strong rules for sums and equality types. Here we are considering theories without equality types. A *model* of such a theory **T** in an rccc **B** is given by an assignment of display morphisms to the constant Types and morphisms to the constant Terms in such a way that the judgements which comprise the axioms of the theory are all satisfied in the rccc; this is a sound notion since one can show that any judgement derivable from the axioms using the rules is also satisfied. Defining an appropriate notion of homomorphism of models, we get a category **Mod(T, B)** of models of **T** in **B**. If $F:\mathbf{B}\longrightarrow\mathbf{B}'$ is an rccc morphism, then it sends **T**-models in **B** to **T**-models in **B'** and gives a functor $F_*:\mathbf{Mod}(\mathbf{T},\mathbf{B})\longrightarrow\mathbf{Mod}(\mathbf{T},\mathbf{B}')$; similarly each natural transformation $\phi:F\longrightarrow F'$ between rccc morphisms induces a natural transformation $\phi_*:F_*\longrightarrow F'_*$. In this way we get a 2-functor **Mod(T, -)** from **RCCC** into the 2-category of categories, **CAT.** The fundamental observation is that, in an appropriately bicategorical sense, this 2-functor is representable:

Theorem. *For each theory* **T** *there is an rccc* **B(T)**, *called the* classifying category *of* **T**, *and a model* $M_{\mathbf{T}}$ *of* **T** *in* **B(T)**, *called the* generic model *of* **T**, *with the property that for any rccc* **B** *the functor*

$$(-)_*(M_{\mathbf{T}}):\mathbf{RCCC}(\mathbf{B}(\mathbf{T}),\mathbf{B})\longrightarrow\mathbf{Mod}(\mathbf{T},\mathbf{B})$$

is an equivalence of categories.

□

The construction of **B(T)** is simplified by the presence of the rules for dependent sums, since as we remarked in 2.3(*iii*), it follows that we can take every morphism to be represented by a term. So we can take the objects of **B(T)** to be the closed types in **T** and the morphisms to be eqivalence classes of closed terms, under the equivalence relation of provable equality in **T**; the display morphisms are those which are isomorphic to first projections, $\mathrm{Fst}:\sum a\in A.B(a)\longrightarrow A$.

Classifying categories enable us to give a very general notion of *interpretation* of one theory in another—by defining an interpretation of **T** in **T'** to be a model of **T** in **B(T')**; similarly we can give a notion of *modification* between interpretations in terms of homomorphisms between models in the classifying category. This gives the collection of theories the structure of a bicategory (*cf*. [B3]), **MLTT**, in such a way that $\mathbf{B}(-):\mathbf{MLTT}\longrightarrow\mathbf{RCCC}$ is a fully faithful homomorphism of bicategories. The essential image of this bicategory homomorphism consists of those rccc's which are equivalent to the classifying category of some theory: this does not include every rccc for the following trivial reason. Suppose that **B** is an rccc and **A** is the class of display morphisms in **B**. If for each $n\in\mathbb{N}$ we introduce symbols for n-ary constant Types for each morphism $A\longrightarrow I_1\times\cdots\times I_n$ in **A** and symbols for n-ary constant Terms for each morphism $I_1\times\cdots\times I_n\longrightarrow J$ in **B**, then there is an evident theory $\mathbf{T_B}$ whose derived judgements are just those satisfied in the rccc. The canonical model of this theory in **B** induces a full (because of Remark 2.3(*iii*)) and faithful rccc morphism $\mathbf{B}(\mathbf{T_B}) \longrightarrow \mathbf{B}$. Its image consists of those objects of **B** which denote closed types: these are the objects B for which the unique morphism $B \longrightarrow 1$ is in **A**. We therefore impose a further condition on **A**, namely:

Display. *For each object B of **B**, the unique morphism $B \longrightarrow 1$ is in **A**.*

For the rccc's satisfying this added condition we have that $\mathbf{B}(\mathbf{T_B}) \simeq \mathbf{B}$ and thus have:

Corollary. *The classifying category construction $\mathbf{B}(-)$ induces an equivalence of bicategories between **MLTT** and the 2-category of rccc's which satisfy the **Display** condition.*

□

Remark. The condition **Display** in the presence of **Stability**, renders redundant some of our other assumptions about **B** and **A**. For one thing it implies that **B** has binary products, since these are given by pullbacks over the terminal object. It also says in particular that the unique morphism $1 \longrightarrow 1$ is in **A**: but that morphism is necessarily the identity on 1 and so by **Stability**, all isomorphisms are in **A**—in other words the **Unit** condition holds automatically.

2.8. Absoluteness of indexed products. In our discussion of indexed sums we justified the assumption that these sums should agree with the corresponding sums in $\mathbf{B^2} \longrightarrow \mathbf{B}$ (i.e. be given by composition in **B**). We will shortly consider a case (in 2.12) where this assumption is not justified. It is instructive to see why the corresponding question does not arise for indexed products.

Suppose that **B** has finite products and that

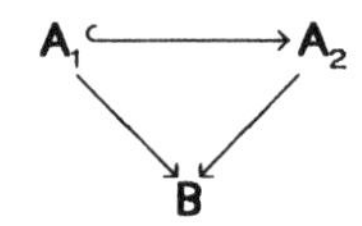

is a full embedding of full subcategories of $\mathbf{B^2}$ over **B** which contain the terminal object (so that $\mathbf{A}_1$ and $\mathbf{A}_2$ as classes of morphisms of **B** contain the isomorphisms). Suppose also that **C** is a class of morphisms in **B** satisfying the **Stability** condition of 2.2. (Thus, although we do not wish to look at it this way, $\mathbf{C} \longrightarrow \mathbf{B}$ is also a full subcategory of $\mathbf{B^2}$ over **B**.) Then we have:

Proposition. *In the above situation, if $\mathbf{A}_1$ and $\mathbf{A}_2$ both have indexed products along maps in **C** satisfying the Beck-Chevalley condition for pullbacks along arbitrary morphisms in **B**, then these indexed products agree (up to isomorphism) on $\mathbf{A}_1$.*

Proof. For $f: J \longrightarrow I$ in **C**, let ${}^1f_{\#}(a): {}^1\Pi_f(A) \longrightarrow I$ and ${}^2f_{\#}(a): {}^2\Pi_f(A) \longrightarrow I$ respectively denote the value at $a: A \longrightarrow J$ of the right adjoints to the pullback functors $f^{\#}: \mathbf{A}_1(I) \longrightarrow \mathbf{A}_1(J)$ and $f^{\#}: \mathbf{A}_2(I) \longrightarrow \mathbf{A}_2(J)$.

Suppose that $y: Y \longrightarrow J$ is in $\mathbf{A}_1(J)$. Then we have a natural comparison morphism

$$\kappa : {}^1\Pi_f(Y) \longrightarrow {}^2\Pi_f(Y)$$

such that for all $z: Z \longrightarrow I$ in $\mathbf{A}_1(I)$, composition with κ induces a bijection

$$\mathbf{B}/I(z,\ {}^1f_\#(y)) \cong \mathbf{B}/I(z,{}^2f_\#(y)).$$

Let us write $h:{}^2\Pi_f(Y)\longrightarrow I$ for ${}^2f_\#(y)$ and form the pullback square:

$$\begin{array}{ccc} L & \xrightarrow{q} & {}^2\Pi_f(Y) \\ {\scriptstyle p}\downarrow & \lrcorner & \downarrow{\scriptstyle h} \\ J & \xrightarrow[f]{} & I \end{array} .$$

We deduce from the Beck-Chevalley condition that $h^\#(\kappa)$ is (isomorphic to) the comparison morphism

$${}^1\Pi_q(p^\#Y)\longrightarrow {}^2\Pi_q(p^\#Y) \ .$$

Hence for all $w:W\longrightarrow {}^2\Pi_f(Y)$ in $\mathbf{A}_1({}^2\Pi_f(Y))$, composition with $h^\#(\kappa)$ induces a bijection

$$\mathbf{B}/{}^2\Pi_f(Y)(w,h^\#({}^1f_\# y)) \cong \mathbf{B}/{}^2\Pi_f(Y)(w,h^\#({}^2f_\# y)),$$

that is, composition with κ induces a bijection

$$\mathbf{B}/I(hw,\ {}^1f_\#(y)) \cong \mathbf{B}/I(hw,{}^2f_\#(y)) \ .$$

Taking w to be the identity on ${}^2\Pi_f(Y)$ (which is in $\mathbf{A}_1$ by hypothesis), we deduce that there is a morphism

$$\lambda:{}^2\Pi_f(Y)\longrightarrow {}^1\Pi_f(Y)$$

such that $\kappa\circ\lambda = id$. Now pull the situation back along $k = {}^1f_\#(y):{}^1\Pi_f(Y)\longrightarrow I$. We find similarly that for $v:V\longrightarrow {}^1\Pi_f(Y)$ in $\mathbf{A}_1({}^1\Pi_f(Y))$, composition with κ induces a bijection

$$\mathbf{B}/I(kv,{}^1f_\#(y)) \cong \mathbf{B}/I(kv,{}^2f_\#(y)) \ .$$

Taking v to be the identity, we find that both id and $\lambda\circ\kappa$ correspond to κ under this bijection, whence $\lambda\circ\kappa=id$.

□

2.9. Remarks.

(*i*) The result we have just proved is an "absoluteness" result for indexed products. The situation should be contrasted with that for indexed sums where no such absoluteness holds. (We will see an example of this in section 5, where the left adjoints (5.9) to pulling back localic-algebraic toposes differ from the left adjoints (4.9) to pulling back algebraic toposes.)

(*ii*) Constructions of a similar (right adjoint) kind—such as finite limits and stable exponentials—are also absolute by essentially the same argument.

2.10. Sums and products of Orders and Types indexed over Types. Let us start by assuming that we are modelling Orders and Operators by a relatively cartesian closed category $\mathbf{B}$ whose class of display morphisms is denoted by $\mathbf{A}$ and satisfies the **Display** condition (see 2.7 *et seq.*).

As we remarked in 1.7, the rules giving the closure of Orders under sums indexed over Types ensure that there is a bijective correspondence between Terms of Type A and Operators of Order $A \times 1_O$. It follows that we may take Types to be special Orders and model them by a distinguished class of morphisms **R** contained in **A** and also satisfying the **Stability** condition (so that we have a cartesian embedding $\mathbf{R} \hookrightarrow \mathbf{A}$ of fibrations over **B**).

We note at once that given this set up, the closure of Orders under sums and products indexed over Types becomes a special case of the closure under sums and products indexed over Orders—which we already have. In order to satisfy the *Unit, Sum* and *Product clauses* for "quantification" of Types over Types (which are the same as those for quantification of Orders over Orders) we apply the analysis of 2.3 to the class **R** itself. Combining this with the absoluteness of products proved in 2.4, we require the following conditions:

Unit'. **R** *contains all isomorphisms.*

Sums'. **R** *is closed under composition.*

Products'. *For all* $f: J \longrightarrow I$ *in* **R**, *the right adjoint* $f_{\#}: \mathbf{A}(J) \longrightarrow \mathbf{A}(I)$ *to* $f^{\#}$ *restricts to a functor* $f_{\#}: \mathbf{R}(J) \longrightarrow \mathbf{R}(I)$.

2.11. Type as an Order. To explain the special role of the Order *Type*, we need to explain the connection between A as it appears in the judgement

$$A \in Type$$

(A as an Operator of Order *Type*) and A as it appears in the judgement

$$a \in A$$

(A as a Type). To do this consider the generic case of the free Type variable X. In the usual way "$X \in Type$" will be interpreted by the identity function $id: [\![Type]\!] \longrightarrow [\![Type]\!]$. On the other hand, to interpret "$s \in X$", we must regard X as a Type (indexed over the Order *Type*) and interpret s as a morphism with that codomain.

This leads to the following situation. We have an object U of **B** which is to interpret *Type* (and so must satisfy that $U \longrightarrow 1$ is in **A**—which is automatic, since we are assuming **A** satisfies the **Display** condition); and we have a morphism $G \longrightarrow U$ in **R** interpreting the Type X indexed over *Type*. Now for any context $\mathbf{\Gamma}$

$$Type \; [\mathbf{\Gamma}]$$

is interpreted by the projection $U \times [\![\mathbf{\Gamma}]\!] \longrightarrow [\![\mathbf{\Gamma}]\!]$ and hence

$$A \in Type \; [\mathbf{\Gamma}]$$

is interpreted by a section $[\![\mathbf{\Gamma}]\!] \longrightarrow U \times [\![\mathbf{\Gamma}]\!]$ of this projection—and this is equivalent to giving a morphism

$$\ulcorner A \urcorner : [\![\mathbf{\Gamma}]\!] \longrightarrow U.$$

Then to interpret A in

$$a \in A \ [\Gamma] \ ,$$

that is, as an object $[\![A]\!] \longrightarrow [\![\Gamma]\!]$ of the category $\mathbf{R}([\![\Gamma]\!])$, we form a pullback square:

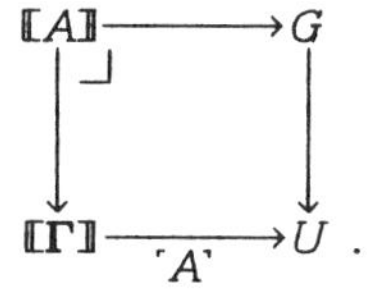

$$\begin{array}{ccc} [\![A]\!] & \longrightarrow & G \\ \downarrow & \lrcorner & \downarrow \\ [\![\Gamma]\!] & \xrightarrow[\ulcorner A \urcorner]{} & U \ . \end{array}$$

In particular the free type variable X is interpreted either as the identity on U, or as the object $G \longrightarrow U$ of $\mathbf{R}(U)$. Since every Type is obtainable from X by substitution and the latter is modelled by pullbacks, we can see that taking $Type$ as an Order amounts to the following assumption on our distinguished class $\mathbf{R}$ of morphisms in $\mathbf{B}$:

> **Generic type.** *There is a morphism $G \longrightarrow U$ in $\mathbf{R}$ which generates the class under pullback: every morphism in $\mathbf{R}$ is obtainable from $G \longrightarrow U$ by pullback along some (not necessarily unique) morphism in $\mathbf{B}$.*

2.12. Sums and products of Types indexed over Orders. We now consider what properties of

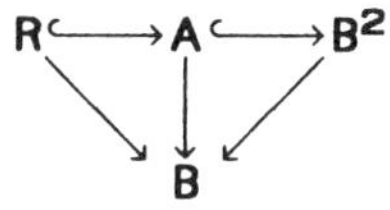

$$\begin{array}{ccccc} \mathbf{R} & \hookrightarrow & \mathbf{A} & \hookrightarrow & \mathbf{B}^2 \\ & \searrow & \downarrow & \swarrow & \\ & & \mathbf{B} & & \end{array}$$

are needed to model the "quantification" of Types over Orders. The main problem is to understand the relevant rules in 1.9 for sums. Of course we still expect indexed sums to provide a left adjoint to substitution. And as in 2.2 we can extract a morphism from the unit of the adjunction: for if we have

$$A(X,Y) \in Type \ [Y \in L(X) > X \in K]$$

then we have $x \in A(X,Y) \longmapsto \langle Y,x \rangle \in \sum Y \in L(X) \, . \, A(X,Y)$, whose interpretation makes the following diagram commute:

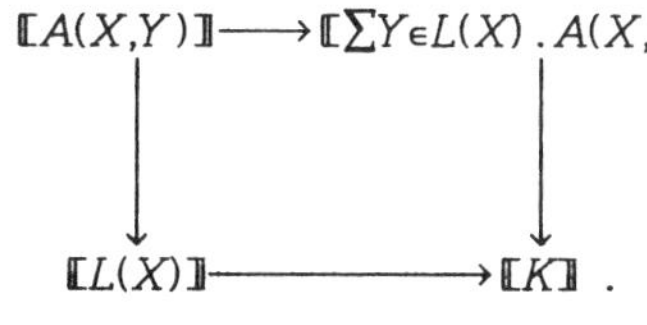

$$\begin{array}{ccc} [\![A(X,Y)]\!] & \longrightarrow & [\![\sum Y \in L(X) \, . \, A(X,Y)]\!] \\ \downarrow & & \downarrow \\ [\![L(X)]\!] & \longrightarrow & [\![K]\!] \ . \end{array}$$

The morphism $[\![A(X,Y)]\!] \longrightarrow [\![L(X)]\!]$ is in $\mathbf{R}$ and the map $[\![L(X)]\!] \longrightarrow [\![K]\!]$ is in $\mathbf{A}$: so their composition $[\![A(X,Y)]\!] \longrightarrow [\![K]\!]$ is in $\mathbf{A}$. It follows that the left adjoint is providing a best possible factorization of morphisms in $\mathbf{A}$ through morphisms in $\mathbf{R}$.

We next have to deal explicitly with substitution: the Beck-Chevalley condition does not come for free anymore (since the left adjoints are not given simply by composition). But it is a standard result that this condition amounts to requiring that the factorization be stable under pullback.

It is tempting to think that that is all, but it is not. What we have suceeded in modelling so far are the *Sum clauses* for Types over Orders from 1.9 with the following restrictions: in the *elimination* and *equality clauses* the B that appears *does not depend on* $z\in\sum X\in K.A$. These are essentially the rules given in [MiPl], which are equivalent to the sum rules in Girard's original version of higher order typed lambda calculus—the system F_ω of [G1]. (We say "essentially" because neither Mitchell and Plotkin, nor Girard give a "second" equality rule.) What then is the added nuance in the Martin-Löf style rule we give in 1.9? It amounts to a condition on the *other* morphism in the factorization. If the morphism $f:L\longrightarrow K$ in **A** has best factorization

$$L \xrightarrow{\ s\ } \textstyle\sum X\in L.1_T \xrightarrow{\ r\ } K$$

with r in **R**, then s is *orthogonal* to **R** in the familiar categorical sense: that is, for any $g:B\longrightarrow A$ in **R** and commutative square

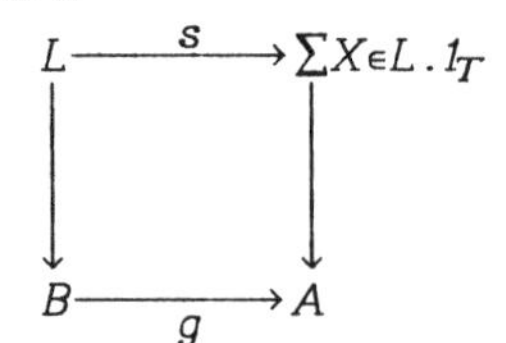

in **B**, there is a unique morphism $\sum X\in L.1_T\longrightarrow B$ making the diagram

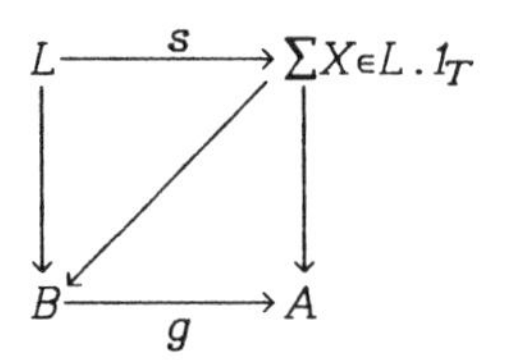

commute. It is easy to see that *any* factorization of f as a morphism orthogonal to **R** followed by a morphism in **R** is necessarily the *universal* factorization of f through a morphism in **R**. Thus the categorical version of the rule for sums of Types over Orders is as follows:

> **Big Sums.** *Any morphism f in* **A** *factors as $f=r\circ s$, where r is in* **R** *and s is orthogonal to* **R**. *Moreover this factorization is stable under pullback along arbitrary morphisms in* **B**.

The orthogonality condition in **Big Sums** remains slightly mysterious. The rest of the condition can be understood as providing a reflection of the fibration **A**$\longrightarrow$**B** into the firbration **R**$\longrightarrow$**B**: that is, a cartesian left adjoint to the inclusion **R**$\hookrightarrow$**A**. In the models of sections 4 and 5 we get the orthogonality condition for free, since for these there is a finite limit preserving inclusion from **B** into a category **G** having all pullbacks (so that $cod:\mathbf{G^2}\longrightarrow\mathbf{G}$ is a fibration) and the reflection of **A** into **R** is the restriction to **A** of a reflection from $\mathbf{G^2}\longrightarrow\mathbf{G}$ into some full subfibration—for which situation orthogonality is automatic.

Turning finally to the case of products of Types indexed over Orders, in view of Proposition 2.8 it is comparatively easy to explain what is needed to interpret the relevant clauses in 1.9. We already have an interpretation of indexed products of Orders over

Orders and since we are interpreting Types as special kinds of Order, the case of products of Types over Orders is necessarily a restriction of this:

> **Big Products.** *For all $f:J\longrightarrow I$ in* **A**, *the right adjoint $f_{\#}:\mathbf{A}(J)\longrightarrow\mathbf{A}(I)$ to $f^{\#}$ restricts to a functor $f_{\#}:\mathbf{R}(J)\longrightarrow\mathbf{R}(I)$.*

Note that this condition directly entails the condition **Products'** of 2.10.

2.13. Summary. We have now dealt with the categorical interpretation of all parts of the syntax of the basic theory of constructions presented in section 1. Let us summarize what we require (at the same time eliminating some of the redundancies we have noted):

(*i*) A category **B** with a terminal object.

(*ii*) Distinguished collections **A** and **R** of morphisms in **B**, with **R** contained in **A**.

(*iii*) The pullback along an arbitrary morphism in **B** of a morphism in **A** (respectively **R**) exists and is again in **A** (respectively **R**).

(*iv*) **R** contains all the isomorphisms in **B**.

(*v*) **R** and **A** are both closed under composition.

(*vi*) For each $f:J\longrightarrow I$ in **A**, the pullback functor $f^{\#}:\mathbf{A}(I)\longrightarrow\mathbf{A}(J)$ has a right adjoint, $f_{\#}:\mathbf{A}(J)\longrightarrow\mathbf{A}(I)$. Moreover, these right adjoints satisfy the Beck-Chevalley condition for pullbacks of f along arbitrary morphisms in **B**; and they send morphisms in $\mathbf{R}(J)$ to morphisms in $\mathbf{R}(I)$.

(*vii*) Every morphism in **A** factors as the composition of a morphism orthogonal to the class **R** followed by a morphism in **R**. Moreover, this factorization is preserved under pullback along arbitrary morphisms in **B**.

(*viii*) There is a morphism $G\longrightarrow U$ in **R** from which all other morphisms in **R** can be obtained by pullback.

(*ix*) For each object B in **B**, the unique morphism from B to the terminal object is in **A**.

The equivalence between **MLTT** and **RCCC** mentioned above extends to one between theories over the basic theory of constructions (as in 1.10) and an appropriate 2-category of categorical structures satisfying (*i*) to (*ix*). It is worth pointing out that the only *structure* required to specify a model is essentially only that of a category **B** and two distinguished classes of morphisms **A** and **R** in **B**—all the other requirements amount to categorical properties of this structure. Thus if (**B**,**A**,**R**) is a model and $I:\mathbf{B}\simeq\mathbf{B}'$ is an equivalence of categories, then (**B'**,**A'**,**R'**) is also a model if we define **A'** to consist of morphisms isomorphic (in $\mathbf{B}^2$) to ones in the image of **A** under I and similarly for **R'**.

We conclude this section by considering the interpretation of the two extensions of the basic theory of constructions given in 1.11 and 1.12.

2.14. Models with "*Order* ∈ *ORDER*". The categorical explanation of the theory "*Order* ∈ *ORDER*" is similar to that given in 2.11 for the modelling of "*Type*" in the basic theory. We have to have an object V in **B** to interpret *Order* and a morphism $h:H\longrightarrow V$ in **A** to interpret $O(X)$ [$X\in$ *Order*]. Then for any $k:K\longrightarrow I$ in **A** (interpreting

$K(\xi)$ $[\xi \in I]$ say) there is a morphism $\ulcorner K \urcorner : I \longrightarrow V$ (interpreting $T_K(\xi)$) so that $(\ulcorner K \urcorner)^{\#}(h) \cong k$ in $\mathbf{A}(I)$ (the isomorphism interpreting $I_K(\xi)$ and its inverse interpreting $J_K(\xi)$). The condition on $\mathbf{A}$ is therefore:

> ***Order* $\in$ *ORDER*.** *There is a morphism $H \longrightarrow V$ in $\mathbf{A}$ from which all other morphisms in $\mathbf{A}$ can be obtained by pullback.*

2.15. Models with "*Type* $\simeq$ *ORDER*". Turning to the theory "*Type* $\simeq$ *ORDER*", our analysis of the interpretation of sum rules in this section shows that if we strengthen the rules for sums of Types indexed over Orders as indicated in 1.12, then in the categorical models we must have:

> when $f : J \longrightarrow I$ is in $\mathbf{A}$, the left adjoint $f_! : \mathbf{R}(J) \longrightarrow \mathbf{R}(I)$ to $f^{\#}$ (whose existence is guarenteed by 2.13(vii)) is given by composition with f; in other words, $f \circ r$ is in $\mathbf{R}$ whenever f is in $\mathbf{A}$ and r is in $\mathbf{R}$.

But taking $r = id$, we get that $\mathbf{A}$ is contained in $\mathbf{R}$ and therefore that $\mathbf{A}$ *and* $\mathbf{R}$ *are equal*. This assumption renders some of the assumptions in 2.13 redundant, and we arrive at the following requirements for a categorical model of the theory of constructions with "*Type* $\simeq$ *ORDER*":

(i) A category $\mathbf{B}$ with a terminal object.

(ii) A distinguished collection $\mathbf{A}$ of morphisms in $\mathbf{B}$.

(iii) The pullback along an arbitrary morphism in $\mathbf{B}$ of a morphism in $\mathbf{A}$ exists and is again in $\mathbf{A}$.

(iv) $\mathbf{A}$ is closed under composition.

(v) For each $f : J \longrightarrow I$ in $\mathbf{A}$, the pullback functor $f^{\#} : \mathbf{A}(I) \longrightarrow \mathbf{A}(J)$ has a right adjoint, $f_{\#} : \mathbf{A}(J) \longrightarrow \mathbf{A}(I)$. Moreover, these right adjoints satisfy the Beck-Chevalley condition for pullbacks of f along arbitrary morphisms in $\mathbf{B}$.

(vi) There is a morphism $G \longrightarrow U$ in $\mathbf{A}$ from which all other morphisms in $\mathbf{A}$ can be obtained by pullback.

(vii) For each object B in $\mathbf{B}$, the unique morphism from B to the terminal object is in $\mathbf{A}$.

3 Lim theories

In this section we will review those parts of (more traditional) categorical logic which we will need in the next two sections to present our topos-theoretic models of the theory of constructions. In specifying these models along the general lines indicated in the previous section, we will use several constructions on *categories with finite limits*—or *lex categories* as we will call them. In order to see that these constructions can be carried out, we need a way of presenting lex categories in terms of "generators and relations". There are a number of ways in which this can be done. For example, one could use *finite projective sketches* : see [BW, 4.4]. Alternatively one could use theories over a fragment of Martin-Löf type theory, such as the *generalized algebraic theories* of [Ca] (see also [Po]). Instead we choose to use the *lim theories* of M.Coste, since their syntax and semantics are part of the familiar formalism of the first order predicate calculus. The following account stresses model-theoretic aspects; for a fuller picture, see the original work of M.Coste [Co1, Co2] on the subject.

For the purposes of this section, a *language* **L** is specified by:

- A collection of *sort symbols*, $S, S', S'', \ldots$
- A collection of *function symbols* of specified *types*. The type of such a function symbol f, is given by a non-empty list of sort symbols $S_1, \ldots, S_n, S$, and this will be indicated by writing $f : S_1 \times \cdots \times S_n \longrightarrow S$. (This includes the case $n=0$, when f is more usually called a *constant symbol* of type S.)
- A collection of *relation symbols* of specified *types*. The type of such a relation symbol R, is given by a (possibly empty) list of sort symbols $S_1, \ldots, S_n$, and this will be indicated by writing $R \rightarrowtail S_1 \times \cdots \times S_n$.

Starting with countably infinite sets of *variables* for each sort symbol, the *terms* of **L** and their associated *types* are defined recursively in the usual way:

- Each variable x of type S is a term of type S.
- If $t_1, \ldots, t_n$ are terms of type $S_1, \ldots, S_n$ and $f : S_1 \times \cdots \times S_n \longrightarrow S$ is a function symbol, then $f(t_1, \ldots, t_n)$ is a term of type S.

(We write $t{:}S$ to indicate that a term t has type S.) We next define the *basic formulas* over **L** recursively as follows:

- If t, t' are terms of the same type, then $t = t'$ is a basic formula.
- If $t_1, \ldots, t_n$ are terms of type $S_1, \ldots, S_n$ and $R \rightarrowtail S_1 \times \cdots \times S_n$ is a relation symbol, then $R(t_1, \ldots, t_n)$ is a basic formula.
- The symbol $\top$ ("true") is a basic formula.
- If ϕ and ψ are basic formulas, then so is $(\phi \wedge \psi)$.

Basic formulas of the first two kinds are usually called *atomic* formulas. Thus a basic formula is a finite (possibly empty) conjunction of atomic formulas.

We now define a *lim theory* **T** to be given by a language **L** and a set of *lim sentences* (the *axioms* of **T**): such a lim sentence is specified formally by two disjoint lists $\overline{x}, \overline{y}$ of distinct variables and two basic formulas ϕ, ψ, the variables involved in the first all appearing in the list $\overline{x}$ and the variables involved in the second appearing either in $\overline{x}$ or in $\overline{y}$. The lim sentence will be written as

$$(3.1) \qquad \forall \overline{x} \, (\phi(\overline{x}) \rightarrow \exists! \overline{y} \, \psi(\overline{x}, \overline{y}))$$

to indicate that its intended meaning is: "for all $\overline{x}$ such that $\phi(\overline{x})$ holds, there exist *unique* $\overline{y}$ so that $\psi(\overline{x}, \overline{y})$ holds". In the case that ψ does not depend on y, then (*3.1*) will be abbreviated to

$$\forall \overline{x} \, (\phi(\overline{x}) \rightarrow \psi(\overline{x}))$$

and if furthermore ϕ is $\top$, then (*3.1*) will be written as

$$\forall \overline{x}\, \psi(\overline{x}).$$

Clearly every (many-sorted) algebraic theory can be regarded as a lim theory. Here are two simple examples of non-algebraic lim theories:

3.1. Example: the lim theory **cat** of *categories*.

The underlying langage of **cat** has two sorts, *Ob* and *Mor*, three function symbols

$$dom : Mor \longrightarrow Ob$$
$$cod : Mor \longrightarrow Ob$$
$$id : Ob \longrightarrow Mor,$$

and one relation symbol

$$comp \rightarrowtail Mor \times Mor \times Mor,$$

whose intended meaning is the graph of the composition partial function on morphisms in a category. Thus the axioms of **cat** are:

- $\forall x{:}Ob(\, dom(id(x))=x \wedge cod(id(x))=x\,)$
- $\forall f,g,h{:}Mor\big(\, comp(f,g,h) \rightarrow (\, dom(f)=dom(h) \wedge cod(f)=dom(g) \wedge cod(g)=cod(h)\,)\big)$
- $\forall f,g{:}Mor(\, cod(f)=dom(g) \rightarrow \exists! h{:}Mor\ comp(f,g,h)\,)$
- $\forall f,g,h,k,l,m,n{:}Mor(\, comp(f,g,k) \wedge comp(g,h,l) \wedge comp(k,h,m) \wedge comp(f,l,n) \rightarrow m=n\,)$
- $\forall f{:}Mor(\, comp(f,id(cod(f)),f) \wedge comp(id(dom(f)),f,f)\,)$.

3.2. Example: the lim theory **lex** of *categories with finite limits*.

We axiomatize the property of having finite limits via those of having a terminal object and pullbacks. Thus **lex** is obtained from **cat** by adding a constant symbol T of type *Ob*, a new relation symbol $pb \rightarrowtail Mor \times Mor \times Mor \times Mor$ and new axioms:

- $\forall x{:}Ob\, \exists! f{:}Mor\, (\, dom(f)=x \wedge cod(f)=\mathrm{T}\,)$
- $\forall f,g,h,k{:}Mor\big(\, pb(f,g,h,k) \rightarrow \exists! l{:}Mor(comp(f,g,l) \wedge comp(h,k,l))\big)$
- $\forall g,k{:}Mor(cod(g)=cod(k) \rightarrow \exists! f,h{:}Mor\ pb(f,g,h,k)\,)$
- $\forall f,f',g,h,h',k,l{:}Mor\big(pb(f,g,h,k) \wedge comp(f',g,l) \wedge comp(h',k,l) \rightarrow$
$\exists! m{:}Mor(comp(m,f,f') \wedge comp(m,h,h'))\big)$.

3.3. Structures and satisfaction. Given a set-valued *structure* for a language **L** (i.e. an assignment of sets for the sorts, functions for the function symbols and relations for the relation symbols, all satisfying the evident typing requirements), the intended meaning of the lim sentence (*3.1*) mentioned above amounts to an informal definition of the notion of *satisfaction* of a lim sentence in a structure. In giving the formal definitions of "structure for a language" and "satisfaction of a lim sentence by a structure" one only needs to use set theoretic operations of a very simple kind: in fact formation of finite limits in the category of sets and functions is all that is needed. Accordingly these definitions can be given for an arbitrary category with finite limits and as such, are a fragment of the categorical interpretation of first order logic given by Makkai and Reyes in [MR, Chapter 2, section 3]. We recall from there the following definitions:

If **L** is a language and **C** is a category with finite limits, then an **L**-*structure* M in **C** assigns

- to each sort symbol S, an object $M(S)$ in **C**,
- to each function symbol $f{:}S_1\times\cdots S_n \longrightarrow S$, a morphism $M(f){:}M(S_1)\times\cdots\times M(S_n)\longrightarrow M(S)$ in **C**,
- to each relation symbol $R\rightarrowtail S_1\times\cdots\times S_n$, a subobject $M(R)\rightarrowtail M(S_1)\times\cdots\times M(S_n)$ in **C**.

If $\overline{x}=x_1,\ldots,x_n$ is a finite list of distinct variables with x_i of type S_i say, and if t is a **L**-term whose variables all occur in $\overline{x}$, then a morphism $[\![t(\overline{x})]\!]{:}M(S_1)\times\cdots\times M(S_n)\longrightarrow M(S)$ in **C** is defined by structural induction, as follows:

- If t is x_i, then $[\![t(\overline{x})]\!]=\pi_i$, the ith product projection.
- If t is $f(t_1,\ldots,t_m)$, then $[\![t(\overline{x})]\!]=M(f)\circ\langle[\![t_1(\overline{x})]\!],\ldots,[\![t_m(\overline{x})]\!]\rangle$ (where $\langle[\![t_1(\overline{x})]\!],\ldots,[\![t_m(\overline{x})]\!]\rangle$ is the unique morphism whose composition with each π_j is $[\![t_j(\overline{x})]\!]$).

Similarly, if ϕ is a basic **L**-formula involving variables from the list $\overline{x}$, then a subobject $[\![\phi(\overline{x})]\!]\rightarrowtail M(S_1)\times\cdots\times M(S_n)$ in **C** is defined by structural induction, as follows:

- If ϕ is $t=t'$, then $[\![\phi(\overline{x})]\!]$ is the equalizer of the pair of morphisms $[\![t(\overline{x})]\!],[\![t'(\overline{x})]\!]$.
- If ϕ is $R(t_1,\ldots,t_m)$, then $[\![\phi(\overline{x})]\!]$ is the pullback of the subobject $M(R)$ along the morphism $\langle[\![t_1(\overline{x})]\!],\ldots,[\![t_m(\overline{x})]\!]\rangle$.
- If ϕ is $\top$, then $[\![\phi(\overline{x})]\!]$ is the *greatest* subobject of $M(S_1)\times\cdots\times M(S_n)$, i.e. that given by the identity morphism for $M(S_1)\times\cdots\times M(S_n)$.
- If ϕ is $\psi\wedge\theta$, then $[\![\phi(\overline{x})]\!]$ is the *meet* of the subobjects $[\![\psi(x)]\!]$ and $[\![\theta(x)]\!]$, i.e. is given by forming a pullback from monomorphisms representing these subobjects.

Now given a lim sentence as in (*3.1*), define the objects X,Y of **C** to be $M(S_1)\times\cdots\times M(S_n)$ and $M(S'_1)\times\cdots\times M(S'_m)$ repectively (where S_i is the type of x_i and S'_j the type of y_j); suppose also that the subobject $[\![\psi(\overline{x},\overline{y})]\!]$ is represented by the monomorphism $\langle a,b\rangle{:}[\![\psi(\overline{x},\overline{y})]\!]\rightarrowtail X\times Y$. Then we say that the **L**-structure M *satisfies* the sentence (*3.1*), and write

$$M\vDash\forall\overline{x}(\phi(\overline{x})\rightarrow\exists!\overline{y}\,\psi(\overline{x},\overline{y})),$$

if on forming the pullback square

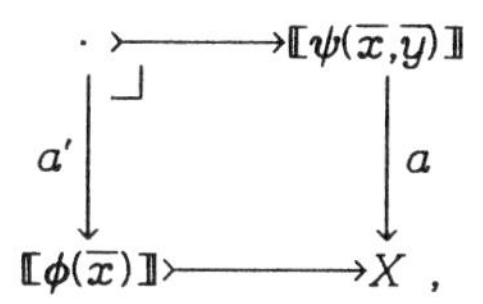

$$\begin{array}{ccc} \cdot & \rightarrowtail & [\![\psi(\overline{x},\overline{y})]\!] \\ {\scriptstyle a'}\downarrow & \lrcorner & \downarrow{\scriptstyle a} \\ [\![\phi(\overline{x})]\!] & \rightarrowtail & X \end{array},$$

in **C**, the morphism a' is an isomorphism.

3.4. Definition. If **T** is a lim theory with underlying language **L**, a *model* of **T** in a category **C** with finite limits is an **L**-structure in **C** which satisfies all the axioms of **T**. Let **T(C)** denote the category whose objects are such models and whose morphisms are *homomorphisms* of **L**-structures in **C**. Such a homomorphism $h{:}M\longrightarrow N$ is specified by a family $h_S:M(S)\longrightarrow N(S)$ of morphisms in **C** indexed by the sort symbols S of **L** such that: for each function symbol $f:S_1\times\cdots\times S_n\longrightarrow S$ in **L** the square

$$\begin{array}{ccc} M(S_1)\times\cdots\times M(S_n) & \xrightarrow{M(f)} & M(S) \\ {\scriptstyle h_{S_1}\times\cdots\times h_{S_n}}\downarrow & & \downarrow{\scriptstyle h_S} \\ N(S_1)\times\cdots\times N(S_n) & \xrightarrow[N(f)]{} & N(S) \end{array}$$

in **C** commutes; and for each relation symbol $R\rightarrowtail S_1\times\cdots\times S_n$ in **L** there is a commutative square in **C** of the form

$$\begin{array}{ccc} M(R) & \rightarrowtail & M(S_1)\times\cdots\times M(S_n) \\ \downarrow & & \downarrow{\scriptstyle h_{S_1}\times\cdots\times h_{S_n}} \\ N(R) & \rightarrowtail & N(S_1)\times\cdots\times N(S_n)\ . \end{array}$$

Composition and identities in **T**(**C**) are given componentwise from those in **C**.

Let us apply this definition to the two examples of lim theories given above. In the case of 3.1 and when **C** = **Set** the category of sets, evidently **cat(Set)** is just the category **Cat** of small categories and functors; and in general, **cat(C)** is the category of *internal categories and functors* in **C**, as defined in [J1, Chapter 2] for example. In the case of 3.2 the situation is more subtle. The objects of **lex(Set)** are small categories equipped with operations specifying terminal object and pullbacks; and the morphisms are functors which exactly preserve these operations. As is well known, the operation of taking a finite limit of any particular shape can be given as a derived operation from terminal object and pullbacks: thus we will refer to the objects of **lex(Set)** simply as *small categories with finite limits*, or *small lex categories*. But whilst a morphism in **lex(Set)** is necessarily a *lex functor* (that is, one which *preserves finite limits* in the usual sense of sending finite limit cones to finite limit cones), the converse is not generally the case since a lex functor need not preserve the given operations for terminal object and pullbacks up to equality. We will call the morphisms in **lex(Set)** *strict lex functors*.

3.5. Classifying categories. Consideration of the models of a lim theory **T** in *all* lex categories rather than just in the category of sets opens up the possibility of constructing a most general, or *generic* model of **T**. First note that models of **T** in different lex categories can be compared by transporting them along lex functors: given lex categories **C**, **D** and a model M of **T** in **C**, then for any lex functor F:**C**$\longrightarrow$**D** there is a model $F(M)$ of **T** in **D**, obtained by applying F to the objects, morphisms and subobjects which comprise the structure M. (That $F(M)$ is a structure follows just from the fact that F preserves finite products and monomorphisms; to see that it is a **T**-model one needs that the satisfaction of lim sentences is preserved by F and this follows precisely from the preservation of finite limits.) Similarly, if $\phi:F\longrightarrow G$ is a natural transformation between lex functors, we get a homomorphism $\phi(M):F(M)\longrightarrow G(M)$ of **T**–models in **D** whose component at a sort symbol S is $\phi_{M(S)}$. In this way one obtains a functor

$$(-)(M) : \mathbf{LEX}(\mathbf{C},\mathbf{D}) \longrightarrow \mathbf{T}(\mathbf{D})$$

from the category of lex functors and natural transformations from $\mathbf{C}$ to $\mathbf{D}$ into the category of $\mathbf{T}$-models in $\mathbf{D}$. We can now state the fundamental result linking lim theories and lex categories:

Theorem. *For each lim theory* $\mathbf{T}$ *there is a small lex category* $\mathbf{C_T}$ *(called the* classifying category *of* $\mathbf{T}$*) and a model* $G_\mathbf{T}$ *of* $\mathbf{T}$ *in* $\mathbf{C_T}$ *(called the* generic model *of* $\mathbf{T}$*) such that for any lex category* $\mathbf{D}$ *the functor* $(-)(G_\mathbf{T}):\mathbf{LEX}(\mathbf{C_T},\mathbf{D}) \longrightarrow \mathbf{T}(\mathbf{D})$ *is an equivalence of categories.*

□

Note that the above property of $\mathbf{C_T}$ and $G_\mathbf{T}$ uniquely determines the former up to equivalence of categories and the latter up to isomorphism of $\mathbf{T}$-models. There are at least two ways of constructing the classifying category. The more elementary way is in terms of the syntax of $\mathbf{T}$: see [Co1, 2.3]. The second way is model theoretic and depends upon the fact that $\mathbf{T}(\mathbf{Set})$ is a (typical) *locally finitely presentable* category in the sense of Gabriel and Ulmer [GU] (see also [MP]). Thus the full subcategory $\mathbf{T}(\mathbf{Set})_{fp} \hookrightarrow \mathbf{T}(\mathbf{Set})$ of finitely presentable set-valued $\mathbf{T}$-models is equivalent to a small category and its opposite category is equivalent to the classifying category:

$$\mathbf{C_T} \simeq (\mathbf{T}(\mathbf{Set})_{fp})^{op}.$$

Any small lex category $\mathbf{D}$ can be presented, up to equivalence, as the classifying category of some lim theory. First define the *internal language* $\mathbf{L}$ of $\mathbf{D}$ to have a sort symbol $\ulcorner X \urcorner$ for each object X of $\mathbf{D}$, a function symbol $\ulcorner f \urcorner : \ulcorner X_1 \urcorner \times \cdots \times \ulcorner X_n \urcorner \longrightarrow \ulcorner X \urcorner$ for each morphism $f:X_1 \times \cdots \times X_n \longrightarrow X$ and a relation symbol $\ulcorner R \urcorner : \rightarrowtail \ulcorner X_1 \urcorner \times \cdots \times \ulcorner X_n \urcorner$ for each subobject $R:\rightarrowtail X_1 \times \cdots \times X_n$. There is an evident $\mathbf{L}$-structure M in $\mathbf{D}$ given by erasing "$\ulcorner$ $\urcorner$". Let $\mathbf{T}$ be the lim theory with underlying language $\mathbf{L}$ and whose axioms are all those lim sentences of $\mathbf{L}$ which are satisfied by M. Then M is by definition a $\mathbf{T}$-model in $\mathbf{D}$; hence by the universal property of the classifying category of $\mathbf{T}$, there is a lex functor $F:\mathbf{C_T} \longrightarrow \mathbf{D}$ and an isomorphism $M \cong F(G_\mathbf{T})$ in $\mathbf{T}(\mathbf{D})$. Finally one can prove that the functor F is necessarily an equivalence, so that $\mathbf{D} \simeq \mathbf{C_T}$, as required.

The construction of classifying categories of lim theories provides a useful way of constructing lex categories with specified properties. We illustrate this with two examples which we will need later:

3.6.Example: *copower* of a lex category by a category.

Let **Cat** denote the 2-category of small categories, functors and natural transformations; and let **Lex** denote the 2-category of small lex categories, lex functors and natural transformations. Given $\mathbf{C}$ in **Cat** and $\mathbf{D}$ in **Lex**, we wish to construct a small lex category $\mathbf{C} \cdot \mathbf{D}$, called the *copower* of $\mathbf{D}$ by the category $\mathbf{C}$, with the property that there is a natural equivalence:

$$\mathbf{Cat}(\mathbf{C},\mathbf{Lex}(\mathbf{D}, -)) \simeq \mathbf{Lex}(\mathbf{C} \cdot \mathbf{D}, -).$$

This amounts to giving a functor $(-)\cdot(-):\mathbf{C}\times\mathbf{D}\longrightarrow\mathbf{C}\cdot\mathbf{D}$ with the properties:

(i) $(-)\cdot(-)$ is *lex in its second variable*, i.e. for each $X\in\mathbf{C}$, $X\cdot(-):\mathbf{D}\longrightarrow\mathbf{C}\cdot\mathbf{D}$ is a lex functor;

(ii) if $B(-,-):\mathbf{C}\times\mathbf{D}\longrightarrow\mathbf{E}$ is any functor into a lex category which is lex in its second variable, then there is a lex functor $\overline{B}:\mathbf{C}\cdot\mathbf{D}\longrightarrow\mathbf{E}$, unique up to unique isomorphism, and a natural isomorphism $\overline{B}((-)\cdot(-))\cong B(-,-)$.

Therefore to construct $\mathbf{C}\cdot\mathbf{D}$ we define a language $\mathbf{L}$ as follows:

For each $X\in\mathbf{C}$ and $Y\in\mathbf{D}$ take a sort symbol $X\cdot Y$. For each $f:X\longrightarrow X'$ in $\mathbf{C}$ and $g:Y\longrightarrow Y'$ in $\mathbf{D}$ take a function symbol $f\cdot g:X\cdot Y\longrightarrow X'\cdot Y'$. There are no relation symbols.

And over this language we take the lim theory $\mathbf{T}$ with the following axioms:

(i) For each $X\in\mathbf{C}$ and $Y\in\mathbf{D}$, the axiom $\forall z:X\cdot Y(\ id_X\cdot id_Y(z)=z\ )$.

(ii) For $f:X\longrightarrow X'$, $f':X'\longrightarrow X''$ in $\mathbf{C}$ and $g:Y\longrightarrow Y'$, $g':Y'\longrightarrow Y''$ in $\mathbf{D}$, the axiom $\forall z:X\cdot Y(\ f'f\cdot g'g(z)=f'\cdot g'(f\cdot g(z))\ )$.

(iii) For each $X\in\mathbf{C}$, the axiom $\exists!z:X\cdot 1(\ z=z\ )$ (where 1 denotes the terminal object in $\mathbf{D}$).

(iv) For each $X\in\mathbf{C}$ and each pullback square

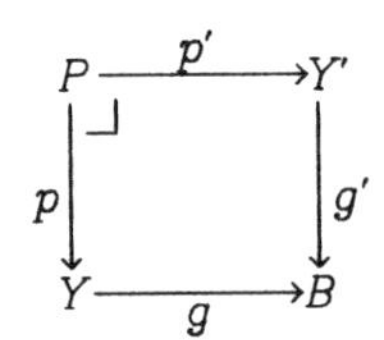

in $\mathbf{D}$, the axiom

$$\forall z:X\cdot Y, z':X\cdot Y'\big(id_X\cdot g(z)=id_X\cdot g'(z')\rightarrow\exists!u:X\cdot P(id_X\cdot p(u)=z\wedge id_X\cdot p'(u)=z')\big).$$

Clearly an $\mathbf{L}$-structure in a lex category $\mathbf{E}$ satisfying the lim sentences of types (i) and (ii) is precisely a functor $\mathbf{C}\times\mathbf{D}\longrightarrow\mathbf{E}$; and this functor is lex in its second variable if and only if the structure satisfies the lim sentences of types (iii) ("preserves terminal object in its second variable") and (iv)("preserves pullbacks in its second variable"). Note also that a homomorphism of $\mathbf{L}$-structures is precisely a natural transformation between the corresponding functors; and the functor transporting $\mathbf{T}$-models along a lex functor $F:\mathbf{E}\longrightarrow\mathbf{E}'$ is given by composing the corresponding functors with F. Thus $\mathbf{T}(-)\cong\mathbf{Cat}(\mathbf{C},\mathbf{Lex}(\mathbf{D},-))$ and hence the copower $\mathbf{C}\cdot\mathbf{D}$ in $\mathbf{Lex}$ is given by the classifying category $\mathbf{C}_{\mathbf{T}}$.

3.7. Example: *oplax colimits* of lex categories.

Let $\mathbf{C}$ be a small category. A *pseudofunctor* $\mathbf{D}:\mathbf{C}^{op}\longrightarrow\mathbf{Lex}$ is specified by the following information:

- for each object U of $\mathbf{C}$, a small lex category $\mathbf{D}(U)$,
- for each morphism $\alpha:U\longrightarrow V$ in $\mathbf{C}$, a lex functor $\alpha^*:\mathbf{D}(V)\longrightarrow\mathbf{D}(U)$,
- for each $U\in\mathbf{C}$, a natural isomorphism $\iota_U:id_{\mathbf{D}(U)}\cong(id_U)^*$,
- for each composable pair of morphisms $\alpha:U\longrightarrow V,\beta:V\longrightarrow W$ in $\mathbf{C}$, a natural

isomorphism $\kappa_{\alpha,\beta}:\alpha^*\circ\beta^* \cong (\beta\circ\alpha)^*$,

satisfying the coherence conditions that the diagrams

$$\begin{array}{ccc} \alpha^*\beta^*\gamma^* & \xrightarrow{\alpha^*\kappa_{\beta,\gamma}} & \alpha^*(\gamma\beta)^* \\ {\scriptstyle \kappa_{\alpha,\beta}\gamma^*}\downarrow & & \downarrow{\scriptstyle \kappa_{\alpha,\gamma\beta}} \\ (\beta\alpha)^*\gamma^* & \xrightarrow[\kappa_{\beta\alpha,\gamma}]{} & (\gamma\beta\alpha)^* \end{array} \quad \text{and} \quad \begin{array}{ccccc} id_{\mathbf{D}(U)}\alpha^* & = & \alpha^* & = & \alpha^*\, id_{\mathbf{D}(V)} \\ {\scriptstyle \iota_U\alpha^*}\downarrow & & \downarrow{\scriptstyle id} & & \downarrow{\scriptstyle \alpha^*\iota_V} \\ (id_U)^*\alpha^* & \xrightarrow[\kappa_{id,\alpha}]{} & \alpha^* & \xleftarrow[\kappa_{\alpha,id}]{} & \alpha^*(id_V)^* \end{array}$$

commute.

If $\mathbf{E}\in\mathbf{Lex}$, then an *oplax cone M under* $\mathbf{D}:\mathbf{C}^{op}\longrightarrow\mathbf{Lex}$ *with vertex* $\mathbf{E}$ is specified by:

- for each $U\in\mathbf{C}$, a lex functor $M_U:\mathbf{D}(U)\longrightarrow\mathbf{E}$,
- for each $\alpha:U\longrightarrow V$ in $\mathbf{C}$, a natural transformation $M_\alpha:M_V\longrightarrow M_U\circ\alpha^*$,

satisfying the coherence conditions

$$M_{\beta\alpha}=(M_U\kappa_{\alpha,\beta})\circ(M_\alpha\beta^*)\circ M_\beta \text{ and } M_{id_U}=M_U\iota_U .$$

Then the *oplax colimit* of the pseudofunctor $\mathbf{D}:\mathbf{C}^{op}\longrightarrow\mathbf{Lex}$, denoted $\underrightarrow{oplax}_{\mathbf{C}^{op}}\mathbf{D}$, is the small lex category which is the vertex of a universal oplax cone I under $\mathbf{D}$—that is, I should have the property that for any other oplax cone M with vertex $\mathbf{E}$ there exists a lex functor $\overline{M}:\underrightarrow{oplax}_{\mathbf{C}^{op}}\mathbf{D}\longrightarrow\mathbf{E}$ unique up to unique isomorphism with the property that there are natural isomorphisms $\mu_U:\overline{M}\circ I_U\cong M_U$ $(U\in\mathbf{C})$ satisfying

$$M_\alpha\circ\mu_V=(\mu_U\alpha^*)\circ(\overline{M}I_\alpha) \qquad (\alpha:U\longrightarrow V \text{ in } \mathbf{C}).$$

One can construct the oplax colimit $\underrightarrow{oplax}_{\mathbf{C}^{op}}\mathbf{D}$ as the classifying category of a suitable lim theory $\mathbf{T}$. The underlying language of $\mathbf{T}$ has

- sort symbols $U\cdot X$ for each $U\in\mathbf{C}$ and $X\in\mathbf{D}(U)$
- function symbols $\alpha\cdot f:V\cdot Y\longrightarrow U\cdot X$ for each $\alpha:U\longrightarrow V$ in $\mathbf{C}$ and $f:\alpha^*Y\longrightarrow X$ in $\mathbf{D}(U)$

and no relation symbols. The axioms of $\mathbf{T}$ are:

(*i*) For each sort symbol $U\cdot X$, the axiom $\forall z{:}U\cdot X(\,id\cdot\iota^{-1}(z)=z)$.

(*ii*) For $\alpha:U\longrightarrow V$, $\beta:V\longrightarrow W$ in $\mathbf{C}$, $f:\alpha^*Y\longrightarrow X$ in $\mathbf{D}(V)$ and $g:\beta^*Z\longrightarrow Y$ in $\mathbf{D}(W)$, the axiom $\forall z{:}W\cdot Z(\beta\alpha\cdot(f\circ\alpha^*g\circ\kappa^{-1})(z)=\alpha\cdot f(\beta\cdot g(z)))$.

(*iii*) For each $U\in\mathbf{C}$ and pullback square

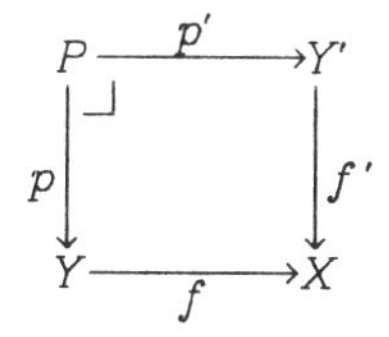

in $\mathbf{D}(U)$, the axiom

$$\forall z{:}U\cdot Y,z'{:}U\cdot Y'\big(id\cdot f\iota^{-1}(z)=id\cdot f'\iota^{-1}(z')\rightarrow \exists! u{:}U\cdot P(id\cdot p\iota^{-1}(u)=z\wedge id\cdot p'\iota^{-1}(u)=z')\big).$$

If G denotes the generic model of **T** in its classifying category $\underrightarrow{oplax}_{\mathbf{C}^{op}}\mathbf{D}$, then the colimiting oplax cone cone I is given by

$$I_U(X) = G(U\cdot X)\ ,\ I_U(f) = G(id_U\cdot f\iota^{-1}) \text{ and } (I_\alpha)_X = G(\alpha\cdot id_X).$$

The proof that this works is a similar, but more complicated version of the argument in 3.6. Indeed the copower $\mathbf{C}^{op}\cdot\mathbf{D}$ is a special case of the oplax colimit construction, since we can take $\mathbf{C}^{op}\cdot\mathbf{D} = \underrightarrow{oplax}_{\mathbf{C}^{op}}\mathbf{D}$ where $\mathbf{D}\in\mathbf{Lex}$ is regarded as the constant pseudofunctor $\mathbf{C}^{op}\longrightarrow\mathbf{Lex}$ with value $\mathbf{D}$.

We will see in 4.20(*ii*) that for one form of the model of the theory of constructions developed in section 4, the constant Orders K are denoted by small lex categories $[\![K]\!]\in\mathbf{Lex}$ and that more generally Orders L dependent on $X\in K$ are denoted by pseudofunctors $[\![L]\!]:[\![K]\!]^{op}\longrightarrow\mathbf{Lex}$. Then the lex category denoting the product $\prod X\in K.L$ is in fact obtained by taking the oplax colimit of $[\![L]\!]:[\![K]\!]^{op}\longrightarrow\mathbf{Lex}$; hence in particular the lex category denoting $K\to K'$ is given by the copower $[\![K]\!]^{op}\cdot[\![K']\!]$.

4 Algebraic toposes

In this section we will describe the first of our topos-theoretic models of the theory of constructions. For what follows, the basic reference is Johnstone's book [J1], especially chapters 2 and 4. We will denote the 2-category of Grothendieck toposes, geometric morphisms and natural transformations (between inverse image functors) by **GTOP**. Given a Grothendieck topos **E**, **GTOP(E)** will denote the (pseudo)slice 2-category whose objects are Grothendieck **E**-toposes, $f:\mathbf{F}\longrightarrow\mathbf{E}$, whose morphisms are triangles in **GTOP** commuting up to a given isomorphism and whose 2-cells are 2-cells in **GTOP** compatible with the given isomorphisms. In the case **E**=**Set**, since **Set** is terminal in **GTOP** we can identify **GTOP(Set)** with **GTOP**; however, even in the general case we will often confuse a Grothendieck **E**-topos with its domain topos **F** when the particular geometric morphism $f:\mathbf{F}\longrightarrow\mathbf{E}$ is clear from the context.

GTOP(E) is the same as the 2-category **BTOP/E** of [J1, Chapter 4]. Just as in that reference, we will loosely refer to certain contructions in **GTOP(E)** as (finite) *limits* even though they are actually *bilimits*, i.e. given by universal properties which involve equivalences rather than isomorphisms of hom-categories. Thus for example, we said in the previous paragraph that **Set** is terminal in **GTOP**, meaning that for any **E**∈**GTOP** the category of geometric morphisms **GTOP(E,Set)** is *equivalent* to **1** (the one object, one morphism category) rather than isomorphic to it.

If **E**∈**GTOP** and **C** is an internal category in **E**, then the **E**-*toposes of* (*internal*) *presheaves on* **C** will be denoted $[\mathbf{C}^{op},\mathbf{E}]$. Thus the objects of $[\mathbf{C}^{op},\mathbf{E}]$ are essentially the discrete fibrations over **C** in **E**: see [J1, 2.15]. Lifting the name from Johnstone's paper [J2], we make the following definition:

4.1. Definition. An **E**-topos $f:\mathbf{F}\longrightarrow\mathbf{E}$ is *algebraic* if it is equivalent in **GTOP(E)** to $[\mathbf{C}^{op},\mathbf{E}]$ for some internal lex category **C**, i.e. for some model in **E** of the lim theory **lex** of 3.2.

ATOP(**E**) will denote the full sub-2-category of **GTOP**(**E**) whose objects are the algebraic **E**-toposes. (And in case **E**=**Set**, we speak simply of *algebraic toposes* and write **ATOP** for the corresponding full sub-2-category of **GTOP**.)

4.2. Proposition. *Algebraic* **E**-*toposes are stable under change of base: if* f:**F**⟶**E** *is a geometric morphism between Grothendieck toposes and* **A**⟶**E** *is an algebraic* **E**-*topos, then on forming the pullback square*

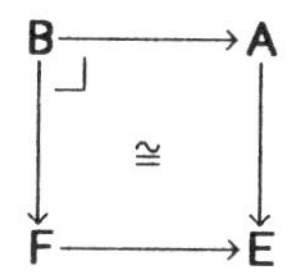

in **GTOP**, *it is the case that* **B**⟶**F** *is an algebraic* **F**-*topos.*

Proof. Suppose that $\mathbf{A}\simeq[\mathbf{C}^{op},\mathbf{E}]$ with $\mathbf{C}\in\mathbf{lex}(\mathbf{E})$. The inverse image functor $f^*:\mathbf{E}\longrightarrow\mathbf{F}$ certainly preserves finite limits and hence as in 3.5 applying it to **C** yields $f^*(\mathbf{C})$, a model of **lex** in **F**. But by [J1, Corollary 4.35] there is a pullback square in **GTOP** of the form

$$(4.1)\qquad \begin{array}{ccc} [(f^*\mathbf{C})^{op},\mathbf{F}] & \longrightarrow & [\mathbf{C}^{op},\mathbf{E}] \\ \big\downarrow & \cong & \big\downarrow \\ \mathbf{F} & \xrightarrow[f]{} & \mathbf{E}\ . \end{array}$$

Hence $\mathbf{B}\simeq[(f^*\mathbf{C})^{op},\mathbf{F}]$ is an algebraic **F**-topos.

□

4.3. Notation. If f:**F**⟶**E** is a geometric morphism between Grothendieck toposes, then

$$f^{\#}:\mathbf{GTOP}(\mathbf{E})\longrightarrow\mathbf{GTOP}(\mathbf{F})$$

will denote the operation of change of base, i.e. of pulling back **E**-toposes along f. (As we remarked above, pullbacks of toposes are strictly speaking "bipullbacks" and consequently $f^{\#}$ is a bicategory homomorphism—it preserves identities and composition only up to coherent isomorphism.) By the previous proposition, we can restrict $f^{\#}$ along the full inclusions $\mathbf{ATOP}(-)\hookrightarrow\mathbf{GTOP}(-)$ to get $f^{\#}:\mathbf{ATOP}(\mathbf{E})\longrightarrow\mathbf{ATOP}(\mathbf{F})$.

4.4. Classifying toposes. For any small lex category **C** and Grothendieck topos **E** there is a natural equivalence of categories:

$$(4.2)\qquad \mathbf{GTOP}(\mathbf{E},[\mathbf{C}^{op},\mathbf{Set}])\simeq\mathbf{LEX}(\mathbf{C},\mathbf{E})$$

where the right hand side denotes the category of lex functors and natural transformations from **C** to **E**. This equivalence is given in one direction by sending a geometric morphism $f:\mathbf{E}\longrightarrow[\mathbf{C}^{op},\mathbf{Set}]$ to the lex functor obtained by restricting f^* along the Yoneda embedding $\mathbf{C}\hookrightarrow[\mathbf{C}^{op},\mathbf{Set}]$; and it is given in the other direction by sending a lex fuctor $F:\mathbf{C}\longrightarrow\mathbf{E}$ to the geometric morphism whose inverse image part is the left Kan extension of F along the

Yoneda embedding; see [J1, Proposition 7.13].

If **T** is a lim theory with classifying lex category $\mathbf{C_T}$, we can combine (*4.2*) with the equivalence of 3.5 to obtain:

$$\mathbf{GTOP}(\mathbf{E},[\mathbf{C_T}^{op},\mathbf{Set}]) \simeq \mathbf{T}(\mathbf{E}).$$

Thus $[\mathbf{C_T}^{op},\mathbf{Set}]$ is the *classifying topos for the lim theory* **T**—meaning that the category of **E**-points of the topos is naturally equivalent to the category of **T**-models in **E**. As we noted in 3.5, any small lex category is the classifying category of some lim theory. Thus *the algebraic toposes are precisely the toposes which classify lim theories.*

Applying the Yoneda embedding $H:\mathbf{C_T} \hookrightarrow [\mathbf{C_T}^{op},\mathbf{Set}]$ to the generic model $G_\mathbf{T} \in \mathbf{T}(\mathbf{C_T})$, we obtain a **T**-model $U_\mathbf{T} = H(G_\mathbf{T})$ in the classifying topos *which is generic amongst models of* **T** *in Grothendieck toposes*. This means in particular that for any $M \in \mathbf{T}(\mathbf{E})$ there is a geometric morphism $m:\mathbf{E} \longrightarrow [\mathbf{C_T}^{op},\mathbf{Set}]$ with $m^*(U_\mathbf{T}) \cong M$ in $\mathbf{T}(\mathbf{E})$. As we mentioned in 3.5, $\mathbf{C_T}$ is equivalent to $(\mathbf{T}(\mathbf{Set})_{fp})^{op}$, the opposite of the full subcategory of $\mathbf{T}(\mathbf{Set})$ whose objects are the finitely presentable **T**-models. Using this fact, we can identify $U_\mathbf{T}$ concretely: since limits are calculated pointwise in functor categories

$$\mathbf{T}[\mathbf{C_T}^{op},\mathbf{Set}] \cong \mathbf{CAT}(\mathbf{C_T}^{op},\mathbf{T}(\mathbf{Set})) \simeq \mathbf{CAT}(\mathbf{T}(\mathbf{Set})_{fp},\mathbf{T}(\mathbf{Set}))$$

and under this equivalence, $U_\mathbf{T}$ corresponds to the inclusion $\mathbf{T}(\mathbf{Set})_{fp} \hookrightarrow \mathbf{T}(\mathbf{Set})$.

Lim theories and small lex categories correspond via the classifying category construction; but as we saw in 3.2, the latter are themselves the models of the particular lim theory **lex** and we can apply the above considerations to this theory:

4.5. Proposition. *There is a Grothendieck topos* Υ *and an algebraic* Υ*-topos* $\Sigma \longrightarrow \Upsilon$ *with the property that for any other Grothendieck topos* **E** *and any algebraic* **E***-topos* $\mathbf{A} \longrightarrow \mathbf{E}$ *there is a pullback square in* **GTOP** *of the form*:

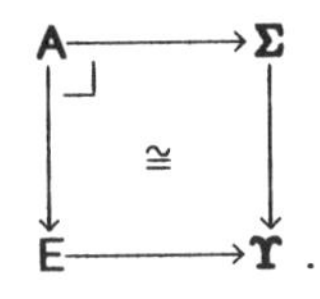

Moreover, Υ *is itself an algebraic topos.*

Proof. Define Υ to be $[\mathbf{C_{lex}}^{op},\mathbf{Set}]$, the classifying topos of the lim theory **lex**. Then Υ is certainly an algebraic topos. The generic model $U_\mathbf{lex}$ is an internal lex category in Υ: define Σ to be the algebraic Υ-topos $[U_\mathbf{lex}^{op},\Upsilon]$. If **E** is a Grothendieck topos and $\mathbf{A} \longrightarrow \mathbf{E}$ an algebraic **E**-topos, then $\mathbf{A} \simeq [\mathbf{C}^{op},\mathbf{E}]$ for some $\mathbf{C} \in \mathbf{lex}(\mathbf{E})$. Let $f:\mathbf{E} \longrightarrow \Upsilon$ be a geometric morphism classifying the model **C**, i.e. for which there is an isomorphism $\mathbf{C} \cong f^*(U_\mathbf{lex})$ in $\mathbf{lex}(\mathbf{E})$. Then by (*4.1*), $\mathbf{A} \simeq f^\#(\Sigma)$ in $\mathbf{GTOP}(\mathbf{E})$, as required.

□

We now turn our attention to the cartesian closed structure of algebraic toposes. Recall that the *exponential* in **GTOP(E)** of two **E**-toposes **F**,**G** if it exists is an **E**-topos, denoted $\mathbf{F}\to_{\mathbf{E}}\mathbf{G}$, together with a geometric morphism $ev{:}(\mathbf{F}\to_{\mathbf{E}}\mathbf{G})\times_{\mathbf{E}}\mathbf{F}\longrightarrow\mathbf{G}$ inducing equivalences

$$\mathbf{GTOP(E)}(-,\mathbf{F}\to_{\mathbf{E}}\mathbf{G}) \simeq \mathbf{GTOP(E)}((-)\times_{\mathbf{E}}\mathbf{F},\mathbf{G}).$$

$\mathbf{F}$ is called *exponentiable* if $\mathbf{F}\to_{\mathbf{E}}\mathbf{G}$ exists for all $\mathbf{G}$. Not every topos is exponentiable; we refer the interested reader to the paper of Johnstone and Joyal [JJ] for a detailed analysis of this property. Luckily for us, the situation for algebraic toposes is quite simple: [J1, Remark 7.49] implies that every algebraic topos is exponentiable. Moreover, as the next proposition shows, **ATOP** is closed in **GTOP** under exponentiation:

4.6. Proposition. *If* $\mathbf{C}$ *and* $\mathbf{D}$ *are small lex categories, then the exponential of* $[\mathbf{D}^{op},\mathbf{Set}]$ *by* $[\mathbf{C}^{op},\mathbf{Set}]$ *in* **GTOP** *is* $[(\mathbf{C}^{op}\cdot\mathbf{D})^{op},\mathbf{Set}]$, *where* $\mathbf{C}^{op}\cdot\mathbf{D}$ *is the copower in* **Lex** *of* $\mathbf{D}$ *by the category* $\mathbf{C}^{op}$ *(as defined in Example 3.6).*

Proof. Let $\mathbf{E}$ be a Grothendieck topos and let $\Delta{:}\,\mathbf{Set}\longrightarrow\mathbf{E}$ denote the constant-sheaf functor (the inverse image part of the (essentially) unique geometric morphism from $\mathbf{E}$ to **Set**). Then from (*4.1*), the product of $\mathbf{E}$ and $[\mathbf{C}^{op},\mathbf{Set}]$ in **GTOP** is $[(\Delta\mathbf{C})^{op},\mathbf{E}]$. An internal presheaf on the constant internal category $\Delta\mathbf{C}$ can be identified with an external presheaf valued in $\mathbf{E}$: thus $\mathbf{E}\times[\mathbf{C}^{op},\mathbf{Set}]$ is the functor category $\mathbf{CAT}(\mathbf{C}^{op},\mathbf{E})$. A geometric morphism out of $\mathbf{CAT}(\mathbf{C}^{op},\mathbf{E})$ is determined by its inverse image part—which is precisely a functor preserving finite limits and small colimits. Then since limits and colimits in such a functor category are calculated pointwise from $\mathbf{E}$, we have

$$\mathbf{GTOP}(\mathbf{E}\times[\mathbf{C}^{op},\mathbf{Set}],-) \simeq \mathbf{CAT}(\mathbf{C}^{op},\mathbf{GTOP}(\mathbf{E},-)). \tag{4.3}$$

Combining (*4.3*) with (*4.2*) and using the universal property of the copower $\mathbf{C}^{op}\cdot\mathbf{D}$ (which we note from its construction in Example 3.6, is valid for lex functors valued in any lex category and not just a small one), we have:

$$\begin{aligned}\mathbf{GTOP}(\mathbf{E}\times[\mathbf{C}^{op},\mathbf{Set}],[\mathbf{D}^{op},\mathbf{Set}]) &\simeq \mathbf{CAT}(\mathbf{C}^{op},\mathbf{GTOP}(\mathbf{E},[\mathbf{D}^{op},\mathbf{Set}]))\\ &\simeq \mathbf{CAT}(\mathbf{C}^{op},\mathbf{LEX}(\mathbf{D},\mathbf{E}))\\ &\simeq \mathbf{LEX}(\mathbf{C}^{op}\cdot\mathbf{D},\mathbf{E})\\ &\simeq \mathbf{GTOP}(\mathbf{E},[(\mathbf{C}^{op}\cdot\mathbf{D})^{op},\mathbf{Set}]).\end{aligned}$$

These equivalences are natural in $\mathbf{E}$ and show that $[(\mathbf{C}^{op}\cdot\mathbf{D})^{op},\mathbf{Set}]$ has the correct universal property to be the exponential $[\mathbf{C}^{op},\mathbf{Set}]\to[\mathbf{D}^{op},\mathbf{Set}]$.

□

The proof of Proposition 4.6 is constructive and in a form admitting *relativization* to the category theory of any topos $\mathbf{E}$ with natural number object (and in particular to any Grothendieck topos), where the internal categories of $\mathbf{E}$ play the role of small categories and categories fibred over $\mathbf{E}$ play the role of large ones. (See [B2] for a general discussion of category theory relative to a base other than **Set** and [PS] for a development of aspects of this theory using indexed categories (pseudofunctors) rather than fibrations.) We therefore have:

4.7. Proposition. *For any Grothendieck topos* $\mathbf{E}$, *the algebraic* $\mathbf{E}$-*toposes are exponentiable objects of* $\mathbf{GTOP(E)}$ *and* $\mathbf{ATOP(E)}$ *is closed in* $\mathbf{GTOP(E)}$ *under exponentiation: for* $\mathbf{C},\mathbf{D}\in\mathbf{lex(E)}$ *the exponential of* $[\mathbf{D}^{op},\mathbf{E}]$ *by* $[\mathbf{C}^{op},\mathbf{E}]$ *is* $[(\mathbf{C}^{op}\cdot\mathbf{D})^{op},\mathbf{E}]$ *where* $\mathbf{C}^{op}\cdot\mathbf{D}$ *is the copower of the internal lex category* $\mathbf{D}$ *by the internal category* $\mathbf{C}^{op}$.

□

4.8. Corollary. *Each* $\mathbf{ATOP(E)}$ *is a cartesian closed bicategory and finite products and exponentials are preserved by the inclusion* $\mathbf{ATOP(E)}\hookrightarrow\mathbf{GTOP(E)}$. *They are also preserved by the operation* $f^{\#}:\mathbf{ATOP(E)}\longrightarrow\mathbf{ATOP(F)}$ *of pullback along a geometric morphism* $f:\mathbf{F}\longrightarrow\mathbf{E}$.

Proof. In view of the previous proposition, for the first sentence of the corollary we just have to see that $\mathbf{ATOP(E)}$ is closed in $\mathbf{GTOP(E)}$ under finite products. Proposition 4.2 implies that $\mathbf{ATOP(E)}$ is closed in $\mathbf{GTOP(E)}$ under taking binary products; indeed by [J1, Corollary 4.36], we can take the product $[\mathbf{C}^{op},\mathbf{E}]\times_{\mathbf{E}}[\mathbf{D}^{op},\mathbf{E}]$ to be $[(\mathbf{C}\times\mathbf{D})^{op},\mathbf{E}]$. The terminal object of $\mathbf{GTOP(E)}$ is certainly algebraic, since $\mathbf{E}\simeq[\mathbf{1}^{op},\mathbf{E}]$ where $\mathbf{1}$ is the trivial internal lex category.

For the second sentence, we just have to show that $f^{\#}$ preserves exponentials, since clearly it preserves finite limits. Since the inclusions $\mathbf{ATOP}(-)\hookrightarrow\mathbf{GTOP}(-)$ preserve exponentials, it is sufficient to prove that $f^{\#}:\mathbf{GTOP(E)}\longrightarrow\mathbf{GTOP(F)}$ preserves any existing exponentials. This is so because $f^{\#}$ has a left adjoint $f_!:\mathbf{GTOP(F)}\longrightarrow\mathbf{GTOP(E)}$ ("compose with f") satisfying the condition of "Frobenius reciprocity": for $\mathbf{G}\in\mathbf{GTOP(E)}$ and $\mathbf{H}\in\mathbf{GTOP(F)}$, simple properties of pullbacks give that $f_!(\mathbf{H}\times_{\mathbf{F}}f^{\#}\mathbf{G})\simeq(f_!\mathbf{H})\times_{\mathbf{E}}\mathbf{G}$. Thus if the exponential $\mathbf{G}\to_{\mathbf{E}}\mathbf{G}'$ exists in $\mathbf{GTOP(E)}$, then

$$\begin{aligned}\mathbf{GTOP(F)}(-,f^{\#}(\mathbf{G}\to_{\mathbf{E}}\mathbf{G}')) &\simeq \mathbf{GTOP(E)}(f_!(-),\mathbf{G}\to_{\mathbf{E}}\mathbf{G}')\\ &\simeq \mathbf{GTOP(E)}(f_!(-)\times_{\mathbf{E}}\mathbf{G},\mathbf{G}')\\ &\simeq \mathbf{GTOP(E)}(f_!(-\times_{\mathbf{F}}f^{\#}\mathbf{G}),\mathbf{G}')\\ &\simeq \mathbf{GTOP(F)}(-\times_{\mathbf{F}}f^{\#}\mathbf{G},f^{\#}\mathbf{G}'),\end{aligned}$$

so that the exponential $(f^{\#}\mathbf{G})\to_{\mathbf{F}}(f^{\#}\mathbf{G}')$ exists in $\mathbf{GTOP(F)}$ and is given by $f^{\#}(\mathbf{G}\to_{\mathbf{E}}\mathbf{G}')$.

□

In the case that $f:\mathbf{F}\longrightarrow\mathbf{E}$ is algebraic, the left adjoint $f_!$ mentioned in the above proof restricts to give a left adjoint to $f^{\#}:\mathbf{ATOP(E)}\longrightarrow\mathbf{ATOP(F)}$, as we now show:

4.9. Proposition. *If* $f:\mathbf{F}\longrightarrow\mathbf{E}$ *is an algebraic* $\mathbf{E}$-*topos and* $g:\mathbf{G}\longrightarrow\mathbf{F}$ *is an algebraic* $\mathbf{F}$-*toposes, then* $gf:\mathbf{G}\longrightarrow\mathbf{E}$ *is an algebraic* $\mathbf{E}$-*topos.*

Proof. Consider first the special case when f is an equivalence. Then $\mathbf{G}\longrightarrow\mathbf{F}\simeq\mathbf{E}$ is the pullback of $\mathbf{G}\longrightarrow\mathbf{F}$ along the inverse equivalence $\mathbf{E}\simeq\mathbf{F}$; hence the composition is algebraic by Proposition 4.2.

Now in the general case, suppose that $\mathbf{F}\simeq[\mathbf{C}^{op},\mathbf{E}]$ with $\mathbf{C}\in\mathbf{lex(E)}$. Then by the previous paragraph, $\mathbf{G}\longrightarrow\mathbf{F}\simeq[\mathbf{C}^{op},\mathbf{E}]$ is an algebraic $[\mathbf{C}^{op},\mathbf{E}]$-topos; hence $\mathbf{G}\simeq[\mathbf{D}^{op},[\mathbf{C}^{op},\mathbf{E}]]$ for some $\mathbf{D}\in\mathbf{lex}([\mathbf{C}^{op},\mathbf{E}])$. By [J1, Exercise 2.7], $[\mathbf{D}^{op},[\mathbf{C}^{op},\mathbf{E}]]\simeq[(Gr\mathbf{D})^{op},\mathbf{E}]$ where $Gr\mathbf{D}\in\mathbf{cat(E)}$ is the result of applying the *Grothendieck construction* to $\mathbf{D}\in\mathbf{cat}([\mathbf{C}^{op},\mathbf{E}])$. In the case $\mathbf{E}=\mathbf{Set}$, this construction can be described as follows: identifying $\mathbf{D}$ with a functor $\mathbf{C}^{op}\longrightarrow\mathbf{Cat}$, then $Gr\mathbf{D}$ is the category whose objects are pairs (U,X) with $U\in\mathbf{C}$ and $X\in\mathbf{D}(U)$, and whose morphisms $(U,X)\longrightarrow(V,Y)$ are pairs (α,f) where $\alpha:U\longrightarrow V$ in $\mathbf{C}$ and $f:X\longrightarrow\mathbf{D}(\alpha)(Y)$ in $\mathbf{D}(U)$. For a general (Grothendieck) topos $\mathbf{E}$, the construction is just that obtained by translating the above recipe for $Gr\mathbf{D}$ into the internal logic of $\mathbf{E}$. It is straightforward to show that when $\mathbf{D}\in\mathbf{cat}[\mathbf{C}^{op},\mathbf{E}]$ is actually an internal *lex* category, then so is $Gr\mathbf{D}$. Therefore $\mathbf{G}\simeq[(Gr\mathbf{D})^{op},\mathbf{E}]$ is an algebraic $\mathbf{E}$-topos.

□

4.10. Corollary. *If $f:\mathbf{F}\longrightarrow\mathbf{E}$ is an algebraic $\mathbf{E}$-topos, then the left adjoint to $f^{\#}:\mathbf{GTOP(E)}\longrightarrow\mathbf{GTOP(F)}$, viz. the operation of composing with f, restricts to give a left adjoint to $f^{\#}:\mathbf{ATOP(E)}\longrightarrow\mathbf{ATOP(F)}$, denoted $f_!:\mathbf{ATOP(F)}\longrightarrow\mathbf{ATOP(E)}$.*

These left adjoints satisfy a Beck-Chevalley condition *with respect to pullback squares in* $\mathbf{GTOP}$: *if*

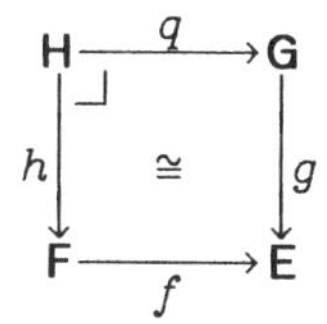

is a pullback with f (and hence also q) algebraic, then the canonical natural transformation $q_!\circ h^{\#}\longrightarrow g^{\#}\circ f_!$ is an equivalence.

Proof. The first paragraph is a consequence of Propositition 4.9; and the second follows from the fact that the Beck-Chevalley condition holds for the left adjoints to pulling back in $\mathbf{GTOP}$, due to the usual elementary properties of pullbacks with respect to composition.

□

So far, the results in this section have all been obtained by marshalling well known facts about presheaf toposes and classifying toposes. We are now going to show that the dual of Corollary 4.10 holds, namely that $f^{\#}:\mathbf{ATOP(E)}\longrightarrow\mathbf{ATOP(F)}$ possesses a right adjoint when $f:\mathbf{F}\longrightarrow\mathbf{E}$ is an algebraic $\mathbf{E}$-topos (and that these right adjoints satisfy the Beck-Chevalley condition). The method we employ is (the bicategorical version of) that in Proposition 2.6: we show that exponentiation by an algebraic topos preserves geometric morphisms which are algebraic, and then construct the right adjoints to pulling back using exponentials. To carry out this plan we have to delve a little more deeply into the structure of internal lex categories.

4.11. Definition. A lex functor $P:\mathbf{C}\longrightarrow\mathbf{B}$ between (small) lex categories is a *lex fibration over* $\mathbf{B}$ if it possesses a right adjoint $P_*:\mathbf{B}\longrightarrow\mathbf{C}$ and the counit of the adjunction, $\varepsilon:P\circ P_*\longrightarrow 1_{\mathbf{B}}$, is an isomorphism. (Note that P_* is necessarily a lex functor.)

If $Q:\mathbf{D}\longrightarrow\mathbf{B}$ is another lex fibration, then a morphism

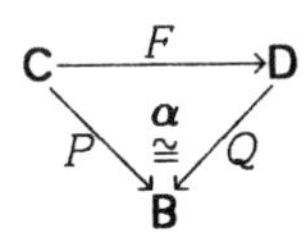

in **Lex/B** is *cartesian* if the natural transformation $F \circ P_* \longrightarrow Q_*$ (obtained from α by transposing across the adjunctions $P \dashv P_*$, $Q \dashv Q_*$) is an isomorphism.

4.12. Remark. A notion of (cloven) fibration can be given in any bicategory with finite (bi)limits: see Street [S]. When the bicategory is **Lex**, this notion reduces to the above, particularly simple one. (*Cf*. [RW1] and [RW2].) Because it is the *bicategorical* rather than 2-categorical notion of fibration, it is not the case that a lex fibration is the same thing as a cloven fibration of categories (in the classical sense [Gr], [B2]) all of whose fibres and pullback functors are lex. However the two concepts only differ up to equivalence in **Lex/B**. To see this, note that each lex fibration P:**C**⟶**B** determines a pseudofunctor **C**(-):**B**^{op}⟶**Lex** (*cf*. 3.7), where each **C**(U) ($U \in$ **B**) is the full subcategory of **C**/$P_*(U)$ whose objects are those $x:X \longrightarrow P_*(U)$ with $P(x)$ an isomorphism; and for $\alpha:U \longrightarrow V$ in **B**, α^*:**C**(V)⟶**C**(U) is given by pullback in **C** along $P_*(\alpha)$. Applying the Grothendick construction (*cf*. the proof of Proposition 4.9) to **C**(-):**B**^{op}⟶**Lex**, one obtains a cloven fibration in the classical sense which is equivalent over **B** to the original functor P.

It is not hard to show that the above assignment of pseudofunctors to lex fibrations extends to give an equivalence between the 2-category of lex fibrations over **B**, cartesian lex functors and natural transformations on the one hand and the 2-category of pseudofunctors **B**^{op}⟶**Lex**, pseudonatural transformations and modifications on the other. Note that if **D**:**B**^{op}⟶**Lex** is a pseudofunctor, then the corresponding lex fibration P:Gr(**D**)⟶**B** has P equal to the projection $(U,X) \longmapsto U$, with right adjoint P_* sending $U \in$ **B** to $(U,1)$ where $1 \in$ **D**(U) is the terminal object.

4.13. Notation. If **C** is a small lex category, we will denote by $\hat{\mathbf{C}}$ the algebraic topos [**C**^{op},**Set**]. The assignment **C** $\longmapsto \hat{\mathbf{C}}$ extends to a bicategory homomorphism $(\hat{-})$:**Lex**^{op}⟶**GTOP** via the natural equivalence of (*4.2*). In particular a lex morphism F:**C**⟶**D** determines a geometric morphism $\hat{F}:\hat{\mathbf{D}} \longrightarrow \hat{\mathbf{C}}$ whose inverse image functor is left Kan extension along F and whose direct image functor is precomposition with F.

4.14. Proposition. *Let* **B** *be a small lex category. A Grothendieck* $\hat{\mathbf{B}}$*-topos* **E**$\longrightarrow \hat{\mathbf{B}}$ *is algebraic iff it is equivalent to* $\hat{P}_*:\hat{\mathbf{C}} \longrightarrow \hat{\mathbf{B}}$ *for some lex fibration* P:**C**⟶**B**.

Proof. Suppose **E** is an algebraic $\hat{\mathbf{B}}$-topos—say **E** ≃ [**D**^{op},$\hat{\mathbf{B}}$] with **D** ∈ **lex**($\hat{\mathbf{B}}$). Then just as in the proof of Proposition 4.9, we can regard **D** as a functor **B**^{op}⟶**lex**(**Set**), hence as a functor **B**^{op}⟶**Lex**, apply the Grothendieck construction to it and obtain **E** ≃ [(Gr**D**)^{op},**Set**] in **GTOP**($\hat{\mathbf{B}}$). The geometric morphism which defines [(Gr**D**)^{op},**Set**] as a topos over [**B**^{op},**Set**] is just that whose inverse image functor is precomposition with the projection P:Gr(**D**)⟶**B**; but as we noted above, P is indeed a lex fibration and (-)·P:[**B**^{op},**Set**]⟶[(Gr**D**)^{op},**Set**] is naturally isomorphic to the functor given by left Kan extension along the right adjoint P_*, i.e. to the inverse image part of the geometric morphism $\hat{P_*}$.

Conversely, given a lex fibration $P:\mathbf{C}\longrightarrow\mathbf{B}$, to show that $\hat{\mathbf{C}}$ is an algebraic $\hat{\mathbf{B}}$-topos we have to find a functor $\mathbf{D}:\mathbf{B}^{op}\longrightarrow\mathbf{lex(Set)}$ with $Gr(\mathbf{D})\longrightarrow\mathbf{B}$ equivalent to $P:\mathbf{C}\longrightarrow\mathbf{B}$ in $\mathbf{Lex/B}$. Using the correspondence between lex fibrations over $\mathbf{B}$ and pseudofunctors $\mathbf{B}^{op}\longrightarrow\mathbf{Lex}$ remarked upon in 4.12, this amounts to proving:

4.15. Lemma. *Every pseudofunctor* $\mathbf{B}^{op}\longrightarrow\mathbf{Lex}$ *is pseudonaturally equivalent to a* functor $\mathbf{B}^{op}\longrightarrow\mathbf{Lex}$ *whose value at any morphism of* $\mathbf{B}$ *is a* strict *lex functor.*

Proof. The lemma can be viewed as a corollary of recent work of G.M.Kelly and J.Power showing how to turn "pseudo" structures in 2-categories into equivalent "strict" structures. Here we shall give a direct proof of this particular result.

With notation as in Remark 4.12, we can assume that the pseudofunctor is of the form $\mathbf{C}(-):\mathbf{B}^{op}\longrightarrow\mathbf{Lex}$ for some lex fibration $P:\mathbf{C}\longrightarrow\mathbf{B}$. For each $\alpha:U\longrightarrow V$ in $\mathbf{B}$, let $P_*\alpha:\mathbf{C}/P_*U\longrightarrow\mathbf{C}/P_*V$ denote the functor between slice categories given by composition with $P_*\alpha$. Then because we can calculate limits in categories of presheaves pointwise from **Set**, we get that $(-)\circ(P_*\alpha)^{op}:[(\mathbf{C}/P_*V)^{op},\mathbf{Set}]\longrightarrow[(\mathbf{C}/P_*U)^{op},\mathbf{Set}]$ is a strict lex functor. Moreover, the assignment $\alpha\longmapsto(-)\circ(P_*\alpha)^{op}$ preserves identities and compositions. Thus we have a *functor*

$$\tilde{\mathbf{C}}(-) =_{def} [(\mathbf{C}/P_*(-))^{op},\mathbf{Set}]:\mathbf{B}^{op}\longrightarrow \mathbf{LEX}$$

whose values at morphisms are strict lex. The composition of the inclusion $\mathbf{C}(U)\hookrightarrow\mathbf{C}/P_*(U)$ with the Yoneda embedding yields a full and faithful lex functor $I_U:\mathbf{C}(U)\hookrightarrow\tilde{\mathbf{C}}(U)$ which is pseudonatural in U (because the functors $P_*\alpha:\mathbf{C}/P_*U\longrightarrow\mathbf{C}/P_*V$ are left adjoint to the pullback functors $(P_*\alpha)^*:\mathbf{C}/P_*V\longrightarrow\mathbf{C}/P_*U$). Then defining $\mathbf{C}'(U)\hookrightarrow\tilde{\mathbf{C}}(U)$ to be the least full subcategory containing all objects of the form $\tilde{\mathbf{C}}(\alpha)(I_V(Y))$ and closed under the given operations for terminal object and pullbacks (inherited from those for **Set**), we get a functor $\mathbf{C}'(-):\mathbf{B}^{op}\longrightarrow\mathbf{Lex}$ whose values on morphisms are strict; and I restricts to a pseudonatural transformation $\mathbf{C}(-)\longrightarrow\mathbf{C}'(-)$ whose components are not only full and faithful but also essentially surjective (since $\mathbf{C}(U)$ and I_U are lex)—and hence I yields a pseudonatural equivalence, as required.

□

Using the above lemma, we can complete the second half of the proof of Proposition 4.14: starting with a lex fibration $P:\mathbf{C}\longrightarrow\mathbf{B}$, form $\mathbf{C}':\mathbf{B}^{op}\longrightarrow\mathbf{lex(Set)}$ as in 4.15; then since $\mathbf{C}(-)\simeq\mathbf{C}'(-)$, the Grothendieck construction yields $\mathbf{C}\simeq Gr(\mathbf{C}')$ in $\mathbf{Lex/B}$ and hence $\hat{\mathbf{C}}\simeq(Gr\mathbf{C}')^{\wedge}\simeq[(\mathbf{C}')^{op},\hat{\mathbf{B}}]$ in $\mathbf{GTOP}(\hat{\mathbf{B}})$ with $\mathbf{C}'\in\mathbf{lex}(\hat{\mathbf{B}})$—so that $\hat{\mathbf{C}}$ is indeed an algebraic $\hat{\mathbf{B}}$-topos.

□

4.16. Corollary. *Let* $\mathbf{E}$ *be a Grothendieck topos and* $f:\mathbf{F}\longrightarrow\mathbf{E}$ *an algebraic* $\mathbf{E}$*-topos. Exponentiating by any algebraic topos* $\mathbf{G}$ *yields an algebraic* $(\mathbf{G}\to\mathbf{E})$*-topos,* $(\mathbf{G}\to f):(\mathbf{G}\to\mathbf{F})\longrightarrow(\mathbf{G}\to\mathbf{E})$.

Proof. Consider first the case in which $\mathbf{E}$ is itself algebraic—say $\mathbf{E}=\hat{\mathbf{B}}$ with $\mathbf{B}\in\mathbf{Lex}$. Then by Proposition 4.14, we can take f to be $\hat{P}_*:\hat{\mathbf{C}}\longrightarrow\hat{\mathbf{B}}$ with $P:\mathbf{C}\longrightarrow\mathbf{B}$ a lex fibration. Supposing that $\mathbf{G}=\hat{\mathbf{D}}$ with $\mathbf{D}\in\mathbf{Lex}$, then we have from Proposition 4.6 that $(\mathbf{G}\to\mathbf{E})$ is $(\mathbf{D}^{op.}\mathbf{B})^{\wedge}$ and that $(\mathbf{G}\to\mathbf{F})$ is $(\mathbf{D}^{op.}\mathbf{C})^{\wedge}$. Moreover the calculations in the proof of that proposition imply that $(\mathbf{G}\to f)$ is the geometric morphism $(\mathbf{D}^{op.}P_*)^{\wedge}$. But $\mathbf{D}^{op.}(-):\mathbf{Lex}\longrightarrow\mathbf{Lex}$ is a homomorphism of bicategories; therefore $\mathbf{D}^{op.}P_*$ is right adjoint to $\mathbf{D}^{op.}P$ with counit an isomorphism, i.e. $\mathbf{D}^{op.}P$ is a lex fibration. Hence by Proposition 4.14 again, $(\mathbf{G}\to f):(\mathbf{G}\to\mathbf{F})\longrightarrow(\mathbf{G}\to\mathbf{E})$ is an algebraic $(\mathbf{G}\to\mathbf{E})$-topos.

Now consider the general case in which $\mathbf{E}$ is an arbitrary Grothendieck topos. We can always find a small site of definition for $\mathbf{E}$ whose underlying category has finite limits (see [J1, Corollary 0.46]). In other words, we can find $\mathbf{B}\in\mathbf{Lex}$ and a geometric inclusion $i:\mathbf{E}\hookrightarrow\hat{\mathbf{B}}$. Since $\mathbf{F}\longrightarrow\mathbf{E}$ is algebraic there is $\mathbf{C}\in\mathbf{lex}(\mathbf{E})$ with $\mathbf{F}\simeq[\mathbf{C}^{op},\mathbf{E}]$ in $\mathbf{GTOP}(\mathbf{E})$. The direct image functor $i_*:\mathbf{E}\longrightarrow\hat{\mathbf{B}}$ is lex and so we can transport $\mathbf{C}$ along it to get $i_*(\mathbf{C})\in\mathbf{lex}(\hat{\mathbf{B}})$. Then as in (4.1) there is a pullback square in $\mathbf{GTOP}$ of the form:

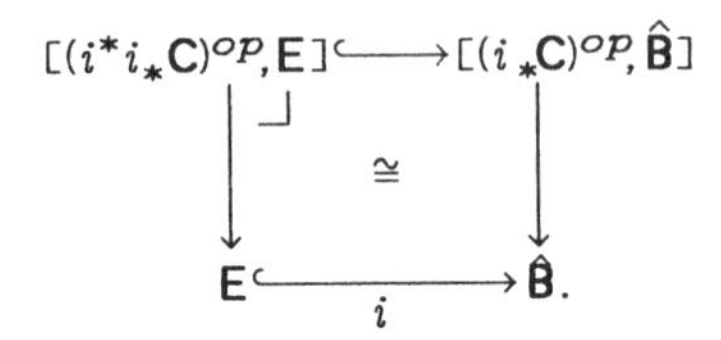

Since i is an inclusion, $i^*i_*\mathbf{C}\cong\mathbf{C}$ in $\mathbf{lex}(\mathbf{E})$; hence there is a pullback square of the form

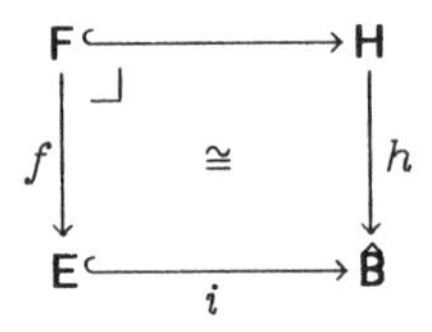

with $\hat{\mathbf{B}}$ algebraic and $h:\mathbf{H}\longrightarrow\hat{\mathbf{B}}$ an algebraic $\hat{\mathbf{B}}$-topos. Now $(\mathbf{G}\to -):\mathbf{GTOP}\longrightarrow\mathbf{GTOP}$ preserves pullbacks (since it has a left (bi)adjoint). Applying it to the above square therefore gives that $(\mathbf{G}\to f)$ is the pullback of $(\mathbf{G}\to h)$ along $(\mathbf{G}\to i)$; but $(\mathbf{G}\to h)$ is algebraic by the special case considered above and hence by Proposition 4.2, the pullback $(\mathbf{G}\to f)$ is also algebraic. □

As with the previous results on exponentiation, the above corollary admits of relativization from **Set** to the category theory of an arbitrary Grothendieck topos $\mathbf{E}$:

4.17. Corollary. *If $\mathbf{F}\longrightarrow\mathbf{E}$ is a Grothendieck $\mathbf{E}$-topos, $\mathbf{A}\longrightarrow\mathbf{E}$ an algebraic $\mathbf{E}$-topos and $b:\mathbf{B}\longrightarrow\mathbf{F}$ an algebraic $\mathbf{F}$-topos, then $(\mathbf{A}\to_{\mathbf{E}}b):(\mathbf{A}\to_{\mathbf{E}}\mathbf{B})\longrightarrow(\mathbf{A}\to_{\mathbf{E}}\mathbf{F})$ is an algebraic $(\mathbf{A}\to_{\mathbf{E}}\mathbf{F})$-topos.* □

We are now in a position to prove the dual version of Corollary 4.10:

4.18. Proposition.

(*i*) *If $f:\mathbf{F}\longrightarrow\mathbf{E}$ is an algebraic $\mathbf{E}$-topos, then $f^{\#}:\mathbf{ATOP}(\mathbf{E})\longrightarrow\mathbf{ATOP}(\mathbf{F})$ has a right adjoint,* *denoted* $f_{\#}$:**ATOP(F)**$\longrightarrow$**ATOP(E)**.

(*ii*) *The adjoints of* (*i*) *satisfy a Beck-Chevalley condition: given a pullback square*

$$\begin{array}{ccc} \mathbf{H} & \xrightarrow{q} & \mathbf{G} \\ {\scriptstyle h}\downarrow & \cong & \downarrow{\scriptstyle g} \\ \mathbf{F} & \xrightarrow[f]{} & \mathbf{E} \end{array}$$

in **GTOP** *with* f *(and hence* q*) algebraic, the canonical natural transformation* $g^{\#}\circ f_{\#}\longrightarrow q_{\#}\circ h^{\#}$ *is an equivalence.*

Proof. We construct the right adjoint as in Proposition 2.6. Thus given b:**B**$\longrightarrow$**F** in **ATOP(F)**, define $f_{\#}$(**B**)$\longrightarrow$**E** via the pullback

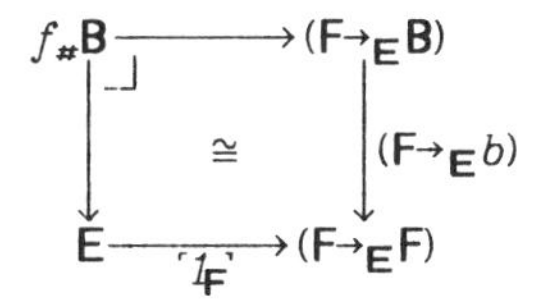

in **GTOP(E)**, where $\ulcorner 1_{\mathbf{F}} \urcorner$ denotes the exponential transpose of $\mathbf{F}\times_{\mathbf{E}}\mathbf{E}\simeq\mathbf{F}\xrightarrow{1}\mathbf{F}$. Combining Corollary 4.17 with Proposition 4.2, we certainly have that $f_{\#}$**B** is an algebraic **E**-topos. To see that it has the right universal property, first note that from the universal property of pullbacks, morphisms

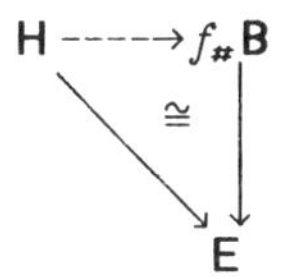

in **GTOP(E)** correspond to diagrams in **GTOP** of the form

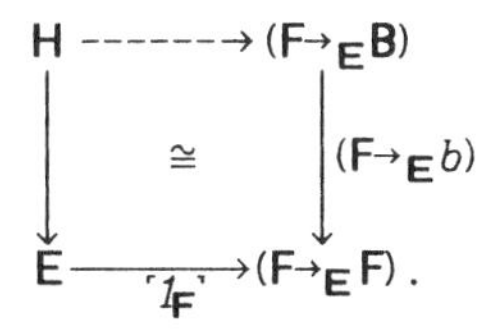

Transposing such a diagram across the exponential adjunction gives

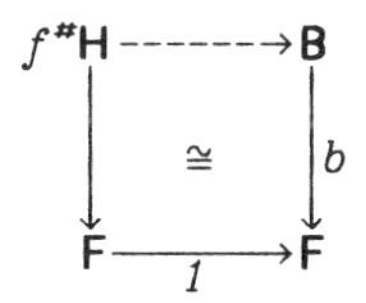

in **GTOP(F)**. Thus **GTOP(E)**(**H**,$f_{\#}$**B**) $\simeq$ **GTOP(F)**($f^{\#}$**H**,**B**) and the equivalence is evidently natural in **H**$\longrightarrow$**E**. This proves not just (*i*), but in fact something slightly more, namely: *when* f *is algebraic then the value of the right adjoint to* $f^{\#}$:**GTOP(E)**$\longrightarrow$**GTOP(F)** *exists at any algebraic* **E**-*topos*. This also suffices to prove (*ii*), since whenever both $f_{\#}$**B**$\in$**GTOP(E)** and $q_{\#}(h^{\#}\mathbf{B}) \in \mathbf{GTOP(G)}$ are defined, then elementary properties of pullbacks with respect to composition imply that $g^{\#}(f_{\#}\mathbf{B})$ and $q_{\#}(h^{\#}\mathbf{B})$ are canonically equivalent in **GTOP(G)**.

□

4.19. Algebraic topos model of the theory of constructions. Collecting together the results of this section, we present our first example of an instance of the categorical structure set out in section 2. In fact it is an example of the theory of constructions with "$Type \simeq ORDER$" (see 1.12) and so we show how to fulfil the conditions listed in 2.15:

(i) The category **B** is obtained from the 2-category **ATOP** (*cf* 4.1) by taking *isomorphism classes* of 1-cells. Thus **B** has for its objects the algebraic toposes and for its morphisms isomorphism classes of geometric morphisms: in other words we identify two geometric morphisms $f,g:\mathbf{F}\longrightarrow\mathbf{E}$ between algebraic toposes if they are isomorphic objects in **GTOP(F,E)**. Composition and identities in **B** are those inherited from **GTOP**. This category **B** certainly has a terminal object, namely **Set**—*cf*. 4.8.

(ii) The class of morphisms **A** is that determined by the geometric morphisms $f:\mathbf{F}\longrightarrow\mathbf{E}$ between algebraic toposes which make **F** an algebraic **E**-topos.

(iii) Note that by Proposition 4.9, if **E** is in **B** and $f:\mathbf{F}\longrightarrow\mathbf{E}$ is an algebraic **E**-topos, then **F** is also in **B**. Consequently Propositions 4.2 implies that the pullback of a morphism in **A** along an arbitrary morphism of **B** exists and is again in **A**.

(iv) **A** is closed under composition by Proposition 4.9.

(v) If $f:\mathbf{F}\longrightarrow\mathbf{E}$ is in **A**, then the pullback functor $f^{\#}:\mathbf{A}(\mathbf{E})\longrightarrow\mathbf{A}(\mathbf{F})$ has a right adjoint satisfying the Beck-Chevalley condition by Proposition 4.18.

(vi) The topos $\mathbf{\Upsilon}$ of Proposition 4.5 is in **B** and the algebraic $\mathbf{\Upsilon}$-topos $\mathbf{\Sigma}\longrightarrow\mathbf{\Upsilon}$ of that proposition determines a morphism in **A** with the property that any other morphism in **A** can be obtained from it by pullback.

(vii) Finally, for each object **E** in **B**, the unique morphism $\mathbf{E}\longrightarrow\mathbf{Set}$ is in **A** because we chose **B** to consist only of *algebraic* toposes.

4.20. Equivalent descriptions of the model. We state without proofs two examples of the categorical structure in 2.15 which are both equivalent to that given in 4.19. These equivalent versions (especially the second) have advantages when it comes to making *calculations* in the model. As we remarked in 2.13, we can specify these equivalent forms of the model by giving equivalent versions of the underlying category **B** (and taking the essential image of the class **A** under the equivalence).

(i) A version in the style of domain theory. Instead of looking at the algebraic toposes **E** themselves, we can look at their *categories of points*, **GTOP(Set,E)**. Supposing that $\mathbf{E}\simeq[\mathbf{C}^{op},\mathbf{Set}]$ with $\mathbf{C}\in\mathbf{Lex}$, then by (*4.2*) this category of points is equivalent to **LEX(C,Set)**, which is a typical *locally finitely presentable* (*lfp*) category. (See [GU] and [MP].) If $\mathbf{F}\simeq[\mathbf{D}^{op},\mathbf{Set}]$ is another algebraic topos, then it is the case that a functor

$$\mathbf{GTOP(Set,E)}\simeq\mathbf{LEX(C,Set)}\longrightarrow\mathbf{LEX(D,Set)}\simeq\mathbf{GTOP(Set,F)}$$

is induced by composition with a geometric morphism if and only if the functor preserves filtered colimits. In this way we get *an equivalence between* **ATOP** (defined in 4.1) *and the 2-category consisting of lfp categories, functors preserving filtered colimits and natural transformations.* Using this equivalence we get an alternative description of the structure in 4.19 in terms of lfp categories and filtered colimit preserving functors. The class of display morphisms is perhaps most simply described as comprising those functors between lfp categories which preserve limits and filtered colimits and have filtered colimit preserving right adjoints with counit of the adjunction an isomorphism.

The model in this form has been studied by Coquand (see [CE, section 5]), although not in terms of the systematic framework developed in section 2. From the point of view of domain theory, it is very natural to approach the model in this way, since lfp categories directly generalize algebraic lattices and filtered colimit preserving functors generalize continuous maps beteen cpo's. Perhaps the main advantage of the topos-theoretic treatment we have given in this section is the extremely simple way in which the existence of a generic family of Orders (4.19(vi)) is demonstrated using standard properties of classifying toposes; we hope that the use of this technique will lead to the discovery of other models.

(ii) A version in the style of Scott's information systems. Instead of dealing with lfp categories, we can work directly with the (essentially) small lex categories which determine them. (Recall that an lfp category **A** is equivalent to **LEX**(**C**, **Set**) when $\mathbf{C} \simeq (\mathbf{A}_{fp})^{op}$.) Our calculations in the proof of Proposition 4.6 imply that specifying a filtered colimit preserving functor **LEX**(**C**, **Set**) $\longrightarrow$ **LEX**(**D**, **Set**) is equivalent to giving a functor $\mathbf{C}^{op} \times \mathbf{D} \longrightarrow \mathbf{Set}$ which is lex in its second variable ($cf.$ 3.6(i)): we will call such a thing a *lex module from* **C** *to* **D**. Recall that a *module* (or *profunctor* or *distributeur*) from **C** to **D** is a functor $\mathbf{C}^{op} \times \mathbf{D} \longrightarrow \mathbf{Set}$; modules can be composed and small categories, modules and natural transformations form a bicategory—see [J1, section 2.4] and [CKW] for example. Restricting the objects to be lex categories and the morphisms to be lex modules, we obtain a sub-bicategory which we call **Lexmod**. Then it is the case that **ATOP** *and* **Lexmod** *are equivalent bicategories*. This forms the basis for a second equivalent description of the model in 4.19—one that has similarities with the "information system" approach in domain theory. The (opposites of) small lex categories are to lfp categories as information systems are to domains; lex modules are to filtered colimit preserving functors as "approximable maps" are to continuous maps between domains.

With a little calculation (which we do not give here) the structures in **Lexmod** which witness the fact that it is a model of the theory of constructions with "$Type \simeq ORDER$" take a rather pleasant form. First note that the terminal object **1** in **Lexmod** is given by the trivial lex category; and the product of **C** and **D** is given by their product in **Lex**. The class of display morphisms **A** in **Lexmod** corresponding under the equivalence to that in 4.19(ii), consists of those lex modules of the form $\mathbf{C}(P(-),+) : \mathbf{E}^{op} \times \mathbf{C} \longrightarrow \mathbf{Set}$ where $P : \mathbf{E} \longrightarrow \mathbf{C}$ is a lex fibration ($cf.$ 4.11). Thus for each **C**, these are the objects of the category **A**(**C**) (notation as in 2.2); the morphisms from $P : \mathbf{E} \longrightarrow \mathbf{C}$ to $Q : \mathbf{F} \longrightarrow \mathbf{C}$ in **A**(**C**) are (isomorphism classes of) lex modules $M:\mathbf{E}^{op}\times\mathbf{F}\longrightarrow\mathbf{Set}$ for which the composition with $\mathbf{C}(Q(-),+)$ is isomorphic to $\mathbf{C}(P(-),+)$, that is, for which we have $M(-,Q_*(+))\cong\mathbf{E}(-,P_*(+))$.

We can use the remarks in 4.12 to describe the objects of $\mathbf{A}(\mathbf{C})$ equivalently as pseudofunctors $\mathbf{C}^{op}\longrightarrow\mathbf{Lex}$. Further calculation shows that from this point of view, a morphism in $\mathbf{A}(\mathbf{C})$ between pseudofunctors $\mathbf{E}:\mathbf{C}^{op}\longrightarrow\mathbf{Lex}$ and $\mathbf{F}:\mathbf{C}^{op}\longrightarrow\mathbf{Lex}$ is specified by the following data:

- a lex module $M_U(-,+):\mathbf{E}(U)^{op}\times\mathbf{F}(U)\longrightarrow\mathbf{Set}$ for each $U\in\mathbf{C}$,
- a natural transformation $M_\alpha:M_V(-,+)\longrightarrow M_U(\alpha^*(-),\alpha^*(+))$ for each $\alpha:U\longrightarrow V$ in $\mathbf{C}$,

satisfying the coherence conditions

$$M_{id_U}=M_U(\iota^{-1},\iota) \quad\text{and}\quad M_{\beta\alpha}=M_U(\kappa,\kappa^{-1})\circ(M_\alpha)_{\beta^*\times\beta^*}\circ M_\beta$$

(where ι and κ are the canonical isomorphisms for pseudofunctors defined in 3.7).

We next describe the adjoints $M_!$ and $M_\#$ to the pullback functor $M^\#:\mathbf{A}(\mathbf{C})\longrightarrow\mathbf{A}(\mathbf{D})$ for a morphism $M:\mathbf{D}\longrightarrow\mathbf{C}$ in $\mathbf{A}$. First consider the special case when $\mathbf{C}$ is the terminal object $\mathbf{1}$. We can identify $\mathbf{A}(\mathbf{1})$ with **Lexmod** and $M^\#$ with the functor $\Delta_{\mathbf{D}}:\mathbf{Lexmod}\longrightarrow\mathbf{A}(\mathbf{D})$ which sends a lex category to the constant pseudofunctor with that value and acts similarly on modules. Then it is the case that

- *the left adjoint to* $\Delta_{\mathbf{D}}$ *sends a pseudofunctor* $\mathbf{E}:\mathbf{D}^{op}\longrightarrow\mathbf{Lex}$ *to the lex category* $Gr(\mathbf{E})$ *obtained by performing the* Grothendieck construction (*cf.* 4.9) *on* $\mathbf{E}$;
- *the right adjoint to* $\Delta_{\mathbf{D}}$ *sends a pseudofunctor* $\mathbf{E}:\mathbf{D}^{op}\longrightarrow\mathbf{Lex}$ *to the lex category* $\underrightarrow{oplax}_{\mathbf{D}^{op}}\mathbf{E}$ *obtained by taking the* oplax colimit *of* $\mathbf{E}$ (*cf.* 3.7).

The Beck-Chevalley condition implies that we can calculate the adjoints to an arbitrary morphism in $\mathbf{A}$ *fibrewise* using the above special case. Thus if $M:\mathbf{D}\longrightarrow\mathbf{C}$ in $\mathbf{A}$ is given as $\mathbf{C}(P(-),+)$ with $P:\mathbf{D}\longrightarrow\mathbf{C}$ a lex fibration, we have:

$$(M_!\mathbf{E})(U)=Gr(\mathbf{E}|_U)$$

and

$$(M_\#)(U)=\underrightarrow{oplax}_{\mathbf{D}(U)^{op}}(\mathbf{E}|_U)$$

where $\mathbf{D}(U)$ is the fibre of P over $U\in\mathbf{C}$ (*cf.* 4.12) and $\mathbf{E}|_U$ is the restriction of $\mathbf{E}$ to that fibre.

The last part of the structure of **Lexmod** which needs describing is that corresponding to $\boldsymbol{\Sigma}\longrightarrow\boldsymbol{\Upsilon}$ in 4.19(*vi*), namely the interpretation of *Order* and the generic family of Orders. The lex category underlying $\boldsymbol{\Upsilon}$ is the classifying category $\mathbf{C}_{\mathbf{lex}}$ of the lim theory **lex**, which we saw in 4.4 can be described equivalently as the opposite of the category of finitely presented lex categories and strict lex functors, $(\mathbf{lex}(\mathbf{Set})_{fp})^{op}$. The generic family in $\mathbf{A}(\mathbf{C}_{\mathbf{lex}})$, regarded as a pseudofunctor $\mathbf{lex}(\mathbf{Set})_{fp}\longrightarrow\mathbf{Lex}$, is just the forgetful functor (not an inclusion because the morphisms in $\mathbf{lex}(\mathbf{Set})$ are *strict* lex).

5 Localic algebraic toposes

We now describe a second topos-theoretic model of the theory of constructions—and moreover one in which the classes **A** and **R** required in 2.13 are distinct. The underlying category and the class **A** are as in 4.19, but we restrict **R** by imposing the condition of being localic. In terms of the alternative description 4.20(i), this corresponds to modelling (families of) Orders by (continuous fibrations of) locally finitely presentable categories, but modelling (families of) Types by (continuous fibrations of) *algebraic lattices*.

A geometric morphism f:**F**⟶**E** is *localic* if the terminal object $1 \in$ **E** is an object of generators for **F** over **E**, i.e. if for every $Y \in$ **F** there is some $X \in$ **E** and a diagram of the form

$$(5.1) \qquad f^*(X) \xleftarrow{m} \cdot \xrightarrow{e} Y$$

in **F** with m a monomorphism and e an epimorphism. An alternative characterization, and the one which gives rise to the name "localic", is: f:**F**⟶**E** is localic iff **F** is equivalent in **GTOP(E)** to an **E**-topos of sheaves on an internal locale of **E**. We refer the reader to [J3] and [JT] for expositions of the basic properties of localic toposes. In particular we will need to use the fact that the localic geometric morphisms form one half of a factorization system on **GTOP**. The other half is given by the *hyperconnected* morphisms—those f:**F**⟶**E** for which f^* is faithful and such that the objects in the image of f^* are closed under taking subobjects in **F**. These morphisms are orthogonal to the localic morphisms (*cf*. 2.12). An arbitrary geometric morphism f:**F**⟶**E** factors uniquely up to equivalence as $f \cong \ell \circ h$:**F**⟶**L**⟶**E**, with h:**F**⟶**L** hyperconnected and ℓ:**L**⟶**E** localic. (**L** can be taken to be the **E**-topos of sheaves on the internal locale $f_*(\Omega_{\mathbf{F}})$, where $\Omega_{\mathbf{F}}$ is the subobject classifier in **F**; more elementarily, it is equivalent to the full subcategory of **F** whose objects are those Y for which there exists a diagram of the form (*5.1*) with m mono and e epi.)

5.1. Meet semilattices. These are the models of the lim theory **msl** having a single sort Ob, a constant symbol $\top{:}Ob$, a function symbol $\wedge{:}Ob{\times}Ob \longrightarrow Ob$ and axioms

- $\forall x,y,z{:}Ob(x\wedge(y\wedge z) = (x\wedge y)\wedge z)$
- $\forall x,y{:}Ob(x\wedge y = y\wedge x)$
- $\forall x{:}Ob(x\wedge\top = x)$
- $\forall x{:}Ob(x\wedge x = x)$.

Clearly a model of **msl** is a partially ordered set (via the relation: $x \le y$ iff $x\wedge y = x$) with all finite meets (including the empty one, $\top$)—and hence can be regarded as a lex category; similarly a homomorphism of meet semilattices is in particular a strict lex functor. Thus **msl(Set)** is a full subcategory of **lex(Set)**. Indeed it is a reflective subcategory: the left adjoint to the inclusion **msl(Set)**$\hookrightarrow$**lex(Set)** sends a lex category **C** to the meet semilattice Po**C** obtained by first forming the pre-order reflection (i.e. the set Ob**C** pre-ordered via: $X \le Y$ iff **C**(X,Y) is inhabited) and then quotienting by the associated equivalence relation

($X \equiv Y$ iff $X \leq Y$ and $Y \leq X$) to get a poset.

Relativizing the above to any Grothendieck topos **E**, the *internal meet semilatices* in **E**, **msl**(**E**), form a full reflective subcategory of **lex**(**E**), the left adjoint $Po:\mathbf{lex}(\mathbf{E})\longrightarrow\mathbf{msl}(\mathbf{E})$ being given by the internal version of the poset reflection of a category. The unit of the adjunction at $\mathbf{C}\in\mathbf{lex}(\mathbf{E})$, the quotient functor $\mathbf{C}\longrightarrow Po\mathbf{C}$, is full and surjective on objects; hence by [J3, Proposition 3.1(i)] the geometric morphism $[\mathbf{C}^{op},\mathbf{E}]\longrightarrow[(Po\mathbf{C})^{op},\mathbf{E}]$ it induces is hyperconnected. Similarly, since $Po\mathbf{C}$ is an internal poset, the unique internal functor $Po\mathbf{C}\longrightarrow\mathbf{1}$ is faithful and hence by [J3, Proposition 3.1(ii)], the induced geometric morphism $[(Po\mathbf{C})^{op},\mathbf{E}]\longrightarrow[\mathbf{1}^{op},\mathbf{E}]\simeq\mathbf{E}$ is localic. Therefore these two geometric morphisms give the hyperconnected-localic factorization of their composition, which is the geometric morphism $[\mathbf{C}^{op},\mathbf{E}]\longrightarrow\mathbf{E}$ defining the topos of internal presheaves as an **E**-topos. If the algebraic **E**-topos $[\mathbf{C}^{op},\mathbf{E}]$ is already known to be localic, then it is equivalent to its localic factorization and so $[\mathbf{C}^{op},\mathbf{E}]\simeq[(Po\mathbf{C})^{op},\mathbf{E}]$ in **GTOP**(**E**). We have thus proved:

5.2. Proposition. *A Grothendieck* **E**-*topos* $\mathbf{F}\longrightarrow\mathbf{E}$ *is both localic and algebraic iff it is equivalent in* **GTOP**(**E**) *to* $[M^{op},\mathbf{E}]\longrightarrow\mathbf{E}$ *for some internal meet semilattice* M.

□

In view of this result we can use the material in 4.4 on classifying toposes applied to the lim theory **msl** rather than to **lex** to obtain analogues of Propositions 4.2 and 4.5:

5.3. Proposition.

(i) *Geometric morphisms which are both localic and algebraic are stable under pullback in* **GTOP**.

(ii) *There is an algebraic topos* $\mathbf{\Upsilon}'$ *and a localic-algebraic* $\mathbf{\Upsilon}'$*-topos* $\mathbf{\Sigma}'\longrightarrow\mathbf{\Upsilon}'$ *with the property that any other* $\mathbf{F}\longrightarrow\mathbf{E}$ *in* **GTOP** *which is both algebraic and localic can be obtained from* $\mathbf{\Sigma}'\longrightarrow\mathbf{\Upsilon}'$ *by pullback.*

Proof. For (i) we can either use the proof of 4.2 applied to internal meet semilattices or recall that localic morphisms are pullback stable—*cf*. [J3, Proposition 2.1] or [JT, Proposition VI.4].

For (ii), take $\mathbf{\Upsilon}'=[\mathbf{C}^{op}_{\mathbf{msl}},\mathbf{Set}]$, the classifying topos of the lim theory **msl** and $\mathbf{\Sigma}'=[U^{op}_{\mathbf{msl}},\mathbf{\Upsilon}']$, where $U^{op}_{\mathbf{msl}}\in\mathbf{msl}(\mathbf{\Upsilon}')$ is the generic meet semilattice—then argue as in the proof of 4.5.

□

5.4. Remark. In Proposition 4.5, the "generic" algebraic topos is defined over a topos $\mathbf{\Upsilon}$ which is itself algebraic. On the other hand, in Proposition 5.3 the generic localic-algebraic topos is defined over a topos $\mathbf{\Upsilon}'$ which is algebraic, but is not localic; for if it were, then $\mathbf{C}_{\mathbf{msl}}$ would have to be equivalent to $Po\mathbf{C}_{\mathbf{msl}}$ and hence be a preorder—but $\mathbf{C}_{\mathbf{msl}}$ is equivalent to the opposite of the category of finitely presented meet semilattices, which is certainly not preordered.

However, the property 5.3(ii) does not determine $\mathbf{\Sigma}'\longrightarrow\mathbf{\Upsilon}'$ uniquely. Perhaps there is another choice of localic-algebraic $\Sigma'' \longrightarrow \Upsilon''$ satisfying 5.3(*ii*) but with $\Upsilon'' \longrightarrow$ **Set** itself both algebraic and localic? In fact no such choice is possible. For if we had such, then we could find pullback squares in **GTOP** of the form

Composing these two squares to give a single pullback and recalling the definition of Σ', we must have $(r \circ i)^* U_{\mathbf{msl}} \cong U_{\mathbf{msl}}$ in $\mathbf{msl}(\Upsilon')$; but Υ' is the classifying topos of meet semilattices and under the equivalence $\mathbf{msl}(\Upsilon') \simeq \mathbf{GTOP}(\Upsilon', \Upsilon')$ this isomorphism corresponds to some isomorphism $r \circ i \cong 1_{\Upsilon'}$. Therefore i and r make Υ' a retract of Υ'' in **GTOP**—from which it follows easily that Υ' is localic if Υ'' is. But we observed above that Υ' is not localic; so no such $\Sigma'' \longrightarrow \Upsilon''$ can exist.

5.5. Notation. For each Grothendieck topos **E**, let **LATOP(E)** denote the full sub-2-category of **GTOP(E)** whose objects are those $\mathbf{A} \longrightarrow \mathbf{E}$ which are both localic and algebraic. Given a geometric morphism $f{:}\mathbf{F} \longrightarrow \mathbf{E}$,

$$f^{\#}{:}\mathbf{LATOP(E)} \longrightarrow \mathbf{LATOP(F)}$$

will denote the restriction to localic-algebraic toposes of the operation of pulling back (which by Proposition 5.3(*i*), is well defined).

We next examine how the additional assumption of being localic effects the properties of algebraic toposes with respect to exponentiation and right adjoints to change of base. To do so, we use the following topos-theoretic result which as far as we know has not appeared elsewhere:

5.6. Proposition. *Let* $\mathbf{A} \longrightarrow \mathbf{E}$ *be an exponentiable* **E**-*toposes. Then* $(\mathbf{A} \to_{\mathbf{E}} -){:}\mathbf{GTOP(E)} \longrightarrow \mathbf{GTOP(E)}$ *preserves localic morphisms.*

Proof. We make use of the following characterization of localic geometric morphisms, which is proved in [Pi1, Proposition 3.5] (for the case **E**=**Set**, but in a way that relativizes to arbitrary **E**):

Given $g{:}\mathbf{G} \longrightarrow \mathbf{F}$ in **GTOP(E)**, form the pullback of g against itself

$$\begin{array}{ccc} \mathbf{G} \times_{\mathbf{F}} \mathbf{G} & \xrightarrow{p_1} & \mathbf{G} \\ {\scriptstyle p_0} \downarrow & \overset{\pi}{\cong} & \downarrow {\scriptstyle g} \\ \mathbf{G} & \xrightarrow[g]{} & \mathbf{F} \end{array}$$

and let $d{:}\mathbf{G} \longrightarrow \mathbf{G} \times_{\mathbf{F}} \mathbf{G}$ be the *diagonal* geometric morphism, i.e. the morphism for which there are natural isomorphisms $\delta_i{:}p_i \circ d \cong 1_{\mathbf{G}}$ $(i=0,1)$ satisfying $g\delta_1 \circ \pi d = g\delta_0$. Then: *$g$ is localic iff d is an inclusion.*

Since $(\mathbf{A}\to_{\mathbf{E}} -):\mathbf{GTOP(E)}\longrightarrow\mathbf{GTOP(E)}$ has a left biadjoint, it preserves the above pullback square and diagonal geometric morphism. Consequently to prove the proposition it is sufficient to prove that $(\mathbf{A}\to_{\mathbf{E}} -)$ *preserves geometric inclusions.* This is the case because inclusions can be characterized using limits in the bicategory **GTOP(E)** (and these are preserved by $(\mathbf{A}\to_{\mathbf{E}} -)$). Indeed, the geometric inclusions are precisely the *inverters* in **GTOP(E)**, i.e. those $i:\mathbf{F}\longrightarrow\mathbf{G}$ for which there are $g,h:\mathbf{G}\longrightarrow\mathbf{H}$ and $\phi:g\longrightarrow h$ with the property that for any $\mathbf{K}$, the functor $i\circ(-):\mathbf{GTOP(E)}(\mathbf{K},\mathbf{F})\longrightarrow\mathbf{GTOP(E)}(\mathbf{K},\mathbf{G})$ is full and faithful with essential image those $k:\mathbf{K}\longrightarrow\mathbf{G}$ for which $\phi k:gk\longrightarrow hk$ is an isomorphism. We briefly indicate why this is the case, giving the argument for $\mathbf{E}=\mathbf{Set}$, but in a form admitting relativization:

First note that inverters can be constructed as sheaf subtoposes: if $(H_u \mid u\in U)$ is a small family of generators for $\mathbf{H}$ and j is the least Lawvere-Tierney topology on $\mathbf{G}$ which forces each $\phi_{H_u}:g^*(H_u)\longrightarrow h^*(H_u)$ to be iso, then $sh_j(\mathbf{G})\hookrightarrow\mathbf{G}$ is necessarily the inverter of ϕ. Conversely, if $i:\mathbf{F}\longrightarrow\mathbf{G}$ is an inclusion—say $\mathbf{F}=sh_j(\mathbf{G})$—then define $\mathbf{H}$ to be the full subcategory of the arrow category $\mathbf{G}^2$ whose objects are those $a:X\longrightarrow Y$ in $\mathbf{G}$ which are j-bidense. One can show that $\mathbf{H}$ is a Grothendieck topos. (This is just a question of exhibiting generators, since it has the right exactness properties automatically.) Then the inclusion $\mathbf{H}\hookrightarrow\mathbf{G}^2$ is the inverse image part of a geometric surjection $q:\mathbf{G}^2\longrightarrow\mathbf{H}$. The two functors and non-identity natural transformation in $\mathbf{Cat}(\mathbf{1},\mathbf{2})$ induce geometric morphisms $l_0,l_1:\mathbf{G}\longrightarrow\mathbf{G}^2$ and a natural transformation $\lambda:l_0\longrightarrow l_1$. Then the definition of $\mathbf{H}$ and the characterization [J1, 3.4] of sheaf subtoposes in terms of categories of fractions imply that $sh_j(\mathbf{G})\hookrightarrow\mathbf{G}$ is the inverter of $q\lambda:ql_0\longrightarrow ql_1$.

□

5.7. Corollary.

(i) *If* $a:\mathbf{A}\longrightarrow\mathbf{E}$ *is an algebraic* $\mathbf{E}$*-topos and* $b:\mathbf{B}\longrightarrow\mathbf{E}$ *a localic-algebraic* $\mathbf{E}$*-topos, then their exponential* $(\mathbf{A}\to_{\mathbf{E}}\mathbf{B})\longrightarrow\mathbf{E}$ *is localic-algebraic. So in particular each* **LATOP(E)** *is cartesian closed and for each geometric morphism* $f:\mathbf{F}\longrightarrow\mathbf{E}$, *the pullback operation* $f^{\#}:\mathbf{LATOP(E)}\longrightarrow\mathbf{LATOP(E)}$ *preserves the cartesian closed structure.*

(ii) *If* $f:\mathbf{F}\longrightarrow\mathbf{E}$ *is algebraic and* $b:\mathbf{B}\longrightarrow\mathbf{F}$ *is a localic-algebraic* $\mathbf{F}$*-topos, then the algebraic* $\mathbf{E}$*-topos* $f_{\#}(\mathbf{B})$ *of Proposition 4.18 is also localic. So when* f *is algebraic,* $f_{\#}$ *gives a right adjoint for* $f^{\#}:\mathbf{LATOP(E)}\longrightarrow\mathbf{LATOP(E)}$ *satisfying the Beck-Chevalley condition.*

Proof. In view of Corollary 4.8, for (i) we just need to see that $(\mathbf{A}\to_{\mathbf{E}}\mathbf{B})\longrightarrow\mathbf{E}$ is localic; but this morphism is the composition of the equivalence $(\mathbf{A}\to_{\mathbf{E}}\mathbf{E})\simeq\mathbf{E}$ with $(\mathbf{A}\to_{\mathbf{E}}b):(\mathbf{A}\to_{\mathbf{E}}\mathbf{B})\longrightarrow(\mathbf{A}\to_{\mathbf{E}}\mathbf{E})$, which is localic by Proposition 5.6.

For (ii), note that the construction of $f_{\#}(\mathbf{B})$ in 4.18 is in terms of exponentiation and pullbacks—so that the result follows from 5.6 and the fact that localic morphisms are pullback stable.

□

5.8. Remark. Using Proposition 5.2 and the formula for exponentials of algebraic toposes given in 4.7, we can rephrase 5.7(i) in more concrete terms: if $\mathbf{C}\in\mathbf{lex(E)}$ and $M\in\mathbf{msl(E)}$, then the copower $\mathbf{C}^{op}\cdot M\in\mathbf{lex(E)}$ is equivalent to an internal meet semilattice and hence is an internal preorder.

We now turn to the construction of left adjoints to pulling back localic-algebraic toposes. In contrast to the case for exponentials and right adjoints (for which we have the absoluteness result 2.8), the left adjoints are not the same as in the algebraic case:

5.9. Proposition. *If* f:**F**⟶**E** *is an algebraic* **E**-*toposes, then* $f^{\#}$:**LATOP(E)**⟶**LATOP(F)** *possesses a left adjoint, denoted* $f_!$:**LATOP(F)**⟶**LATOP(E)**, *which satisfies the Beck-Chevalley condition* (*cf.* *4.10*).

Proof. Let **LTOP(E)** denote the full sub-2-category of **GTOP(E)** whose objects are localic **E**-toposes. (The morphisms of **LTOP(E)** are also localic since **B**⟶**A** is localic when both **A**⟶**E** and **B**⟶**A**⟶**E** are.) For an arbitrary geometric morphism f:**F**⟶**E**, the hyperconnected-localic factorization of geometric morphisms mentioned at the beginning of this section provides a left adjoint $f_!$:**LTOP(F)**⟶**LTOP(E)** to the pullback operation $f^{\#}$:**LTOP(E)**⟶**LTOP(F)**. Indeed, given b:**B**⟶**F** in **LTOP(F)**, $f_!$(**B**)⟶**E** is the localic factorization of the composition $f \circ b$:**B**⟶**E**. These left adjoints satisfy the Beck-Chevalley condition for pullback squares in **GTOP** simply because both localic and hyperconnected morphisms are stable under pullback: see [JT, VI.5] and [J3, section 2].

Therefore the proposition will be proved if we can show that $f_!$:**LTOP(F)**⟶**LTOP(E)** takes algebraic toposes to algebraic toposes in the case that f is itself algebraic. In view of 4.9, this amounts to showing that the localic part of an algebraic **E**-topos is again algebraic. But we saw above that for **C**∈**lex(E)**, the localic factorization of [$\mathbf{C}^{op}$,**E**]⟶**E** is [$(Po\mathbf{C})^{op}$,**E**]⟶**E**, where Po**C**∈**msl(E)** is the poset reflection of **C**.

□

5.10. Localic-algebraic model of the theory of constructions. We now organize the results of this section to give a second topos-theoretic model of the theory of constructions. It is a model of the theory "*Order*∈*ORDER*" of 1.11 (but not of the stronger theory "*Type*≃*ORDER*" of 1.12). So we will show how to fulfil the conditions in 2.13 and the condition in 2.14.

For (*i*), we take the category **B** to be just as in 4.19(*i*): so **B** consists of algebraic toposes and isomorphism classes of geometric morphisms.

For (*ii*), we take **A** as in 4.19(*ii*): so it is determined by those geometric morphisms f:**F**⟶**E** making **F** an algebraic **E**-topos; and for the subclass **R** we take those morphisms determined by geometric morphisms which are not only algebraic but also localic.

(*iii*) follows from 4.19(*iii*) and 5.3(*i*).

(*iv*) holds because of the fact that equivalences between toposes are trivially localic-algebraic.

(*v*) holds because of 4.19(*iv*) and the fact (easily deduced from the definition at the beginning of this section) that the composition of localic geometric morphisms is again localic.

(vi) holds because of 4.19(v) and 5.7(ii).

(vii) follows as in 5.9: the hyperconnected-localic factorization is pullback stable, hyperconnected morphisms are orthogonal to localic ones and the factorization applied to an algebraic geometric morphism yeilds a localic-algebraic morphism.

($viii$) is a consequence of 5.3(ii).

(ix) holds by definition of **B**.

Finally, the condition ***Order*∈*ORDER*** of 2.14 is just 4.19(vi).

5.11. Equivalent descriptions of the model. In 4.20 we gave two equivalent forms of the category **B**. The first was in terms of locally finitely presentable categories and functors preserving filtered colimits. The second was in terms of small lex categories and lex modules. We explain briefly what the class of morphisms **R** looks like in these equivalent formulations.

Recall from 4.20(i) that a filtered colimit preserving functor P:**F**⟶**E** between lfp categories is in the class **A** if it preserves limits (this is equivalent to its having a left adjoint) and has a right-adjoint-right-inverse which also preserves filtered colimits. In particular P is a fibration with lfp fibres; and then P is in **R** simply if its fibres are in fact pre-ordered. Note that pre-ordered lfp categories are equivalent to algebraic lattices. Thus the category **R**(**1**), whose objects model the constant Types, is equivalent to the category of algebraic lattices and continuous (i.e. directed sup preserving) maps. More generally, one can show that for **E** locally finitely presentable, **R**(**E**) is equivalent to a category whose objects are filtered colimit preserving functors from **E** into the lfp category of meet semilattices, **msl**(**Set**). The latter category is equivalent to the category of algebraic lattices and continuous maps possessing continuous right adjoints—so each object of **R**(**E**) in particular determines a functor into algebraic lattices and continuous maps; then the morphisms in **R**(**E**) are given by "lax natural" families of continuous maps. This is the form in which Coquand, Gunter and Winskel have studied the model—see [CE, section 5].

Turning to the formulation of the model in terms of lex categories and lex modules, the modules **D**⟶**C** which are in **R** are those induced by lex fibrations P:**D**⟶**C** whose fibres are pre-ordered and hence are equivalent to meet semi-lattices. Thus we can take the objects of **R**(**C**) to be those pseudofunctors $\mathbf{C}^{op}\longrightarrow\mathbf{Lex}$ which are in fact functors $\mathbf{C}^{op}\longrightarrow\mathbf{msl(Set)}$. Restricting the description in 4.20(ii) of the morphisms of **A**(**C**) to these objects, we find that specifying a morphism from $\mathrm{E}:\mathbf{C}^{op}\longrightarrow\mathbf{msl(Set)}$ to $\mathrm{F}:\mathbf{C}^{op}\longrightarrow\mathbf{msl(Set)}$ in **R**(**C**) amounts to giving a family of *relations* $M_U\subseteq \mathrm{E}(U)\times\mathrm{F}(U)$ ($U\in\mathbf{C}$) satisfying:

- $x'\leq x$ and $M_U(x,y)$ and $y\leq y'$ $\Rightarrow$ $M_U(x',y')$
- $M_U(x,\top)$
- $M_U(x,y)$ and $M_U(x,y')$ $\Rightarrow$ $M_U(x,y\wedge y')$
- $\alpha:U\longrightarrow V$ and $M_V(x,y)$ $\Rightarrow$ $M_U(\alpha^*(x),\alpha^*(y))$.

(One can use relations, since if $M:\mathbf{C}^{op}\times\mathbf{D}\longrightarrow\mathbf{Set}$ is a lex module and $\mathbf{D}$ is a meet semilattice, then $y\leq\top$ induces a monomorphism $M(X,y)\rightarrowtail M(X,\top)\cong 1$, so that each $M(X,y)$ has at most one element.) Describing the categories $\mathbf{R}(\mathbf{C})$ in this way also makes it easy to describe the reflection of $\mathbf{R}$ into $\mathbf{A}$ (i.e. 5.10(*vii*)): it is given on objects of $\mathbf{A}(\mathbf{C})$ by composing a pseudofunctor $\mathbf{C}^{op}\longrightarrow\mathbf{Lex}$ with the poset reflection functor $Po:\mathbf{Lex}\longrightarrow\mathbf{msl}(\mathbf{Set})$ of 5.1. Finally, *Type* is modelled in this setting simply by the opposite of the category of *finite* meet semilattices (since finitely presentable implies finite in this case); and the generic family of Types is modelled by the inclusion of this category into **msl(Set)**.

REFERENCES

[B1] J.Bénabou, *Théories relatives à un corpus,* C. R. Acad. Sc. Paris, Série A 281(1975) 831-834.

[B2] J.Bénabou, *Fibred categories and the foundations of naive category theory,* J. Symbolic Logic 50(1985) 10-37.

[B3] J.Bénabou, *Introduction to bicategories,* Lecture Notes in Math. No.47 (Springer-Verlag, Berlin - Heidelberg, 1967), pp1-77.

[BW] M.Barr and C.Wells, *Toposes, Triples and Theories,* Grundlehren der mathematischen Wissenschaften 278 (Springer-Verlag, New York, 1985).

[CKW] A.Carboni, S.Kasangian and R.F.C.Walters, *An axiomatics for bicategories of modules,* J. Pure Appl. Algebra 45(1987) 127-141.

[Cd] L.Cardelli, *A polymorphic λ-calculus with Type:Type,* SRC Report (DEC, Palo Alto, 1986).

[Ca] J.Cartmell, *Generalised algebraic theories and contextual categories,* Annals Pure Appl. Logic 32(1986) 209-243.

[C] A.Church, *A formulation of the simple theory of types,* J. Symbolic Logic 5(1940) 56-68.

[CE] T.Coquand and T.Ehrhard, *An equational presentation of higher order logic.* In: D.H.Pitt *et al* (eds), *Category Theory and Computer Science,* Lecture Notes in Computer Science No.283 (Springer-Verlag, Berlin - Heidelberg, 1987), pp40-56.

[CGW] T.Coquand, C.Gunter and G.Winskel, *Polymorphism and domain equations,* Computer Laboratory Technical Report No.107 (Cambridge University, 1987).

[CH] T.Coquand and G.Huet, *The Calculus of Constructions,* Information and Computation 76(1988) 95-120.

[Coq1] T.Coquand, *Une théorie des constructions,* Thèse de troisième cycle, Univ. Paris VII, 1985.

[Coq2] T.Coquand, *An analysis of Girard's paradox.* In: *Proc 1st Annual Symp. Logic in Computer Science* (IEEE Computer Society Press, Washington, 1986), pp227-236.

[Co1] M.Coste, *Localization, spectra and sheaf representation.* In: M.P.Fourman *et al* (eds), *Applications of Sheaves,* Lecture Notes in Math. No.753 (Springer-Verlag, Berlin - Heidelberg, 1979) pp212-238.

[Co2] M.Coste, *Localisation dans les catégories de modèles.* Thesis, Univ. Paris-Nord, 1977.

[G1] J.-Y.Girard, *Intérpretation fonctionelle et élimination des coupures de l'arithmétique d'ordre supérieur.* Thèse de doctorat d'état, Univ. Paris VII, 1972.

[G2] J.-Y.Girard, *The system F of variable types, fifteen years later,* Theoretical Computer Science 45(1986) 159-192.

[GU] P.Gabriel and F.Ulmer, *Lokal präsentierbare Kategorien,* Lecture Notes in Math. No.221 (Springer-Verlag, Berlin - Heidelberg, 1971).

[Gr] J.W.Gray, *Fibred and cofibred categories.* In: S.Eilenberg *et al* (eds), *Proc. LaJolla Conference on Categorical Algebra* (Springer-Verlag, New York, 1966) pp21-83.

[HHP] R.Harper, F.Honsell and G.Plotkin, *A framework for defining logics.* In: *Proc. 2nd Annual Symp. Logic in Computer Science,* (IEEE Computer Society Press, Washington, 1987), pp194-204.

[HRR] J.M.E.Hyland, E.P.Robinson and G.Rosolini, *The discrete objects in the effective topos,* preprint, Cambridge, 1987.

[J1] P.T.Johnstone, *Topos Theory,* L.M.S. Monographs No.10 (Academic Press, London, 1977).

[J2] P.T.Johnstone, *Injective toposes.* In: B.Banaschewski and R.-E.Hoffmann (eds), *Continuous Lattices,* Lecture Notes in Math. No.871 (Springer-Verlag, Berlin - Heidelberg, 1981), pp284-297.

[J3] P.T.Johnstone, *Factorization theorems for geometric morphisms, I,* Cahiers Top. et Géom. Diff. 22(1981) 3-17.

[JJ] P.T.Johnstone and A.Joyal, *Continuous categories and exponentiable toposes,* J. Pure Appl. Algebra 25(1982) 255-296.

[JT] A.Joyal and M.Tierney, *An extension of the Galois theory of Grothendieck,* Mem. Amer. Math. Soc. 51(1984), No.309.

[La] F.Lamarche, *A simple model of the theory of constructions.* In this volume.

[LS] J.Lambek and P.J.Scott, *Introduction to higher order categorical logic,* Cambridge Studies in Advanced Math. 7 (Cambridge University Press, 1986).

[M-L1] P.Martin-Löf, *Constructive mathematics and computer programming.* In: L.J.Cohen *et al* (eds), *Sixth International Congress for Logic Methodology and Philosophy of Science* (North-Holland, Amsterdam, 1982), pp153-175.

[M-L2] P.Martin-Löf, *Intuitionistic Type Theory. (Notes by G.Sambin of a series of lectures given in Padua, June 1980)* (Bibliopolis, Napoli, 1984).

[MiPl] J.C.Mitchell and G.D.Plotkin, *Abstract types have existential type.* In: *Proc. 12th Annual Symp. Principles of Programming Languages,* (ACM, New York, 1985) pp237-51.

[MP] M.Makkai and A.M.Pitts, *Some results on locally finitely presentable categories,* Trans Amer. Math. Soc. 299(1987) 473-496.

[MR] M.Makkai and G.E.Reyes, *First Order Categorical Logic,* Lecture Notes in Math. No.611 (Springer-Verlag, Berlin - Heidelberg, 1977).

[Ob] A.Obtułowicz, *Categorical and algebraic aspects of Martin-Löf type theory,* preprint, 1987.

[Pi1] A.M.Pitts, *Conceptual completeness for first order intuitionistic logic: an application of categorical logic,* Annals Pure Appl. Logic, *to appear.*

[Pi2] A.M.Pitts, *Polymorphism is set theoretic, constructively.* In: D.H.Pitt *et al* (eds), *Category Theory and Computer Science,* Lecture Notes in Computer Science No.283 (Springer-Verlag, Berlin - Heidelberg, 1987), pp12-39.

[Po] A.Poigné, *Tutorial on algebra categorically.* In: D.H.Pitt *et al* (eds), *Category Theory and Computer Programming,* Lecture Notes in Computer Science No.240 (Springer-Verlag, Berlin - Heidelberg, 1986) pp76-102.

[PS] R.Paré and D.Schumacher, *Abstract families and the adjoint functor theorems.* In:

P.T.Johnstone and R.Paré (eds), *Indexed Categories and their Applications*, Lecture Notes in Math. No.661 (Springer-Verlag, Berlin - New York, 1978), pp1-125.

[RW1] R.Rosebrugh and R.J.Wood, *Cofibrations in the bicategory of topoi*, J. Pure Appl. Algebra 32(1984) 71-94.

[RW2] R.Rosebrugh and R.J.Wood, *Cofibrations II: left exact right actions and composition of gamuts*, J. Pure Appl. Algebra 39(1986) 283-300.

[Se1] R.A.G.Seely, *Locally cartesian closed categories and type theory*, Math. Proc. Camb. Phil. Soc. 95(1984) 33-48.

[Se2] R.A.G.Seely, *Categorical models for higher order polymorphic lambda calculus*, J. Symbolic Logic 52(1987) 969-989.

[Sc] D.S.Scott, *Data types as lattices*, SIAM J. Computing 5(1976) 522-587.

[S] R.Street, *Fibrations in bicategories*, Cahiers Top. et Géom. Diff. 21(1980) 111-160 and *(corrections)* 28(1987) 53-56.

[Ta] P. Taylor, *Recursive domains, indexed category theory and polymorphism*, Thesis, University of Cambridge, 1986.

[Tr] A.S.Troelstra, *On the syntax of Martin-Löf's theories*, Theoretical Computer Science 51(1987) 1-26.

DEPARTMENT OF PURE MATHEMATICS,
UNIVERSITY OF CAMBRIDGE,
CAMBRIDGE CB2 1SB, ENGLAND

and

MATHEMATICS DIVISION,
UNIVERSITY OF SUSSEX,
BRIGHTON BN1 9QH, ENGLAND.

Contemporary Mathematics
Volume **92**, 1989

A SIMPLE MODEL OF THE THEORY OF CONSTRUCTIONS

François Lamarche

The theory of semigranular categories and aggregates [6] , [7] has been developped as a tool to provide models for polymorphic type theories, following the lead of [5] . In this paper we present a simple model (meaning that one can calculate simple types) for the theory of constructions [2] which uses only the restriction of the theory to posets (the formal system therein is not amenable to a semantical treatment; it has been modified for this purpose by more than one author, see below). We do not give the proofs in full details, but we feel we provide enough information for the dedicated reader to fill in the gaps without too much trouble.

During the last year a general categorical framework for interpreting type theories "à la Martin-Löf" has emerged. This framework is characterized by a remarkable simplicity, in comparison with the formal system (judgements, and the like) which is interpeted in it. A detailed account is in Hyland-Pitts [9] , in this volume, which we will use as a reference. The authors give a large formal system, which they call "the" theory of constructions, and then an axiomatic setting (a category with added structure and some universal properties) for interpreting it. The formal system differs from the original [2] in essentially three respects:

a) Reduction is controlled by a typing discipline, which guarantees soundness of semantics. The controlling is done by introducing equality judgements, which are a little bit more than a "typing discipline". Their categorical counterpart is the idea of presenting a category with generators and relations.
   η-reduction is added to β-reduction, since adjunctions always obey this rule.

b) Sum types are added, i.e. Σ .

c) Unit types are added, corresponding to one-element sets.

In the model we present it is possible to interpret this formal system except the sum types. A study of [9] shows that all mentions of sum types can be removed from the syntax (and from the categorical interpretation) without affecting the other parts. It goes without saying that any subsystem of this theory of constructions can be interpreted in our model.

The terminology we will use is guaranteed to be nonstandard since the subject is much too recent to have any standards (this paper was written independently from [9]). In the two paragraphs that follow we will use square brackets to give the corresponding term used by Hyland-Pitts.

Definition

A Martin-Löf category $\mathfrak{C}$ is a (locally small) category with finite products and a distinguished subclass $\mathfrak{F} \subset \mathfrak{C}$ of arrows, the fibrations [display maps] , such that the pullback of any fibration by any morphism of $\mathfrak{C}$ exists and is a fibration [Stability] . For any object $X \in \mathfrak{C}$ we define $\mathfrak{F}_X$ to be the full subcategory of $\mathfrak{C}/X$ whose objects are fibrations. For any $f{:}X \to Y$ in $\mathfrak{C}$ the condition above defines a pullback functor $f^*{:}\mathfrak{F}_Y \to \mathfrak{F}_X$. To be able to interpret Π in $\mathfrak{C}$ we require that if f is a fibration then f^* has a right adjoint Π_f , and that the Beck condition hold: if

$$\begin{array}{ccc} Z & \xrightarrow{z} & X \\ g\downarrow & & \downarrow f \\ W & \xrightarrow[w]{} & Y \end{array}$$

is a pullback, with $f,g \in \mathfrak{F}$ and if $\varphi \in \mathfrak{F}_X$ then the canonical morphism $w^*\Pi_f\varphi \to \Pi_g z^*\varphi$ in $\mathfrak{F}_X$ is an isomorphism [Products] .

We will say an object X of $\mathfrak{C}$ is a $\mathfrak{C}$-class if the unique $X \to 1$ to the terminal object is a fibration. The condition [Display] corresponds to saying that every object is a $\mathfrak{C}$-class; it is necessary only if one wants a back-and forth correspondence between theories and categories, and will not be satisfied in our model. One should think of a fibration $Z \to Y$ as an indexed family $(Z_y)_{y \in Y}$ of classes Z_y . Then, for any morphism $f{:}X \to Y$ the pullback should be thought of as the family $(Z_{f(x)})_{x \in X}$. For $\varphi, \psi \in \mathcal{F}_X$ we define the exponentiation $\varphi \Rightarrow \psi$ to be $\Pi_\varphi \varphi^* \psi$. Therefore the operator $\Rightarrow$ is defined only on pairs of fibrations having a common codomain. The Beck condition ensures that pullback preserves exponentiation. The definition of $\Rightarrow$ implies that whenever $\mathcal{F}_X$ has products, it is cartesian-closed, and in particular $\mathcal{F}_1$ always is cartesian-closed. In model categories that one meets in practice, $\mathcal{F}_X$ has all finite products for any X , and in particular [Unit] isomorphisms are fibrations. This allows us to interpret a version of Martin-Löf theory where Σ can be applied to constant families, giving binary products of variable types. A stronger condition, not always met in practice, is to require that $\mathcal{F}$ be a subcategory of $\mathcal{C}$, allowing us to interpret Σ in full [Sums] . All that we need to add to a Martin-Löf category to interpret the theory of constructions is:

Definition

A Coquand category is a Martin-Löf category $(\mathcal{C},\mathcal{F})$, along with a distinguished fibration $\lambda:\Lambda \to S$ [Generic Type for $\mathcal{F}^s$] such that S is a $\mathcal{C}$-class and if $\mathcal{F}^s_X$ is the full subcategory of $\mathcal{F}_X$ composed of objects isomorphic to fibrations of the form $x^*\lambda$, for some $x:X \to S$, then for any $f:X \to Y$ in $\mathcal{F}$ the restriction of Π_f to $\mathcal{F}^s_X$ lands in $\mathcal{F}^s_Y$. We call $\mathcal{F}^s_X$ the category of small fibrations above X . An object X such that $X \to 1$ is a small fibration is called a small $\mathcal{C}$-class , or a $\mathcal{C}$-set . Obviously, pulling back by any morphism sends a small fibration to a small fibration. One should think of a small fibration as an indexed family of $\mathcal{C}$-sets, and of the object S as the $\mathcal{C}$-class of all $\mathcal{C}$-sets.

We can now start describing the model per se.

Definitions

Let X be a partially ordered set. X is said to be multi cocomplete [4] , or alternately to have multisups , if the following holds: for any $A \subset X$ there exists $A^+ \subset X$ such that all $a \in A^+$ are upper bounds for A , and for any upper bound x to A there exists a unique $a \in A^+$ with $a \leq x$. We will call A^+ , which can be empty, the multisup family of A . This amounts to saying that every connected component of the subposet of upper bounds of A has a bottom element, and thus shows that A^+ is uniquely defined. We call an $a \in A^+$ a sup candidate for A (or simply an A-candidate, or even a candidate). If x is an upper bound of A then the unique

$a \in A^+$ with $a \le x$ is called the A-candidate determined by x. In particular, if we take $A = \emptyset$, we get that every component of X itself has a bottom element; we will call these the empty elements of X. An example of a multicocomplete poset is one which is consistently (co)complete [10] : every bounded subset has a supremum. This is equivalent to saying that multisup families are either empty or singletons. This forces the poset to be connected if nonempty. We say that $x \in X$ is prime if whenever $A \subset X$ is such that $x \in A^+$, then $x \in A$. Notice this prevents x from being empty. We say x is atomic if it is not empty, and $y < x$ implies that y is empty. Clearly, atoms are prime.

Proposition

Let X be a multicocomplete poset. Let $A \subset X$ be such that $A \cap A^+ \neq \emptyset$. Then A^+ is a singleton, i.e. A has a sup.

This is because if $a \in A \cap A^+$ then a is smaller than any other element of A^+ and then the uniqueness clause in the definition of A^+ forces any element of it to be equal to a. ///

Proposition

Let $A \subset X$ non empty be contained in a unique component of X (equivalently, A has some lower bound). Then A has an infimum. In particular, X has pullbacks.

Look at B the set of lower bounds for A. B^+ is obviously nonempty, since elements of A bound B. Since every element of A is above some element of B^+ we have $B \cap B^+ \neq \emptyset$. But then $\sup B$ is the inf of A, as usual. ///

Suppose now X has directed (filtered) sups. Recall that $x \in X$ is compact (or finitely presented, or isolated) if whenever $A \subset X$ directed is such that $x \le \sup A$ then there exists $a \in A$ such that $x \le a$ (remember that a directed set is always nonempty). We say x is finitary if the set $x{\downarrow} = \{y \in X \mid y \le x\}$ is finite.

Definition

An aggregate poset (which in this paper shall simply be called an aggregate) is a poset X such that

a) X has directed sups and multisups.
b) Every prime of X is both finitary and compact, and every $x \in X$ is a sup candidate for a set of primes.

c) If x is a sup candidate for A, and $p \le x$ a prime, then there is $a \in A$ with $p \le a$.

An aggregate is said to be _granular_ if every one of its primes is atomic.

We list some of the consequences of that definition:

Proposition

i) An element is compact iff it is finitary (this warrants calling a compact element a _finite_ one).
ii) For every $x \in X$ the set of compact elements below it is directed and x is the sup of this set (X is sometimes said to be _algebraic_).
iii) If x is a sup candidate for $(x_i)_{i \in I}$ and $y \le x$ then y is a sup candidate for $(y \wedge x_i)_i$ (candidates are preserved under pullbacks).
iv) If X is granular, then the subposet $x\downarrow$ is a complete atomic boolean algebra for every $x \in X$. ///

In the light of this it follows easily that an aggregate which is consistently cocomplete is the same as a DI domain [3]. Also, it is easy to see (exercise) that a (non-empty) consistently cocomplete granular poset is always isomorphic to a qualitative domain [5] seen as a poset. We will denote the set of finite elements of an aggregate X by X_f, and its set of connected components by X^*. Notice that $(-)^*$ defines a functor from the category of aggregates and stable functions (see below) to the category of sets. If $\alpha \in X^*$ we denote the bottom element of α by 0_α. If X is granular we denote its set of atoms by $|X|$.

Proposition

Let X be granular. Then for any $a \in X$ there is a unique $|a| \subset |X|$ such that $a \in |a|^+$. ///

Definition

If $\mathbb{C},\mathbb{D}$ are categories with filtered colimits and pullbacks, then an _entire functor_ $F:\mathbb{C} \to \mathbb{D}$ is a functor which preserves both. If $\mathbb{C}$ and $\mathbb{D}$ are now posets considered as categories we will call an entire functor a _stable function_.

Definition

Let X,Y be posets, and $f:X\to Y$ be a morphism of posets. Let $x\in X$, $a\in Y$ be such that $a\le f(x)$. We say (a,x) is a <u>generic pair</u> (or simply <u>$a\le f(x)$ is generic</u>) if whenever we have $y\in X$ such that $a\le f(y)$ and $\{x,y\}$ has an upper bound, then $x\le y$.

Proposition

Let X,Y be posets with pullbacks, and let $f:X\to Y$ preserve them. Then $a\le f(x)$ is generic iff $z\le x$ and $a\le f(z)$ implies that $z=x$. ///

Proposition

Let X,Y be aggregates and $f:X\to Y$ a morphism of posets. Then f is stable iff for any prime $p\in Y$ and any $y\in X$ such that $p\le f(y)$ there exists an $x\le y$ such that $p\le f(x)$ is generic and x finite.

The proof of "if" is exactly like that of [5] , in view of the preceding proposition and the fact that every element of X is the sup of a directed set of finite elements, and that finite elements are finitary. ///

If $f,g:X\to Y$ are stable functions between aggregates, then the Berry ordering is defined just as in [5] , i.e. it is a natural transformation $f\to g$ such that the usual commutative squares defining naturality are pullbacks. We also get that $f\le g$ iff $x\le f(a)$ generic $\Rightarrow$ $x\le g(a)$ generic for all x prime in Y . We will prove later on that the category of granular posets and stable functions is cartesian-closed, with the exponential object Y^X being the poset $\mathit{Stab}(X,Y)$ of stable functions $X\to Y$ with the Berry order. It suffices now to notice (exercise) that if X,Y are aggregates then the product poset $X\times Y$ is also an aggregate, and is the categorical product in the category of aggregates and stable functions. If X,Y are granular, then $X\times Y$ is granular too. The terminal object for both granular and aggregates is the one-element poset 1 .

Notation

If X is a set, we denote the set of all two-element subsets of X by $\mathbb{P}^2(X)$

Definition

Let X be an aggregate. A _2-network_ in X is a pair (A,α), where $A \subset X$, and α is a function $\alpha: \mathcal{P}^2(A) \to X$ sending any $\{a,b\} \in \mathcal{P}^2(X)$ to a candidate $\alpha_{a,b}$ for $\{a,b\}$. We say X is _binary_ if given any 2-network (A,α) there exists a unique A-candidate $z \in X$ such that for any $\{a,b\} \in \mathcal{P}^2(X)$ the $\{a,b\}$-candidate determined by z is $\alpha_{a,b}$. We will give a structure theorem for nonempty connected binary semigranular posets.

Definition

A (non oriented, antireflexive) _multigraph_ Q is a triple (Q_1,Q_2,ξ), where Q_1, Q_2 are sets, called the set of _nodes_ and the set of _edges_ respectively, and ξ is a function $Q_2 \to \mathcal{P}^2(Q_1)$. Every multigraph Q gives rise to a poset $\hat{Q}$ as follows: an element of $\hat{Q}$ is a pair (M,μ) where $M \subset Q_1$ and $\mu: \mathcal{P}^2(M) \to Q_2$ is such that $\xi \circ \mu$ is the identity on $\mathcal{P}^2(M)$. We say that $(M,\mu) \le (N,\nu)$ if $M \subset N$ and μ is the restriction of ν on $\mathcal{P}^2(M)$.

Theorem 1

A nonempty granular poset X is connected and binary iff it is isomorphic to $\hat{Q}$ for a multigraph Q.

Let us sketch the proof. If $Q = (Q_1,Q_2,\xi)$ is a multigraph, it is easy to see that $\hat{Q}$ is a connected binary granular poset. For example, if $(M_i,\mu_i)_{i \in I}$ is a family of elements of $\hat{Q}$, and $M = \bigcup_i M_i$, then a candidate for $(M_i)_i$ is an element of $\hat{Q}$ of the form (M,ν) where the restriction of ν to M_i is μ_i. Given X binary, granular and connected, we construct Q such that $\hat{Q} \cong X$ as follows: $Q_1 = |X|$. If $x,y \in |X|$, $x \neq y$, then the set $\xi^{-1}\{x,y\}$ is $\{x,y\}^+$, and this suffices to define both Q_2 and ξ. ///

This generalizes the Girard approach to coherent domains: a _coherent domain_ is a nonempty consistently cocomplete binary granular poset. It is easy to see that $\hat{Q}$ is a coherent domain iff ξ is an injective function, i.e. the multigraph Q is simply a symmetric, antireflexive relation on Q_1. We can simply call such a multigraph a _graph_. Therefore, if X is a coherent domain, we define its associated graph to be the pair $(|X|, \tilde{X})$, where $\tilde{X} \subset \mathcal{P}^2(|X|)$ is the set of all $\{x,y\}$ which have an upper bound in X.

Definition

An aggregate X is finitary if for any finite $A \subset X_f$ the set A^+ is finite.

Proposition

If Q is a multigraph, then $\hat{Q}$ is finitary iff for any $K \in \mathcal{P}^2(Q_1)$ the set $\xi^{-1}(K)$ is finite. In other words, to prove finitaryness for a binary granular poset, one can check the finite multisup property for finite families of atoms only. ///

We are now ready to construct Martin-Löf categories. For the time being $\mathfrak{C}$ will be the category of all aggregates and stable functions. First let us take $\mathfrak{C}$-classes to be granular posets. We want to define a notion of fibration $E \rightarrow X$ corresponding to an X-indexed family of granular posets:

Definition

Let X,Y be granular. A strong morphism $f:X \rightarrow Y$ is a morphism of posets sending atoms of X to atoms of Y such that the function $|f|:|X| \rightarrow |Y|$ thus defined is injective and has the following property: for any $A \subset |X|$, $f(A^+) = f(A)^+$. This implies that f has a right adjoint f^-, for if $a \in Y$ there is $|a| \in Y$ with $a \in |a|^+$. Then if $B = |a| \cap |f|(|X|)$ let $b \leq a$ be the B-candidate determined by a. Since $|b|$ is contained in the image of $|f|$ there is a unique $f^-(a) \in X$ such that $f(f^-(a)) = b$. It follows that for any $c \in X$, $a \in Y$, $f^-(f(c)) = c$ and $f(f^-(a)) \leq a$ and that is enough to show that f^- is a right adjoint to f. This shows that X is isomorphic to the subposet Z of Y which is generated by $|f|(|X|)$, in the sense that $C \subset Z$ implies $C^+ \subset Z$. We leave it to the reader to show that f, f^- are stable and that $f \circ f^- \leq 1_Y$ for the Berry order (we recommend the use of generic elements).

Let *Poset* be the category of posets and morphisms, X a poset, and let $H:X \rightarrow$ *Poset* be some functor. That is, for every $s \in X$ there is a poset $H(s)$ and for $s \leq t$ there is a morphism $H^s_t:H(s) \rightarrow H(t)$ with $H^s_t \circ H^r_s = H^r_t$ when $r \leq s \leq t$. Let el(H) be the following poset: an element of el(H) is a pair (s,a) where $s \in X$ and $a \in H(x)$. We say $(s,a) \leq (t,b)$ iff $s \leq t$ and $H^s_t(a) \leq b$ (this is a special case of the Grothendieck construction; the verification that el(H) is indeed a poset is left to the reader). There is a morphism $h: el(H) \to X$ of posets defined by $h(s,a) = s$. A remarkable fact is that the full functor H can be recovered from the order structure of $el(H)$ and from h. First, it is obvious that for any $s \in X$, $H(s) \cong h^{-1}(s)$. Then, if $(s,a) \in el(H)$, $t \geq s$, we want $(t, H^s_t(a))$ to be entirely defined by a property intrinsic to the order structure. This property exists:

> $(t, H^s_t(a))$ is the unique $w \in el(H)$ such that $(s,a) \leq w$, $h(w) = t$, and for any $u \geq t$, $(u,b) \in el(H)$ such that $(s,a) \leq (u,b)$, we have $(t, H^s_t(a)) \leq (u,b)$.

The inexperienced reader is urged to verify this property. It entails the following definition:

Definition

Let $e: E \to X$ be some morphism of posets. We say a pair $v \leq w$ of elements of E is cocartesian if for any $y \geq v$ such that $e(w) \leq e(y)$ we have that $w \leq y$. We say e is an opfibration if for any $x \in E$, $t \geq e(x)$ there is $w \in E$ with $e(w) = t$ and $x \leq w$ is cocartesian. When $v \leq w$ is cocartesian we will write $v \preccurlyeq w$ or $w \succcurlyeq v$.

Let $e: E \to X$ be an opfibration. It is easy to prove that for v, w, w' in E, $v \preccurlyeq w$, $v \preccurlyeq w'$ and $e(w) = e(w')$ implies $w = w'$. If, for $s \in X$ we denote $e^{-1}(s)$ by E^s, a very slight additional effort shows that for $s \leq t$ in X the cocartesian pairs define a morphism $E^s_t: E^s \to E^t$, and that E^s_t varies functorially with s, t, because $v \preccurlyeq w$ and $w \preccurlyeq x$ implies $v \preccurlyeq x$. We have thus constructed a functor $X \to$ *Poset*, and this construction and the Grothendieck construction are inverse to each other. Notice that for $H: X \to$ *Poset* and $f: Y \to X$ some morphism of posets, the opfibration $el(H \circ f) \to Y$ is isomorphic to the pullback of $el(H) \to X$ by f.

Let X be an aggregate. We are interested in functors $X \to \mathcal{G}$, where $\mathcal{G}$ is the category of granular posets and strong morphisms, and in the opfibrations corresponding to them.

Definition

A granular fibration is a morphism $e: E \to X$ of posets where X is an aggregate such that

a) e is an opfibration, and for any $s \in X$ the fiber E^s above s is granular.

b) For any $s \leq t$ in X the morphism $E_t^s : E^s \to E^t$ is a strong morphism of granular posets. If $a \in E^s$, $b \in E^t$ with $s \leq t$ we will denote the application of the right adjoint of E_t^s to b by $b_{|s}$. Also, we will sometimes write $E_t a$ for $E_t^s(a)$, i.e. $E_t c$ is defined for any $c \in E$ such that $e(c) \leq t$. We will call an element x of E which is atomic in its fiber $E^{e(x)}$ a <u>semiatom</u>.

Here are some consequences of the definition of a granular fibration.

Proposition

Let $e: E \to X$ be a granular fibration. Then

i) E has filtered sups, multisups, and e preserves all the stucture that exists in E : filtered sups, existing infs, and if x is a candidate for $(x_i)_{i \in I}$ then $e(x)$ is a candidate for $(e(x_i))_i$.

ii) The inclusion of any fiber $E^s \to E$ also preserves all the structure as above.

iii) If $a \in E^s$ is such that a is finite in E^s and s is finite in X, then a is finite in E. ///

The definition of a granular fibration is not entirely satisfactory. For example, we cannot show yet that E is an aggregate. The category $\mathcal{G}$ can be shown to have filtered colimits and pullbacks, and it turns out that the right notion for us is a functor $X \to \mathcal{G}$ which is *entire*. This can be described by a form of generic element.

Definition

Let $e: E \to X$ be a granular fibration. We say that $x \in E$ is <u>E-generic</u> if whenever $x \lessdot y$ and $z \lessdot y$ then $x \leq z$. We say that e is an <u>entire fibration</u> if for every semiatom y in E there exists an E-generic x (necessarily a semiatom) such that $x \lessdot y$ and $e(x)$ is finite in X.

Proposition

Let $\varphi: \Phi \to X$ be an entire fibration. Then

i) Φ is an aggregate.

ii) If $Y \to X$ is a stable function then the pullback $\Phi' \to Y$ is an entire fibration, and is the pullback in $\mathfrak{C}$. In particular, if Z is granular then the projection $Z \times X \to X$ is an entire fibration ($Z \to 1$ is obviously an entire fibration).

iii) If $\psi: \Psi \to X$ is another entire fibration then the pullback $\Phi \times_X \Psi \to X$ is an entire fibration.

To prove i) we first show that if $x \in E$ is E-generic and semiatomic (implying that $e(x)$ is finite), or it is empty in $E^{e(x)}$ where $e(x)$ is prime in X, then x is prime in E. It follows easily that those primes are finite in E, that every element of E is a candidate for a set of primes, and that axiom c) for aggregates holds. ///

For X an aggregate, let Fib_X be the full subcategory of $\mathcal{C}/X$ whose objects are entire fibrations. Part iii) of the proposition above says that Fib_X has finite products; it is well known that (notation as above) for $s \in X$, $(\Phi \times_X \Psi)^s = \Phi^s \times \Psi^s$. Also, $Fib_X(\varphi,\psi)$ can be made into a poset via the Berry order, since $Fib_X(\varphi,\psi)$ is a subset of the set of stable functions $\Phi \to \Psi$.

Definition

Let $h \in Fib(\varphi,\psi)$. Let $a \in \Phi$, $x \in \Psi$ be such that $\varphi(a) = \psi(x)$, x is semiatomic and $x \le h(a)$. We say $x \le h(a)$ is <u>ultrageneric</u> if for any semiatom $y \lessdot x$, $y \le h(a)$ is generic. In particular, whenever $x \le h(a)$ is ultrageneric, x is semiatomic and $x \le h(a)$ is generic itself.

Proposition

Let φ,ψ,h as above. Then h is stable iff for any $c \in \Phi$, $y \in \Psi$ semiatomic such that $\varphi(c) = \psi(y)$ and $y \le h(c)$ in Ψ there exist $a \le c$ and $x \lessdot y$ with $x \le h(a)$ ultrageneric and a finite in its fiber. If k is another element of $Fib(\varphi,\psi)$ then $h \le k$ iff for all $x \in \Psi$ and $a \in \Phi$ such that $x \le h(a)$ is ultrageneric we have that $x \le k(a)$ and is ultrageneric. ///

Theorem 2

Let X be an aggregate, $\varphi:\Phi \to X$ and $\psi:\Psi \to X$ be entire fibrations. Then $Fib_X(\varphi,\psi)$ is a granular poset.

Let $q \in \mathit{Set}/X^*(\varphi^*,\psi^*)$. That is, q is a function $\Phi^* \to \Psi^*$ such that $\psi^* \circ q = \varphi^*$. We can define a function $g_q:\Phi \to \Psi$ as follows: if $a \in \Phi$ is in connected component α then $g_q(a) = 0_\alpha$. It is easy to check that g_q is actually an element of $Fib_X(\varphi,\psi)$. Also, if h is any element of $Fib_X(\varphi,\psi)$ then we have $g_{h^*} \le h$. Therefore, the set of components of $Fib_X(\varphi,\psi)$ is isomorphic to $\mathit{Set}/X^*(\varphi^*,\psi^*)$ and every component has a bottom element of the form g_q. For $q \in \mathit{Set}/X^*(\varphi^*,\psi^*)$, $a \in \Phi$, $x \in \Psi$ with $\varphi(a) = \psi(x)$, a finite in $|\Phi^{\varphi(x)}|$

and x semiatomic, define $[q,a,x]:\Phi\to\Psi$ by

$$[q,a,x](b) = \Psi_{e(b)}x \text{ if } a\le b \text{ , otherwise}$$
$$= 0_{q(\alpha)} \text{ if } b \text{ is in component } \alpha \text{ .}$$

We claim $[q,a,x]$ is an element of $\mathrm{Fib}_X(\varphi,\psi)$, and that all atoms therein are of this form. The first statement is easy to prove, and for the second we just have to show that if $y\le[q,a,x](c)$ is ultrageneric then $y=x$ and $c=a$. This will imply that if $h\in\mathrm{Fib}_X(\varphi,\psi)$ and if $x\le h(a)$ is ultrageneric, with $x\in|\Psi|$, then $[h^*,a,x]\le h$. It is then a formality to check that the axioms of a semigranular poset are verified. The multisups are calculated pointwise, i.e., k is a candidate for $(h_i)_{i\in I}$ iff $k\ge h_i$ and for any $a\in\Phi$, $k(a)$ is a candidate for $(h_i(a))_i$. ///

Theorem 3

If $f:Y\to X$ is an entire fibration then the pullback functor $f^*:\mathrm{Fib}_X\to\mathrm{Fib}_Y$ has a right adjoint Π_f. The Beck condition holds for any pullback square with two parallel entire fibrations.

If $\varphi:\Phi\to Y$ is an entire fibration we define $\Pi_f\varphi$ to be the granular fibration $e:E\to X$ such that for $s\in X$,

$$E^s = \{h:Y^s\to\Phi \mid h \text{ is stable and } \forall_{x\in Y^s}\ \varphi(h(x))=x\} \text{ .}$$

E^s has the Berry order induced by its inclusion in $\mathit{Stab}(Y^s,\Phi)$. The preceding proposition shows that E^s is granular, since it is isomorphic to the poset $\mathrm{Fib}_{Y^s}(1_{Y^s},I^*\varphi)$, where $I^*\varphi$ is the pullback of φ by the inclusion $I:Y^s\to Y$. Also, we know that

$$|E^s| = \{[q,a,x] \mid q\in \mathit{Set}/Y^{s*}(1_{Y^{s*}},I^*\varphi)\ ,a\in Y^s_f\ ,x\in|\Phi^a|\} \text{ .}$$

If $s,t\in X$ and $s\le t$ we define $E^s_t:E^s\to E^t$ to send $h:Y^s\to\Phi$ to the stable $Y^t\to\Phi$ such that

$$(E^s_t(h))(a) = \Phi_a(h(a|_s)) \text{ .}$$

To show that E^s_t is a strong morphism, and then that e is an entire fibration, we use the characterization of the semiatoms of E given in the preceding theorem. The order structure in E can be shown to be as follows: for $h\in E^s$ and $k\in E^t$ we have $h\le k$ iff

i) $s\le t$,
ii) for all $a\in Y^s$, $h(a)\le k(Y^s_t(a))$ in Φ,

iii) for $a \le b$ in Y^s the square

$$\begin{array}{ccc} h(a) & \rightarrow & k(Y^s_t(a)) \\ \downarrow & & \downarrow \\ h(b) & \rightarrow & k(Y^s_t(b)) \end{array}$$

is a pullback.

This is equivalent to having $s \le t$, and for all $c \in Y^t$ the inequality $h(c_{|s}) \le k(c)$ in Φ, along with the corresponding pullback condition. We can now describe the co-unit $\varepsilon : f^*e \rightarrow \varphi$. This is just evaluation: if f^*e is denoted by $d:D \rightarrow Y$, an element of D is a pair (a,h), where $a \in Y$ and $h \in E^{f(a)}$. Since $h: Y^{f(a)} \rightarrow \Phi$ we can put

$$\varepsilon(a,h) \quad = \quad h(a) \in \Phi^a .$$

Showing that $\varepsilon \in Fib_Y(f^*e,\varphi)$ is done using ultrageneric elements. One can then verify the universal property of ε and the Beck condition, which is closely related to the fact that Π_f is calculated by taking the product (the poset of all splittings $Y^s \rightarrow \Phi$) fiberwise.

///

If P is a property of granular posets (say, of being binary, finitary, coherent, etc) we say an entire fibration $e: E \rightarrow X$ is a <u>P fibration</u> if all the fibers of E have property P. We say the aggregate X is <u>separable</u> if it is finitary, has a countable or finite set of primes and a finite set of connected components. If X is separable a separable fibration is the same as an entire fibration $E \rightarrow X$ where E is separable.

Proposition

Let P be one of the properties below, $f:Y \rightarrow X$ an entire fibration between aggregates. Then Π_f sends any P fibration to a P fibration.

i) binary

ii) binary and finitary

iii) binary and consistently cocomplete, i.e. "coherent domain or the empty poset". We will call such a fibration a Coherent fibration

In addition, if f is a separable fibration then Π_f sends separable fibrations to separable fibrations and one can add the word "separable" to the three clauses above.

The proof that Π_f preserves binary fibrations and consistent cocompleteness is a consequence of the fact that multisups are calculated pointwise in each fiber, as is said in Theorem 2 . To prove the preservation of finitariness, we need the following observation: Let Z be an aggregate, $\varphi:\Phi\to Z$ and $\psi:\Psi\to Z$ binary entire fibrations. Let $h=[q,a,x]$ and $k=[q,b,y]$ be atomic morphisms of fibrations $\varphi\to\psi$ belonging to the same component . Let $(W_c)_{c\in\{a,b\}^+}$ be the family of sets

$$W_c = \{h(c),k(c)\}^+ \text{ if } h(c)\neq k(c)\ ,\ \emptyset \text{ otherwise.}$$

Then the set $\{h,k\}^+$ is isomorphic to the set of all functions $g:\{a,b\}^+\to\amalg_c W_c$ such that $g(c)\in W_c$. The verification that the separability condition is conserved follows directly from the definitions. ///

We now have a variety of notions of fibration we can use to define $\mathfrak{F}$ and obtain a Martin-Löf category on $\mathfrak{C}$. We can also get a Coquand structure, but in order to do so we have to redefine $\mathfrak{C}$ as the category of separable aggregates and stable functions. We will define the object S as a poset derived from a multigraph.

Let $Q=(Q_1,Q_2,\xi)$ be the following multigraph. Q_1 the set of nodes is $\mathbb{N}$, the set of natural numbers. $Q_2=\mathfrak{P}^2(\mathbb{N})\times\{0,1\}$ and $\xi:Q_2\to\mathfrak{P}^2(Q_1)$ is the projection, that is, we get a multigraph for which every two different nodes are connected by two edges labeled 0 and 1 .

We define a functor $H:\hat{Q}\to$ *Coh* , where *Coh* is the category of coherent domains and strong morphisms. It sends $(M,\mu)\in\hat{Q}$ to the coherent domain whose graph has M for set of nodes, and for which $\{m,n\}$ is an edge iff $\mu(\{m,n\})=(\{m,n\},1)$. It is quite easy to see that H is a functorial construction, and therefore, taking the Grothendieck construction we get a granular fibration $\ell:L\to\hat{Q}$. It is also easy to prove that ℓ is entire, and it is obviously a coherent fibration, and finally a separable one. If we take $S=\hat{Q}+1$ (disjoint sum) and $\lambda:\Lambda\to S$ to be the fibration which will have the same fibers as ℓ on $\hat{Q}$ and the empty fiber on 1 , then λ will classify all the Coherent separable fibrations, and we thus get a Coquand category on our smaller $\mathfrak{C}$. Let us sketch the proof of this claim: let $e:E\to X$ be a Coherent fibration. Since the subposet of all $s\in X$ such that E^s is empty is a union of connected components, and must go to $1 \subset S$, we can assume all the fibers of E are inhabited. Let $A \subset E$ be the subset of all elements which are atomic in their fiber. In other words $A = \coprod_{s \in X} |E^s|$. Let $\sim$ be the equivalence relation obtained by taking the symmetric closure of $\leqslant$. The relation

$R(x,y)$ iff x,y are in the same fiber and have a sup in it

is compatible with $\sim$ in the sense that given $x \sim x'$, $y \sim y'$ such that $Ex = Ey$, $Ex' = Ey'$, then $R(x,y)$ iff $R(x',y')$; this is just a simple consequence of the definition of a strong morphism. Choose an injective function $\varsigma : A/\sim \to Q_1$. Let $f : X \to \hat{Q}$ send $s \in X$ to the element (M, μ) where $M = \varsigma(|E^s|/\sim)$ and $\mu\{\varsigma(x/\sim), \varsigma(y/\sim)\} = 1$ iff $R(x,y)$. An easy calculation will show f is stable and e is the pullback of ℓ by f.

The desirability of disconnected posets is now apparent: it allows us to have an empty $\mathfrak{C}$-set, and thus give a model which is "consistent" according to the Curry-Howard isomorphism.

$\mathfrak{C}$ can be made even smaller: If X is a binary aggregate and $E \to X$ a binary fibration, then E is also binary. Hence one can take $\mathfrak{C}$ to be the category of separable binary aggregates.

References

[1] J. CARTMELL Generalized algebraic theories and contextual categories, *Ann. Pure Appl. Logic* 32 (1986) 209-243.

[2] T. COQUAND Une théorie des constructions, *Thèse de 3ème cycle*, Paris VII (1985).

[3] T. COQUAND, C. GUNTER, G. WINSKELL DI-domains as a model of polymorphism, *Technical report, University of Cambridge Computer Laboratory*, N° 107

[4] Y. DIERS Familles universelles de morphismes, *Ann. Soc. Sci. Bruxelles*, 93-III (1979) 175-195.

[5] J.-Y. GIRARD The system F of variable types fifteen years later, *Theoret. Comp. Sci.* 45 (1986) 159-192.

[6] F. LAMARCHE A model of Coquand's theory of constructions, *Comptes Rendus Soc. Roy. Canada*, Vol X, No. 2 (1988) 89-94.

[7] F. LAMARCHE Modelling polymorphism with categories, *Ph. D Thesis*, McGill, (1988).

[8] G. HUET A uniform approach to type theory, in *Proceedings of the University of Texas Programming institute on logical foundations of functional programming, June 1987*, G. Huet, ed.

[9] M. HYLAND, A. M. PITTS The theory of constructions: categorical semantics and topos-theoretic models, *This volume.*

[10] D. S. SCOTT Domains for denotational semantics, *Lect. Notes Comp. Sci.* 140, Springer (1982).

Department of Mathematics
University of Pennsylvania

Contemporary Mathematics
Volume **92**, 1989

Multicategories Revisited

J. LAMBEK[1]

Abstract: This is an account of the origin of multicategories and their relation to algebra, logic and linguistics. It is shown that a multicategory is completely characterized by its internal language and how this language may be exploited to obtain properties of tensor product and internal *hom*. There is a discussion of what happens when some or all of Gentzen's structural rules are imposed.

1. SYNTACTIC CALCULUS

This article is largely historical, describing my interest in a common thread to algebra, logic and linguistics over the last thirty years or so; but I hope that it contains at least one new idea. It all began when George Findlay and I were trying to study homological algebra, just prior to the appearance of the book by Cartan and Eilenberg. We were looking at bimodules ${}_RA_S$, ${}_SB_T$ and ${}_RC_T$, R, S and T being associative rings with unity, and became obsessed with the natural isomorphisms

$$Hom(B, A\backslash C) \cong Hom(A \otimes B, C) \cong Hom(A, C/B),$$

where $A \otimes B$ is the tensor product $A \otimes_R B$, $A\backslash C$ (read A under C) is the $S - T$ - bimodule $Hom_R(A, C)$ and C/B (read C over B) is the $R - S$ - bimodule $Hom_T(B, C)$. We had noticed an analogous relationship for two-sided ideals A, B and C of a ring R, namely

$$B \subseteq A \cdot .C \Leftrightarrow A \cdot B \subseteq C \Leftrightarrow A \subseteq C. \cdot B,$$

where $A \cdot C$ is the product ideal, and $A \cdot .C = \{r \in R \mid \forall_{a \in A} ar \in C\}$ and $C. \cdot B = \{r \in R \mid \forall_{b \in B} rb \in C\}$ are two kinds of residual quotients, distinct when R is not commutative. Although our ideas on this topic were never published, I tried to push the notational distinction between left and right division in my book on ring theory (1966), but most ring theorists rejected this notation.

1980 Mathematics Subject Classification (1985): 03B40; 03F05; 18D15; 68S05; 68S20.

[1]This research was supported in part by the Natural Sciences and Engineering Research Council of Canada and the Quebec Department of Education. The author is attached to the Category Research Center in Montreal. He is indebted to Dr. Franz Guenthner for the hospitality of the Forschungsstelle für natürlich-sprachliche Systeme in Tübingen, where most of this was written.

The manuscript with Findlay had already pointed out a possible application to linguistics, which first appeared in print in 1958. The idea was this: given syntactic types N for "name" and S for "statement", and perhaps other basic types, one can build compound types $A \cdot B, A\backslash C$ and C/B, where $A \cdot B$ is the type of expressions consisting of expressions of type A followed by expressions of type B, $A\backslash C$ is the type of expressions which when preceded by expressions of type A yield expressions of type C and C/B is the type of expressions which when followed by expressions of type B yield expressions of type C. If we write $A \to B$ to mean that every expression of type A may also be assigned the type B, we notice that

$$B \to A\backslash C \Leftrightarrow A \cdot B \to C \Leftrightarrow A \to C/B.$$

In my 1958 paper I also assumed associativity in the form

$$(A \cdot B) \cdot C \leftrightarrow A \cdot (B \cdot C),$$

but I had second thoughts about this in 1961.

To illustrate these ideas, note that the name *John* has type N and that the statement *John works* has type S. Since *John* can here be replaced by any other name, it follows that *works* has type $N\backslash S$, as has any other predicate. Indeed, if we work in the so-called *associative syntactic calculus*, a formal system which also postulates the reflexive and transitive laws for the binary relation $\to$ between types, we can formally prove that $N \cdot (N\backslash S) \to S$, as follows:

$$\frac{A\backslash C \to A\backslash C}{A \cdot (A\backslash C) \to C}.$$

Similarly, in *John (likes Jane)*, the predicate *likes Jane* has type $N\backslash S$, hence *likes* has type $(N\backslash S)/N$. Indeed, $((N\backslash S)/N) \cdot N \to N\backslash S$ and, more generally,

$$\frac{C/B \to C/B}{(C/B) \cdot B \to C}.$$

Had we bracketed the above sentence differently as *(John likes) Jane*, we would have obtained the type $N\backslash(S/N)$ for *likes*. Fortunately, this does not matter, as one can prove that $(N\backslash S)/N \leftrightarrow N\backslash(S/N)$, using associativity. For example, abbreviating $(A\backslash B)/C$ as D, we have the following proof in tree form of this result in one direction:

$$\cfrac{(A \cdot D) \cdot C \to A \cdot (D \cdot C) \qquad \cfrac{\cfrac{D \to (A\backslash B)/C = D}{D \cdot C \to A\backslash B}}{A \cdot (D \cdot C) \to B}}{\cfrac{\cfrac{(A \cdot D) \cdot C \to B}{A \cdot D \to B/C}}{(A\backslash B)/C = D \to A\backslash(B/C)}}$$

Readers with appropriate background will recall an analogous isomorphism in homological algebra, which in our notation is written

$$(A \obslash B) \oslash C \cong A \obslash (B \oslash C).$$

As it happened, the linguistic application discussed so far had been anticipated by Bar–Hillel (1953), based on still earlier work by Ajdukiewicz (1937). As opposed to the latter, Bar– Hillel distinguished between the rules $A \cdot (A \backslash C) \to C$ and $(C/B) \cdot B \to C$, although not in this notation. My methods went a bit further, as will be seen from the following example.

Consider the statement *he works* of type S. We cannot give *he* the type N, as **Jane likes he* is ungrammatical, but we can assign to it the type $S/(N\backslash S)$, as $(S/(N\backslash S)) \cdot (N\backslash S) \to S$. On the other hand, *him* in *(Jane likes) him* has type $(S/N)\backslash S$, as $(S/N) \cdot ((S/N)\backslash S) \to S$. Of course, as *John* can be written in place of *he* or *him, John* should also have types $S/(N\backslash S)$ and $(S/N)\backslash S$ in addition to type N. (Indeed, this becomes mandatory in coordinate structures such as *John and he*.) A case can be made for assigning infinitely many types to *John* by iterating this process. Must all these types be listed under *John* in the dictionary? Not at all, because we can prove that $N \to S/(N\backslash S)$ and $N \to (S/N)\backslash S$. For example, the former is proved as follows:

$$\cfrac{\cfrac{A\backslash C \to A\backslash C}{A \cdot (A\backslash C) \to C}}{A \to C/(A\backslash C)}$$

Linguists now call this process "type raising".

Another example of this sort of thing arises when we bracket *John (likes him)* in the more usual way. Since *likes* has type $N\backslash(S/N)$ and *likes him* has type $N\backslash S$, *him* can also be assigned type $(N\backslash(S/N))\backslash(N\backslash S)$, in addition to the type $(S/N)\backslash S$ determined earlier. Again, these two types are not independent, in view of the law $C\backslash A \to (C\backslash B)\backslash(C\backslash A)$, an easy consequence of the associative law, now known to linguists as "Geach's Law".

Let this suffice for illustrating what can be done in applying the associative syntactic calculus to natural languages. There are some obvious limitations to this method and my paper was ignored for many years. (The only thing that was ever quoted was an example pointing out that *time flies* can be construed in two different ways: as a statement and as a command.) It is only recently that linguists have become interested in the associative syntactic calculus, as witnessed by the work of van Benthem (1983, 1988), Buszkowski (1986) and others.

Let me now turn to another question. As long as only the rules $(C/B) \cdot B \to C$ and $A \cdot (A\backslash C) \to C$, already known to Bar–Hillel, were used, it was easy to decide whether, for given types A and B, one could prove that $A \to B$. For, as one proceeded from left to right, the types were getting progressively simpler. Such a decision procedure was not so obvious in the presence of such rules as $A \to B/(A\backslash B)$ and $A/B \to (A/C)/(B/C)$.

2. DECISION PROCEDURE

A decision procedure for the associative syntactic calculus was obtained by modifying a method by Gentzen (see, e.g., Kleene 1952 or Szabo 1969), who had replaced simple deductions $A \to B$ by so–called "sequents" $A_1 A_2 \ldots A_n \to B$.

Using capital Greek letters to denote strings of types, we may present the syntactic calculus in Gentzen–style by means of the following axiom scheme and rules of inference:

$$A \to A;$$

$$\frac{\Gamma \to A \qquad \Delta \to B}{\Gamma\Delta \to A \cdot B};$$

$$\frac{\Gamma A B \Delta \to C}{\Gamma A \cdot B \Delta \to C};$$

$$\frac{\Gamma B \to A}{\Gamma \to A/B}; \qquad \frac{B\Gamma \to A}{\Gamma \to B\backslash A}$$

$$\frac{\Theta \to B \quad \Gamma A \Delta \to C}{\Gamma A/B \Theta \Delta \to C}; \qquad \frac{\Theta \to B \quad \Gamma A \Delta \to C}{\Gamma \Theta B\backslash A \Delta \to C}.$$

Note that each of the rules of inference introduces one of the symbols $\cdot, / or \backslash$ on the right or on the left of the arrow. These rules are easily verified if we interpret ABC, for example, as another way of writing $(A \cdot B) \cdot C$. The transivity of the arrow in the syntactic calculus gives rise to the so–called "cut":

$$\frac{\Theta \to A \qquad \Gamma A \Delta \to B}{\Gamma \Theta \Delta \to B}.$$

The *cut elimination theorem* says that any proof with cut can be replaced by one without cut. This is just as well, as the cut would cause difficulty when we are trying to look for a proof of $\Gamma\Theta\Delta \to B$, as we might have to try infinitely many choices for A. Here is an example of a cut– free proof:

$$\frac{\dfrac{N \to N \qquad\qquad S \to S}{N(N\backslash S) \to S}}{N \to S/(N\backslash S)}$$

The reader will also notice that the reflexive law $A \to A$ is used only for basic types.

The proof of the cut elimination theorem for the associative syntactic calculus is even easier than Gentzen's original proof for intuitionistic propositional logic, as for the latter he had to postulate three so–called structural rules, which caused some complications:

$$\frac{\Gamma A B \Delta \to C}{\Gamma B A \Delta \to C} \quad \text{(interchange)},$$

$$\frac{\Gamma A A \to B}{\Gamma A \to B} \quad \text{(contraction)},$$

$$\frac{\Gamma \to B}{\Gamma A \to B} \quad \text{(thinning)}.$$

While all three structural rules are required for intuitionistic logic, only the first two are required for relevance logic (as pointed out by Minc 1977) and only interchange is required for a version of Girard's linear logic (1987). We repeat that none of Gentzen's structural rules are used for the syntactic calculus.

3. MULTICATEGORIES

I had pointed out in 1961 that Gentzen's sequent calculus was essentially the same as Bourbaki's (1958) method of bilinear maps. For argument's sake, let us assume we are dealing with $R-R$–bimodules for a given ring R. Bourbaki had described a canonical bilinear map $AB \to A \otimes B$ which induced a natural isomorphism between *bilinear* maps $AB \to C$ and *linear* maps $A \otimes B \to C$. This was exploited to obtain all properties of the tensor product.

More generally, if $\Gamma = A_1 \ldots A_m$ and $\Delta = B_1 \ldots B_n$ are sequences of $R-R$–bimodules (we leave out the commas), there is a natural isomorphism

$$Mult(\Gamma AB\Delta, C) \cong Mult(\Gamma A \otimes B\Delta, C)$$

between $m+n+2-$ linear and $m+n+1-$ linear maps. In the same spirit, we have natural isomorphisms

$$Mult(\Gamma B, C) \cong Mult(\Gamma, C\emptyset B),$$
$$Mult(A\Gamma, C) \cong Mult(\Gamma, A\backslash C).$$

Once the analogy between the methods of Gentzen and Bourbaki has been realized, it is quite natural to ask that the sequents in the Gentzen style presentation of the syntactic calculus be interpreted as some kind of multilinear maps. This was done in my 1968 paper, to be followed by a formal definition of " multicategories" in 1969.

A multicategory is supposed to be like a category, except that arrows $f : A \to B$ have been replaced by arrows $f : A_1 \ldots A_n \to B$, sometimes called "multiarrows".

Formally, we first introduce a *multigraph* consisting of a class of arrows (also called "sequents") and a class of objects (also called "types") together with two mappings

$$\text{source: } \{\text{arrows}\} \longrightarrow \{\text{objects}\}^*,$$
$$\text{target: } \{\text{arrows}\} \longrightarrow \{\text{objects}\},$$

where $\{\text{objects}\}^*$ is the free monoid generated by the class of objects, its elements are strings $\Gamma = A_1 \ldots A_n$ of objects. Note that n may be zero, in which case Γ is the empty string, which will here be denoted by a blank. An arrow $f : \ \to B$ is also called an *element* of B.

To obtain a Gentzen–style sequent calculus (but without his three structural rules) we impose the axiom

$$1_A : A \to A$$

and the rule of inference

$$\frac{f : \Theta \to A \qquad g : \Gamma A \Delta \to B}{g < f >: \Gamma\Theta\Delta \to B},$$

Gentzen's "cut". (Out of context, the notation $g < f >$ is ambiguous, as it does not indicate where f is "substituted" into g, but we shall use it nonetheless.) What we obtained up to now is a contextfree production grammar, but with arrows reversed, we shall call it a *contextfree recognition grammar*.

To obtain a *multicategory* it remains to impose an appropriate equivalence relation between arrows $f, g : \Gamma \to B$; we shall denote it by the equality symbol. We impose the following equations:

$$(1a) \qquad \frac{f : \Gamma \to A \quad 1_A : A \to A}{1_A < f >: \Gamma \to A} = f : \Gamma \to A;$$

$$(1b) \qquad \frac{1_A : A \to A \quad g : \Gamma A \Delta \to B}{g < 1_A >: \Gamma A \Delta \to B} = g : \Gamma A \Delta \to B;$$

$$(2) \qquad \frac{\dfrac{f : \Theta \to A \quad g : \Gamma A \Delta \to B}{g < f >: \Gamma \Theta \Delta \to B} \quad h : \Phi B \Psi \to C}{h < g < f \gg: \Phi \Gamma \Theta \Delta \Psi \to C}$$

$$= \frac{f : \Theta \to A \qquad \dfrac{g : \Gamma A \Delta \to B \qquad h : \Phi B \Psi \to C}{h < g >: \Phi \Gamma A \Delta \Psi \to C}}{h < g >< f >: \Phi \Gamma \Theta \Delta \Psi \to C};$$

$$(3) \qquad \frac{f : \Gamma \to A \qquad \dfrac{g : \Delta \to B \quad h : \Phi A \Theta B \Psi \to C}{h < g >: \Phi A \Theta \Delta \Psi \to C}}{h < g >< f >: \Phi \Gamma \Theta \Delta \Psi \to C}$$

$$= \frac{g : \Delta \to B \qquad \dfrac{f : \Gamma \to A \quad h : \Phi A \Theta B \Psi \to C}{h < f >: \Phi \Gamma \Theta B \Psi \to C}}{h < f >< g >: \Phi \Gamma \Theta \Delta \Psi \to C}.$$

These equations were called *identity* laws, *associative* law and *commutative* law respectively in my 1969 paper. (The inadequacy of the notation is apparent from the different proof–trees labelled $h < g >< f >$ in (2) and (3) above.)

Examples of multicategories are categories, trivially, but also monoidal categories and closed categories (Bénabou 1963), as we shall see below.

As a matter of notation: we shall write $Mult(\Gamma, A)$ for the set of (multi)arrows $\Gamma \to A$.

4. STRUCTURED MULTICATEGORIES

Instead of defining monoidal categories, closed categories etc., it is often more convenient to introduce operatives $\otimes, \emptyset$ and o into a multicategory directly.

A *tensor product* in a multicategory is a binary operation $\otimes$ on objects together with a canonical family of (multi)arrows $m_{AB} : AB \to A \otimes B$ inducing a bijection

$$Mult(\Gamma A \otimes B \Delta, c) \xrightarrow{\sim} Mult(\Gamma A B \Delta, C).$$

Thus, to each arrow $f : \Gamma AB\Delta \to C$ there corresponds a unique arrow $f^{\S} : \Gamma A \otimes B\Delta \to C$ such that $f^{\S} < m_{AB} >= f$. The uniqueness of f may also be expressed equationally by saying that, for each $g : \Gamma A \otimes B\Delta \to C$, $(g < m_{AB} >)^{\S} = g$.

A *right internal hom* in a multicategory is a binary operation $\oslash$ on objects together with a canonical family of arrows $e_{CB} : (C\oslash B)B \to C$ inducing a bijection

$$Mult(\Gamma, C\oslash B) \xrightarrow{\sim} Mult(\Gamma B, C).$$

Thus, to each arrow $f : \Gamma B \to C$ there corresponds a unique arrow $f^* : \Gamma \to C\oslash B$ such that $e_{CB} < f^* >= f$. The uniqueness of f^* may be expressed equationally by saying that, for each $g : \Gamma \to C\oslash B$, $(e_{CB} < g >)^* = g$.

A *left internal hom* is defined similarly by an operation $\obslash$ on objects, a family of arrows $e'_{AC} : A(A\obslash C) \to C$ and by assigning to each arrow $f : A\Gamma \to C$ an arrow $f^+ : \Gamma \to A\obslash C$ satisfying appropriate equations to ensure a bijection

$$Mult(\Gamma, A\obslash C) \xrightarrow{\sim} Mult(A\Gamma, C).$$

A *unity object* in a multicategory is an object I together with an arrow $i : \ \to I$ (an empty string on the left of the arrow) inducing a bijection

$$Mult(\Gamma I\Delta, C) \xrightarrow{\sim} Mult(\Gamma\Delta, C).$$

Thus, to each arrow $f : \Gamma\Delta \to C$, there corresponds a unique arrow $f^{\#} : \Gamma I\Delta \to C$ such that $f^{\#} < i >= f$. The uniqueness of $f^{\#}$ may be expressed by saying that, for each $g : \Gamma I\Delta \to C$, $(g < i >)^{\#} = g$.

We list some of the structured multicategories that have appeared in the literature: a *monoidal* multicategory has $\otimes$ and I, a *right closed* multicategory has $\oslash$, a *left closed* multicategory has $\obslash$, a *biclosed* multicategory has $\oslash$ and $\obslash$, a *residuated* multicategory has $\otimes$, $\oslash$ and $\obslash$.

If one wants to add the logical operations $\wedge, \top, \vee, \bot$ to the syntactic calculus, one may of course do so. For multicategories, these symbols are better replaced by $\times, 1, +, 0$, the *Cartesian product, terminal object, coproduct* and *initial object* respectively. The Cartesian product comes equipped with canonical projections $\pi_{AB} : A \times B \to A$, $\pi'_{AB} : A \times B \to B$ and the coproduct with canonical injections $\kappa_{AB} : A \to A + B$, $\kappa'_{AB} : B \to A + B$. The following bijections are assumed to be induced:

$$Mult(\Gamma, A \times B) \xrightarrow{\sim} \quad Mult(\Gamma, A) \times Mult(\Gamma, B),$$

$$Mult(\Gamma, 1) \xrightarrow{\sim} \{*\},$$

$$Mult(\Gamma A + B\Delta, C) \xrightarrow{\sim} \quad Mult(\Gamma A\Delta, C) \times Mult(\Gamma B\Delta, C),$$

$$Mult(\Gamma 0\Delta, C) \xrightarrow{\sim} \{*\}.$$

Here $\{*\}$ is taken to be a prototypical one–element set.

For expository purposes, let us now confine attention to residuated multicategories (as in my paper of 1968), almost the same as biclosed monoidal multicategories (as in my 1969 paper) but without a unity object. A residuated category is nothing else than a sequent presentation of the syntactic calculus with a suitable equivalence relation between sequents.

Given a multicategory $\mathcal{X}$, one may consider the *free* residuated multicategory $F(\mathcal{X})$ generated by $\mathcal{X}$. By now everybody knows how to construct this, so let us immediately turn to the problem of calculating $Mult(\Gamma, B)$ for given $\Gamma = A_1 \dots A_n$ and B in $F(\mathcal{X})$. The problem breaks into two parts:

(I) Find all sequents $\Gamma \to B$ in the Gentzen–style syntactic calculus freely generated by $\mathcal{X}$.
(II) Decide when two sequents $f, g : \Gamma \to B$ determine the same arrow in $F(\mathcal{X})$.

(I) is answered by the cut elimination theorem discussed above. My answer to (II) in 1968 was only partly correct, namely when $\mathcal{X}$ is *discrete*, that is, a multicategory with no arrows except the obligatory $1_x : x \to x$ for each object x. What I should have done was to reduce f and g to normal form. This approach was in fact followed by various authors who have been working on related multicategories since, to my knowledge Szabo, Mann, Minc and his students. I hope to return to this problem on another occasion.

Incidentally, the cut elimination theorem for residuated multicategories says that every arrow of the free residuated multicategory constructed with cut is actually equal to one constructed without cut, not just that it can be replaced by the latter. Similar results hold for other structured multicategories.

5. TYPE LANGUAGES

It has become fashionable in the past few years to look at internal languages of various categories, e.g., toposes and Cartesian closed categories (Lambek and Scott 1986) and, most recently, monoidal categories (Barry Jay 1987). One may just as easily consider internal languages of multicategories and even define multicategories linguistically, as we propose to do here.

DEFINITION 5.1. A *type language* consists of two classes, the class of *types* and the class of *terms*, together with a mapping from the latter to the former assigning a unique type to each term. Among the terms there are countably many *variables* of each type A, say $x_1^A, x_2^A, \dots$. (Instead of specifying $x = x_i^A$, we usually just say: let x be a variable of type A.)

The class of terms is built up inductively from the variables by means of a given class of *operation symbols* $f : A_1 \dots A_n \to B$, the A_i and B being types, as follows: if a_i is a term of type $A_i (i = 1, \dots, n)$ and f is an operation symbol as shown, then $fa_1 \dots a_n$ is a term of type B. (If $n = 0$, then f is just a *constant* term of type B.) Moreover, for any finite set X of variables, there is a *congruence relation* $=_X$ between terms which contain no (free) variables other than members of X. By "congruence relation" we mean an equivalence relation satisfying the following substitution rule (substituting equals for equals):
(EO) if $f : A_1 \dots A_n \to B$ is any operation symbol and a_i, a_i' are terms of type A_i such that $a_i =_X a_i'$ $(i = 1, \dots, n)$, then $fa_1 \dots a_n =_X fa_1' \dots a_n'$.
Finally, $=_X$ is supposed to satisfy the obvious rule:
(E1) if $X \subseteq Y$, then from $a =_X b$ one may infer $a =_Y b$,
as well as the following substitution rule (substituting terms for variables):
(E2) if $z \notin X$, then from $a =_{X \cup \{z\}} b$ one may infer $[c/z]a =_X [c/z]b$.

Here $[c/z]a$ denotes the result of substituting c of type C for each (free) occurrence of the variable z of type C in a, and z is assumed to be outside X.

The rules (E1) and (E2) permit, in particular, change of (free) variable. Thus, for example, from $a =_z b$ we may obtain $a =_{\{z,z'\}} b$ by (E1), whence $[z'/z]a =_{z'} [z'/z]b$ by (E2), assuming that z and z' are variables of the same type.

Is it really necessary to write $=_X$ in place of $=$? Indeed, suppose a and b to be terms of type A not containing any free variables. Certainly, from $a = b$ we can infer $a =_z b$, where z is a variable of type C, in view of (E1). Conversely, from $a =_z b$ we can infer $a = b$, by (E2), by substituting for z some term c of type C, assuming, of course, that there are terms of type C. (Here $=$ means $=_\emptyset$.)

Thus, if there exist constant terms of each type, it is not necessary to place the subscript X on the equality sign (hence rule (E1) need not be stated explicitly). There may be other cases where it is not necessary to write $=_X$ in place of $=$.

For future reference, it will be convenient to make the following definition.

DEFINITION 5.2. Two operation symbols $f, g : A_1 \ldots A_n \to B$ are said to denote the same *operations* if

$$(*) \qquad f x_1 \ldots x_n =_X g x_1 \ldots x_n,$$

where $X = \{x_1, \ldots, x_n\}$, the x_i being distinct variables of type $A_i (i = 1, \ldots, n)$. (Even if $A_i = A_j$, we insist that $x_i \neq x_j$, as long as $i \neq j$.) We write $f = g$.

The reader will be able to think of many examples of type languages. The main example we shall be concerned with here is the internal language of a multicategory or, more generally, the type language associated with a contextfree recognition grammar.

DEFINITION 5.3. The *internal language* of a contextfree recognition grammar (or of a multicategory) is defined as follows. Its types are the objects of the grammar and its operation symbols are the arrows (sequents). The identity arrow $1_A : A \to A$ counts as an operation symbol and the cut

$$\frac{f : \Theta \to A \qquad g : \Gamma A \Delta \to B}{g < f >: \Gamma \Theta \Delta \to B}$$

produces a new operation symbol $g < f >$ from two given operation symbols f and g. Let $=_X$ be the smallest congruence relation (actually function from finite sets of variables to congruence relations) between terms of the same type; containing at most the variables in the set X, satisfying (E1) and (E2) as well as the following:

$$\text{(E3)} \qquad 1_A x =_x x,$$

where x is a variable of type A, and

$$\text{(E4)} \qquad g < f > u_1 \ldots u_m x_1 \ldots x_k v_1 \ldots v_n =_X g u_1 \ldots u_m f x_1 \ldots x_k v_1 \ldots v_n,$$

where $X = \{u_1, \ldots, u_m,\ x_1, \ldots x_k,\ v_1, \ldots, v_n\}$, the u_i, x_l and v_j being variables of the types occuring in the strings Γ, Θ and Δ respectively.

Note that the cut is thus interpreted as substituting $f x_1 \ldots x_k$ for x in $g u_1 \ldots u_m x v_1 \ldots v_n$. (It appears that here is one situation where we might as well have written $=$ in place of $=_X$. Anyway, from now on, subscripts will be omitted.)

6. RESTRICTED FUNCTIONAL COMPLETENESS

Recalling how the terms of a type language are defined inductively from the variables by means of operation symbols, we infer the following completeness property for the internal language of a contextfree grammar.

THEOREM 6.1 The internal language of a contextfree recognition grammar, hence of a multicategory, satisfies *restricted functional completeness*: given any term $\varphi(x_1, \ldots, x_n)$ which contains no free variables other than x_i of type $A_i (i = 1, \ldots, n)$, assume that these variables occur

(a) in this order,

(b) without repetition,

(c) explicitly,

then there exists an operation symbol (sequent) $f : A_1 \ldots A_n \to B$ such that

$$(+) \qquad f x_1 \ldots x_n = \varphi(x_1, \ldots, x_n),$$

where the subscript $X = \{x_1, \ldots, x_n\}$ has been omitted. Moreover, the operation defined by f is uniquely determined.

The restrictions (a),(b) and (c) on functional completeness are due to the absence of Gentzen's three structural rules. We shall see later what happens when some or all of these restrictions are removed. In any case, restriction (b) can be relaxed somewhat for the existence of f. In view of (E2) of Section 5, we may substitute the same variable for two variables of the same type. Hence repetition of variables is permitted in (+) as long as each occurrence is counted separately.

Proof: Let $a \equiv \varphi(x_1, \ldots, x_n)$ be any term of type A in which the variables $x_1, \ldots, x_n$ occur explicitly in this order and different occurrences of the same variable are listed separately. To each such term a we assign a sequent $\bar{a} : A_1 \ldots A_n \to A, A_i$ being the type of $x_i (i = 1, \ldots, n)$, defined by induction on the length of a as follows:

if $a \equiv x$, a variable of type $A, \bar{a} \equiv 1_A$;

if $a \equiv f a_1 \ldots a_m, \quad \bar{a} \equiv f < \bar{a}_1 > \ldots < \bar{a}_m >$.

We prove by induction that

$$\bar{a} x_1 \ldots x_n = \varphi(x_1, \ldots, x_n).$$

Indeed, if $a \equiv x$,

$$\bar{a} x = 1_A x = x,$$

and, if $a \equiv f a_1 \ldots a_m$,

$$\begin{aligned} \bar{a} x_1 \ldots x_n &= f \bar{a}_1 x_1 \ldots x_{r_1} \bar{a}_2 x_{r_1+1} \ldots x_{r_2} \ldots \bar{a}_k x_{r_{k-1}+1} \ldots x_m \\ &= f a_1 a_2 \ldots a_m, \end{aligned}$$

by inductional assumption.

Finally, if f satisfies (+), then the operation defined by f is uniquely determined in view of Definition 5.2.

Let us also remark that in a language which satisfies restricted functional completeness we need not distinguish between operations $f : \quad \to B$ whose source is the empty string and constant terms b of type B. Indeed, it follows from (+) with $n = 0$ and X empty that, given any term b of type B, there is an operation symbol f such that $f = b$.

Moreover, equality of terms coincides with equality of operations in this case.

For future reference we require a lemma.

LEMMA 6.2. In the internal language of a multicategory, if a' results from a by replacing all k free occurrences of the variable z by the closed term c of type C, so that $a' \equiv [c/z]a$, then $\bar{a}' = \bar{a} < \bar{c} >^k$, the notation indicating k cuts.

Proof: If $a \equiv z$ then $a' \equiv c$, so $\bar{a'} = \bar{c} = 1_A < \bar{c} >= \bar{a} < \bar{c} >$. Suppose that $a \equiv f a_1 \ldots a_m$, put $a'_i \equiv [c/z]a_i$, then $\bar{a'_i} = \bar{a_i} < \bar{c_i} >^{k_i}$, by inductional assumption, where k_i is the number of occurrences of z in a_i. Therefore

$$\begin{aligned} \bar{a'} &= f\bar{a'_1} \ldots \bar{a'_m} \\ &= f\bar{a_1} < \bar{c} >^{k_1} \ldots \bar{a_m} < \bar{c} >^{k_m} \\ &= f\bar{a_1} \ldots \bar{a_m} < \bar{c} >^k \\ &= \bar{a} < \bar{c} >^k, \end{aligned}$$

using the equations of a multicategory.

7. THE MULTICATEGORY GENERATED

DEFINITION 7.1. A type language clearly generates a multigraph, its arrows being the operation symbols $f : A_1 \ldots A_m \to B$. If the type language satisfies restricted functional completeness, this multigraph is a contextfree recognition grammar: the sequent $1_A : A \to A$ is the operation symbol 1_A defined by $1_A x = x$ and the cut

$$\frac{f : \Theta \to A \quad g : \Gamma A \Delta \to B}{g < f >: \Gamma \Theta \Delta \to B}$$

is defined by

$$g < f > u_1 \ldots u_m \; x_1 \ldots x_k \; v_1 \ldots v_n$$

$$= \; g \; u_1 \ldots u_m \; f \; x_1 \ldots x_k \; v_1 \ldots v_n.$$

In fact, this contextfree recognition grammar is a multicategory.

Indeed, the equations of a multicategory are easily checked. For example, the equation $g < 1_A >= g$ holds because

$$g < 1_A > u_1 \ldots u_m \; x \; v_1 \ldots v_n \quad = \; g \; u_1 \ldots u_m \; 1_A \; x \; v_1 \ldots v_n$$

and $1_A x \; = \; x$.

What if the type language was already the internal language of a multicategory, which satisfies restricted functional completeness by Theorem 6.1?

THEOREM 7.2. The multicategory generated by the internal language of a multicategory $\mathcal{M}$ is $\mathcal{M}$.

Proof: It suffices to show that, if two arrows $f, g : A_1 \ldots A_n \to B$, when regarded as operation symbols of the internal language, define the same operations (see Definition 5.2.), then $f = g$ in $\mathcal{M}$.

We begin by defining an equivalence relation $\cong$ between terms of the same type in the internal language: $a \cong b$ means that a and b contain the same variables taken from the set X in the same order (repetition is allowed as long as each occurrence counts separately) and that $\bar{a} = \bar{b}$, according to the definition of $\bar{a}$ in the proof of Theorem 6.1, that is, a and b determine the same operations.

We show that $\cong$ is a congruence relation,that is, it satisfies (E0) of Section 5. For, if f, a_i, b_i are given such that $a_i \cong b_i$, then a_i and b_i contain the same variables in the same order and $\bar{a_i} = \bar{b_i}$. Therefore also $a = fa_1 \ldots a_n$ and $b = fb_1 \ldots b_n$ contain the same variables and

$$\bar{a} = f < \bar{a_1} > \ldots < \bar{a_n} >= f < \bar{b_1} > \ldots < \bar{b_n} >= \bar{b}.$$

(E1) is vacuous as long as we omit the subscripts X and Y.

To check (E2), suppose $a \cong b$. Then, a given variable $z \notin X$ will occur as often in a as in b, say k times, and in the same places. Moreover, $\bar{a} = \bar{b}$. Let $a' = [c/z]a$ and $b' = [c/z]b$, where c does not contain z. We claim that $a' \cong b'$. Indeed, a' and b' contain the same variables in the same order and, by Lemma 6.2,

$$\bar{a'} = \bar{a} < \bar{c} >^k = \bar{b} < \bar{c} >^k = \bar{b'}.$$

Finally, we must show that $\cong$ satisfies (E3) and (E4), namely

$$1_A x \cong x, \quad g < f > u_1 \ldots u_m \, x_1 \ldots x_k \, v_1 \ldots v_n \cong g \, u_1 \ldots u_m \, f \, x_1 \ldots x_k \, v_1 \ldots v_n.$$

To check the former, for example, we note that

$$\overline{1_A x} = 1_A < \bar{x} >= 1_A < 1_A >= 1_A = \bar{x}.$$

The latter is verified similarly.

To complete the proof of the theorem, suppose that $f, g : A_1 \ldots A_n \to B$ define the same operation, that is

$$f \, x_1 \ldots x_n = g \, x_1 \ldots x_n,$$

x_i being of type A_i, without repetition. Recall that $=$ (actually $=_X$) was the smallest congruence relation satisfying (E1) to (E4) and that we have just proved that $\cong$ is also such. Therefore,

$$b \equiv f x_1 \ldots x_n \cong g x_1 \ldots x_n \equiv b'.$$

In particular, $\bar{b} = \overline{b'}$, that is,

$$f < 1_{A_1} > \ldots < 1_{A_n} >= g < 1_{A_1} > \ldots < 1_{A_n} >,$$

that is, $f = g$, as was to be proved.

At first sight, there is something puzzling about the above argument. Suppose $f, g : AA \to B$ and $fxx = gxx$, then the same argument will show that $f = g$. How come? The answer is that, from the manner in which $=$ was defined, it follows that any proof of $fxx = gxx$ will generalize to a proof of $fxx' = gxx'$.

To sum up, let $L(\mathcal{M})$ be the internal language of a multicategory $\mathcal{M}$ and $M(\mathcal{L})$ the multicategory generated by the type language $\mathcal{L}$, which is assumed to satisfy restricted functional completeness. Then Theorem 7.2 asserts that

$$M(L(\mathcal{M})) = \mathcal{M}.$$

In other words, a multicategory is completely determined by its internal language. It would be nice if we could also assert that

$$L(M(\mathcal{L})) = \mathcal{L}.$$

This is certainly the case if $\mathcal{L}$ is the internal language of some multicategory. The following appear to be necessary and sufficient conditions on $\mathcal{L}$ for this to be so:

(a) $\mathcal{L}$ satisfies restricted functional completeness;
(b) equality between terms (of the same type) is the smallest congruence relation respecting (E2) to (E4);
(c) operation symbols defining the same operations are equal.

8. INTERNAL LANGUAGE AND TENSOR PRODUCT

Recall that a tensor product in a multicategory is a binary operation $\otimes$ between objects together with a family of arrows $m_{AB} : AB \to A \otimes B$ inducing a bijection

$$Mult(\Gamma A \otimes B\Delta, C) \xrightarrow{\sim} Mult(\Gamma AB\Delta, C)$$

for given strings Γ, Δ and objects A, B and C. Thus, for each arrow $f : \Gamma AB\Delta \to C$ there is given an arrow $f^{\S} : \Gamma A \otimes B\Delta \to C$ satisfying

$$f^{\S} < m_{AB} >= f, \quad (g < m_{AB} >)^{\S} = g,$$

for each $g : \Gamma A \otimes B\Delta \to C$.

In the internal language, it is convenient to abbreviate $m_{AB}xy$ as (x, y). (Some people write $x \otimes y$, but this may lead to confusion with the tensor product of arrows introduced below.)

The above bijection may now be expressed by saying that to each arrow $f : \Gamma AB\Delta \to C$ there is associated a unique arrow $f^{\S} : \Gamma A \otimes B\Delta \to C$ such that

$$f^{\S} x_1 \ldots x_m(x, y)y_1 \ldots y_n = f x_1 \ldots x_m xy\ y_1 \ldots y_n.$$

(As above, we omit the subscript on the equality sign when there is no likelihood of confusion.) From this the usual properties of the tensor product are easily deduced.

For example, if $f : A \to A'$ and $g : B \to B'$ are given arrows, we may define $f \otimes g : A \otimes B \to A' \otimes B'$ by putting $f \otimes g \equiv h^{\S}$, where $h : AB \to A' \otimes B'$ is the unique arrow such that $h\ xy = mfxgy$, whose existence is assured by restricted functional completeness. In other words, we define $f \otimes g$ by the equation

$$(f \otimes g)(x, y) = (fx,\ gy).$$

To show that $\otimes$ is a bifunctor we require, for example, that

$$(f' \otimes g')(f \otimes g) = f'f \otimes g'g,$$

whenever $f' : A' \to A''$ and $g' : B' \to B''$. We prove this by calculating

$$\begin{aligned}(f' \otimes g')(f \otimes g)(x, y) &= (f' \otimes g')(fx,\ gy) \\ &= (f'fx,\ g'gy) \\ &= (f'f \otimes g'g)(x,\ y).\end{aligned}$$

The associativity arrow $\alpha_{ABC} : (A \otimes B) \otimes C \to A \otimes (B \otimes C)$ is defined by the equation

$$\alpha_{ABC}((x, y), z) = (x, (y, z)).$$

It is easily checked that α is a natural transformation. Similarly one defines $\alpha^{-1}_{ABC} : A \otimes (B \otimes C) \to (A \otimes B) \otimes C$ and checks that $\alpha_{ABC}\alpha^{-1}_{ABC} = 1_{A\otimes(B\otimes C)}, \alpha^{-1}_{ABC}\alpha_{ABC} = 1_{(A\otimes B)\otimes C}$. MacLane's pentagonal coherence condition is the commutativity of the following diagram:

$$\begin{array}{ccccc} ((A\otimes B) \otimes C) \otimes D & \to & (A \otimes B) \otimes (C \otimes D) & \to & A \otimes (B \otimes (C \otimes D)) \\ \downarrow & & & & \uparrow \\ (A\otimes(B \otimes C)) \otimes D & & \longrightarrow & & A \otimes ((B \otimes C) \otimes D) \end{array}$$

The way to prove this in the internal language is to point out that there is a unique arrow $f : ((A \otimes B) \otimes C) \otimes D \to A \otimes (B \otimes (C \otimes D))$ satisfying the equation

$$f(((x, y), z), t) = (x, (y, (z, t))).$$

We thus see that the usual properties of the tensor product in a category follow from the canonical bijection between arrows $\Gamma A \otimes B\Delta \to C$ and $\Gamma AB\Delta \to C$ in a multicategory. Of course, the traditional procedure is the reverse: one postulates that $\otimes$ is a bifunctor with a coherent natural associativity defined in a category. One then turns the category into a multicategory by defining $f : ABCD \to E$ as $f : ((A\otimes B)\otimes C)\otimes D \to E$ (to take the case $n = 4$ for example). The equations of a multicategory are then easily checked. However, in the spirit of Gentzen and Bourbaki, I prefer to go in the opposite direction, beginning with a multicategory.

We leave it as an exercise to the reader to explore the properties of a unity object in the internal language as follows:

$$i : \ \to I\ , \quad \frac{f : \Gamma\Delta \to C}{f^{\#} : \Gamma I\Delta \to C}\ , \quad f^{\#} < i >= f,\ (g < i >)^{\#} = g.$$

Let me draw attention here to the internal language of a monoidal category recently developed by Barry Jay (1987) and to the interesting applications made by him. His approach is somewhat different from ours, as the terms of his language are not allowed to contain repetitions of the same variable.

9. INTERNAL LANGUAGE AND INTERNAL HOM

We recall that a right internal hom in a multicategory is a binary operation $\emptyset$ between objects together with a family of arrows $e_{CB} : (C\emptyset B)B \to C$ inducing a bijection

$$Mult(\Gamma, C\emptyset B) \xrightarrow{\sim} Mult(\Gamma B, C).$$

Thus, to each arrow $f : \Gamma B \to C$ one assigns an arrow $f^* : \Gamma \to C\not{o}B$ satisfying

$$e_{CB} < f^* >= f, \quad (e_{CB} < g >)^* = g$$

for each $g : \Gamma \to C\not{o}B$.

In the internal language it is convenient to abbreviate $e_{CB}uy$ as $u`y$, read "u of y", where u and y are variables of types $C\not{o}B$ and B respectively. It is also customary to write $f^*x_1 \dots x_n$ as $\lambda_{y\epsilon B} f x_1 \dots x_n y$, where y is a so- called "bound" variable, assumed to be distinct from the x_i. The above equations then become:

$$(\lambda_{y\epsilon B} f x_1 \dots x_n y)\, `y' = f x_1 \dots x_n y',$$

$$\lambda_{y\epsilon B}(g x_1 \dots x_n\, `y) = g\; x_1 \dots x_n.$$

For the reader's convenience, we have here distinguished the free variable y' of type B from the bound variable y of the same type, but this was not necessary. In any case, rule (E2) of Section 5 allows us to replace y' by any term b of type B, so the first equation may be generalized to

$$(\lambda_{y\epsilon B} f x_1 \dots x_n y)`b = f x_1 \dots x_n b.$$

While this resembles the usual λ-calculus, it differs from it inasmuch as λ -abstraction applies only to the rightmost free variable occurring in an expression and that this variable is assumed to be distinct from all the others.

We can now show that $\not{o}$ is a bifunctor, contravariant in its second argument. Thus, given $f : A \to A'$ and $g : B' \to B$, we define $f\not{o}g : A\not{o}B \to A'\not{o}B'$ by the equation

$$((f\not{o}g)u)`y' = f(u`(gy')),$$

where u and y' are variables of types $A\not{o}B$ and B' respectively. In other words,

$$(f\not{o}g)u = \lambda_{y'\epsilon B'} u`(g\; y').$$

If furthermore $f' : A' \to A''$ and $g' : B'' \to B'$, one easily checks that

$$(f'\not{o}g')(f\not{o}g)u = ((f'f)\not{o}(gg'))u.$$

We can also obtain a canonical arrow $\beta_{ABC} : A\not{o}B \to (A\not{o}C)\not{o}(B\not{o}C)$, expressing Geach's law, by putting

$$\beta_{ABC}u = \lambda_{v\epsilon B\not{o}C}\; \lambda_{z\epsilon C} u`(v`z),$$

where u is a variable of type $A\not{o}B$. In this way, one can obtain all the properties of $\not{o}$ that may turn out to be useful.

Conversely, given a category with a bifunctor $\not{o}$, covariant in the first and contravariant in the second argument, it is an exercise, albeit somewhat tedious, to write down all the properties of $\not{o}$ which are needed to show that one obtains a multicategory by defining $f : ABCD \to E$ as $f : A \to ((E\not{o}D)\not{o}C)\not{o}B$, to take the case $n = 4$ for example.

If one is given a multicategory with both right internal hom and tensor product, it is immediate that one has a natural bijection

$$Mult(A \otimes B, C) \cong Mult(AB, C) \cong Mult(A, C\not{o}B),$$

so that $- \otimes B$ is left adjoint to $-\not{o}\; B$.

A left internal hom may be introduced dually by an arrow $e'_{AC} : A(A\backslash C) \to C$ which induces a bijection

$$Mult(\Gamma, A\backslash C) \xrightarrow{\sim} Mult(A\Gamma, C).$$

We write $e'_{AC}xv$ as $x`v$, think of "x substituted into v", and $f^+x_1 \ldots x_n$ as $\lambda'_{x\epsilon A} fx\, x_1 \ldots x_n$. λ'–abstraction thus applies to the leftmost free variable in an expression.

In a multicategory which has both right and left internal hom, one obtains canonical arrows $\tau_{AB} : A \to B/(A\backslash B)$ and $\tau'_{AB} : A \to (B/A)\backslash B$, type raising, given by

$$\tau_{AB}x = \lambda_{u\epsilon A\backslash B}(x`u),$$

for example. Also one can define an associativity isomorphism $\gamma_{ABC} : (A\backslash B)/C \to A\backslash(B/C)$ by

$$\gamma_{ABC}\omega = \lambda'_{x\epsilon A}\lambda_{z\epsilon C}(x`(\omega`z)),$$

where ω is a variable of type $(A\backslash B)/C$.

10. GENTZEN MULTICATEGORIES

We shall take another look at Gentzen's structural rules, which we had discussed earlier. (See Szabo 1974.)

DEFINITION 10.1. A *Gentzen deductive system* is a contextfree recognition grammar which satisfies the three standard rules:

$$\frac{f : \Gamma AB\Delta \to C}{f^i : \Gamma BA\Delta \to C} \quad \text{(interchange)},$$

$$\frac{f : \Gamma AA \to B}{f^c : \Gamma A \to B} \quad \text{(contraction)},$$

$$\frac{f : \Gamma \to B}{f^t : \Gamma A \to B} \quad \text{(thinning)}.$$

By a *Gentzen multicategory* we shall understand a multicategory with operation symbols f^i, f^c, f^t, introduced as above, satisfying the following equations:

$$\begin{aligned} f^i x_1 \ldots x_m yx\, y_1 \ldots y_n &= fx_1 \ldots x_m\, xy\, y_1 \ldots y_n, \\ f^c x_1 \ldots x_m x &= fx_1 \ldots x_m x\, x, \\ f^t x_1 \ldots x_m x &= fx_1 \ldots x_m. \end{aligned}$$

In particular, in a multicategory, from fxy, gxx' and hx, where x and x' have the same type, we can form

$$f^i yx = fxy, \quad g^c x = gxx, \quad h^t xy = hx$$

thus allowing permutations of arguments, diagonal maps and projections. One might point out that the possibility to construct diagonal maps lies at the bottom of many interesting arguments in mathematics: for example, Cantor's diagonal argument, Gödel's incompleteness proof and the construction of a non–calculable numerical function.

PROPOSITION 10.2. The internal language of a Gentzen multicategory satisfies unrestricted *functional completeness*:

Given any term $\varphi(x_1, \ldots, x_n)$ which contains no variables other than the x_i of type $A_i (i = 1, \ldots, n)$, there exists a unique operation $f : A_1 \ldots A_n \to B$ such that

$$f x_1 \ldots x_n = \varphi(x_1, \ldots, x_n).$$

Here the x_i need not occur in this order, they may occur at several places and they need not occur explicitly.

Thus we have dropped the restrictions (a), (b) and (c) of Section 6 completely. We leave the proof to the reader.

A Gentzen deductive system is of course the underlying deductive system of intuitionistic logic. A Gentzen multicategory is nothing else than a *manysorted algebraic theory*, in the sense of Lawvere (1963) and Bénabou (1968). (See also Birkhoff and Lipson, 1970.) In retrospect, Michael Barr's (1978) use of algebraic theories as analogues of grammars in his illustration of syntax acquisition was not so farfetched.

PROPOSITION 10.3 If a Gentzen multicategory is equipped with a tensor product, this will be a Cartesian product. If a Gentzen multicategory is equipped with a unity object, this will be a terminal object. If a Gentzen multicategory is right closed, it is also left closed; in fact $B \circ A$ may be taken to be $A \not{o} B$ and viewed as exponentiation A^B.

Proof: Given a tensor product in a multicategory, we may define $\pi_{AB} : A \otimes B \to A$ and $\pi'_{AB} : A \otimes B \to B$ by

$$\pi_{AB}(x, y) = x \quad , \pi'_{AB}(x, y) = y.$$

where x and y are variables of types A and B respectively. Also, given $f : \Gamma \to A$ and $g : \Gamma \to B$, we may define $< f, g >: \Gamma \to A \otimes B$ by

$$< f, g > x_1 \ldots x_n =< f x_1 \ldots x_n, \; g x_1 \ldots x_n >,$$

where $\Gamma = A_1 \ldots A_n$ and the x_i are variables of type $A_i (i = 1, \ldots, n)$. Then

$$\pi_{AB} < f, g > x_1 \ldots x_n = f x_1 \ldots x_n,$$

so that $\pi_{AB} < f, g >= f$, by Definition 5.2, and similarly $\pi'_{AB} < f, g >= g$. Also

$$\begin{aligned} < \pi_{AB}, \pi'_{AB} > (x, y) &= (\pi_{AB}(x, y), \pi'_{AB}(x, y)) \\ &= (x, y), \end{aligned}$$

so that $<\pi_{AB},\pi'_{AB}>=1_{A\otimes B}$. This is so in view of Section 8, because there is a unique arrow $m^{\S}: A\otimes B\to A\otimes B$ such that $m^{\S}(x,y)=mxy=(x,y)$.

This shows that $A\otimes B$ is a Cartesian product with projections π_{AB} and π'_{AB}. We leave the proof that a unity object is terminal to the reader.

Suppose our multicategory is right closed. Define $B\backslash A\equiv A/B$ and $e'_{BA}: B(B\backslash A)\to A$ by

$$e'_{BA}yu=e_{AB}uy,$$

where y and u are variables of types B and A/B respectively. (Thus $e'_{BA}=e^{i}_{AB}$.) Also put

$$\lambda'_{y\epsilon B}fy\ x_1\ldots x_n\equiv\quad\lambda_{y\epsilon B}f^{i}x_1\ldots x_ny,$$

where f^i results from f by interchange. Then

$$\begin{aligned}e'_{BA}y'\lambda'_{y\epsilon B}fy\ x_1\ldots x_n&=e_{AB}\lambda'_{y\epsilon B}fy\ x_1\ldots x_ny'\\&=e_{AB}\lambda_{y\epsilon B}f^{i}x_1\ldots x_ny\ y'\\&=f^{i}x_1\ldots x_ny'\\&=f\ y'x_1\ldots x_n.\end{aligned}$$

The other equation involving λ' may be shown similarly.

11. SEMANTICS OF BICLOSED MULTICATEGORIES.

The internal language of a *closed* Gentzen multicategory (right closed suffices, in view of interchange) is the usual λ–calculus. For example, we may define

$$\begin{aligned}\lambda_{y\epsilon B}fxyz&\equiv\lambda_{y\epsilon B}f^{i}xzy,\\\lambda_{y\epsilon A}gxx&\equiv\lambda_{x\epsilon A}g^{c}x,\\\lambda_{y\epsilon B}hx&\equiv\lambda_{y\epsilon B}h^{t}xy.\end{aligned}$$

While intuitionistic logic and the associated Cartesian closed multicategories make use of all three structural rules (and the associated equations), the syntactic calculus and the associated residuated (or even biclosed monoidal) multicategories use none of them. Two intermediate cases have appeared in the literature.

The linear logic of Girard (1987) and the associated symmetric monoidal multicategories allow the interchange rule but neither contraction nor thinning. This connection had been noticed by Minc (1977), who called it "relevance logic" in place of "linear logic". However, it seems that the usual relevance logic allows not only interchange, but also contraction, forbidding the thinning rule only (see Anderson and Belnap 1975).

Incidentally, what has been called "λ–calculus" here was called "λK–calculus" by Church (1941). He reserved "λ–calculus" for what corresponds to relevance logic: the thinning rule is absent and the definition of $\lambda_{y\epsilon B}hx$ is consequently not permitted, if x is distinct from y.

Let us now take another look at the linguistic application discussed in Section I, namely the attempt to describe English syntax by means of the syntactic calculus. We shall consider types built up from basic types S, N and perhaps others with the help of two binary operations "over" and "under". It will do no harm to forget about the operation "times", as long as we stick to the Gentzen–style presentation of Section 2. If we take account of the actual derivations of sentencehood and if we specify when two derivations are to be regarded as equivalent, we are led to study a biclosed multicategory. To be precise, we are interested in the *biclosed multicategory freely generated by the dictionary* (of English).

The *dictionary* is assumed to list certain arrows, whose source is the empty string:

$$\begin{aligned} &John\colon && \longrightarrow N,\\ &works\colon && \longrightarrow N\backslash S,\\ &likes\colon && \longrightarrow (N\backslash S)/N, \end{aligned}$$

etc. For argument's sake, we shall here ignore the ambiguity of words such as *works*, which could just as well be the plural of a count noun, as this problem can easily be taken care of with the help of subscripts.

By the *syntactic* λ*–calculus* of English we shall mean the internal language of the biclosed multicategory freely generated by the dictionary. English words appear as constant terms in the syntactic λ–calculus. For example, *John* is of type N and *works* is of type $N\backslash S$, from which we infer that *John' works* has type S.

So much about English syntax. Its semantics is best described by a closed Gentzen multicategory or, equivalently, its internal language, the usual λ–calculus, also known as combinatory logic. This had been advocated for a long time by Curry (see e.g. Curry and Feys 1958) and was carried out more recently by Montague (see his selected papers 1974).

Since both syntax and semantics are biclosed multicategories, an *interpretation* of English should be a "biclosed multifunctor" (for a definition of this term see below). As the syntax is freely generated by the dictionary, an interpretation is completely determined by the interpretation of the basic types and the English words listed in the dictionary.

In a first attempt to find an interpretation of English we might take the semantics to be the multicategory of sets. Then S is interpreted by the set $\overline{S} = \{\top, \bot\}$ of truth values and N is interpreted by a given set $\overline{N}$ of individuals. Thus *John* will be interpreted as a certain individual who bears that name and *works* as a function from $\overline{N}$ to $\overline{S}$, that is, as a property of individuals.

This attempt at interpreting English is however not very realistic. Thus, whether John works is not just true or false but depends on time and other circumstances. It may be more reasonable to take for the semantics not the multicategory of sets, but some topos, for example Sets $^{\mathcal{T}^{op}}$, where $\mathcal{T}$ is a tree (or even a forest), allowing the present to develop into a number of possible futures, the past being fixed.

But even this attempt at interpreting English ignores one aspect of Montague semantics, as he insists that the language in which the interpretation takes place is an intensional language. We shall not pursue this matter any further here. Instead we shall end this section with the definition that had been promised.

DEFINITION 11.1 A *multifunctor* F between multicategories $\mathcal{M}$ and $\mathcal{M}'$ is a mapping from the class of objects of $\mathcal{M}$ to that of $\mathcal{M}'$ together with a mapping from the class of arrows of $\mathcal{M}$ to that of $\mathcal{M}'$ such that, when $f : A_1 \dots A_n \to B$ in $\mathcal{M}$ then $F(f) : F(A_1)\dots F(A_n) \to F(B)$ in $\mathcal{M}'$ and $F(1_A) = 1_{F(A)}$ as well as $F(g < f >) = F(g) < F(f) >$. If both $\mathcal{M}$ and $\mathcal{M}'$ are biclosed multicategories, F is said to be a *biclosed multifunctor* if it preserves the biclosed structure on the nose, thus

$$F(A \phi B) = F(A) \phi F(B),$$
$$F(e_{AB}) = e_{F(A)F(B)},$$
$$F(f^*) = F(f)^*, \qquad \text{etc.}$$

12. SOME FINAL REMARKS

An imminent deadline won't allow me to carry out everything I would have liked to do here, for example, to re–examine cut elimination and coherence from the point of view of the internal language, the latter along the lines pioneered by Minc. Instead, I shall consider some of the points raised in the discussion after my talk at the Boulder conference and make some observations about the recent literature related to multicategories, although I cannot claim to have completely digested all of this material.

More general than contextfree recognition grammars are *production grammars*, also called "rewriting systems" or "semi–Thue systems", which deal with arrows (productions)

$$A_1 \dots A_m \to B_1 \dots B_n.$$

In fact, my own recent work concerning natural languages makes use of production grammars rather than categorial grammars.

If a suitable equivalence relation is introduced between derivations of a production grammar, one obtains a strictly monoidal category, which may also be described as a 2–category whose underlying 1–category is a monoid, in fact a free monoid. Something like this was done by Hotz (1966) and pursued by Benson in several papers. In particular, Benson (1970) pointed out that an interpretation should be viewed as a functor from a category called "syntax" to a category called "semantics", where the former was a strict monoidal category, thus anticipating the ideas of Section 11 (and those in Section 4 of my paper in Oehrle et al, 1988).

Inasmuch as contextfree recognition grammars are special kinds of production grammars, one should be able to view multicategories as special kinds of strict monoidal categories. This was in fact pointed out by Linton and Reynolds, in developing ideas which Linton (1971) had first presented in his customary inimitable style. The trick was to use sequences of arrows $f_i : \Gamma_i \to B_i \quad (i = 1, \dots, n)$ of the multicategory to construct arrows $(f_1, \dots, f_k) : \Gamma_1 \dots \Gamma_n \to B_1 \dots B_n$ of the strict monoidal category, which Linton calls a "multilinear" category.

Incidentally, the productions $A_1 \dots A_m \to B_1 \dots B_n$ should not be confused with the sequents used by Gentzen in his treatment of classical logic, which happen to look the same. In Gentzen's sequents, juxtaposition on the left denotes conjunction, while juxtaposition on the right denotes disjunction. Gentzen's classical sequents have also been lifted to the level of categories in Szabo's "polycategories" (1975). They reappear in Girard's linear logic, where they may be interpreted as arrows

$$A_1 \otimes \dots \otimes A_m \to (B_1^* \otimes \dots \otimes B_n^*)^*,$$

the star representing some kind of involution.

The first time I came across the use of lambda calculus in connection with text processing was in a preprint by Masterman et al (1956). Van Benthem (1983,1988) studied a version of the syntactic calculus admitting the structural rule of interchange and considered the restricted form of the lambda calculus that goes with it. Probably his system is the same as that studied by Minc (1977), who established a cut elimination theorem for his system and used the restricted form of the lambda calculus to solve the coherence problem for symmetric monoidal closed categories (see also MacLane 1982). The same calculus appears in the linear logic of Girard (1987), among many other interesting ingredients, and a cut elimination theorem may also be found in Girard and Lafont (1987).

A completeness theorem for the syntactic calculus was proved by Buszkowski (1986): a sequent $A_1 \ldots A_n \rightarrow A_{n+1}$ is provable if and only if it is valid under every interpretation of the A_i as downward closed subsets of a partially ordered semigroup, much like the ideals of a ring in Section 1. A similar result for linear logic was obtained by Girard (1987).

Bibliography

Ajdukiewicz, K.: Die syntaktische Konnexität, Studia Philosophica I (1937), 1-27; translated in S. McCall, Polish Logic 1920 - 1939, Clarendon Press, Oxford 1967.

Anderson, A.R., and N.D.Belnap: Entailment: the logic of relevance and necessity, Princeton University Press, Princeton, N.J., 1975.

Bar–Hillel, Y.: A quasiarithmetical notation for syntactic description, Language 29 (1953), 47-58.

Barr, M.: The theory of theories as a model of syntactic acquisition, Theoretical Linguistics 5 (1978), 261-274.

Bénabou, J.: Catégories avec multiplications, C.R. Acad. Sci. Paris 256 (1963), 1887-1890.

Bénabou, J.: Introduction to bicategories, Report on the Midwest Category Seminar, Springer Lecture Notes in Mathematics 47 (1967), 1-77.

Bénabou, J.: Structures algébriques dans les catégories, Cahiers Topologie Géom. Differentielle 10 (1968), 1-126.

Benson, D.B.: Syntax and semantics: a categorical view, Information and Control 17 (1970), 145-160.

Benson, D.B.: The basic algebraic structures in categories of derivations, Information and Control 28 (1975), 1-29.

Birkhoff, G. and J.D. Lipson: Heterogeneous algebras, J. Combinatorial Theory 8 (1970), 115-133.

Bourbaki, N.: Algèbre multilinéaire, Hermann, Paris, 1948.

Buszkowski, W.: Completeness results for Lambek syntactic calculus, Zeitschr. f. math. Logik und Grundlagen d. Math. 32 (1986), 13-28.

Cartan, H. and S. Eilenberg: Homological Algebra, Princeton University Press, Princeton, N.J., 1956.

Church, A.: The calculi of lambda–conversion, Annals of Mathematics Studies 6, Princeton University Press, Princeton, N.J., 1941.

Curry, H.B. and R. Feys: Combinatory Logic I, North Holland, Amsterdam, 1958.

Eilenberg, S. and G.M. Kelly: Closed categories, in: Proc. Conf. Categorical Algebra La Jolla 1965, Springer Verlag, Berlin, 421- 562.

Findlay, G.D. and J. Lambek: Calculus of bimodules, McGill University, manuscript 1955.

Girard, J.Y.: Linear logic, Theoretical Computer Science 50 (1987), 1-102.

Girard, J.Y. and Y. Lafont: Linear logic and lazy computation, preprint 1987.

Hotz, G.: Eindeutigkeit und Mehrdeutigkeit formaler Sprachen, Elektronische Informationsverarbeitung und Kybernetik 2 (1966), 235-247.

Jay, C.B.: Languages for monoidal categories, Dalhousie University, preprint 1987.

Kleene, S.C.: Introduction to metamathematics, Van Nostrand, New York and Toronto, 1952.

Lambek, J.: The mathematics of sentence structure, Amer. Math. Monthly 65 (1958), 154-69.

Lambek, J.: On the calculus of syntactic types, Amer. Math. Soc. Proc. Symposia Appl. Math. 12 (1961), 166-78.

Lambek, J.: Lectures on rings and modules, Ginn and Co. Waltham Mass., 1966; Chelsea Publ. Co., New York, 1986.

Lambek, J.: Deductive systems and categories I, J. Math. Systems Theory 2 (1968), 278-318.

Lambek, J.: Deductive systems and categories II, Springer LNM 86 (1969), 76-122.

Lambek, J.: Categorial and categorical grammars, in: Oehrle et al, 1988.

Lambek, J. and P.J. Scott: Introduction to higher order categorical logic, Cambridge studies in advanced mathematics 7 (1986).

Lawvere, F.W.: Functional semantics of algebraic theories, Proc. Nat. Acad. Sci. U.S.A. 50 (1963), 869-872.

Linton, F.E.J.: The multilinear Yoneda lemmas, Lecture Notes in Mathematics 195 (1971), Springer Verlag.

Linton, F.E.J. and G. Reynolds: Pro–U–categories, undated manuscript.

MacLane, S.: Natural associativity and commutativity, Rice University Studies 49 (1963), 28-46.

MacLane, S.: Categories for the working mathematician, Graduate Texts in Mathematics 5 (1971), Springer Verlag.

MacLane, S.: Why commutative diagrams coincide with equivalent proofs, Contemporary Mathematics 13 (1982), 387-401.

Mann, C.R.: The connection between equivalence of proofs and cartesian closed categories, Proc. London Math. Soc. 31 (1975), 289-310.

Masterman, M., A.F. Parker Rhodes and M.T. Hoskyns: An account of the pilot project of the Cambridge Language Research Unit, Progress Report II, Annex V (1956), 1-58.

Minc, G.E.: Theory of proofs (arithmetic and analysis), translated from Itogi Nauki i Tekhniki (Algebra, topologiya, geometriya), 13 (1975), 5-49.

Minc, G.E.: Closed categories and the theory of proofs, translated from Zapiski Nauchnych Seminarov Leningradskogo Otdeleniya Mat. Instituta im. V.A. Stuklova AN SSSR 68 (1977), 83-114.

Montague, R.: Formal philosophy, selected papers of Richard Montague, edited by R.H. Thomason, Yale University Press, New Haven, 1974.

Oehrle, R.T., E. Bach and D. Wheeler (editors): Categorial grammars and natural language structures, Reidel, Dordrecht, 1988.

Szabo, M.E.(ed.): The collected papers of Gerhard Gentzen, Studies in Logic and the Foundations of Mathematics, North Holland, Amsterdam 1969.

Szabo, M.E.: A categorical equivalence of proofs, Notre Dame J. Formal Logic 15 (1974), 177-91.

Szabo, M.E.: Polycategories, Communications in Algebra 3 (1975), 663-689.

Szabo, M.E.: Algebra of Proofs, Studies in Logic and the Foundations of Mathematics 88, North Holland, Amsterdam, 1978.

Van Benthem, J.: The semantics of variety in categorial grammar, Simon Fraser University, Report 83-26, 1983.

Van Benthem, J.: The Lambek calculus, in: Oehrle et al. 1988.

Mathematics Department
McGill University
Montreal, Quebec H3A 2K6

Contemporary Mathematics
Volume **92**, 1989

An Application of Minimal Context-free Intersection Partitions to Rewrite Rule Consistency Checking

Dana May Latch *

Abstract

We describe a method of showing that a set of rewrite rules specifying the semantics of a functional language is consistent (confluent). Given a set of rewrite rules, the procedure begins by extracting a subset of rules known to be consistent, because the domain of each of the rules of the subset is disjoint from that of all other rules. The remaining rules are examined in groups based on the minimal intersection partitions of the domains. Two of these rules are consistent if, when applied to all elements of the partition, the final results can be shown to be equal using (a) the rewrite rules themselves or rewrite rules from the set already shown to be consistent, (b) narrowing substitutions, and (c) substitution of variables for innermost applications.

1 Introduction

In this paper, we develop a consistency checking procedure for a class of formal functional programming (FFP) languages [Ba78]. The input to the consistency checker is an operational semantics in the form of a collection of rewrite rules that implement, within a FFP language, the semantic specifications of the primitive functions and combining forms (henceforth, primitives) of the language. A set of operational rewrite rules is *consistent* whenever rewrite evaluation is confluent [Hu80]; that is, whenever all innermost evaluation sequences of a primitive, either are nonterminating or terminate in the same value. The consistency checker partitions each set of operational rewrite rules into a consistent set and its complement.

The semantics describes the transformational effect of a primitive on its operand(s) (possibly resulting in recursive applications of the language primitives). The performance of this transformation might be considered a "macro-operation," or machine instruction,

*On leave 1985-1988 at CUNY; Work partially supported by NSF Grants #MCS81-04217; #DCR83-02897 and by CUNY Grants #PSC-CUNY-667920; #PSC-CUNY-668293.

of the interpreter, occurring in a single "instruction cycle." The operational rewrite rules describe the "micro-operations" that implement a language primitive, where a basic "machine cycle" is occupied by a single rewrite operation.

Variously defined rewrite rule collections can reflect the operational nature of diverse target architectures, while the specification of the semantics hides this level of detail from the language user. Typically, an operational semantics will assign a fixed recursive structure to compound objects. But, at the operational level, it may be appropriate to consider lists variously as LISP-like lists, tail-constructed lists, or concatenated sublists, depending on the context. On a single sequential processor, it might be perfectly adequate to unwind a list from front to back, one element at a time, but, on a processor with n simultaneously addressible neighbors, it seems sensible to provide rewrite rules that can capture up to n elements in a single rewrite cycle, resorting to recursion only for lists of length greater than n.

Different structural interpretations, reflected in distinct operational semantics, could require separate validations. Our system, however, permits ambiguity in the operational "deconstruction" of compound structures. Thus, for example, lists can be treated variously by different rewrite rules as left-constructed, right-constructed or a mixture of the two. Validation of an operational semantics that permits this kind of ambiguity may be considered the certification of a class of deterministic interpreter models, or of one or more nondeterministic models, one for each subset of the rewrite rules adequate to implement the semantic specification.

The user of the consistency checker must provide a set of rewrite rules based on a syntactic term algebra [Ru87]. The consistency checker must be able to determine a set of most general unifiers [Si84] for any pair of left-hand-sides and there must be only a finite number. Each left-hand-side of a rewrite rule is a term in a typed syntactic term algebra which is generated from a context-free grammar. Moreover, each left-hand-side corresponds to a sentential form derivation in the grammar, and represents the set of all expressions that match it. Describing all expressions that match more than one left-hand-side is equivalent both to giving a unification algorithm for the term algebra and to finding a minimal partition of the intersection of left-hand-side expression sets [La88]. Thus we are interested in most general unifiers (mgu's), or equivalently minimal intersection partitions (mip's). Unfortunately, since we are not doing simple syntactic unification (where unifiable terms must have the same structure), two left-hand-sides may have more than one, indeed an infinite number of, mgu's [Si84]. Each mgu or equivalent intersection component represents a separate case that must be examined by the consistency checker. More specifically, then, each pair of terms must have only a finite number of mgu's (that is, there exist an equational unification algorithm of finite type [Si84]). The equational unification theory identifies two differently constructed expression lists whenever their components are identified. In [La88], sufficient conditions for the existence of such unification algorithms are given. However, there is no hope of determining necessary conditions, because finding a unification algorithm for any context-free term algebra is equivalent to determining whether the intersection of two context-free languages is empty, an undecidable problem [HoUl79].

Sections 2 through 4 illustrate and develop these concepts in the context of a set of rewrite rules for a class $\mathbf{L}_{\mathrm{FP}}$ of FFP languages. Each language $\mathcal{L} \in \mathbf{L}_{\mathrm{FP}}$ has as basis a specific Turing-complete language $\mathcal{L}_\beta \in \mathbf{L}_{\mathrm{FP}}$ [Wi81]. In Section 2, we present a universal algebraic description of a context-free term algebra for each language $\mathcal{L}$. Section 3 defines the concept of a "good" set of rewrite rules and exhibits a good, but syntactically ambiguous, set of rewrite rules for $\mathcal{L}_\beta$. In Section 4, a unification algorithm of finite type is presented for the term algebra of each language $\mathcal{L}$. Section 5 includes an outline of the consistency checking procedure, an outline of a proof of its existence, and a detailed proof of the consistency of the rewrite rules for the basis language $\mathcal{L}_\beta$.

Acknowledgments: The author would like to thank Don Stanat, Geoffrey Frank, Quan Nguyen, John Cherniavsky, and Ron Sigal for fundamental contributions to the consistency checker. She also extends her appreciation to Ellis Cooper, Melvin Fitting, and Anil Nerode for many useful conversations.

2 Syntactic FFP Algebras.

This section includes (**a**) a review, similar to the development in [D²Qu78], of a context-free grammar (CFG) as a derivation system for expressions, along with an example class $\mathbf{G}_{\mathrm{FP}}$ of CFG's for expressions in FFP languages [Ba78]; and (**b**) a universal algebraic definition of typed context-free algebras [Ru87], along with the corresponding FFP algebras.

2.1 A Traditional Presentation of Context-free Languages

Let T be a finite set of symbols, and let T^{m} denote the set of all strings of m symbols from T. T^0 represents the singleton set containing only the empty string ϵ. The set T^* of all finite strings of symbols over T is the disjoint union $\bigsqcup \mathrm{T}^{\mathrm{m}}, \mathrm{m} \geq 0$. The set T^* together with the associative operation of concatenation is a free monoid with the empty string ϵ as the identity.

Definition 2.1.1: Let S be a finite set of symbols. A *language* L over S is any subset of T^*; that is, $\mathrm{L} \subseteq \mathrm{T}^*$. An element $\mathrm{e} \in \mathrm{L}$ is called an *expression* (or a *sentence*) of L. □

Definition 2.1.2: A *context-free grammar* G is a four-tuple (T, N, P, σ) where:

(a) T is a finite set of *terminal symbols*;

(b) N is a finite set of *nonterminal symbols*;

(c) T and N are disjoint; that is $\mathrm{T} \bigcap \mathrm{N} = \emptyset$;

(d) P is the finite set of *productions* or syntactic rewrite rules; $\mathrm{P} \subset \mathrm{N} \times (\mathrm{T} \bigcup \mathrm{N})^*$;

(e) $\sigma \in \mathrm{N}$ is the *starting nonterminal*. □

Notation 2.1.3: If $p \in \mathrm{P}$ and $p = (\mathrm{A}, \mathrm{S}(p))$, then we write $\mathrm{A} \rightarrow_p \mathrm{S}(p)$. The left-hand-side (lhs) of p is A and the right-hand-side (rhs) is $\mathrm{S}(p)$. The rhs $\mathrm{S}(p)$ can assume the following "shape":

$$\mathrm{S}(p) = \mathrm{s}_1\, \mathrm{A}_1\, \mathrm{s}_2 \ldots \mathrm{s}_k\, \mathrm{A}_k\, \mathrm{s}_{k+1}\,,$$

with $k \geq 0$, $\mathrm{A}_i \in \mathrm{N}$, and $\mathrm{s}_j \in \mathrm{T}^*$. □

The process of generating an expression according to a CFG G is the successive rewriting of *sentential forms* (that is, elements of $(\mathrm{T} \bigcup \mathrm{N})^*$) through the use of productions of the grammar, starting with the start nonterminal σ. The sequence of sentential forms and productions required to generate an expression constitutes a *derivation* of the expression according to the grammar.

Definition 2.1.4: Let $G = (T, N, P, \sigma)$ be a CFG.

(a) If $A \rightarrow_p S(p)$ is a production in G, and $w_0 = u\, A\, v$ and $w_1 = u\, S(p)\, v$ are sentential forms, we say that w_1 is *immediately derived* from w_0 in G, and we indicate this relation by writing $w_0 * \!\rightarrow_p w_1$.

(b) If $(w_0, w_1, \ldots w_n)$ is a sequence of sentential forms such that

$$w_0 * \!\rightarrow_{p_1} w_1 * \!\rightarrow_{p_2} \cdots * \!\rightarrow_{p_n} w_n$$

we say that w_n is *derivable* from w_0 and indicate this relation by writing $w_0 * \!\rightarrow_d w_n$, where

$$d = (w_0, p_1, w_1, p_2, \cdots, p_n, w_n).$$

The sequence d is called a *derivation* of w_n from w_0 according to G. The beginning sentential form w_0 is called the *domain* of d, while the last sentential form w_n, denoted by $S(d)$, is called *codomain* of d.

(c) A derivation $w_0 * \!\rightarrow_d w$ is *leftmost*, whenever, for each j, $1 \le j \le n$, the nonterminal rewritten at the jth step is the leftmost nonterminal in the sentential form w_{j-1}.

(d) The grammar G is said to be *ambiguous* if there exists a terminal string $e \in T^*$ and distinct leftmost derivations $\sigma * \!\rightarrow_{d_1} e$ and $\sigma * \!\rightarrow_{d_2} e$.

(e) The *length* $len(d)$ of derivation d is the number of productions in d. □

Definition 2.1.5: Let $G = (T, N, P, \sigma)$ be a CFG and $u * \!\rightarrow_d w$.

(a) The set $|d| = \{u * \!\rightarrow_d w * \!\rightarrow_{d_1} y \mid y \in (T \bigcup N)^*\}$ consists of all *derivation extensions* of d.

(b) $|d|^T = \{u * \!\rightarrow_d w * \!\rightarrow_{d_1} e \mid e \in T^*\}$ is the set of all *terminal* derivation extensions of d.

(c) Let $|w|$ denote the set of all derivations that begin with $w \in (T \bigcup N)^*$. If $A \in N$, $|A|$ represents the set of all derivations with domain A.

(d) If $A \in N$, let $\mathbf{L}_G(A)$ denote the *language* of all terminal strings derivable from A; that is, $\mathbf{L}_G(A)$ is the set of all derivation codomains of derivations in $|A|^T$.

(e) The language $\mathbf{L}_G(\sigma)$ is called the *language generated by* G and is denoted by $\mathbf{L}(G)$. If $e \in \mathbf{L}(G)$, we say that e is a *string*, an *expression*, or a *word* generated by G.

(f) If $A * \!\rightarrow_d S(d)$, then $\mathbf{L}_G(S(d))$ is the *sublanguage* of $\mathbf{L}_G(A)$ of sentential form $S(d)$. □

An FFP language contains three basic kinds of expressions: atoms, applications, and sequences. *Atomic expressions* include numbers and Boolean constants, as well as a collection of "names" for the language primitives (that is, primitive functions and functional forms of the language). An atomic expression is treated as a single symbol. An expression consisting of an application of a function to its argument is called an *application*. The application $(f : e)$ represents the application of the function (or program) component f

to the operand (or input) component e; both f and e can be any expressions in the FFP language. *Sequences* are defined recursively as bracketed lists of FFP expressions. The following example develops a context-free description of the syntactic structure of a class of FFP languages.

Example 2.1.6 The following table describes a class $\mathbf{G}_{\mathrm{FP}}$ of CFG's for FFP languages [Ba78], by listing the principal productions appearing in any CFG in $\mathbf{G}_{\mathrm{FP}}$. This class is partially specified; the remaining productions which generate atomic expression collections such as numbers, Boolean constants, primitive program names, functional form names, etc., are not given. The set of atoms is the language $\mathbf{L}_{\mathbf{G}_{\mathrm{FP}}}(\mathrm{AT})$ of nontermial AT and is denoted by $|\mathrm{AT}|$. (When no confusion arises, we refer to this class of CFG's as "the grammar $\mathbf{G}_{\mathrm{FP}}$".)

Because the consistency checking procedure assumes an innermost model for evaluation of expressions, the grammar $\mathbf{G}_{\mathrm{FP}}$ has the following properties. There are only two productions from the start symbol E; E $\rightarrow$ O and E $\rightarrow$ X with $\mathbf{L}(\mathrm{O})\bigcap\mathbf{L}(\mathrm{X}) = \emptyset$. The sublanguage L(X) represents expressions containing at least one application, while L(O) represents expressions with no applications. Expressions in L(O) are called *objects*. Since E is the start nonterminal, $\mathbf{G}_{\mathrm{FP}}$; that is

$$\mathbf{L}(\mathrm{E}) = \mathbf{L}(\mathbf{G}_{\mathrm{FP}}) = \mathbf{L}(\mathrm{O})\bigsqcup\mathbf{L}(\mathrm{X}),$$

where $\bigsqcup$ denotes disjoint union.

Because the consistency checking procedure takes as input a given set of rewrite rules that contain variables for objects and object lists, the grammar $\mathbf{G}_{\mathrm{FP}}$ contains an arbitrary number of (terminal) object variables $\mathrm{V}(\mathrm{O}) = \{x_i\}$ and object list variables $\mathrm{V}(\mathbf{L}_{\mathrm{O}}) = \{l_j\}$. A list variable l_j represents a (possibly empty) sublist of a list. For example[1], if θ is a substitution such that the instanitiation of l_j, denoted by $\theta\, l_j$, is the empty string ϵ, then the $\theta\ < x_i,\ l_j > = < \theta\, x_i >$; similarly for $< l_j,\ x_i >$. If $\theta\, l_j = \mathrm{ob}_1, \mathrm{ob}_2$ with ob_1 and ob_2 in $\mathbf{L}(\mathrm{O})$, then $\theta\ < x_i,\ l_j > = < \theta\, x_i,\ \mathrm{ob}_1,\ \mathrm{ob}_2 >$.

The grammar $\mathbf{G}_{\mathrm{FP}}$ is ambiguous, because lists can be constructed by gluing a single component onto the left-end or the right-end of an already constructed list. For example, the distinct length 3 leftmost derivations:

$$\begin{aligned} d_l &= \mathrm{O} \ast\!\!\rightarrow_{bro} \ < \mathrm{L_O} > \ \ast\!\!\rightarrow_{glo} \ < \mathrm{O},\ \mathrm{L_O} > \ \ast\!\!\rightarrow_{gro} \ < \mathrm{O},\ \mathrm{L_O},\ \mathrm{O} > \\ d_r &= \mathrm{O} \ast\!\!\rightarrow_{bro} \ < \mathrm{L_O} > \ \ast\!\!\rightarrow_{gro} \ < \mathrm{L_O},\ \mathrm{O} > \ \ast\!\!\rightarrow_{glo} \ < \mathrm{O},\ \mathrm{L_O},\ \mathrm{O} > . \end{aligned}$$

derive the same sentential form, $\mathrm{S}(d_l) =< \mathrm{O},\ \mathrm{L_O},\ \mathrm{O} >= \mathrm{S}(d_r)$. In fact, $| < \mathrm{O},\ \mathrm{L_O},\ \mathrm{O} > |^{\mathrm{T}}$ is the set of all derivations to sequences (bracketed list expressions) having at least two components. □

[1]Our presentation uses commas to separate components in a list, but the commas are only for readability; they are not actually part of a list expression.

Table 1: A Grammar for FFP Languages

Terminals: $\mathrm{T} = |\mathrm{AT}| \cup \mathrm{V(O)} \cup \mathrm{V(L_O)} \cup \{\ <\ >\ \ ,\ \ (\ )\ \ :\ \}$

Nonterminals: $\mathrm{N} = \{\ \mathrm{E, O, X, AT, V_O, L_O, V_L, L_X}\ \}$

Start Nonterminal: E

Productions:

Production Type	Labeled Production	Production Type	Labeled Production
EXPRESSION		OBJECT	
• object	$\mathrm{E} \rightarrow_{eo} \mathrm{O}$	LISTS	
• applicative	$\mathrm{E} \rightarrow_{ex} \mathrm{X}$	• empty	$\mathrm{L_O} \rightarrow_{em} \epsilon$
OBJECT		• singleton	$\mathrm{L_O} \rightarrow_{lio} \mathrm{O}$
• atom	$\mathrm{O} \rightarrow_{at} \mathrm{AT}$	• glue left	$\mathrm{L_O} \rightarrow_{glo} \mathrm{O} , \mathrm{L_O}$
• sequence	$\mathrm{O} \rightarrow_{bro}\ < \mathrm{L_O} >$	• glue right	$\mathrm{L_O} \rightarrow_{gro} \mathrm{L_O} , \mathrm{O}$
• variable	$\mathrm{O} \rightarrow_{vo} \mathrm{V_O}$	• variable	$\mathrm{L_O} \rightarrow_{vl} \mathrm{V_L}$
ATOMIC	$\mathrm{AT} \rightarrow \mathrm{a}$	VARIABLE	$\mathrm{V_L} \rightarrow l_j$
VARIABLE	$\mathrm{V_O} \rightarrow x_i$	APPLICATIVE	
APPLICATIVE		LISTS	
EXPRESSIONS		• singleton	$\mathrm{L_X} \rightarrow_{lix} \mathrm{X}$
• application	$\mathrm{X} \rightarrow_{ap} (\mathrm{E}\ :\ \mathrm{E})$	• glue left	$\mathrm{L_X} \rightarrow_{glx} \mathrm{E} , \mathrm{L_X}$
• sequence	$\mathrm{X} \rightarrow_{brx}\ < \mathrm{L_X} >$	• glue right	$\mathrm{L_X} \rightarrow_{grx} \mathrm{L_X} , \mathrm{E}$

2.2 Universal Syntactic Algebras.

Definition 2.2.1: Let G = (T, N, P, σ) be a CFG. The *free universal algebra* generated by a CFG G [Ru87] is a typed syntactic algebra **F**(G).

(a) The *carrier sets* (that is, the sets used to define the algebraic operations) correspond to the nonterminal derivation sets

$$|\mathrm{A}| = \{A \ast\!\!\rightarrow_d \mathrm{S}(d)\},$$

where A $\in$ N. The nonterminal collection N represents the *types*.

(b) The *operations* correspond to the productions as follows:

(b1) Each terminal production $\mathrm{A} \rightarrow_p \mathrm{e}_p,\ \ \mathrm{e}_p \in \mathrm{T}^*$ corresponds to a *nullary* operation

$$\hat{p} : \{\ast\} \ \rightarrow\ |\mathrm{A}|.$$

The collection $\mathcal{A}$(G) of all nullary operations is the set of *algebraic atoms*.

(b2) If $\mathrm{A} \rightarrow_p \mathrm{s}_1\, \mathrm{A}_1\, \mathrm{s}_2 \ldots \mathrm{s}_k\, \mathrm{A}_k\, \mathrm{s}_{k+1},\ k \geq 1$, is a production, then

$$\hat{p} : |\mathrm{A}_1| \times \cdots \times |\mathrm{A}_k| \ \rightarrow\ |\mathrm{A}|$$

is the operation, which we call a *p-composition*, defined by:

$$\begin{aligned}
\hat{p}\ \ (\mathrm{A}_1 * \!\!\rightarrow_{d_1} \mathrm{S}(d_1),\ \ldots,\ \mathrm{A}_k * \!\!\rightarrow_{d_k} \mathrm{S}(d_k)) = \\
\mathrm{A} \rightarrow_p \mathrm{s}_1\, \mathrm{A}_1\, \mathrm{s}_2 \ldots \mathrm{s}_k\, \mathrm{A}_k\, \mathrm{s}_{k+1} \\
* \!\!\rightarrow_{d_1} \mathrm{s}_1\, \mathrm{S}(d_1)\, \mathrm{s}_2 \ldots \mathrm{s}_k\, \mathrm{A}_k\, \mathrm{s}_{k+1} \\
* \!\!\rightarrow_{d_2} \mathrm{s}_1\, \mathrm{S}(d_1)\, \mathrm{s}_2\, \mathrm{S}(d_2) \ldots \mathrm{A}_{k-1}\, \mathrm{s}_k\, \mathrm{A}_k\, \mathrm{s}_{k+1} \\
\vdots \\
* \!\!\rightarrow_{d_{k-1}} \mathrm{s}_1\, \mathrm{S}(d_1)\, \mathrm{s}_2\, \mathrm{S}(d_2) \ldots \mathrm{S}(d_{k-1}) \mathrm{s}_k\, \mathrm{A}_k\, \mathrm{s}_{k+1} \\
* \!\!\rightarrow_{d_k} \mathrm{s}_1\, \mathrm{S}(d_1)\, \mathrm{s}_2\, \mathrm{S}(d_2) \ldots \mathrm{S}(d_{k-1}) \mathrm{s}_k\, \mathrm{S}(d_k)\, \mathrm{s}_{k+1}\ .\ \square
\end{aligned}$$

Each element d of $\mathbf{F}(\mathrm{G})$ is a derivation from a nonterminal A to the sentential form $\mathrm{S}(d)$. In the free algebra $\mathbf{F}(\mathbf{G}_{\mathrm{FP}})$, the leftmost O-derivation (a derivation with domain O):

$$\begin{aligned}
d_l \ = \ & \mathrm{O} * \!\!\rightarrow_{bro} < \mathrm{L_O} > * \!\!\rightarrow_{glo} < \mathrm{O},\, \mathrm{L_O} > * \!\!\rightarrow_{gro} < \mathrm{O},\, \mathrm{L_O},\, \mathrm{O} > \\
& * \!\!\rightarrow_{vo} < \mathrm{V_O},\, \mathrm{L_O},\, \mathrm{O} > * \!\!\rightarrow < x_1,\, \mathrm{L},\, \mathrm{O} > * \!\!\rightarrow_{vl} < x_1, \mathrm{V_L}, \mathrm{O} > \\
& * \!\!\rightarrow < \mathrm{V_O},\, l_1,\, \mathrm{O} > * \!\!\rightarrow_{vo} < x_1,\, l_1,\, \mathrm{V_O} > * \!\!\rightarrow < x_1, l_1, x_2 >
\end{aligned}$$

corresponds to the *algebraic element*

$$\widehat{d_l} = \widehat{bro}\ \widehat{glo}\ \widehat{gro}\ \widehat{vo}\ \widehat{x_1}\ \widehat{vl}\ \widehat{l_1}\ \widehat{vo}\ \widehat{x_2}\ .$$

Definition 2.2.2: For grammar $\mathrm{G} = (\mathrm{T},\ \mathrm{N},\ \mathrm{P},\ \sigma)$, there are two important syntactic congruence relations for the free algebra $\mathbf{F}(\mathrm{G})$.

(a) The first is the *interchange congruence* $\mathcal{I}(\mathrm{G})$. Two interchange congruent derivations differ only in the order that productions are applied. The interchange congruence is generated by the interchange derivation pairs; that is, if $\mathrm{w}_i \in (\mathrm{T} \bigcup \mathrm{N})^*$, $i = 1,2,3$, $\mathrm{A}_i \in \mathrm{N}$, $i = 1,2$, and $\mathrm{A}_i * \!\!\rightarrow_{p_i} \mathrm{S}(p_i)$, $i = 1,2$, then the following derivation pair is a generating pair in the interchange congruence:

$$\begin{aligned}
(\ \ & \mathrm{w}_1\, \mathrm{A}_1\, \mathrm{w}_2\, \mathrm{A}_2\, \mathrm{w}_3 * \!\!\rightarrow_{p_1} \mathrm{w}_1\, \mathrm{S}(p_1)\, \mathrm{w}_2\, \mathrm{A}_2\, \mathrm{w}_3 * \!\!\rightarrow_{p_2} \mathrm{w}_1\, \mathrm{S}(p_1)\, \mathrm{w}_2\, \mathrm{S}(p_2)\, \mathrm{w}_3\ , \\
& \mathrm{w}_1\, \mathrm{A}_1\, \mathrm{w}_2\, \mathrm{A}_2\, \mathrm{w}_3 * \!\!\rightarrow_{p_2} \mathrm{w}_1\, \mathrm{A}_2\, \mathrm{w}_2\, \mathrm{S}(p_2)\, \mathrm{w}_3 * \!\!\rightarrow_{p_1} \mathrm{w}_1\, \mathrm{S}(p_1)\, \mathrm{w}_2\, \mathrm{S}(p_2)\, \mathrm{w}_3\ \ )\ .
\end{aligned}$$

Each congruence class has a unique leftmost derivation which corresponds to a unique *parse tree*; hence, the quotient universal algebra $\mathbf{F}(\mathrm{G})/\mathcal{I}(\mathrm{G})$ is called the algebra $\mathbf{PR}(\mathrm{G})$ of *parse trees* generated by G. Under the interchange congruence, concatenation of derivations is *monoidal* (see, for example, [Be75] or [Ne80]).

(b) The second congruence on $\mathbf{F}(\mathrm{G})$ is the the *pre-order congruence* $\mathcal{PO}(\mathrm{G})$ Two derivations $d_1 : \mathrm{A}_1 * \!\!\rightarrow\ \mathrm{S}(d_1)$ and $d_2 : \mathrm{A}_2 * \!\!\rightarrow\ \mathrm{S}(d_2)$ are pre-order congruent, denoted $d_1 \equiv d_2$, if and only if $\mathrm{A}_1 = \mathrm{A}_2$ and $\mathrm{S}(d_1) = \mathrm{S}(d_2)$. This pre-order relation includes the interchange congruence, and identifies any two derivations that have the same domains and codomains. The quotient algebra $\mathbf{F}(\mathrm{G})/\mathcal{PO}(\mathrm{G})$ is isomorphic to the typed algebra $\mathbf{S}(\mathrm{G})$ of *sentential forms* generated by G □.

Example 2.2.2: The free syntactic algebra $\mathbf{F}(\mathbf{G}_{\mathrm{FP}})$ for the grammar $\mathbf{G}_{\mathrm{FP}}$ has the *p*-compositions that are given below in Table 2.

Table 2: FFP p-Compositions

Production Type	Production Composition	Production Type	Production Composition
EXPRESSION		OBJECT LISTS	
• object	$\|O\| \to_{eo} \|E\|$	• empty	$\{\epsilon\} \to_{em} \|L_O\|$
• applicative	$\|E\| \to_{ex} \|X\|$	• singleton	$\|O\| \to_{lio} \|L_O\|$
OBJECT		• glue left	$\|O\| \times \|L_O\| \to_{glo} \|L_O\|$
• atom	$\|AT\| \to_{at} \|O\|$	• glue right	$\|L_O\| \times \|O\| \to_{gro} \|L_O\|$
• sequence	$\|L_O\| \to_{bro} \|O\|$	• variable	$\|V_L\| \to_{vl} \|L_O\|$
• variable	$\|V_O\| \to_{vo} \|O\|$	VARIABLE	$\{l_j\} \to \|V_L\|$
ATOMIC	$\{a\} \to \|AT\|$	APPLICATIVE LISTS	
VARIABLE	$\{x_i\} \to \|V_O\|$	• singleton	$\|X\| \to_{lix} \|L_X\|$
APPLICATIVE EXPRESSIONS		• glue left	$\|E\| \times \|L_X\| \to_{glx} \|L_X\|$
• application	$\|E\| \times \|E\| \to_{ap} \|X\|$	• glue right	$\|L_X\| \times \|E\| \to_{grx} \|L_X\|$
• sequence	$\|L_X\| \to_{brx} \|X\|$		

The nullary operations in $\mathbf{F}(\mathbf{G}_{\mathrm{FP}})$ correspond to variable productions ($V_O \to x_i$ or $V_L \to l_j$) or to the atomic productions ($AT \to a_k$). Let $\mathbf{F}^{T}(\mathbf{G}_{\mathrm{FP}})$ denote the subalgebra of all terminal string derivations in $\mathbf{F}(G_{\mathrm{FP}})$. Each terminal string derivation congruence class

$$[\,A *\!\!\to_d t\,], \quad t \in T^*,$$

in the sentential form algebra $\mathbf{S}(G_{\mathrm{FP}})$ can be viewed as a *term* t in a typed term algebra $\mathbf{T}(\mathbf{G}_{\mathrm{FP}})$ that contains only two kinds of variables, object and object list variables. This subalgebra $\mathbf{T}(G_{\mathrm{FP}})$ is called the *FFP term algebra.* For example, the terminal string derivation class

$$[\,O *\!\!\to_{d_l}\ < x_1,\ l_1,\ x_2 >]$$

corresponds to the O-term (that is, a term of type O)

$$< x_1, l_1, x_2 >$$

that "forgets" the derivation of $< x_1, l_1, x_2 >$ from O.

For each A-term $t = [\,A *\!\!\to_d t\,]$ in $\mathbf{T}(G_{\mathrm{FP}})$, there exists a *retraction* $r(t)$ onto the pre-order congruence class

$$[\,A *\!\!\to_{r(d)} S(t)\,]$$

of the A-*sentential form* S(t) of t constructed from t by replacing each x_i by nonterminal O and each l_j by nonterminal L_O. For the O-term $< x_1, l_1, x_2 >$,

$$r(< x_1, l_1, x_2 >) = [\,O *\!\!\to_{r(d_l)}\ < O,\ L_O,\ O >].$$

This retraction $r : \mathbf{T}(\mathbf{G}_{\mathrm{FP}}) \to \mathbf{S}(\mathbf{G}_{\mathrm{FP}})$ is monoidal.

A *substitution* in $\mathbf{T}(G_{\mathrm{FP}})$ is a finite set θ of variable-term pairs:

$$\theta = \{(x_i \Rightarrow \mathrm{ob}_i)\} \bigcup \{(l_j \Rightarrow \mathrm{list}_j)\}$$

where, for each i, ob_i is an O-term and, for each j, list_j is an L_O-term. For any A-term $t = [A *\!\!\to_d t]$ in $\mathbf{T}(\mathbf{G}_{\mathrm{FP}})$ and substitution θ, the *instantiated term* θt corresponds to a derivation extension $d(\theta)$ of the retract $r(t) = [A *\!\!\to_{r(d)} S(t)]$. For example, if

$$\theta = \{(x_1 \Rightarrow \mathrm{true}),\ (l_1 \Rightarrow < >,\ x_3),\ (x_2 \Rightarrow < l_2,\ \mathrm{false} >)\}\,,$$

then the instantiated O-term $\theta < x_1,\ l_1,\ x_2 >$ is defined by

$$\theta < x_1,\ l_1,\ x_2 > \ = \ [\, \mathrm{O} *\!\!\longrightarrow_{r(d_l)} < \mathrm{O},\ \mathrm{L_O},\ \mathrm{O} > \\ *\!\!\longrightarrow_{d(\theta)} < \mathrm{true}, < >,\ x_3, < l_2,\ \mathrm{false} >>].$$

Composition of substitutions corresponds to composition of derivation pre-order congruence classes, viewed as composition of paths in the monoidal graph of immediate derivations (see, for example [Ne80], [Si84], or [Ru87]). We use the notation $\theta_2 \circ \theta_1$ to represent the composition of substitutions. If $\mathrm{t} = [\mathrm{A} *\!\!\longrightarrow_d \mathrm{t}]$ is a term in $\mathbf{T}(\mathbf{G}_{\mathrm{FP}})$ with retract $r(\mathrm{t}) = [\mathrm{A} *\!\!\longrightarrow_{r(d)} S(\mathrm{t})]$, then the instantiated term $(\theta_2 \circ \theta_1)\,\mathrm{t}$ corresponds to the composition

$$[\, \mathrm{A} *\!\!\longrightarrow_{r(d)} S(\mathrm{t}) *\!\!\longrightarrow_{r(d(\theta_1))} S(\theta_1 \mathrm{t}) *\!\!\longrightarrow_{d(\theta_2)} (\theta_2 \circ \theta_1)\mathrm{t}\ ].$$

Composition of substitutions is associative. □

3 Good Sets of Rewrite Rules for FFP Languages.

For each FFP language $\mathcal{L}$, the input to the consistency checking procedure is a set $\mathbf{RR}(\mathcal{L})$ of rewrite rules that specifies an operational semantics for $\mathcal{L}$. For each language primitive f, the corresponding rewrite rule set $\mathbf{RR}(f)$ specifies how each variable-free f-application $(f : \mathrm{e})$ whose value is not "bottom" (denoted by $\perp$) can be transformed using rewrite evaluation. Utilization of the rules in $\mathbf{RR}(f)$ may be nondeterministic in that an f-application may match both left-recursive and right-recursive rules. If an f-application does not match any rewrite rule, then it is assigned the value $\perp$.

The evaluation of expressions in the FFP sublanguage $\mathbf{L}(\ (< \mathrm{L_O} > : \mathrm{O})\ )$ is done using special rules, called *meta-composition rules* [Ba78] . Any FFP semantics is required to contain meta-composition rules, and they are included as part of the definition of a good set of rewrite rules. (Within the category of sets, meta-composition is an example of the unit natural transformation for the adjoint functor pair $(Cartesian - product,\ Hom - set)$ [Mac71;IV].)

Definition 3.1:

(a) Suppose that f is a language primitive in $\mathcal{L}$. A *rewrite rule* is of the form

$$(f : \mathrm{u}) \ \rightarrow\ rhs(f)$$

where:

- u is a O-term with no repeated variables ;
- the set of variables occurring in the $rhs(f)$ is a subset of the variables in u.

(b) The set $\mathbf{RR}(meta)$ of meta-composition rewrite rules consists of:

- $\mathrm{RR_O}(meta) = [\ (< > : x_0) \ \rightarrow\ x_0\ ]$
- $\mathrm{RR}_L(meta) = [\ (< x_1,\ l_1 > : x_2) \ \rightarrow\ (x_1 : << x_1,\ l_1 >,\ x_2 >)\ ].$

(c) An operational semantics $\mathbf{RR}(\mathcal{L})$ is *good*, whenever $\mathbf{RR}(meta) \subseteq \mathbf{RR}(\mathcal{L})$. □

Example 3.2: All the language primitives described in [Ba78] have good rewrite rule implementations. The following rules are included in a good set $\mathbf{RR}(\mathcal{L}_\beta)$ of rewrite rules for the basis FFP language $\mathcal{L}_\beta$. This listing includes rules for the primitives functions *append-left* and *append-right* and for the combining forms *composition*, *apply*, and *construction*. In [Wi81], it was shown that any FFP language that contained these combining forms is Turing complete. The format of the FFP lhs's is taken from [Ba78].

Composition:

- $\mathrm{RR}_0(comp) = [\,(comp :<< comp >,\, x_1 >) \;\rightarrow\; x_1\,]$
- $\mathrm{RR}_{>0}(comp) = [\,(comp :<< comp,\, x_2,\, l_1 >,\, x_1 >) \;\rightarrow\; (x_2 : (< comp,\, l_1 > : x_1))\,]$

Apply:

- $\mathrm{RR}(app) = [\,(app :<< app,\, x_2 >,\, x_1 >) \;\rightarrow\; (x_2 : x_1)\,]$

Construction:

$$
\begin{aligned}
\bullet\, \mathrm{RR}_0(con) &= [\,(con :<< con >,\, x_1 >) \rightarrow <>\,] \\
\bullet\, \mathrm{RR}_L(con) &= [\,(con :<< con,\, x_2,\, l_1 >,\, x_1 >) \rightarrow \\
&\qquad (apnd_L :< (x_2 : x_1),\, (< con,\, l_1 > : x_1) >)\,] \\
\bullet\, \mathrm{RR}_R(con) &= [\,(con :<< con,\, l_2,\, x_3 >,\, x_1 >) \rightarrow \\
&\qquad (apnd_R :< (< con,\, l_2 >: x_1),\, (x_3 : x_1) >)\,]
\end{aligned}
$$

Append-left:

- $\mathrm{RR}(apnd_L) = [\,(apnd_L :< x_1,\, < l_1 >>) \rightarrow < x_1,\, l_1 >\,]$

Append-right:

- $\mathrm{RR}(apnd_R) = [\,(apnd_R :<< l_1 >,\, x_1 >) \rightarrow < l_1,\, x_1 >\,].$

The rewrite rules $\mathrm{RR}_L(con)$ and $\mathrm{RR}_R(con)$ are well-formed *recursive* rules because the lhs's contain no repeated variables and the rhs's contain no new variables.

Note that the collection $\mathbf{RR}(con)$ is nondeterministic. For example, the meta-composition rewrite

$$
\bullet\quad (< con,\, and,\, or > : < \text{true},\, \text{false} >) \rightarrow_{\mathrm{RR}_L(meta)} (con :<< con,\, and,\, or >,\, < \text{true},\, \text{false} >>),
$$

results in an application which matches both $\mathrm{RR}_L(con)$ and $\mathrm{RR}_R(con)$. On the otherhand, the set $\mathbf{RR}(con)\backslash\{\mathrm{RR}_R(con)\}$ is deterministic; that is, any application $(con : \mathrm{e})$ can match at most one rewrite rule, because each term

$$
(con :< con,\, \text{list} >,\, \mathrm{e}_1 >),
$$

that matches a *con*-rule, has “list” with zero, or one or more top-level components. □

4 A Minimal Intersection Partition Algorithm.

A good set $\mathbf{RR}(\mathcal{L})$ of rewrite rules for an FFP language $\mathcal{L}$ may be used nondeterministically and it must be possible to recognize when two or more rewrite rules (for the same primitive) can be used to evaluate the same applicative expression. Each left-hand-side corresponds to a class of derivations in the grammar $\mathbf{G}_{\mathrm{FP}}$, and represents the set of all expressions that match it. Describing all expressions that match more than one left-hand-side is equivalent both to giving a unification algorithm for the term algebra $\mathbf{T}(\mathbf{G}_{\mathrm{FP}})$ and to finding a minimal partition of the intersection of left-hand-side expression sets [La88]. Thus we are interested in most general unifiers (mgu's), or equivalently minimal intersection partitions (mip's). Unfortunately, since we are not doing simple syntactic unification (where unifiable terms must have the same structure), two left-hand-sides may have more than one, indeed an infinite number of, mgu's [Si84]. Each mgu or corresponding intersection component represents a separate case that must be examined by the consistency checker. In this section, we describe an equational unification algorithm of finite type [Si84] for the term algebra $\mathbf{T}(\mathbf{G}_{\mathrm{FP}})$ and show that the sentential form algebra $\mathbf{S}(\mathbf{G}_{\mathrm{FP}})$ satisfies sufficient conditions, presented in [La88], which guarantee the existence of such unification algorithms. The equational unification theory for $\mathbf{T}(\mathbf{G}_{\mathrm{FP}})$ identifies two differently constructed expression lists whenever their components are identified. There is no hope of determining necessary conditions for the existence of such unification algorithms, because finding a unification algorithm for arbitrary context-free term algebras is equivalent to determining whether the intersection of two context-free languages is empty which is undecidable [HoUl79].

First, for a CFG G = (T, N, P, σ), A $\in$ N, and A-derivations A $*\!\!\rightarrow_{d_1}$ S(d_1) and A $*\!\!\rightarrow_{d_2}$ S(d_2) we define the concept of *minimal intersection partition* (mip) for the intersection $\mathbf{L}_{\mathrm{G}}(\mathrm{S}(d_1)) \bigcap \mathbf{L}_{\mathrm{G}}(\mathrm{S}(d_2))$ of the sentential form sublanguages $\mathbf{L}_{\mathrm{G}}(\mathrm{S}(d_1))$ and $\mathbf{L}_{\mathrm{G}}(\mathrm{S}(d_2))$ of $\mathbf{L}_{\mathrm{G}}(\mathrm{A})$. Next, for the term algebra $\mathbf{T}(\mathbf{G}_{\mathrm{FP}})$, we show that, for terms t_1 and t_2 of nonterminal type A, determining a finite set $mgu(\mathrm{t}_1, \mathrm{t}_2)$ of *most general unifiers* for t_1 and t_2 is equivalent to giving a mip for the "retract" sublanguage intersection $\mathbf{L}_{\mathrm{G}}(\mathrm{S}(\mathrm{t}_1)) \bigcap \mathbf{L}_{\mathrm{G}}(\mathrm{S}(\mathrm{t}_2))$. Lastly, we show that mip's exist for the sentential form algebra $\mathbf{S}(\mathbf{G}_{\mathrm{FP}})$, by verifying that $\mathbf{S}(\mathbf{G}_{\mathrm{FP}})$ satisfies the sufficient conditions presented in [La88].

Definition 4.1: Suppose that CFG G = (T, N, P, σ), A $*\!\!\rightarrow_{d_1}$ S(d_1) and A $*\!\!\rightarrow_{d_2}$ S(d_2) are derivations in |A|.

(a) A finite (possibly empty) set of derivation pairs

$$\{(\,\mathrm{A}*\!\!\rightarrow_{d_1} \mathrm{S}(d_1)\ *\!\!\rightarrow_{d_{1k}} \mathrm{S}(d_{1k}),\ \ \mathrm{A}*\!\!\rightarrow_{d_2} \mathrm{S}(d_2)\ *\!\!\rightarrow_{d_{2k}} \mathrm{S}(d_{2k})\,) \mid 1 \le k \le K\,\}$$

with $0 \le K$, is said to be an *intersection partition* of $\mathbf{L}_{\mathrm{G}}(\mathrm{S}(d_1)) \bigcap \mathbf{L}_{\mathrm{G}}(\mathrm{S}(d_2))$, denoted by $ip(\mathrm{S}(d_1),\ \mathrm{S}(d_2))$, whenever

- for each k, $1 \le k \le K$, as sentential forms, $\mathrm{S}(d_{1k}) = \mathrm{S}(d_{2k})$. We call each $\mathrm{z}_k = \mathrm{S}(d_{1k}) = \mathrm{S}(d_{2k})$ an *intersection component*;
- $\mathbf{L}_{\mathrm{G}}(\mathrm{S}(d_1)) \bigcap \mathbf{L}_{\mathrm{G}}(\mathrm{S}(d_2)) = \bigsqcup \mathbf{L}_{\mathrm{G}}(\mathrm{z}_k)$.

(b) An intersection partition $ip(\mathrm{S}(d_1), \mathrm{S}(d_2))$ is said to be a *minimal intersection partition* of $\mathbf{L}_{\mathrm{G}}(\mathrm{S}(d_1)) \bigcap \mathbf{L}_{\mathrm{G}}(\mathrm{S}(d_2))$, denoted by $mip(\mathrm{S}(d_1),\ \mathrm{S}(d_2))$, whenever

- the number K of intersection components is smallest;

- for each k, the length pair $(len(d_{1k}),\ len(d_{2k}))$ is smallest for each entry, among length pairs $(len(e_1),\ len(e_2))$ of all derivation extension pairs

$$(\mathrm{A}{*\!\!\rightarrow_{d_1}}\ \mathrm{S}(d_1) *\!\!\rightarrow_{e_1} \mathrm{S}(e_1),\ \ \mathrm{A}{*\!\!\rightarrow_{d_2}}\ \mathrm{S}(d_2) *\!\!\rightarrow_{e_2} \mathrm{S}(e_2))$$

with common sentential form $\mathrm{S}(e_1) = \mathrm{z} = \mathrm{S}(e_2)$. □

Example 4.2: For the grammar $\mathbf{G}_{\mathrm{FP}}$, consider the length two derivations

$$\begin{aligned} d_L &= \mathrm{O} *\!\!\rightarrow_{bro} \ < \mathrm{L_O} > \ *\!\!\rightarrow_{glo} \ < \mathrm{O},\ \mathrm{L_O} > \\ d_R &= \mathrm{O} *\!\!\rightarrow_{bro} \ < \mathrm{L_O} > \ *\!\!\rightarrow_{gro} \ < \mathrm{L_O},\ \mathrm{O} > . \end{aligned}$$

The minimal intersection partition

$$mip(\,\mathrm{O} *\!\!\rightarrow_{d_L} \ < \mathrm{O},\ \mathrm{L_O} >,\ \mathrm{O} *\!\!\rightarrow_{d_R} \ < \mathrm{L_O},\ \mathrm{O} >\,)$$

consists of the following two derivation pairs

- $(\,\mathrm{O} *\!\!\rightarrow_{d_L} \ < \mathrm{O},\ \mathrm{L_O} > *\!\!\rightarrow_{em} \ < \mathrm{O} >,\ \ \mathrm{O} *\!\!\rightarrow_{d_R} \ < \mathrm{L_O},\ \mathrm{O} > *\!\!\rightarrow_{em} \ < \mathrm{O} >\,)$
- $(\,\mathrm{O} *\!\!\rightarrow_{d_L} \ < \mathrm{O},\ \mathrm{L_O} > *\!\!\rightarrow_{gro} \ < \mathrm{O},\ \mathrm{L_O},\ \mathrm{O} >,$
  $\mathrm{O} *\!\!\rightarrow_{d_R} \ < \mathrm{L_O},\ \mathrm{O} > *\!\!\rightarrow_{glo} \ < \mathrm{O},\ \mathrm{L_O},\ \mathrm{O} >\,)$.

It is an intersection partition, because any O-sequence in

$$\mathbf{L}(< \mathrm{O},\ \mathrm{L_O} >) \bigcap \mathbf{L}(< \mathrm{L_O},\ \mathrm{O} >)$$

has either one component or two or more components. There are two length pairs

- $(\,len(< \mathrm{O},\ \mathrm{L_O} > *\!\!\rightarrow_{em} \ < \mathrm{O},\ \mathrm{O} >),\ len(< \mathrm{L_O},\ \mathrm{O} > *\!\!\rightarrow_{em} \ < \mathrm{O},\ \mathrm{O} >\,)) = (1,\ 1)$
- $(\,len(< \mathrm{O},\ \mathrm{L_O} > *\!\!\rightarrow_{gro} \ < \mathrm{O},\ \mathrm{L_O},\ \mathrm{O} >),\ len(< \mathrm{L_O},\ \mathrm{O} > *\!\!\rightarrow_{glo} \ < \mathrm{O},\ \mathrm{L_O},\ \mathrm{O} >\,))$
  $= (1,\ 1)$.

These length pairs are minimal. Since

$$\mathrm{S}(d_L) = < \mathrm{O},\ \mathrm{L_O} > \neq < \mathrm{L_O},\ \mathrm{O} > = \mathrm{S}(d_R),$$

two is the smallest possible number of intersection components. □

Definition 4.3: Suppose that t_1 and t_2 are terms in the term algebra $\mathbf{T}(\mathbf{G}_{\mathrm{FP}})$.

(a) A substitution pair $(\theta_1,\ \theta_2)$ is said to be a *unifier* of t_1 and t_2, whenever $\theta_1\,\mathrm{t}_1 = \theta_2\,\mathrm{t}_2$.

(b) A set $\{\ (\mu_{1k},\ \mu_{2k})\ \}$ of unifiers of t_1 and t_2 is a set of *most general unifiers*, denoted by $mgu(\mathrm{t}_1,\ \mathrm{t}_2)$, whenever the following property is satisfied:

- If $(\theta_1,\ \theta_2)$ is a unifier of t_1 and t_2, then there is a *unique* mgu $(\mu_{1k},\ \mu_{2k})$ and (not necessarily unique) substitution pair $(\omega_{1k},\ \omega_{2k})$ so that

$$\theta_1 = \omega_{1k} \circ \mu_{1k} \quad \text{and} \quad \theta_2 = \omega_{2k} \circ \mu_{2k}\,.$$

(c) For each k, we use the notation mgu_k to represent the *most general unified term* $\mu_{1k}\,\mathrm{t}_1 = \mu_{2k}\,\mathrm{t}_2$. □

Example 4.4: For the term algebra $\mathbf{T}(\mathbf{G}_{\mathrm{FP}})$, consider the O-term pair

$$\begin{aligned} \mathrm{t}_L &= [\,\mathrm{O} *\!\!\longrightarrow_{d_L} \;<\mathrm{O},\, \mathrm{L_O}> \;*\!\!\longrightarrow\; <x_1,\, l_1>] \\ \mathrm{t}_R &= [\,\mathrm{O} *\!\!\longrightarrow_{d_R} \;<\mathrm{L_O},\, \mathrm{O}> \;*\!\!\longrightarrow\; <l_2,\, x_2>] \end{aligned}$$

The $mgu(\mathrm{t}_L,\, \mathrm{t}_R)$ consists of the following two unifiers:

$$mgu(\mathrm{t}_L,\, \mathrm{t}_R) = \{\, (\mu_{L\,1},\, \mu_{R\,1}),\; (\mu_{L\,\geq 2},\, \mu_{R\,\geq 2})\,\}$$

where

- $\mu_{L\,1} = \{\,(l_1 \Rightarrow \epsilon)\,\}$ and $\mu_{R\,1} = \{\,(l_2 \Rightarrow \epsilon)\,\}$;
- $\mu_{L\,\geq 2} = \{\,(l_1 \Rightarrow l_3,\, x_2)\,\}$ and $\mu_{R\,\geq 2} = \{\,(l_2 \Rightarrow x_1,\, l_3)\,\}$.

The most general unified terms are

- $\mathrm{mgu}_1 =< x_1 >$ and $\mathrm{mgu}_{\geq 2} =< x_1,\, l_3,\, x_2 >$.

That $mgu(\mathrm{t}_L,\, \mathrm{t}_R) = \{\, (\mu_{L\,1},\, \mu_{R\,1}),\; (\mu_{L\,\geq 2},\, \mu_{R\,\geq 2})\,\}$ follows from the observation that any term t that equals both $\theta_L\, \mathrm{t}_L$ *and* $\theta_L\, \mathrm{t}_R$ must have either one component, or two or more components. □

Using the monoidal retraction $r : \mathbf{T}(\mathbf{G}_{\mathrm{FP}}) \to \mathbf{S}(\mathbf{G}_{\mathrm{FP}})$, a careful translation of the definition of mip for the sentential form algebra $\mathbf{S}(\mathbf{G}_{\mathrm{FP}})$ to the definition of mgu for the term algebra $\mathbf{T}(\mathbf{G}_{\mathrm{FP}})$ results in the following proposition:

Proposition 4.5: Let $(\mathrm{t}_1, \mathrm{t}_2)$ be a term pair in $\mathbf{T}(\mathbf{G}_{\mathrm{FP}})$. Each mgu (μ_{1k}, μ_{2k}) in $mgu(\mathrm{t}_1, \mathrm{t}_2)$ corresponds to a unique intersection component z_k of a minimal intersection partition $mip(\mathrm{S}(\mathrm{t}_1),\, \mathrm{S}(\mathrm{t}_2))$ of the retract sentential forms $\mathrm{S}(\mathrm{t}_1)$ and $\mathrm{S}(\mathrm{t}_2)$. □

Example 4.6: For the O-term pair $(\mathrm{t}_L,\, \mathrm{t}_R)$

$$\begin{aligned} \mathrm{t}_L &= [\,\mathrm{O} *\!\!\longrightarrow_{d_L} \;<\mathrm{O},\, \mathrm{L_O}> \;*\!\!\longrightarrow\; <x_1,\, l_1>] \\ \mathrm{t}_R &= [\,\mathrm{O} *\!\!\longrightarrow_{d_R} \;<\mathrm{L_O},\, \mathrm{O}> \;*\!\!\longrightarrow\; <l_2,\, x_2>], \end{aligned}$$

the retract $r(\mathrm{mgu}_1)$ of the most general unified term $\mathrm{mgu}_1 =< x_1 >$ is the intersection component

- $(\,(\mathrm{O} *\!\!\longrightarrow_{d_L} \;<\mathrm{O},\, \mathrm{L_O}> *\!\!\longrightarrow_{em} \;<\mathrm{O}>),\; (\mathrm{O} *\!\!\longrightarrow_{d_R} \;<\mathrm{L_O},\, \mathrm{O}> *\!\!\longrightarrow_{em} \;<\mathrm{O}>)\,),$

while the retract $r(\mathrm{mgu}_{\geq 2})$ of $\mathrm{mgu}_{\geq 2} = < x_1,\, l_3,\, x_2 >$ is the intersection component

- $(\,(\mathrm{O} *\!\!\longrightarrow_{d_L} \;<\mathrm{O},\, \mathrm{L_O}> *\!\!\longrightarrow_{gro} \;<\mathrm{O},\, \mathrm{L_O},\, \mathrm{O}>),$
  $(\mathrm{O} *\!\!\longrightarrow_{d_R} \;<\mathrm{L_O},\, \mathrm{O}> *\!\!\longrightarrow_{glo} \;<\mathrm{O},\, \mathrm{L_O},\, \mathrm{O}>)\,).$ □

Definition 4.7: Let $\mathrm{G} = (\mathrm{T},\, \mathrm{N},\, \mathrm{P},\, \sigma)$

(a) Suppose that $\mathrm{A} \to_p \mathrm{S}(p)$ is a production in P with $\mathrm{S}(p) = \mathrm{s}_1\, \mathrm{A}_1\, \mathrm{s}_2 \ldots \mathrm{s}_k \mathrm{A}_k\, \mathrm{s}_{k+1}$. An (algebraic) p-composition

$$\hat{p}\,(\mathrm{A}_1 *\!\!\longrightarrow_{d_1} \mathrm{S}(d_1),\; \ldots,\; \mathrm{A}_k *\!\!\longrightarrow_{d_k} \mathrm{S}(d_k))$$

is said to be *simple*, whenever, for each m, $1 \leq m \leq k$, the derivation $\mathrm{A}_m *\!\!\longrightarrow_{d_m} \mathrm{S}(d_m)$ is either of length 0 or length 1.

(b) A minimal intersection partition

$$mip(\mathrm{A} \to_{p_1} \mathrm{S}(p_1),\ \mathrm{A} \to_{p_2} \mathrm{S}(p_2))$$

for productions $\mathrm{A} \to_{p_1} \mathrm{S}(p_1)$ and $\mathrm{A} \to_{p_2} \mathrm{S}(p_2)$ in G is said to be *simple*, if each intersection component derivation pair is a pair of simple p-compositions. □

Definition 4.8:Let $\mathrm{G} = (\mathrm{T, N, P}, \sigma)$ and $\mathrm{A} \to_p \mathrm{S}(p)$ be a production in P with $\mathrm{S}(p) = \mathrm{s}_1\, \mathrm{A}_1\, \mathrm{s}_2 \ldots \mathrm{s}_k \mathrm{A}_k\, \mathrm{s}_{k+1}$. Then the p-composition *preserves* mip's if and only if

$$\begin{aligned} mip\ \ (&\hat{p}\,(\mathrm{A}_1 *\!\to_{d_{11}} \mathrm{S}(d_{11}),\ \ldots,\ \mathrm{A}_k *\!\to_{d_{1k}} \mathrm{S}(d_{1k})),\\ &\hat{p}\,(\mathrm{A}_1 *\!\to_{d_{21}} \mathrm{S}(d_{21}),\ \ldots,\ \mathrm{A}_k *\!\to_{d_{2k}} \mathrm{S}(d_{2k}))\ )\\ = \hat{p}\,(\ &mip\ (\mathrm{A}_1 *\!\to_{d_{11}} \mathrm{S}(d_{11}),\ \mathrm{A}_1 *\!\to_{d_{21}} \mathrm{S}(d_{21})),\\ &\ldots,\ mip\ (\mathrm{A}_k *\!\to_{d_{1k}} \mathrm{S}(d_{1k}),\ \mathrm{A}_k *\!\to_{d_{2k}} \mathrm{S}(d_{2k}))\ ). \ \square \end{aligned}$$

In [La88], it is shown that a p-composition preserves mip's if and only if, as a *function* in the sentential form algebra, the p-composition is one-to-one. When p-compositions preserve mip's, then the mip of a derivation pair can be structurally constructed from simple production pair mip's [La 88].

Proposition 4.9:Let $\mathrm{G} = (\mathrm{T, N, P}, \sigma)$ and $\mathrm{A} \to_p \mathrm{S}(p)$ be a production in P with $\mathrm{S}(p) = \mathrm{s}_1\, \mathrm{A}_1\, \mathrm{s}_2 \ldots \mathrm{s}_k \mathrm{A}_k\, \mathrm{s}_{k+1}$. Then the p-composition

$$\hat{p} : \prod |\mathrm{A}_m| \to |\mathrm{A}|$$

is a one-to-one function in the sentential form algebra $\mathbf{S}(\mathrm{G})$ if and only if the p-composition preserves mip's. □

Theorem 4.10:Let $\mathrm{G} = (\mathrm{T, N, P}, \sigma)$. Suppose that

(A) the p-composition for each production $p \in \mathrm{P}$ is a one-to-one function in the sentential form algebra $\mathbf{S}(\mathrm{G})$;

(B) for each production pair $(\ \mathrm{A} \to_{p_1} \mathrm{S}(p_1),\ \mathrm{A} \to_{p_2} \mathrm{S}(p_2)\ )$ in G, there exists a simple minimal intersection partition $mip(\ \mathrm{A} \to_{p_1} \mathrm{S}(p_1),\ \mathrm{A} \to_{p_2} \mathrm{S}(p_2)\ )$.

Then, for each derivation pair $(\ \mathrm{A} \to_{d_1} \mathrm{S}(d_1),\ \mathrm{A} \to_{d_2} \mathrm{S}(d_2)\ )$ in G, there exists a minimal intersection partition

$$mip(\ \mathrm{A} \to_{d_1} \mathrm{S}(d_1),\ \mathrm{A} \to_{d_2} \mathrm{S}(d_2)\ ). \ \square$$

Determining whether p-compositions are one-to-one and whether production pair mip's are simple are both undecidable problems. By Proposition 4.9, the first is equivalent to determining whether the intersection $\mathbf{L}_1 \bigcap \mathbf{L}_2$ of any two context-free languages is nonempty which is undecidable [HoUl79]. The second problem is equivalent to determining whether an arbitrary CFG is ambiguous which is also undecidable [HoUl79]. For the FFP grammar $\mathbf{G}_{\mathrm{FP}}$, each p-composition is a one-to-one operation in the sentential form algebra $\mathbf{S}(\mathbf{G}_{\mathrm{FP}})$, and each production pair mip is either empty or consists of two simple intersection components. Each p-composition is one-to-one, because at most one list nonterminal, $\mathrm{L_O}$ or $\mathrm{L_X}$, is introduced by each list production and because the function component e_1 of an applicative expression $(\mathrm{e}_1 : \mathrm{e}_2)$ is "separated" from the operand component e_2 by a top-level colon. The only nonempty production mip's are those for the glue-left and glue-right list productions; that is, the simple mip

$$mip(\ (\mathrm{L_O} *\!\!\rightarrow_{glo} \mathrm{O},\ \mathrm{L_O}),\ (\mathrm{L_O} *\!\!\rightarrow_{gro} \mathrm{L_O},\ \mathrm{O})\ )$$

consists of the following two derivation pairs:

- $(\ (\mathrm{L_O} *\!\!\rightarrow_{glo} \mathrm{O},\ \mathrm{L_O} *\!\!\rightarrow_{em} \mathrm{O}),\ (\mathrm{L_O} *\!\!\rightarrow_{gro} \mathrm{L_O},\ \mathrm{O} *\!\!\rightarrow_{em} \mathrm{O})\ )$
- $(\ (\mathrm{L_O} *\!\!\rightarrow_{glo} \mathrm{O},\ \mathrm{L_O} *\!\!\rightarrow_{gro} \mathrm{O},\ \mathrm{L_O},\ \mathrm{O}),\ (\mathrm{L_O} *\!\!\rightarrow_{gro} \mathrm{L_O},\ \mathrm{O} *\!\!\rightarrow_{glo} \mathrm{O},\ \mathrm{L_O},\ \mathrm{O})\ )$,

while the simple mip

$$mip(\ (\mathrm{L_X} *\!\!\rightarrow_{glx} \mathrm{E},\ \mathrm{L_X}),\ (\mathrm{L_X} *\!\!\rightarrow_{grx} \mathrm{L_X},\ \mathrm{E})\ )$$

consists of the following two derivation pairs:

- $(\ (\mathrm{L_X} *\!\!\rightarrow_{glx} \mathrm{E},\ \mathrm{L_X} *\!\!\rightarrow_{lix} \mathrm{E},\ \mathrm{X} *\!\!\rightarrow_{ex} \mathrm{X},\ \mathrm{X}),\ (\mathrm{L_X} *\!\!\rightarrow_{grx} \mathrm{L_X},\ \mathrm{E} *\!\!\rightarrow_{lix} \mathrm{X},\ \mathrm{E} *\!\!\rightarrow_{ex} \mathrm{X},\ \mathrm{X})\ )$
  $= (\ \widehat{glx}\ (\mathrm{E} *\!\!\rightarrow_{ex} \mathrm{X};\ \mathrm{L_X} *\!\!\rightarrow_{lix} \mathrm{X}),\ \widehat{grx}\ (\mathrm{L_X} *\!\!\rightarrow_{lix} \mathrm{X};\ \mathrm{E} *\!\!\rightarrow_{ex} \mathrm{X})\ )$
- $(\ (\mathrm{L_X} *\!\!\rightarrow_{glx} \mathrm{E},\ \mathrm{L_X} *\!\!\rightarrow_{grx} \mathrm{E},\ \mathrm{L_X},\ \mathrm{E}),\ (\mathrm{L_X} *\!\!\rightarrow_{grx} \mathrm{L_X},\ \mathrm{E} *\!\!\rightarrow_{glx} \mathrm{E},\ \mathrm{L_X},\ \mathrm{E})\ )$.

Corollary 4.11:

(A) A minimal intersection partition

$$mip(\ \mathrm{A} *\!\!\rightarrow_{d_1} \mathrm{S}(d_1),\ \mathrm{A} *\!\!\rightarrow_{d_2} \mathrm{S}(d_2)\ )$$

exists for each derivation pair $(\ \mathrm{A} *\!\!\rightarrow_{d_1} \mathrm{S}(d_1),\ \mathrm{A} *\!\!\rightarrow_{d_2} \mathrm{S}(d_2)\ )$ in the sentential form algebra $\mathbf{S}(\mathbf{G}_{\mathrm{FP}})$.

(B) A most general unifier set

$$mgu(\ \mathrm{t_1},\ \mathrm{t_2}\ )$$

exists for each A-term pair $(\mathrm{t_1},\ \mathrm{t_2})$ in the term algebra $\mathbf{T}(\mathbf{G}_{\mathrm{FP}})$. □

5 An FFP Consistency Checking Procedure.

5.1 Steps of an FFP Consistency Checking Procedure.

In this section, we develop a consistency checking procedure for a class of FFP languages. For each FFP language $\mathcal{L}$, the input to the consistency checker is a good set $\mathbf{RR}(\mathcal{L})$ of rewrite rules. $\mathbf{RR}(\mathcal{L})$ is *consistent* whenever rewrite evaluation is confluent [Hu80]; that is, whenever all innermost evaluation sequences of an expression, either are nonterminating or terminate in the same value. The consistency checker partitions each set of operational rewrite rules into a consistent set and its complement.

Procedure 5.1.1: Starting with a good set $\mathbf{RR}(\mathcal{L})$ of rewrite rules for the FFP language $\mathcal{L}$, the consistency checker will partition $\mathbf{RR}(\mathcal{L})$ into a set $\mathbf{RR}_{\mathrm{CO}}(\mathcal{L})$ of rules known to be consistent and its complement $\mathbf{RR}_{?}(\mathcal{L})$. The consistency checker executes the following procedure.

(1) For each language primitive f, using the unification algorithm, partition $\mathbf{RR}(f)$ into two subsets:

- the *deterministic* set $\mathbf{RR}_\delta(f)$ of rules $\mathrm{RR}_i(f)$ whose left-hand-sides do not overlap with those of any other rules in $\mathbf{RR}(f)$; that is, $\mathrm{RR}_i(f) \in \mathbf{RR}_\delta(f)$, if and only if, for every $\mathrm{RR}_j(f) \in \mathbf{RR}(f)$,

$$mgu(lhs_i(f),\ lhs_j(f)) = \emptyset;$$

- the *nondeterministic* set $\mathbf{RR}_\eta(f)$ is the complement of $\mathbf{RR}_\delta(f)$

$$\mathbf{RR}_\eta(f) = \mathbf{RR}(f) \setminus \mathbf{RR}_\delta(f).$$

The set $\mathbf{RR}_\delta(f)$ represents a set of deterministic rules, because each innermost f-application $(f : \mathrm{e})$ that matches a rule in $\mathbf{RR}_\delta(f)$ can match no other rule in $\mathbf{RR}(\mathcal{L})$. For each primitive f, the set $\mathbf{RR}_\delta(f)$ is consistent. The sets $\mathbf{RR}_\delta(\mathcal{L})$ and $\mathbf{RR}_\eta(\mathcal{L})$ are defined as follows:

$$\mathbf{RR}_\delta(\mathcal{L}) = \bigsqcup \mathbf{RR}_\delta(f) \quad \mathbf{RR}_\eta(\mathcal{L}) = \bigsqcup \mathbf{RR}_\eta(f).$$

(2) The consistent collection $\mathbf{RR}_{\mathrm{CO}}(\mathcal{L})$ is initialized to be the set $\mathbf{RR}_\delta(\mathcal{L})$ of deterministic rewrite rules of $\mathcal{L}$.

(3) For each subset S of $\mathbf{RR}_\eta(f)$, find the mgu's of the f-lhs's of the rules in S. A set S of rules is consistent if it is consistent for each of the most general unified f-lhs terms. For simplicity, we describe the procedure for S where S is a pair $(\mathrm{RR}_1(f),\ \mathrm{RR}_2(f))$ in $\mathbf{RR}_\eta(f)$. For each most general unified term

$$\mu_1\ lhs_1(f) = (f : \mathrm{z}_{12}) = \mu_2\ lhs_2(f)$$

for $(\mu_1,\ \mu_2) \in mgu(lhs_1(f),\ lhs_2(f))$, create new terms t_1 and t_2, by rewriting $(f : \mathrm{z}_{12})$ using $\mathrm{RR}_1(f)$ and $\mathrm{RR}_2(f)$, giving

$$\mathrm{t}_1 = \mu_1 rhs_1(f) \ \text{ and } \ \mathrm{t}_2 = \mu_2 rhs_2(f).$$

(4) Any of the following steps may now be taken any number of times to transform the terms t_1 and t_2 produced in Step (3). If, at any point, the process produces unequal *object* terms, $\mathrm{RR}_1(f)$ and $\mathrm{RR}_2(f)$ are included in the collection $\mathbf{RR}_?(\mathcal{L})$ of rules not known to be consistent. If, at any point, the process produces terms that are equal, the rules are consistent. Three kinds of steps that can be taken are:

(4a) Rewriting of terms using $\mathrm{RR}_1(f)$ and $\mathrm{RR}_2(f)$, or using the rules in the current collection $\mathbf{RR}_{\mathrm{CO}}(\mathcal{L})$;

(4b) Special substitution to replace, with an unused object variable x_i, all occurrences of any application that occurs in both terms;

(4c) Narrowing variables in pairs $((g_1 : \mathrm{w}_1),\ (g_2 : \mathrm{w}_2))$ of innermost applications occurring in both terms, using a process similar to the one presented in [GoMe86]. If $\mathbf{RR}(g_1)$ and $\mathbf{RR}(g_2)$ are in the current collection $\mathbf{RR}_{\mathrm{CO}}(\mathcal{L})$, and the variable y_k is an object or object list variable that occurs in w_1 and w_2, then we use the unification algorithm to *narrow* y_k to match all possible lhs pairs $(lhs_{1i}(g_1),\ lhs_{2j}(g_2))$. This narrowing process determines all rewrite rule pairs

$(\mathrm{RR}_i(g_1), \mathrm{RR}_j(g_2))$ so that there exists a substitution pair $(\theta_{1i}, \theta_{2j})$ that satisfies both of the following conditions:

- $\theta_{1i}\, y_k = \theta_{2j}\, y_k$;
- $\theta_{1i}\,(g_1 : \mathrm{w}_1) = lhs_{1i}(g_1)$ and $\theta_{2j}\,(g_2 : \mathrm{w}_2) = lhs_{2j}(g_2)$.

Narrowing produces one or more pairs; the terms of each pair must be shown to be equal under the operations of Step 4. □

The procedure begins by specifying a set of consistent rewrite rules, no two of which can be applied to the same subexpression. It then considers each subset of the rules that might not be consistent, because they can rewrite a subexpression in more than one way. The rules are applied to each of their most general unified terms, producing a pair of terms that must be shown to be equal. The elements of the pair of terms are then transformed using only:

(1) rewrite rules that are from the consistent set of rules or from the subset under consideration, and therefore are known not to change the meaning of the expression (with respect to the subset under consideration), or

(2) special substitution, which replaces a specific application that occurs in both terms with an object variable denoting the value of the application, or

(3) narrowing of variables, which replaces a variable by an object term or an object list term in both terms so that rewriting can be continued, using rules only from the consistent set.

Therefore, when this process results in equal terms for a most general unified term, the rewrite rules are consistent with respect to the expressions of the form of this most general unified term. These observations form the basis for a proof of the following theorem:

Theorem 5.1.2: For an FFP language $\mathcal{L}$, if $\mathbf{RR}(\mathcal{L})$ is a good set of rewrite rules, then the resulting collection $\mathbf{RR}_{\mathrm{CO}}(\mathcal{L})$ is consistent. □

5.2 Examples of FFP Consistency Checking.

Consider the good set $\mathbf{RR}(\mathcal{L}_\beta)$ of rewrite rules for the basis FFP language $\mathcal{L}_\beta$ presented in Example 3.2. Besides the rewrite rule set $\mathbf{RR}(meta)$ for meta-composition, $\mathbf{RR}(\mathcal{L}_\beta)$ includes the rewrite rule collections:

$$\mathbf{RR}(apnd_L),\ \mathbf{RR}(apnd_R),\ \mathbf{RR}(comp),\ \mathbf{RR}(app),\ \mathbf{RR}(con).$$

(a) The initial consistent collection $(\mathbf{RR}_{\mathrm{CO}}(\mathcal{L}_\beta))^0$ of rewrite rules contains almost all the rules in $\mathbf{RR}(\mathcal{L}_\beta)$. In particular, $\mathbf{RR}_\delta(f) = \mathbf{RR}(f)$ for

$$f \in \{\ meta,\ apnd_L,\ apnd_R,\ comp,\ app\ \}.$$

This is because the mgu sets of the left-hand-sides of distinct rules $\mathrm{RR}_i(f)$ and $\mathrm{RR}_j(f)$ are empty. Furthermore, for $0 \neq j$, the *con*-lhs most general unifier sets $mgu(lhs_0(con), lhs_j(con))$ are empty, where $lhs_0(con) = (con : << con >, x_1 >)$. Hence, $\mathbf{RR}_\delta(con) = \{\mathrm{RR}_0(con)\}$, and

$$\begin{aligned}(\mathbf{RR}_{\mathrm{CO}}(\mathcal{L}_\beta))^0 \ &= \ \mathbf{RR}(meta) \sqcup \mathbf{RR}(apnd_L) \sqcup \mathbf{RR}(apnd_R) \\ &\quad \sqcup \mathbf{RR}(comp) \sqcup \mathbf{RR}(app) \sqcup \mathbf{RR}_\delta(con).\ \square\end{aligned}$$

(b) The only nondeterministic reduction rules in $\mathbf{RR}(\mathcal{L}_\beta)$ are

$$\mathbf{RR}_\eta(con) = \{\mathrm{RR}_R(con),\ \mathrm{RR}_L(con)\},$$

because *con*-lhs most general unifier set $mgu(lhs_L(con),\ lhs_R(con)) \neq \emptyset$, where

- $lhs_L(con) = (con : << con,\ x_2,\ l_1 >,\ x_1 >)$
- $lhs_R(con) = (con : << con,\ l_2,\ x_3 >,\ x_1 >)$.

The most general unified *con*-lhs terms are

- $\mathrm{mgu}_1 = (con : << con,\ x_2 >,\ x_1 >)$
- $\mathrm{mgu}_{\geq 2} = (con : << con,\ x_2,\ l_3,\ x_3 >,\ x_1 >)$.

(c) Using the consistency checking procedure, we generate the following specially substituted sequences for the term $\mathrm{mgu}_1 = (con : << con,\ x_2 >,\ x_1 >)$. Note that these sequences must be generated together, step by step. The same special substitution occurs in Step 4 of both sequences.

$$\begin{aligned}
\mathrm{ev}_L \;=\; & [\,(con : << con,\ x_2 >,\ x_1 >) \rightarrow_{\mathrm{RR}_L(con)} \\
& (apnd_L : < (x_2 : x_1),\ (< con > : x_1) >) \rightarrow_{\mathrm{RR}_1(meta)} \\
& (apnd_L : < (x_2 : x_1),\ (con : << con >,\ x_1 >) >) \rightarrow_{\mathrm{RR}_0(con)} \\
& (apnd_L : < (x_2 : x_1),\ < >>) \Rightarrow_{\mathbf{SS}(con)} \\
& (apnd_L : < x_3,\ < >>) \rightarrow_{\mathrm{RR}(apnd_L)} \ < x_3 >\,]
\end{aligned}$$

$$\begin{aligned}
\mathrm{ev}_R \;=\; & [(con : << con,\ x_2 >,\ x_1 >) \rightarrow_{\mathrm{RR}_R(con)} \\
& (apnd_R : < (< con >: x_1),\ (x_2 : x_1) >) \rightarrow_{\mathrm{RR}_1(meta)} \\
& (apnd_R : < (con : << con >,\ x_1 >),\ (x_2 : x_1) >) \rightarrow_{\mathrm{RR}_0(con)} \\
& (apnd_R : << >,\ (x_2 : x_1) >) \Rightarrow_{\mathbf{SS}(con)} \\
& (apnd_R : << >,\ x_3 >) \rightarrow_{\mathrm{RR}(apnd_R)} \ < x_3 >\,].
\end{aligned}$$

Since the evaluation sequences ev_L and ev_R both terminate in $< x_3 >$, the rewrite rules $\mathrm{RR}_L(con)$ and $\mathrm{RR}_R(con)$ are consistent with respect to mgu_1.

(d) Using the consistency checking procedure, we generate the following specially substituted and narrowed sequences for the term $\mathrm{mgu}_{\geq 2} = (con : << con,\ x_2,\ l_3,\ x_3 >,\ x_1 >)$.

$$\begin{aligned}
\mathrm{ev}_L \;=\; & [\,(con : << con,\ x_2,\ l_3,\ x_3 >,\ x_1 >) \rightarrow_{\mathrm{RR}_L(con)} \\
& (apnd_L : < (x_2 : x_1), (< con, l_3, x_3 >: x_1) >) \rightarrow_{\mathrm{RR}_L(meta)} \\
& (apnd_L : < (x_2 : x_1), (con : << con, l_3, x_3 >, x_1 >) >) \rightarrow_{\mathrm{RR}_R(con)} \\
& (apnd_L : < (x_2{:}x_1), (apnd_R : < (< con, l_3 > : x_1),\ (x_3{:}x_1) >) >) \rightarrow_{\mathrm{RR}_L(meta)} \\
& (apnd_L : < (x_2{:}x_1), (apnd_R : < (con{:} << con, l_3 >, x_1 >),\ (x_3{:}x_1) >) >) \\
& \Rightarrow_{\mathbf{SS}(con)} (apnd_L : < x_4,\ (apnd_R : < x_5,\ x_6 >) >) \\
& \Rightarrow_{\mathbf{NR}(\theta_R)} (apnd_L : < x_4, (apnd_R : << l_4 >, x_6 >) >) \\
& \quad \rightarrow_{\mathrm{RR}(apnd_R)} (apnd_L : < x_4, < l_4, x_6 >>) \\
& \quad \rightarrow_{\mathrm{RR}(apnd_L)} \ < x_4, l_4, x_6 >\,]
\end{aligned}$$

$$
\begin{aligned}
\mathrm{ev}_R \;=\; & [\,(con :<< con,\, x_2,\, l_3,\, x_3 >,\, x_1 >) \rightarrow_{\mathrm{RR}_R(con)} \\
& (apnd_R :< (< con, x_2, l_3 >: x_1), (x_3 : x_1) >) \rightarrow_{\mathrm{RR}_L(meta)} \\
& (apnd_R :< (con :<< con, x_2, l_3 >, x_1 >), (x_3 : x_1) >) \rightarrow_{\mathrm{RR}_L(con)} \\
& (apnd_R :< (apnd_L :< (x_2 : x_1), (< con, l_3 > : x_1) >), (x_3 : x_1) >) \rightarrow_{\mathrm{RR}_L(meta)} \\
& (apnd_R :< (apnd_L :< (x_2 : x_1), (con :<< con, l_3 >, x_1 >) >), (x_3 : x_1) >) \\
& \Rightarrow_{\mathbf{SS}(con)} (apnd_R :< (apnd_L :< x_4,\, x_5 >),\, x_6 >) \\
& \Rightarrow_{\mathbf{NR}(\theta_L)} (apnd_R :< (apnd_L :< x_4, < l_4 >>), x_6 >) \\
& \quad \rightarrow_{\mathrm{RR}(apnd_L)} (apnd_R :<< x_4, l_4 >, x_6 >) \\
& \quad \rightarrow_{\mathrm{RR}(apnd_R)} < x_4, l_4, x_6 >].
\end{aligned}
$$

Since the evaluation sequences ev_L and ev_R both terminate in $< x_4, l_4, x_6 >$, the rewrite rules $\mathrm{RR}_L(con)$ and $\mathrm{RR}_R(con)$ are consistent with respect to $\mathrm{mgu}_{\geq 2}$. □

6 References.

BaDe87 BACHMAIR, L. and DERSHOWITZ, N., 'Completion for rewriting modulo a congruence,' *Proceedings of Rewriting Techniques and Applications*, Bordeaux, France, May 1987, LNICS 256, Springer-Verlag, New York, 1987, 192-203.

Ba78 BACKUS, J., 'Can programming be liberated from the von Neumann style? A functional style and its algebra of programs,' *Communications of the ACM*, 21(1978), 613-641.

BaW²86 BACKUS, J., WILLIAMS, J. H., and WIMMERS, E. L., 'FL Language Manual,' Research Report RJ 5339, IBM Research Laboratory, San Jose, CA, November, 1986.

Be75 BENSON, D. B., 'The basic algebraic structures in categories of derivations,' *Inf. and Contr.*, 28(1975), 1-29.

D²Qu78 DENNING, P. J., DENNIS, J. B., QUALITZ, J. E., *Machines, Languages and Computation*, Prentice-Hall, Englewood Cliffs, NJ, 1978.

De82 DERSHOWITZ, N., 'Orderings for term rewriting systems,' *J. of Theoretical Comp. Sci.*, 17(1982), 279-310.

GoMe86 GOGUEN, J.A., and MESEGUER, J., '*EQLOG*: Equality, types and generic modules for logic programming,' *Logic Programming: Theory and Implementation*, Prentice Hall, Englewood Cliffs, NJ, 1986, 295-363.

HaW³85 HALPERN, J. Y., WILLIAMS, J. H., WIMMERS, E.L. and WINKLER, T. C., 'Denotational semantics and rewrite rules for FP,' *Proceedings of the Twelfth ACM Symposium of Principles of Programming Languages*, January 1985, 108-120.

HoUl79 HOPCROFT, J. E., and ULLMAN, J. D., *Introduction to Automata Theory, Languages, and Computation*, Addison-Wesley, Reading, MA, 1979.

Hu80 HUET, G., 'Confluent resuctions: abstract properties and applications to term rewriting systems,' *J. of Assoc. Comp. Mach.*, 27(1980), 797-821.

KnBe70 KNUTH, D. and BENDIX, P., 'Simple word problems in universal algebras,' *Computational Problems in Abstract Algebra*, Leech, J., ed., Pergamon Press, 1970, 263-297.

La88 LATCH, D. M., 'Finite generation of ambiguity in context-free languages,' to appear in *J. of Pure and Applied Algebra.*

LaSi88 LATCH, D. M. and SIGAL, R., 'Generating evaluation theorems for functional programming languages,' to appear in *Proceedings of the International Symposium on Methodologies for Intelligent Systems*, Torino, Italy, October, 1988.

Mac71 MAC LANE, S., *Categories for the Working Mathematician*, Springer-Verlag, New York, 1971.

McJ75 McJONES, P. A., 'A Church-Rosser property of closed applicative languages,' Research Report RJ 1589, IBM Research Laboratory, San Jose, CA, May, 1975.

Ne80 NELSON, E., 'Categorical and topological aspects of formal languages,' *Mathematical Systems Theory*, 13(1980), 255-273.

Ru87 RUS, T., 'An algebraic model for programming languages,' *Computer Languages*, 12(1987), 173-195.

Si84 SEIKMAN, J. H., 'Universal unification,' *Proceedings of the* 7th *C.A.D.E.*, LNICS 170, Springer-Verlag, New York, 1984, 1-42. *Computer Languages*, 12(1987), 173-195.

Wi81 WILLIAMS, J. H., 'Formal representations for recursively defined functional programs,' *Formalization of Programming Concepts*, LNICS 107, Springer-Verlag, New York, 1981, 460-470.

Department of Mathematics
North Carolina State University
Raleigh, NC 27695-8205

Contemporary Mathematics
Volume **92**, 1989

Qualitative Distinctions Between Some Toposes of Generalized Graphs

F. WILLIAM LAWVERE

I have divided this paper into three sections as follows: Section I sketches, in a somewhat impressionistic way, the general conceptual context which I intend to exemplify in section III; section II reviews some of the basic constructions on presheaf categories which are implicit in nearly all branches of mathematics, with some examples which it is hoped will appeal to computer scientists; section III describes three sequences of "simplest possible" examples of some of the basic phenomena, pointing out some of the peculiarities which arise even within these sequences.

I. This section should become clearer on a second reading, after having studied the rest of the paper.

"Being is doing", and hence particular being is known (at least partly) by what it can do. If B is an object in a "gros" topos $\mathcal{B}$ of cohesive active sets, what it can do is to continuously parameterize and dynamically act on mathematical structures. The zeroth form of mathematical structure is the abstract sets which form a Boolean topos and which can be realized in two opposite ways (chaotic and discrete) $\mathcal{S} \leftrightarrows \mathcal{B}$ as very special extreme cases of "cohesive active" sets. The category $\mathcal{S}(B)$ of all possible ways that B can parameterize and act on abstract sets is again a topos, but of a qualitatively different sort called "petit" by Grothendieck (typically both the gros $\mathcal{B}$ and all the petit $\mathcal{S}(B)$ are proper classes from the mid-century "set-theoretic" point of view; it is not cardinality but another quality which distinguishes the two). There is a geometric morphism of toposes $\mathcal{B}/B \longrightarrow \mathcal{S}(B)$ from the non-petit "comma category", which already exhibits the qualitative difference even in case $B = 1$, the terminal object of $\mathcal{B}$: for $\mathcal{B}/1 = \mathcal{B}$ but $\mathcal{S}(1) = \mathcal{S}$, and the geometric morphism $\mathcal{B} \longrightarrow \mathcal{S}$ is the functor whose right adjoint is the inclusion

Received by the editors December 1987.
1980 *Mathematics Subject Classification*. Primary 54C40; Secondary 46E25.
This work was done with partial support of the NSF (US) and CNR (Italy).

as chaotics and whose left adjoint is the inclusion as discretes; moreover, in this case the discrete inclusion has a further left adjoint $\pi_0 : \mathcal{B} \longrightarrow \mathcal{S}$ which assigns to each cohesive/active set B the corresponding abstract set of its components/orbits with the adjunction morphism $B \longrightarrow \pi_0(B)$ being the universal map to a discrete space; moreover, the truth value object $\Omega_{\mathcal{B}}$ of $\mathcal{B}$ is *connected* $\pi_0(\Omega) = 1$, hence in particular $\Omega_{\mathcal{B}} \neq 1 + 1$ nor any other sum so that $\mathcal{B}$ cannot be Boolean, whereas $\Omega_{\mathcal{S}} = 1 + 1$ (the earmark of Booleanness) and π_0 (being a left adjoint) preserves sums so that in a positive sense $\mathcal{B} \to \mathcal{S}$ is far from an equivalence. In most determinations of $\mathcal{B}$ which I have considered, $\mathcal{S}$ is in fact determined by $\mathcal{B}$ in that the codiscrete inclusion $\mathcal{S} \hookrightarrow \mathcal{B}$ is characterized as the smallest subtopos of $\mathcal{B}$ which contains the empty object 0 of $\mathcal{B}$. Moreover, it seems essential that π_0 has a tendency to preserve finite products for gros $\mathcal{B}$. Thus what I have said puts many stringent restrictions on the topos $\mathcal{B}$.

I have used above the term "space" as short for cohesive/active set; already in Grassmann it was clear that space is generated by, and lays the foundation for, motion and hence that general spaces have aspects of both. These two aspects can be illustrated in somewhat pure form as follows: If B is essentially determined by a poset of regions in some space, ordered by inclusion (pure cohesiveness in one of its simpler manifestations) it can still act in the following sort of way (presheaf): Let UX be the (abstract) set of all X-valued smooth functions on U for any region U of B: then an inclusion $U' \subset U$ in B acts by restriction to give $UX \longrightarrow U'X$, the totality of sets of functions and restriction-actions determining a single object $\underline{X}$ of $\mathcal{S}(B)$. At the opposite extreme (pure activeness) we might consider a B which is determined by a *suitable monoid* (for example the group of all rigid motions of physical space under composition, or the set of non-negative time durations under addition) and take for $\mathcal{S}(B)$ the category of all *right* actions of B on abstract sets. But already here the "pure" activeness passes over into cohesiveness: if the monoid B satisfies the cancellation property which says that *left* multiplication by any element is a monomorphism, and if $T_B \in \mathcal{S}(B)$ denotes the regular representation [the Yoneda-Grothendieck-Dedekind-Cayley embedding gives a *single* object in the case of a monoid] then the topos $\mathcal{S}(B)/T_B$ (whose objects are objects of $\mathcal{S}(B)$ equipped with a morphism to T_B and whose morphisms are commutative triangles over T_B in $\mathcal{S}(B)$) can also be expressed as $\mathcal{S}(B)/T_B \cong \mathcal{S}(B/*)$ where $*$ is the unique object of B, but $B/*$ is in such a cancellative case a *poset*. Thus in such a case the category $\mathcal{S}(B)/T_B$ of "non-autonomous dynamics" derived from pure action $\mathcal{S}(B)$ turns out to be pure cohesiveness. [Caution: If B is a group then $\mathcal{S}(B)/T_B \approx \mathcal{S}$ since then the poset $B/*$ is codiscrete.] Of course already the "orbit set" functor $\mathcal{S}(B) \longrightarrow \mathcal{S}$ reveals some of the cohesiveness induced by dynamical action: if two elements of a state space $X \in \mathcal{S}(B)$ can be moved into a common element by some acts of B, they thereby "stick together" in one sense. If a monoid B_1 acts on a pure space B_0 then there will be an induced action of B_1

on the poset of subregions of B_0 and the resulting semidirect product $B_0 \ltimes B_1$ is a genuinely mixed cohesive/active set, leading to $\mathcal{S}(B_0 \ltimes B_1)$ etc.

Many of the kinds of mathematical structures which need to be parameterized and acted on by B are commonly understood in their constant form as conditioned diagrams in a category $\mathcal{S}$ of abstract sets, for example a group $1 \longrightarrow S \longleftarrow S \times S$, or a poset $E \rightrightarrows S$ (where the induced single $E \hookrightarrow S \times S$ is monomorphic and contains $S \longrightarrow E$ as the diagonal and *is* the "order relation"), etc. Hence the *variable* groups, posets etc. may to a large extent be treated as exactly similar diagrams *in* the topos $\mathcal{S}(B)$ of *variable* abstract sets; that is

$$\begin{aligned} &B\text{-variable } A\text{-structured sets} \\ &= A\text{-structured } B\text{-variable sets} \end{aligned}$$

often holds, where A is any "theory" of quite a general kind. Here B-variable sets $= \mathcal{S}(B) =$ the topos of abstract sets parameterized and acted on by B.

One of my aims will be to bring out properties of the "petit" toposes $\mathcal{S}(B)$ which will distinguish them in a positive way from the "gros" toposes such as $\mathcal{B}$. A typical $\mathcal{S}(B)$, unlike the special case $\mathcal{S}(1) = \mathcal{S}$, will *not* be Boolean in its equational logic

$$\Omega \times \Omega \overset{\wedge}{\underset{\Rightarrow}{\overset{\vee}{\rightrightarrows}}} \Omega,$$

but Heyting like $\mathcal{B}$; on the other hand its *object* Ω of truth values will almost never be connected, unlike $\mathcal{B}$.

Why is it important that a gros topos $\mathcal{E}$ have a *finite product preserving* components functor $\pi_0 \dashv$ discrete $\dashv \mathcal{E}(1, -)$? For one thing it means that the Burnside ring of $\mathcal{E}$ is the monoid ring (of measures under convolution) of the monoid of finite connected objects under cartesian product, avoiding the series of structure constants involved in a petit topos such as G-sets for a group(oid) G. More importantly perhaps, it means that for any category $\mathcal{A}$ enriched in $\mathcal{E}$, (that is a category whose homs are "spaces" $\mathcal{A}(A', A) \in \mathcal{E}$ rather than abstract sets) such as $\mathcal{A} = \mathcal{E}$ itself, we can begin the qualitative "homotopical" classification of the objects of $\mathcal{A}$ without fractions by defining an associated abstract category $\pi\mathcal{A}$ via $(\pi\mathcal{A})(A', A) = \pi_0\mathcal{A}(A', A)$, being assured by the product-preserving property of π_0 that these "homotopy-classes" of $\mathcal{A}$-maps can still be composed. Roughly, if $\mathcal{A}$ is the study of specific processes $a_0 \xrightarrow{\alpha} a_1$ in objects A, then $\pi\mathcal{A}$ is the study of the higher ramifications of the ensemble of conditions "there exists a process $a_0 \dashrightarrow a_1$ of the kind allowed by A"; here by "process" we mean one which proceeds by one simple law at least to the extent that if for example

$$a_k = \alpha\partial_k, \qquad p\partial_k = 1_{A'}$$

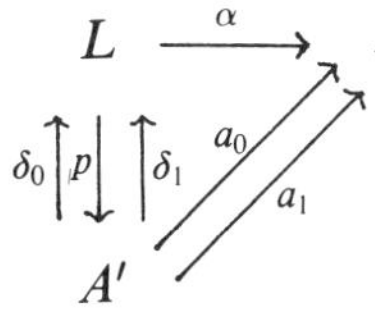

$$\begin{array}{ccc} L & \xrightarrow{\alpha} & A \\ \delta_0 \uparrow \downarrow p \uparrow \delta_1 & \nearrow a_0, \nearrow a_1 & \\ A' & & \end{array}$$

with A' a terminal object in $\mathcal{A}$, we want L also to be "contractible" in $\mathcal{A}$ in the sense that L becomes terminal in $\pi\mathcal{A}$. Such considerations are so compelling that one is driven to believe that if some approximation $\mathcal{E}$ to the notion of "continuous map" fails to have product-preserving π_0, then $\mathcal{E}$ should be changed (as it is often done to achieve cartesian-closure).

II. The next few paragraphs may be simply skimmed (to assimilate my notation) by any reader already familiar with the basics of category theory, of which they are mainly a review.

The examples which we wish to consider are all presheaf toposes, so that Grothendieck topologies do not play a major role although coverings may. Thus for any small abstract category $\mathbf{C}$ (i.e. one whose total set of morphisms constitutes a single object in the category $\mathcal{S}$ of abstract sets, and whose domain, codomain, identity specification, and composition operations are maps in $\mathcal{S}$) we will be considering the category $\mathcal{S}^{\mathbf{C}^{op}}$ of all contravariant functors from $\mathbf{C}$ to $\mathcal{S}$ and all natural transformations between these; an object X of $\mathcal{S}^{\mathbf{C}^{op}}$ may equally well be considered as a *right action* of $\mathbf{C}$ on a set which is partitioned into sorts parameterized by the objects of $\mathbf{C}$ and such that whenever $C' \xrightarrow{\lambda} C$ is a morphism in $\mathbf{C}$ and x is an element of X of sort C, then $x\lambda$ is specified as an element of X of sort C', all this being subject to the conditions

$$\boxed{\begin{array}{l} x1_C = x, \\ x(\lambda\mu) = (x\lambda)\mu \quad \text{whenever} \quad C'' \xrightarrow{\mu} C' \xrightarrow{\lambda} C \quad \text{in} \quad \mathbf{C}. \end{array}}$$

Such an action X is also often referred to as a $\mathbf{C}$-set when there is little danger of confusion with other possible uses of that term. Similarly the $\mathbf{C}$-naturality of any morphism $X \xrightarrow{f} Y$ in the category $\mathcal{S}^{\mathbf{C}^{op}}$ is really just the homogeneity condition

$$f(x\lambda) = f(x)\lambda$$

wherein, of course, the first action of $\lambda \in \mathbf{C}$ is the one given by X and the second by Y. The Yoneda-Grothendieck-Dedekind-Cayley embedding

$$\mathbf{C} \xrightarrow[T_{\mathbf{C}}]{} \mathcal{S}^{\mathbf{C}^{op}}$$

is the functor which associates to each object A of C the $\mathbf{C}$-set $T_{\mathbf{C}}(A) = \mathbf{C}(-, A)$ whose C-th sort is the set $\mathbf{C}(C, A)$ of $\mathbf{C}$ morphisms $C \longrightarrow A$, with action by composition: $C' \xrightarrow{\lambda} C \xrightarrow{x} A$; this is a functor because for any $A \xrightarrow{\alpha} A'$ we get an induced $\mathbf{C}$-homogeneous map $\mathbf{C}(-A) \longrightarrow \mathbf{C}(-, A')$ which by the associativity of composition in $\mathbf{C}$ behaves functorially under composition $A \longrightarrow A' \longrightarrow A''$. The famous *lemma* of the four illustrious mathematicians says that $T_{\mathbf{C}}$ is full embedding $\mathbf{C} \hookrightarrow \mathcal{S}^{\mathbf{C}^{op}}$ (so that it is justified and often convenient to confuse A with $T_{\mathbf{C}}(A) = \mathbf{C}(-, A)$) and says much more, namely that for *any* $\mathbf{C}$-set X and for any object A of $\mathbf{C}$, the set of elements of X of sort A is naturally identifiable with the set of $\mathcal{S}^{\mathbf{C}^{op}}$-morphisms from

$\mathbf{C}(-, A)$ to X. It is thus justified, as well as extremely useful, to adjoin to the above parenthetically-introduced abuse still a further abuse of notation and to henceforth regard the elements of X of sort A as morphisms $A \longrightarrow X$ in $\mathcal{S}^{\mathbf{C}^{op}}$; thus the action of $\mathbf{C}$ on any $\mathbf{C}$-set becomes a special case of composition of morphisms

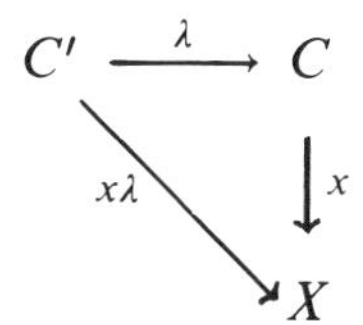

now all in $\mathcal{S}^{\mathbf{C}^{op}}$ as does the application of a morphism f to an element, and the homogeneity (or naturality) property of every $X \longrightarrow Y$ in $\mathcal{S}^{\mathbf{C}^{op}}$ becomes a special case of the associativity of composition in $\mathcal{S}^{\mathbf{C}^{op}}$:

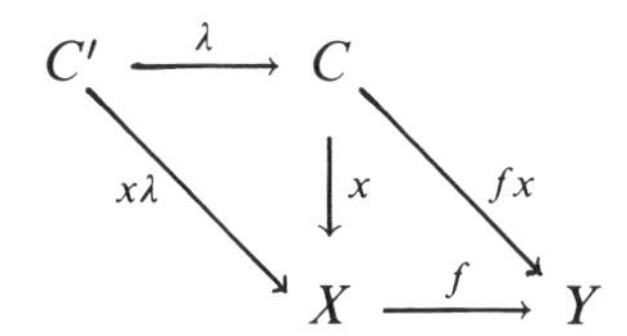

In case the objects and morphisms of $\mathbf{C}$ have some kind of geometrical interpretation, it is often helpful to imagine that the more general objects of $\mathcal{S}^{\mathbf{C}^{op}}$ push that interpretation to a natural limit: an object A of $\mathbf{C}$ may be considered in $\mathcal{S}^{\mathbf{C}^{op}}$ as a generic "figure" and any $A \xrightarrow{x} X$ as a particular figure in X (quite possibly singular, i.e. not necessarily monomorphic) of sort A. Then if $x' = x\lambda$ in X one may consider that λ establishes change of figures in X and that an equation $x_1\lambda_1 = x_2\lambda_2$ is an incidence relation; the naturality or homogeneity of morphisms $X \longrightarrow Y$ is therefore essentially the preservation of incidence relations. One must distinguish between elements (figures of any generic sort) and points. By the latter we understand morphisms $1 \longrightarrow X$ where 1 is the terminal object of $\mathcal{S}^{\mathbf{C}^{op}}$, defined as the $\mathbf{C}$-set which for any A has exactly one element (figure) of sort $A : A \xrightarrow{A} 1$. Then it easily follows that for any X there is exactly one morphism $X \longrightarrow 1$, also denoted by X. [This useful abuse (due to Johnstone) is justified by the fact that for any object X of any category $\mathcal{X}$, the comma category $\mathcal{X}/X$ has a terminal object 1_X which is none other than the identity map of X in $\mathcal{X}$, and that for any object E of $\mathcal{X}/X$ the unique $E \longrightarrow 1_X$ is represented in $\mathcal{X}$ as ϵ

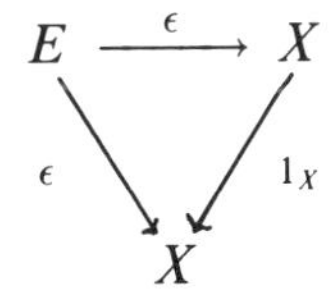

where ϵ is the structural datum in $\mathcal{X}$ that any object in $\mathcal{X}/X$ must have.]

Now a morphism $1 \xrightarrow{x} X$ must operate at each sort, picking an element x_A of X of each sort A; but since the action of $\mathbf{C}$ on 1 by any λ is trivial, and

x is homogeneous, we therefore have

$$x_{C'} = x_C \lambda \quad \text{for all} \quad C' \xrightarrow{\lambda} C \quad \text{in} \quad \mathbf{C}.$$

For example, if $1 \in \mathbf{C}$ [i.e. if $\mathbf{C}$ itself has a terminal object, in which case it is easy to see that $\mathbf{C}(-,1) = 1$ the terminal object of $\mathcal{S}^{\mathbf{C}^{op}}$, a further justification for the Yoneda abuse] then any point x of X is determined by a single figure x_1 of sort 1, by $x_C = x_1 C$ for all $C \xrightarrow{C} 1$ in $\mathbf{C}$. Since in combinatorial geometry there are often many higher-dimensional figures with few specified vertices, and since in dynamical systems there are often no rest states at all, it is not surprising that morphisms $X \longrightarrow Y$ are often not determined by their values on points alone: we may have

$$1 \overset{x}{\dashrightarrow} X \underset{g}{\overset{f}{\rightrightarrows}} Y \quad fx = gx \quad \text{for all such } x$$

$$f \neq g.$$

Of course if $f \neq g$ then there is some A in $\mathbf{C}$ and some figure x of sort A in X with $fx \neq gx$ in Y. It is sometimes interesting to consider the full subcategory $\mathcal{Y}_{\mathbf{C}}$ of $\mathcal{S}^{\mathbf{C}^{op}}$ consisting of all those Y such that for all A in $\mathbf{C}$ and any two figures $A \underset{y_2}{\overset{y_1}{\rightrightarrows}} Y$ with $y_1 \neq y_2$ there exists a point $1 \xrightarrow{a} A$ for which $y_1 a \neq y_2 a$; in other words, there is some point in the generic figure A such that the a-th *vertex* of y_1 differs from the a-th vertex of y_2. Consideration of $\mathcal{Y}_{\mathbf{C}}$ may seem especially pertinent when $1 \in \mathbf{C}$ and when $\mathbf{C} \subset \mathcal{Y}_{\mathbf{C}}$; however, we will *always* have $\mathcal{Y}_{\mathbf{C}} \neq \mathcal{S}^{\mathbf{C}^{op}}$ unless $\mathbf{C} \cong 1$. $\mathcal{Y}_{\mathbf{C}}$ is cartesian closed (see below) but does not have a truth-value object (see below) and is hence never a topos unless $\mathbf{C} \cong 1$; the advantage of being able to treat also the power objects etc. in $\mathcal{S}^{\mathbf{C}^{op}}$ as generalized $\mathbf{C}$-objects, as well as the use of superior set-like exactness properties toposes enjoy, derives from the fact that many conceptual constructions on objects (even if starting from objects of $\mathcal{Y}_{\mathbf{C}}$) will naturally lead to objects of $\mathcal{S}^{\mathbf{C}^{op}}$. Since $\mathcal{Y}_{\mathbf{C}}$ is closed under cartesian products Π and arbitrary subobjects in $\mathcal{S}^{\mathbf{C}^{op}}$, it follows that for every X in $\mathcal{S}^{\mathbf{C}^{op}}$ there is a natural surjective map $X \longrightarrow X^b$ to an object of $\mathcal{Y}_{\mathbf{C}}$ such that

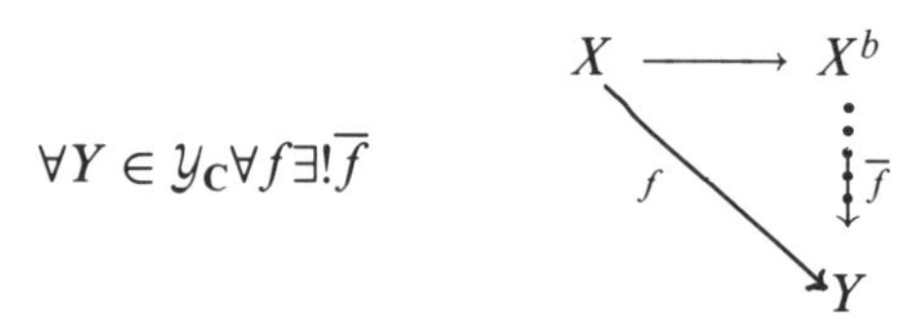

The reader may wish to calculate X^b for some of the kinds of graphs discussed below.

The condition $1 \in \mathbf{C}$ may for some $\mathbf{C}$ not be strictly true yet quasi-true up to "splitting idempotents". There is for any $\mathbf{C}$ a "Cauchy-completion" $\mathbf{C} \subseteq \overline{\mathbf{C}} \subset \mathcal{S}^{\mathbf{C}^{op}}$ consisting of all retracts R of objects of $\mathbf{C}$ in the sense that there exist $R \underset{p}{\overset{i}{\rightleftarrows}} A$ morphisms of $\mathcal{S}^{\mathbf{C}^{op}}$ such that $A \in \mathbf{C}$, $pi = 1_R$. It is always the case that there is an equivalence of categories $\mathcal{S}^{\overline{\mathbf{C}}^{op}} \xrightarrow{\sim} \mathcal{S}^{\mathbf{C}^{op}}$. By $1 \in \overline{\mathbf{C}}$, we thus mean that for some A in $\mathbf{C}$ there is for every C a morphism $C \xrightarrow{e_C} A$ such that $e_C\lambda = e_{C'}$ for *all* λ; thus in particular e_A is an idempotent but of the very special kind sometimes called a "right zero" (unlike two-sided zeroes, such are not necessarily unique, as we will see). In general the idempotents of $\mathbf{C}$ correspond to objects of $\overline{\mathbf{C}}$, and one says that $\mathbf{C}$ is "closed under splitting of idempotents" or "Cauchy-complete as an $\mathcal{S}$-category" if $\mathbf{C} \xrightarrow{\sim} \overline{\mathbf{C}}$ is an equivalence of categories. $\overline{\mathbf{C}}$ can be constructed purely abstractly from $\mathbf{C}$ without reference to actions by defining, for any $A_i \circlearrowleft_{e_i}$, $i = 1, 2$ with $e_i^2 = e_i$,

$$\overline{\mathbf{C}}(e_1, e_2) = \{\lambda \in \mathbf{C}(A_1, A_2) \mid \lambda e_1 = \lambda = e_2\lambda\}$$

noting the *two* equations, and verifying the required properties, including 2-functoriality and $\overline{\overline{\mathbf{C}}} \xrightarrow{\sim} \overline{\mathbf{C}}$ for any $\mathbf{C}$. Note that $1_e \in \overline{\mathbf{C}}(e, e)$ is just e in $\mathbf{C}$. [For example, if $\mathbf{C}$ is the category of all smooth maps between all open subsets of all Euclidean spaces, then $\overline{\mathbf{C}}$ is the category of all smooth manifolds. This powerful theorem justifies bypassing the complicated considerations of charts, coordinate transformations, and atlases commonly offered as a "basic" definition of the concept of manifold. For example the 2-sphere, a manifold but not an open set of any Euclidean space, may be fully specified with its smooth structure by considering any open set A in 3-space E which contains it but not its center (taken to be 0) and the smooth idempotent endomap of A given by $e(x) = x/|x|$. All general constructions (i.e. functors into categories which are Cauchy complete) on manifolds now follow easily (without any need to check whether they are compatible with coverings, etc.) provided they are known on the opens of Euclidean spaces: for example, the tangent bundle of the sphere is obtained by splitting the idempotent e' on the tangent bundle $A \times V$ of A (V being the vector space of translations of E) which is obtained by differentiating e. The same for cohomology groups, etc.]

Even if $\mathbf{C}$ is a monoid, i.e. a category with one object C, $\overline{\mathbf{C}}$ will not be a monoid if $\mathbf{C}$ has idempotent elements other than the identity 1_C. It will often not even be equivalent to $\mathbf{C}$ [the question of equivalence is different since $\mathbf{C}$ could have elements f, g with $fg = 1_C$ but $gf \neq 1_C$ then $e \underset{\text{def}}{=} gf$ determines an object $e \neq 1_C$ in $\overline{\mathbf{C}}$, but $e \xrightarrow{f} 1_C$ is an isomorphism in $\overline{\mathbf{C}}$; in other words, *such* an idempotent e would already be split through C itself, but in a non-trivial way.]

Since we are primarily interested in $\mathcal{S}^{\mathbf{C}^{op}} \simeq \mathcal{S}^{\overline{\mathbf{C}}^{op}}$, we will freely pass back and forth between $\mathbf{C}$ and $\overline{\mathbf{C}}$ in describing objects X: for minimalistic sufficiency, we may prefer $\mathbf{C}$, especially when it is a monoid; but $\overline{\mathbf{C}}$ usually gives a more plastic vision of the kinds of figures which X really has and their relationships.

The most important example of the above, and probably of our whole discussion to follow, is the three element monoid $\mathbf{\Delta}_1$:

$$1_{\mathbf{1}}, e_0, e_1 \qquad e_i e_j = e_i \qquad i, j = 0, 1$$

This is a non-commutative monoid consisting entirely of idempotents (therefore called a "band" by semigroup theorists) but the other two equations tell us that moreover the two new objects in $\overline{\mathbf{\Delta}}_1$ are both *terminal* objects (so we may say $1 \in \overline{\mathbf{\Delta}}_1$) and hence in particular isomorphic. Thus we have

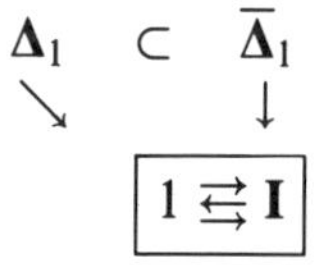

$$\begin{array}{ccc} \mathbf{\Delta}_1 & \subset & \overline{\mathbf{\Delta}}_1 \\ & \searrow & \downarrow \\ & & \boxed{1 \rightleftarrows \mathbf{I}} \end{array}$$

where the vertical functor is an equivalence of categories with two quasi-inverses and all three resulting presheaf toposes are essentially the same. In the two-object category pictured, the original monoid is recovered as all the endomorphisms of $\mathbf{I}$, and

$$e_i = \partial_i \mathbf{I} \qquad i = 0, 1$$

where $1 \xrightarrow{\partial_i} \mathbf{I} \xrightarrow{\mathbf{I}} 1$ is of course 1_1. We may even abbreviate ∂_i to i, in which case we can say that 0,1 are the only *points* of $\mathbf{I}$ even in $\mathcal{S}^{\mathbf{\Delta}_1^{op}}$ but that $\mathbf{I}$ has one further element, namely the figure $1_{\mathbf{I}}$. In general a figure $\mathbf{I} \xrightarrow{X} X$ is often called an "edge" of X and the two points $x\partial_0$, $x\partial_1$ are the initial and terminal vertices of x

$$x_i = x\partial_i \qquad \begin{array}{ccc} 1 & \overset{\delta_0}{\underset{\delta_1}{\rightrightarrows}} & \mathbf{I} \\ & {}_{x_0}\searrow\searrow^{x_1} & \downarrow x \\ & & X \end{array}$$

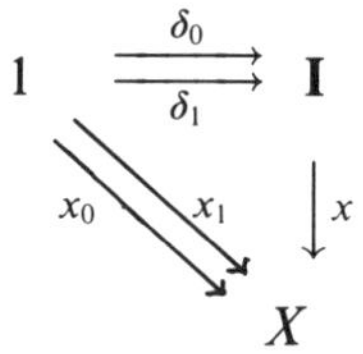

Given any point $1 \xrightarrow{p} X$ of X, there is a corresponding "degenerate edge" at p $\mathbf{I} \xrightarrow{\mathbf{I}} 1 \xrightarrow{p} X$ whose initial and terminal vertices are both p. In general an edge x for which $x\partial_0 = x\partial_1$ is called a "loop" at the corresponding point; there may be several loops at p but the degenerate loop $p\mathbf{I}$ is always among them though it may be the only one. Thus we have in this particular topos a good way of picturing the "inside" of any object X, the general elements as *arrows*, but the degenerate loops as *dots*. Thus for $\mathbf{I}$ itself we have a single non-degenerate arrow:

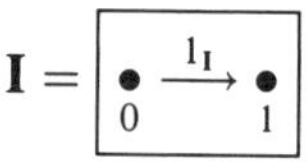

$$\mathbf{I} = \boxed{\underset{0}{\bullet} \xrightarrow{1_{\mathbf{I}}} \underset{1}{\bullet}}$$

and there are morphisms

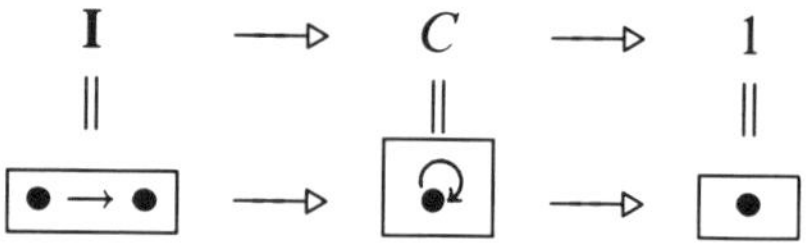

$$\begin{array}{ccccc} \mathbf{I} & \longrightarrow & C & \longrightarrow & 1 \\ \| & & \| & & \| \\ \boxed{\bullet \to \bullet} & \longrightarrow & \boxed{\bullet\circlearrowleft} & \longrightarrow & \boxed{\bullet} \end{array}$$

This illustrates that

1) [in *contrast* to the category $\mathcal{S}^{\mathbf{P}^{op}}$ of "irreflexive" graphs where there are not necessarily any loops and where even if there are loops there is no specific one preserved by morphisms: $\mathbf{P} = \boxed{\bullet \rightrightarrows \bullet}$ is a *category* which happens

to be determined by its underlying reflexive graph structure] in our "reflexive" graphs $\mathcal{S}^{\Delta_1^{op}}$ an edge may become degenerate under the application of a morphism such as $C \to 1$ or $\mathbf{I} \to 1$, and

2) it may be useful to consider quotients (like C) of objects of $\mathbf{C}$ (even when $C \notin \overline{\mathbf{C}}$) as further "generic" figures, so that various types of singular figures (such as loops) also become "representable" at least by objects of $\mathcal{S}^{\mathbf{C}^{op}}$: the surjection $\mathbf{I} \twoheadrightarrow C$ induces the inclusion

$$(C, X) \hookrightarrow (\mathbf{I}, X)$$

of the set of all loops into the set of all edges for any reflexive graph X. Another important figure type in this category is

$E =$

Note that there is a morphism $E \xrightarrow{p} \mathbf{I}$ which has *two* sections $\mathbf{I} \overset{s}{\underset{s'}{\rightrightarrows}} E$ (i.e. $ps = 1_{\mathbf{I}} = ps'$) but that there are two more morphisms $\mathbf{I} \longrightarrow E$ which are not sections of p (in fact they factor across 1) because E has four edges in all. E is not even a quotient of any object of $\overline{\mathbf{C}}$, but it is a quotient of $2\mathbf{I} = \mathbf{I} + \mathbf{I}$ the disjoint sum (coproduct) of two copies of $\mathbf{I}$.

An object of $\mathcal{S}^{\Delta_1^{op}}$ which has several universal significances is

$\Omega =$

which has two points but five edges in the configuration shown (which uniquely specifies the action of ∂_0, ∂_1 on all five). The most basic of these is that it classifies subgraphs. Here by a subgraph of X we mean (not a graph with a property but) a graph A equipped with a specified "inclusion" morphism $A \longrightarrow X$, denoted for example by i_A when it is not understood: for example there are two *different* subobjects of $\mathbf{I}$ which, without their inclusions, are the same 1, and two *different* subobjects of E which without their inclusions are the same $\mathbf{I}$. The only condition for a morphism i (in a topos such as our presheaf toposes; one might well need to complicate this in a non-topos such as the $\mathcal{Y}_{\mathbf{C}}$ mentioned before) to be a subobject inclusion is that it be a *mono*morphism, i.e. satisfy the cancellation property $ia_1 = ia_2 \Rightarrow a_1 = a_2$ for any $T \overset{a_1}{\underset{a_2}{\rightrightarrows}} A$; this $\mathbf{I}$ has essentially five subobjects $0, \mathbf{I}$, $0, 1$, $\{0, 1\}$ where the last is $2 \hookrightarrow \mathbf{I}$ with $2 = 1 + 1$ the "discrete" graph, while E has seven and for the generic loop C we get three. Here "essentially" refers to isomorphism in the comma category $\mathcal{S}^{\Delta_1^{op}}/X$ with $X = \mathbf{I}$ or $X = E$ or $X = C$ for example. In more detail, we write $A \underset{X}{\subseteq} B$ to mean we are given i_A, i_B monomorphisms and can find β so that $i_A = \beta i_B$ in

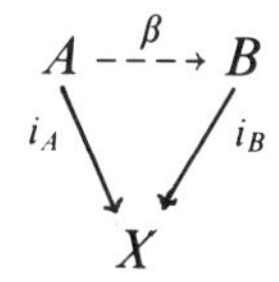

We also write $x \in A$ to mean that x is any morphism with codomain X while we are given a *mono*morphism i_A and can find an α such that $x = i_A\alpha$

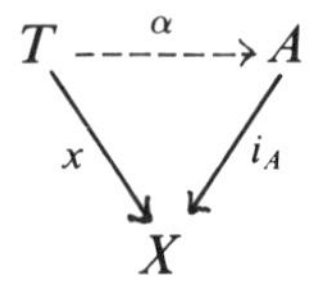

In either case the β or the α which "proves" the inclusion or membership is unique because the second map being compared is monic; in the first case β is also monic since more generally $k\beta$ monic implies β is even if k isn't. The fact that inclusion is a special case of membership, and in the case of points membership a special case of inclusion, of course plays havoc with mid-century axiomatic set-theory but accords well with the naive set theory and geometry which has survived since well before that. For example, we obviously have (by composition in the comma category of "proofs")

$$x \in A \;\&\; A \underset{X}{\subset} B \quad \Rightarrow x \in B$$

and the resulting quantified implication

$$A \underset{X}{\subset} B \Rightarrow \forall x[x \in A \Rightarrow x \in B]$$

can be *reversed* (trivially since we can take $x = i_A$); it can (less trivially) still be reversed in a category like $S^{\mathbf{C}^{op}}$ even if we restrict the universal quantifier in the hypothesis to range only over those x whose domains $T \in \mathbf{C}$. Thus in particular a subgraph of X is determined by some of its edges and some of its points, with the only constraint that both vertices of any selected edge must be selected points. Now the claim that the five-edge graph Ω classifies subgraphs means that Ω has a subgraph $1 \xrightarrow{\text{true}} \Omega$ such that for any graph X and any subgraph $A \xrightarrow{i_A} X$ there is a unique morphism $X \xrightarrow{\varphi_A} \Omega$ such that i_A is the inverse of true under φ_A. That is, for all $T \xrightarrow{x} X$, $x \in A$ iff $\varphi x = \text{true}T$, and the "characteristic" map φ_A of i_A is the *only* morphism with this property. This uniquely determines Ω up to isomorphism. In the case $S^{\Delta_1^{op}}$ of reflexive graphs, Ω is forced to be as claimed, with true the point at which the non-degenerate loop also resides, for the following reason. Picture the inclusion of A in X impressionistically as follows:

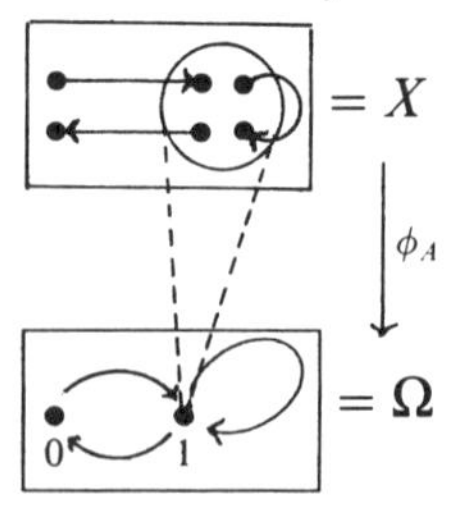

Of course all edges and points in A go by φ to the one true point, and all *points* of X not in A must go to the other place "false"; but there may be edges of X which initiate or terminate in A but not conversely and they will have to go to the appropriate arrow joining true and false since φ must be homogeneous with respect to ∂_0, ∂_1; there may well be paths between points of A which are provided by X but not by A, and they are forced to go by φ to the non-trivial loop at true. Thus all of Ω is needed because of the variety of inclusions that exist in $S^{\Delta_1^{op}}$. On the other hand, taking anything bigger than Ω would ruin the *uniqueness* of φ. Collapsing Ω, for example $\Omega \longrightarrow \Omega^b$ the reflection into the category $\mathcal{Y}_{\Delta_1}$ of graphs with no multiple edges, would induce a closure operation on subgraphs, but could only *classify* certain ones, in the case of Ω^b the "full" subgraphs.

Another "universal" property which happens to be true and interesting in the $S^{\Delta_1^{op}}$ province is that there are enough morphisms $X \longrightarrow \Omega$ to distinguish edges, so that if $F(X)$ denotes the abstract set $F(X) = (X, \Omega)$ of subgraphs of X and $\Omega^{F(X)}$ the $F(X)$-fold cartesian product of copies of Ω, then the canonical morphism

$$X \longrightarrow \Omega^{F(X)}$$
$$x \mapsto [\varphi \longrightarrow \varphi x]$$

is actually a *subgraph* ! (of course in a particular case a much smaller subset $S \subset F(X)$ may suffice).

In any category of the form $S^{\mathbf{C}^{op}}$ there is such a distinguished truth-value object $\Omega_{\mathbf{C}}$, whose elements of sort C "are" just all the ($S^{\mathbf{C}^{op}}/C$-isomorphism types of) subactions of C; such is just any set of arrows $B \longrightarrow C$ for various B in $\mathbf{C}$, subject only to the requirement that for any $B' \longrightarrow B$ in $\mathbf{C}$, and $B \longrightarrow C$ in the set, we must also have the composite $B' \longrightarrow C$ in the set, in other words, any "right ideal" of $\mathbf{C}$ in the case of a monoid. The action of $\mathbf{C}$ on $\Omega_{\mathbf{C}}$ is by inverse image (or "division" or "analysis"): If S is a right ideal in C and if $C' \xrightarrow{\lambda} C$, then

$$t \in S\lambda \quad \text{iff} \quad \lambda t \in S;$$

the requisite property $S(\lambda\mu) = (S\lambda)\mu$ for $C'' \xrightarrow{\mu} C' \xrightarrow{\lambda} C$ then follows. The object Ω thus *constructed* to classify subactions of the $\mathbf{C}(-, C)$ then actually succeeds to classify subactions of any X, for given a subobject A of X, we can define φ by

$$\boxed{t \in \varphi x \quad \text{iff} \quad xt \in A}$$

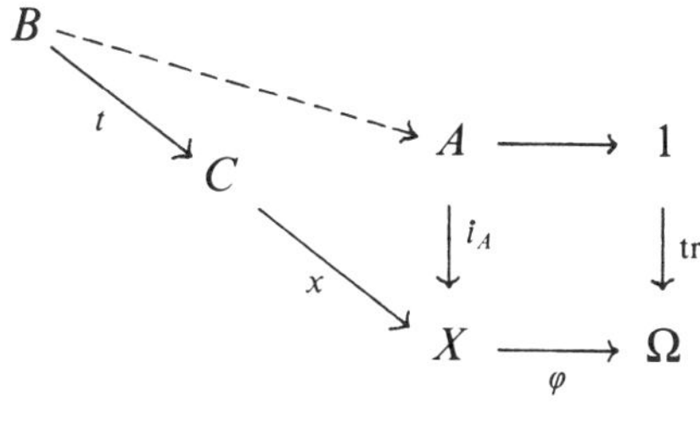

$$\begin{array}{ccccc} B & & & & \\ & \overset{t}{\searrow} & \dashrightarrow & A & \longrightarrow & 1 \\ & & C & \Big\downarrow{\scriptstyle i_A} & & \Big\downarrow{\scriptstyle \text{true}} \\ & & & \overset{x}{\searrow} \; X & \xrightarrow[\varphi]{} & \Omega \end{array}$$

(where of course true$C$ = $\mathbf{C}(-, C)$ the greatest of all right ideals in C). In other words, the truth-value φx of the statement "$x \in A$" is identified engineering-wise with the set of all possible acts t which would bring about its actual truth. [This suggests a notion of numerical "measures" on the $\cup$-lattice $\boldsymbol{\Omega}$, whose theory is so far only fragmentarily developed.]

The functor $\mathcal{S}^{\mathbf{C}^{op}} \xrightarrow{(1,-)} \mathcal{S}$ assigning to every $\mathbf{C}$-set its abstract set of *points* always has a left adjoint assigning to each set the corresponding *discrete* $\mathbf{C}$-set which has S elements of each sort and in which $s_{C'} = s_C\lambda$ for all λ for each $s \in S$, and each $C' \xrightarrow{\lambda} C$ in $\mathbf{C}$. Composing these two adjoints we get (in the case of graphs) the maximal discrete subgraph $|X| \hookrightarrow X$ of any graph, whose characteristic map $X \xrightarrow{\delta} \Omega$ factors through the loop $C \hookrightarrow \Omega$. For any morphism $X \xrightarrow{f} Y$, the composite $\delta_Y f$ classifies the subgraph D_f of X which is the *degeneracy* of f.

On the other hand the "points" functor $\mathcal{S}^{\mathbf{C}^{op}} \xrightarrow{(1,-)} \mathcal{S}$ will have a right adjoint iff $1 \in \overline{\mathbf{C}}$; this assigns to any set S the codiscrete or chaotic action whose figures of C-th sort are the elements of $S^{(1,C)}$. Thus for $\mathbf{C} = \boldsymbol{\Delta}_1$, the chaotic graph on S points has S^2 edges.

To pass to another example, recall that we remarked before the elementary "parallel process" $E = \boxed{\bullet \rightrightarrows \bullet}$ is a reflexive graph which happens to admit only one definition of composition making it into a category $\mathbf{P}$. This also happens to be a self-dual category in the sense that there exists an isomorphism $\mathbf{P}^{op} \xrightarrow{\sim} \mathbf{P}$ of categories. Its actions $\mathcal{S}^{\mathbf{P}^{op}}$ are the *irreflexive* graphs (the negative is in a way appropriate even for those objects which happen to have loops at some point p, for morphisms are allowed to interchange *any* two such loops). It has no sense here to speak of degenerating via a morphism since $\mathbf{P} = \boxed{U \rightrightarrows \mathbf{I}} \hookrightarrow \mathcal{S}^{\mathbf{P}^{op}}$ where $0 \subseteq U \subseteq 1$. Indeed the "points" $1 \longrightarrow X$ should really be pictured here as LOOPS and the "nodes" $U \longrightarrow X$ pictured as dots. Our *new* $\mathbf{I}$ still has five subobjects in $\mathcal{S}^{\mathbf{P}^{op}}$ but $\Omega = \Omega_{\mathbf{P}}$ must be pictured as

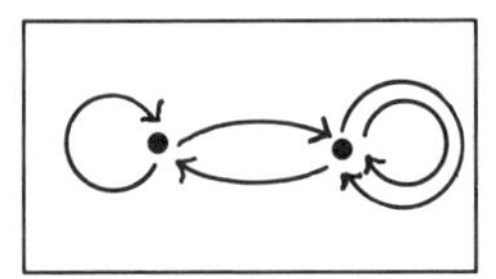

There are three morphisms $1 \longrightarrow \Omega$ (i.e. loops in Ω) since the terminal object 1 itself (the generic loop now) has three subobjects. We can choose either of the two loops at the one node as "true"; the node itself can't be so chosen since in any topos it can be shown that the generic subobject $G \hookrightarrow \Omega$ must have a domain $G = 1$ (i.e. in this case a loop) and not for example $U \subseteq 1$. The unique non-trivial automorphism of Ω induces an involutory modal "negation" operator on subobjects of any object X in $\mathcal{S}^{\mathbf{P}^{op}}$, different from the intuitionistic negation (which is also present) since it preserves false! Since $\mathbf{P}$ has cancellation, the comma category $\mathcal{S}^{\mathbf{P}^{op}}/\mathbf{I} \approx \mathcal{S}^{(\mathbf{P}/\mathbf{I})^{op}}$ is actually the

actions of the *poset*

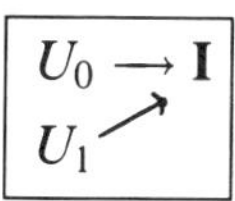

which can usefully be considered as the basic open sets of a three-point topological space

in which one point is not open, making five open sets in all. The "quotient map" $\mathcal{S}^{\mathbf{P}^{op}}/\mathbf{I} \longrightarrow \mathcal{S}^{\mathbf{P}^{op}}$ identifies the two basic regions without thereby coalescing the actions of the corresponding restriction maps from the whole. Note that *discrete* **P** sets are just disjoint sums of single loops, so that the analogue (pullback) of the degeneracy D_f of a map $X \xrightarrow{f} Y$ is not a sub-object of X; the fibers of $|Y| \longrightarrow Y$ disconnect the multiple loops, and nodes which have no loops are not covered at all.

Our three-element example $\mathbf{\Delta}_1$ could even more concretely be realized as the full subcategory of Posets or of Cat consisting of (1 and) 2 in which case we have the adjointness relations

$$\partial_0 \dashv \mathbf{I} \dashv \partial_1$$

uniquely determining all by any one, and hence for any category **A** there is a reflexive graph *in Cat* (rather than in **S**) in which the graph structure is determined by adjointness $\mathbf{A}^2 \rightleftarrows \mathbf{A}$. A more general kind of such "adjoint graph" is

$$\mathcal{S}^{\mathbf{C}^{op}} \rightleftarrows \mathcal{S}$$

in which the initial set of any "edge" action X is its set of (rest) points, the degenerate "edge" at any "point" $S \in \mathcal{S}$ is the trivial action, and the terminal set of any X is its set of components. In an example like $\mathbf{C} = \mathbf{\Delta}_1$, every component contains a point, and the equivalence condition on pairs of points of $|X|$ induced by the surjection

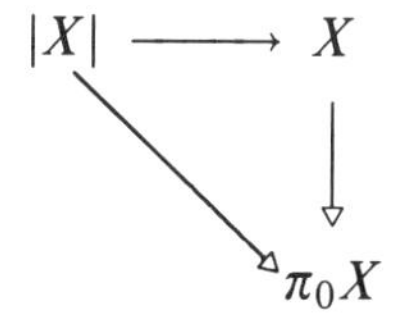

is also the one generated by the existence of two actors $\lambda \in \mathbf{C}$ taking one element to the two points. Thus it is apparent from its picture that $\pi_0\Omega = 1$.

As another example we might consider the four element monoid **F** of all endomaps of a two element set. The two inclusion functors $\mathbf{\Delta}_1 \rightrightarrows \mathbf{F}$ show that any **F**-set X may be viewed as a directed graph in two ways, with the *same* set of *points*. The additional act τ with $\tau^2 = 1_{\mathbf{I}}$ in **F** *reverses* each edge of X to a canonically associated edge of X going the opposite way, and this canonical return trip is preserved by morphisms $X \longrightarrow Y$. There are objects in which some non degenerate loops are fixed by τ, for example the two element C'. This category $\mathcal{S}^{\mathbf{F}^{op}}$ has been extensively used to proving nontrivial theorems about free groups on the basis of geometric intuition by Serre, Bass, Gersten, Duskin, and others, though without ever yet exploiting its topos structure! The connection is as follows: Since $\mathbf{F} \hookrightarrow \mathbf{S}$ is a full subcategory of the category $\mathbf{S}$ of non empty finite sets, we get a full inclusion $\mathcal{S}^{\mathbf{F}^{op}} \hookrightarrow \mathcal{S}^{\mathbf{S}^{op}}$ by taking the right adjoint of the restriction (as well as another full inclusion by taking the left adjoint; it is the left adjoint which extends $\mathbf{F} \hookrightarrow \mathbf{S}$ with respect to the Yoneda embeddings $T_{\mathbf{F}}$, $T_{\mathbf{S}}$, but the right adjoint which is considered the more basic inclusion in sheaf theory). $\mathcal{Y}_{\mathbf{S}} \hookrightarrow \mathcal{S}^{\mathbf{S}^{op}}$ is the classical category of simplicial complexes, which explains the relation to combinatorial topology. A third reflective subcategory of $\mathcal{S}^{\mathbf{S}^{op}}$ is the category $\mathcal{G}$ of groupoids, i.e. categories in which every morphism is an isomorphism, and functors (= homomorphisms) between these; since $1 \in \mathbf{S}$, there is a *codiscrete* (chaotic) inclusion $\mathcal{S} \hookrightarrow \mathcal{S}^{\mathbf{S}^{op}}$ right adjoint to "points" = (1,-) which takes any set S to the action whose elements of sort n are just S^n, for $n \in \mathbf{S}$ and where $n' \xrightarrow{\lambda} n$ in $\mathbf{S}$ acts by composing $n' \xrightarrow{\lambda} n \xrightarrow{x} S$. Now if we consider elements of $(\underline{2}, X)$ as arrows between the points $(1, X)$, X already suggests how to compose these via the elements a of $(\underline{3}, X)$, whose "boundary" elements are deduced via inclusions $2 \hookrightarrow 3$ in $\mathbf{S}$ and might be pictured as

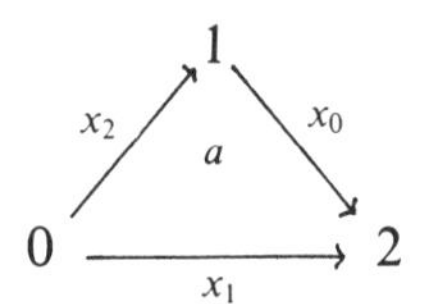

We consider "$a \models x_1 \equiv x_0 x_2$" as a relation to be imposed on the free groupoid generated by words from $(\underline{2}, X)$, for each $a \in (\underline{3}, X)$, thus obtaining the groupoid $\pi_1 X$ and the left adjoint

$$\mathcal{S}^{\mathbf{S}^{op}} \xrightarrow{\pi_1} \mathcal{G}$$

to the full inclusion which to each groupoid $\mathbf{G}$ associates the composable strings of arrows from $\mathbf{G}$. The latter is acted on by *all maps* $n' \xrightarrow{\lambda} n$ since where λ is not surjective, we can compose in $\mathbf{G}$ some segments of an n-string, where it is not injective we can insert some identities into the string to bring it up to length n', and where symbols are "interchanged" we can imagine using the $(\ )^{-1}$ operation of the groupoid. A more unified description of the inclusion is obtained if we note that there is also a codiscrete inclusion $\mathbf{S} \hookrightarrow \mathcal{G}$, since if $\underline{S}(i,j) \underset{\text{def}}{=} 1$ for all $\langle i,j \rangle$ in S^2, there is a unique way to compose these, and $\underline{S}$ is a groupoid with S objects since $\underline{S}(i,j) \xrightarrow{(\ )^{-1}} \underline{S}(j,i)$ just reverses. This restricts to $\mathbf{S} \hookrightarrow \mathcal{S} \hookrightarrow \mathcal{G}$, which then induces $\mathcal{G} \hookrightarrow \mathcal{S}^{\mathbf{S}^{op}}$ immediately by $(\underline{n}, \mathbf{G}) \underset{\text{def}}{=} \mathcal{G}(\underline{n}, \mathbf{G})$; this is full since any homogeneous map $X \xrightarrow{f} Y$ between *such* actions of $\mathbf{S}$ preserves $(\underline{3}, X) \longrightarrow (\underline{3}, Y)$ which are essentially the multiplication tables of the original groupoids, hence f comes from a homomorphism between the latter. The composites $\mathcal{S}^{\mathbf{F}^{op}} \hookrightarrow \mathcal{S}^{\mathbf{S}^{op}} \longrightarrow \mathcal{G}$ and $\mathcal{Y}_{\mathbf{S}} \hookrightarrow \mathcal{S}^{\mathbf{S}^{op}} \longrightarrow \mathcal{G}$ thus provide Poincaré groupoids for **F**-graphs and simplicial schemes respectively. Lemmas showing how to spread out the graphs thus giving rise to free groups lead to geometric proofs of theorems of Schreier, Gersten, etc.

Besides $\Omega_{\mathbf{C}}$, the other crucially topos-theoretic property of the presheaf categories $\mathcal{S}^{\mathbf{C}^{op}}$ is "cartesian closedness", that is the existence of exponential functors $(\ )^A$ characterized by their ("λ-conversion") right adjointness to the cartesian product functor $A \times (\ )$ for each $A \in \mathcal{S}^{\mathbf{C}^{op}}$:

$$\frac{X \longrightarrow Y^A}{A \times X \longrightarrow Y}$$

The object Y^A can be constructed using this adjointness applied to the special case $X = C \in \mathbf{C}$ by invoking Yoneda's lemma: the elements of the sort C in Y^A "are" just the arbitrary morphisms $A \times C \longrightarrow Y$, acted on by

$$A \times C' \xrightarrow{1_A \times \lambda} A \times C \longrightarrow Y \quad \text{for} \quad C' \xrightarrow{\lambda} C.$$

[Somewhat more conceptually, these elements are the morphisms $C^*A \to C^*Y$ in $\mathcal{S}^{\mathbf{C}^{op}}/C$ where C^*X denotes the trivial fibration $X \times C \to C$.] In the case of a monoid $\mathbf{C}$ with its single dominant sort $\mathbf{I}$, the objects of the form $Y^{\mathbf{I}}$ (e.g. for $Y = \mathbf{I}$ itself, or $Y = \Omega_{\mathbf{C}}$) present themselves as the simplest interesting cases for calculation: $Y^{\mathbf{I}}$ has as elements the maps $\mathbf{I} \times \mathbf{I} \xrightarrow{f} Y$ satisfying

$$f(x\lambda, t\lambda) = f(x,t)\lambda$$

remembering that the elements of $\mathbf{I}$ are the elements of the original monoid itself. The action of $\mathbf{C}$ on this set of maps is

$$(f\lambda)(x,t) = f(x,t\lambda)$$

by acting only on the "test" component t, not on the one that is to be exponentiated. There is as yet no very systematic way of calculating even the case $\mathbf{I}^{\mathbf{I}}$, which therefore must be done one $\mathbf{C}$ at a time. For example, if $\mathbf{C}$ is the monoid of all continuous self maps of the unit interval $[0, 1]$, so that the points $1 \longrightarrow \mathbf{I}$ in $\mathcal{S}^{\mathbf{C}^{op}}$ are the points of $[0, 1]$ in the usual sense, then the maps $\mathbf{I} \times \mathbf{I} \longrightarrow \mathbf{I}$ are just the usual continuous functions of two variables, which is mildly surprising, but not too difficult to prove. The corresponding statement for the monoid of *smooth* ($= C^\infty$) self maps of the line is surprising (once one realizes that one has to show that all the higher formally defined partial derivatives are the actual partial derivatives so in particular commute) and rather difficult to prove (Bowman 1966 and forthcoming book by Frölicher and Kriegl). Having calculated $Y^{\mathbf{I}}$, even more interest attaches to the natural path functionals $Y^{\mathbf{I}} \longrightarrow Y$. In both of the examples mentioned, the real line R determines an object of $\mathcal{S}^{\mathbf{C}^{op}}$ with just the reals as points by defining its elements of sort $\mathbf{I}$ to be just the continuous (resp. smooth) paths in R. Since multiplication is continuous it gives rise to a morphism $R \times R \longrightarrow R$ in $\mathcal{S}^{\mathbf{C}^{op}}$ and hence to a left action $R \times R^{\mathbf{I}} \longrightarrow R^{\mathbf{I}}$ by $(af)(x) = a(f(x))$. Thus (in the smooth case) one can look for the object of *linear* functionals

$$\mathrm{Lin}_R(R^R, R) \hookrightarrow R^{(R^R)}$$

whose points are just the morphisms $R^R \xrightarrow{\varphi} R$ which moreover satisfy $\varphi(af) = a\varphi(f)$. Schanuel, Zame, and I showed in 1980 that these are just the Schwartz distributions (of compact support); see the forthcoming book by Frölicher and Kriegl for details.

In the case $\mathbf{C} = \Delta_1$, the reader should be able to compute that the elements of $\mathbf{I}^{\mathbf{I}}$ are the six maps $\mathbf{I} \times \mathbf{I} \longrightarrow \mathbf{I}$ which are the two constants, the two projections, and the maps corresponding to "max" and "min" (when we consider that $0 \leq 1$ in $\mathbf{I}$) but that the action of Δ_1 on this set is such that its standard picture as a graph is

$\mathbf{I}^{\mathbf{I}} =$ 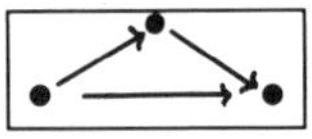

In an interpretation where $\mathbf{I} \longrightarrow X$ are thought of as processes, the following names for the elements of $\mathbf{I}^{\mathbf{I}}$ are suggestive:

$\mathbf{I}^{\mathbf{I}} =$ 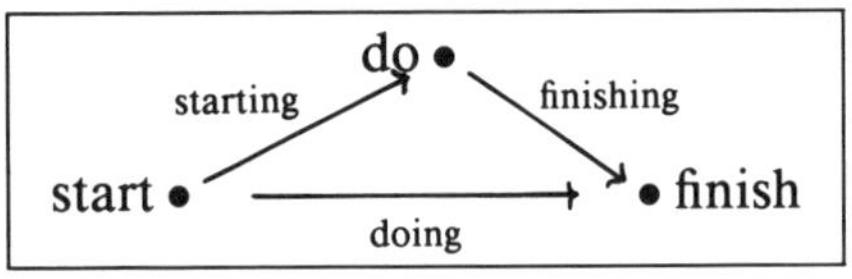

The reader is invited to correlate these six names with the names $0, 1, \mathrm{proj}_1$, proj_2, max, min. Computation of $\mathbf{I}^{\mathbf{I}} \longrightarrow \mathbf{I}$, $\Omega^{\mathbf{I}}$, etc. is further invited.

For presheaf categories $\mathcal{S}^{\mathbf{C}^{op}}$ there is the extremely important formula

$$\mathcal{S}^{\mathbf{C}^{op}}/X \cong \mathcal{S}^{(\mathbf{C}/X)^{op}}$$

showing that all the "slice" or "comma" categories (relative to any $\mathbf{C}$-set X) are again presheaf categories. Here $\mathbf{C}/X$ is the category whose objects are the figures of X and whose morphisms are the incidence relations: $x' \xrightarrow{\lambda} x$ where $x' = x\lambda$, $C' \xrightarrow{\lambda} C$ in $\mathbf{C}$, and x (resp. x') is a figure in X of sort C (resp. C'). The forgetful functor $\mathbf{C}/X \longrightarrow \mathbf{C}$ is what is called a (discrete op-) *fibration.*

For one example of the kind of "cohesiveness" which might be expressible by a directed graph X in $\mathcal{S}^{\Delta_1^{op}}$, consider a paragraph in which there are several concrete or abstract things talked about (and which are taken as points) whereas each sentence (edge) has a subject and an object (represented by the operations ∂_0, ∂_1). Many intransitive verbs can be considered as reflexive versions (loops) of transitive verbs, so that these too can be included in such an analysis. Since the paragraph contains many sentences, which make many interlocking statements about the things, a non-trivial graph structure thus results. (The degenerate loop at each point b may be considered as the sentence "b is b", which is implicit in the paragraph.) A translation or interpretation of one paragraph into another (perhaps in another language) should at least be a morphism of graphs. But it should preserve more than just the "subject of a sentence" and "object of a sentence" incidence relations, and this can be partly expressed by passing to a category of the form $\mathcal{S}^{\Delta_1^{op}}/V$ where V is a fixed graph of "labels" or "values". Frequently V is a graph that consists entirely of loops; then for $X \to V$ to be a graph morphism merely means that every *degenerate* loop in X is mapped to a *point* of V. Thus for example V could be a classification of verbs such as that into "state, activity, achievement" verbs, including the single point "is", and a paragraph X could be given the structure of an $\mathcal{S}^{\Delta_1^{op}}/V$ object by mapping each sentence to the type of its principal verb (it might be reasonable to map some *non*-degenerate sentences such as "x_1 resembles x_2" into the point "is"); or tenses might also be included in V. Then a translation $X \longrightarrow Y$ would be required to preserve the labeling of X and Y

$$\begin{array}{ccc} X & \longrightarrow & Y \\ & \searrow \quad \swarrow & \\ & V & \end{array}$$

that is, to be a morphism in $\mathcal{S}^{\Delta_1^{op}}/V$.

The special role of labeling graphs as objects in $\mathcal{S}^{\Delta_1^{op}}/V$ where V has only loops comes up very frequently. For example, if the edges in X are to be interpreted as processes or as roads between towns and the labeling signifies time or cost or distance, the only *general* requirement is that a trivial process costs nothing, so V could be taken as a set of vectors or real numbers or other abstract quantities, all construed as loops at a single point, with the quantity 0 identified as the degenerate loop.

This special role of loops can be formalized as follows: There is a unique surjective homomorphism of monoids $\Delta_1 \twoheadrightarrow \{0, 1\}$ from our three-element monoid to the multiplicative monoid consisting of the numbers 0, 1. (This is what results if we "force Δ_1 to become commutative".) The only significant feature of the element 0 is that it is a generic idempotent, so the objects of the category $\mathcal{S}^{\{0,1\}^{op}}$ may be identified with diagrams of sets

$$X_0 \underset{i}{\overset{p}{\leftrightarrows}} X, \qquad pi = id_{X_0}$$

the action of 0 being the composite ip on X (one of the great many kinds of examples of such a structure arises when we are given an arbitrary mapping $X_0 \xrightarrow{\alpha} X_1$, $X \underset{\text{def}}{=} X_0 \times X_1$, and i is taken to be the inclusion of the "graph" of α). In this case the discrete inclusion $\mathcal{S} \hookrightarrow \mathcal{S}^{\{0,1\}^{op}}$ not only has a right adjoint "points" functor and a left adjoint "components" functor, but these two functors are *isomorphic,* viz. to $X \mapsto X_0$; hence the components functor trivially preserves products (indeed all limits) and there is trivially a notion of "codiscrete" (which indeed coincides with "discrete"). However, the truth-value object $\Omega_{\{0,1\}}$ is *not* connected since it has three elements and *two* components. We will return to this remarkable topos (which in particular seems to be neither "gros" nor "petit") in the third section, but for the moment let us return to its relationships with the topos $\mathcal{S}^{\Delta_1^{op}}$ of reflexive graphs.

The homomorphism $\mathbf{\Delta}_1 \xrightarrow{q} \{0,1\}$ induces (as does any functor) three adjoint functors between presheaf categories

$$q_! \dashv q^* \dashv q_* \qquad \begin{array}{c} \mathcal{S}^{\Delta_1^{op}} \\ q_! \downarrow \quad \uparrow q^* \quad \downarrow q_* \\ \mathcal{S}^{\{0,1\}^{op}} \end{array}$$

where q^* simply reinterprets $X_0 \underset{i}{\overset{p}{\leftrightarrows}} X$ as a graph with X_0 points and X loops, the location of the loops being specified by p. This is the basic relationship required for the kind of "labeling" applications mentioned above, and in fact suggests considering the comma category (or "glued" topos) $\mathcal{S}^{\Delta_1^{op}}/q^*$ as a category of labeled graphs in which morphisms include the possibility of re-labeling. The right adjoint functor q_* essentially discards all the non-loops in an arbitrary graph, whereas the left adjoint $q!$ forces all edges to become loops by replacing the old set of points with the new set of points which is the set of components of the original graph; the new set of edges is a quotient set of the old in that all the *degenerate* loops in each given component become identified.

How should one internally picture the objects $X_0 \underset{i}{\overset{p}{\leftrightarrows}} X$, $pi = id_{X_0}$ of $\mathcal{S}^{\{(0,1\}^{op}}$? Probably there is no single preferred way, since objects so abstract have many diverse applications; however, one way suggested (by Meloni) is that the generic figure is

$$\mathbf{I}_{\{0,1\}} = \boxed{\;\text{—•—}\;}$$

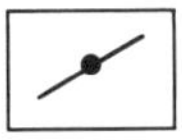

so that in general every object of $\mathcal{S}^{\{0,1\}^{op}}$ is a sum

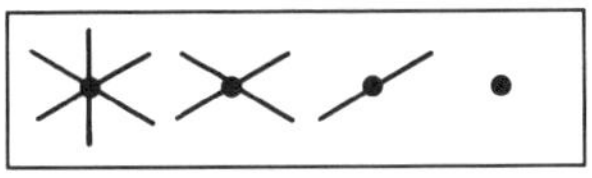

of connected components (the same number as the number of points !) while each component (which is essentially nothing but a pointed set) is indicated by drawing one small path (without endpoints) through the point for each non degenerate element. Then the effect of the functor q^* is to close up each little path into a loop. There are more connections between the two toposes, induced by the two obvious injective homomorphisms

$$\{0,1\} \underset{i_1}{\overset{i_0}{\rightrightarrows}} \mathbf{\Delta}_1$$

where $i_K(0) = e_K$ (recall that 0 is a generic idempotent; of course i_0, i_1 are both sections of q). Each of i_0, i_1 induce three adjoint functors as q does; for example i_0^* redefines every edge in a graph as a loop at its beginning point.

Although $\mathcal{S}^{\Delta_1^{op}}$ is just one topos and the toposes $\mathcal{S}^{\Delta_1^{op}}/B$ are still quite special, it can be shown that for any Grothendieck topos $\mathcal{X}$ there exists a graph B and an idempotent left-exact endofunctor E of $\mathcal{S}^{\Delta_1^{op}}/B$ such that $\mathcal{X}$ is equivalent to the subcategory of $\mathcal{S}^{\Delta_1^{op}}/B$ consisting of the objects fixed by E. As an example, consider the topos $\mathcal{S}^{\mathcal{F}(B)^{op}}$ of right actions of the free category (or path category) determined by a reflexive graph B. This is equivalent to the subcategory of $\mathcal{S}^{\Delta_1^{op}}/B$ consisting of all $X \longrightarrow B$ which satisfy the condition of "fibration" type:

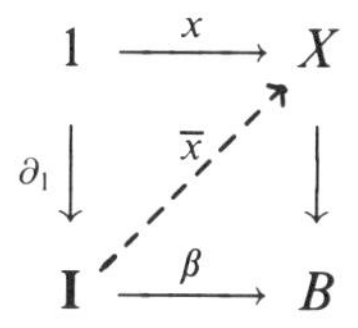

i.e. that for any edge β of B and any point x projecting to $\beta\partial_1$ there is a unique edge $\overline{x}$ which projects to β and has $\overline{x}\partial_1 = x$ (then $x \cdot \beta \underset{\text{def}}{=} \overline{x}\partial_0$ uniquely defines the "action" of β); and this implies that any edge $\overline{x}$ which projects to a degenerate b is itself already degenerate (then in particular we have a further derived "fibration" type condition

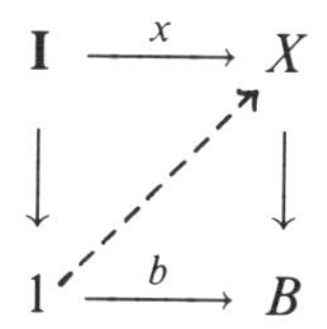

and each degenerate loop b *acts* as the identity). Under the equivalence

$$\mathcal{S}^{\Delta_1^{op}}/B \cong \mathcal{S}^{(\overline{\Delta}_1/B)^{op}}$$

this means that those morphisms in Δ_1/B which are over either $1 \xrightarrow{\partial_1} \mathbf{I}$ or $\mathbf{I} \longrightarrow 1$ in $\overline{\Delta}_1$ are being required to act as *bijections* between the fibers of X, or again that what is really acting on X is the category of fractions

$$\overline{\Delta}_1/B \longrightarrow \mathcal{F}(B)$$

obtained by formally inverting $\mathcal{D}(B)$, the inverse image under $\overline{\Delta}_1/B \longrightarrow \overline{\Delta}_1$ of the subcategory $\mathcal{D}(1) = 1 \underset{I}{\overset{\partial_1}{\rightleftarrows}} \mathbf{I} \circlearrowleft_{e_1}$ of $\overline{\Delta}_1$. Every edge β in B determines a morphism in $\mathcal{F}(B)$, namely the one coming from

$$\sigma(\beta) = \boxed{\begin{array}{ccc} 1 & \overset{\partial_0}{\text{———}} & \mathbf{I} \\ & {\scriptstyle b'} \searrow \quad \swarrow {\scriptstyle \beta} & \\ & B & \end{array}}$$

in Δ_1/B. But composites of these (hence longer words) can be formed in $\mathcal{F}(B)$, since from $b'' \xrightarrow{\alpha} b' \xrightarrow{\beta} b$ in B we get in $\overline{\Delta}_1/B$ the solid arrows

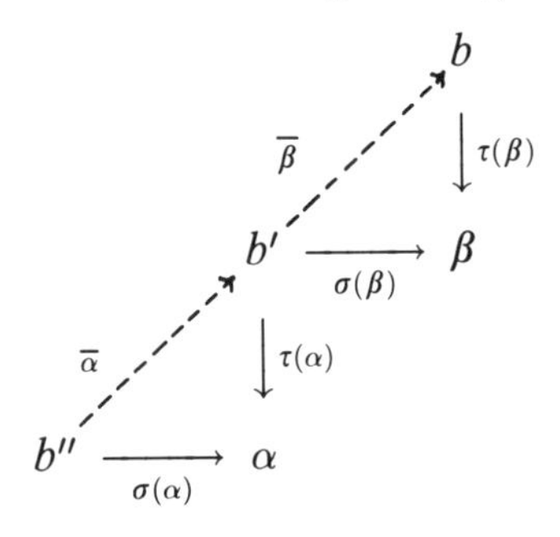

where the "target" morphisms $\tau(\ )$ are well defined like the "source" morphisms $\sigma(\ )$, but using ∂_1 instead of ∂_0; but in $\mathcal{F}(B)$ the τ arrows (with label ∂_1) are invertible, hence there are the dotted arrows, so in particular a well defined $b'' \xrightarrow{\overline{\beta}\,\overline{\alpha}} b$ results in $\mathcal{F}(B)$. On the other hand, for any B there is a unique nondegenerate morphism of graphs $B \longrightarrow C =$ [loop] to the generic loop (defined by the condition that the fiber be discrete, in other words, that *every* nondegenerate edge of B be labeled by the nondegenerate loop of C). Since $\mathcal{F}$ is a functor

$$\mathcal{S}^{\Delta_1^{op}} \xrightarrow{\mathcal{F}} \mathrm{Cat}$$

we thus get an induced functor $\mathcal{F}(B) \longrightarrow \mathcal{F}(C) \cong \mathbf{N}$ to the additive monoid of natural numbers; this functor is the *length* function on the words in $\mathcal{F}(B)$ (Caution: this length function, like $B \longrightarrow C$ itself, is only functorial with respect to *nondegenerate* morphisms $B_1 \longrightarrow B_2$). For example a graph X over the loop $C =$ [loop] in $\mathcal{S}^{\Delta_1^{op}}/C$ satisfies our fibration condition if and only if it is determined by an arbitrary endomap u of the fiber set $X_0 \subset X$, where the rest of edges in X are just the "graph" $\{\langle ux_0, x_0\rangle \mid x_0 \in X_0\}$ of u. Note that usually the free category on B is construed as the skeletal (hence equivalent) subcategory of $\mathcal{F}(B)$ on the objects over 1 in $\overline{\Delta}_1$.

For free categories $\mathcal{F}(B)$ on reflexive graphs, something can occasionally occur (and indeed does in some of our examples such as $B =$ [graph]) which can never occur for the often-studied free categories $W(G)$ on an *irreflexive* G: namely $B \xrightarrow{\sim} \mathcal{F}(B)$ as graphs; of course B would have to be acyclic (i.e. very loop-free) for this to happen.

Since the topos $\mathcal{S}^{\mathcal{F}(B)^{op}}$ is determined as a subcategory of $\mathcal{S}^{\Delta_1^{op}}/B$ via the (epimorphic) functor $\Delta_1/B \xrightarrow{f} \mathcal{F}(B)$, the left-exact idempotent f^*f_* in this case actually preserves even infinite limits, since indeed it has a left adjoint $f^*f_!$ (which is also idempotent)

$$\mathcal{S}^{(\Delta_1/B)^{op}} \underset{f_*}{\overset{f_!}{\rightrightarrows}} \!\!\xleftarrow{f^*}\; \mathcal{S}^{\mathcal{F}(B)^{op}} \qquad f_! \dashv f^* \dashv f_*$$

Now any free category $\mathcal{F}(B)$ has the property that every morphism in it is both monic and epic. Thus, (as will be explained further) $\mathcal{S}(B) \underset{\text{def}}{=} \mathcal{S}^{\mathcal{F}(B)^{op}}$ serves well as one notion of the "petit topos associated to the object B of the gros topos $\mathcal{B} \underset{\text{def}}{=} \mathcal{S}^{\Delta_1^{op}}$".

Before closing this section, let us illustrate the difference between left actions $\mathcal{S}^{\mathbf{C}}$ and right actions $\mathcal{S}^{\mathbf{C}^{op}}$ in the case $\mathbf{C} = \mathbf{\Delta}_1$. In contrast to the infinitely complicated graphs, such left $\mathbf{\Delta}_1$, - sets or "cylinders"

$$A_0 \underset{\overset{\longrightarrow}{d_1}}{\overset{\overset{d_0}{\longrightarrow}}{\xleftarrow{\mathbf{I}}}} A \qquad\qquad \mathbf{I}d_0 = \mathbf{I}d_1 = \mathrm{id}_{A_0}$$

(at least in $\mathcal{S}$!) are all uniquely expressible as disjoint sums of the very special ones in which $A_0 = 1$; these connected cylinders, apart from the cardinality of the fiber A, are determined up to isomorphism by whether or not the two designated points d_0, d_1 are distinct or not.

Note that our cylinders generalize the common notion of cylinders as products as follows: If $1 \rightrightarrows \mathbf{I}$ is any bipointed set (i.e. connected objects of $\mathcal{S}^{\Delta_1}$) and A_0 is any set, then $A \underset{\text{def}}{=} A_0 \times \mathbf{I}$ is an object of $\mathcal{S}^{\Delta_1}$ having the special property that all components are isomorphic. Of course, cylinders *in* a topos other than $\mathcal{S}$ can have highly non-trivial significance; for example when A_0 and A are "contractible" objects of the topos but $d_0(A_0) \cap d_1(A_0) = 0$ in A. In general a $\mathbf{C}^{op}$ action R *in* $\mathcal{S}^{\mathbf{C}}$ gives rise to an adjoint pair of contravariant functors

$$(\mathcal{S}^{\mathbf{C}})^{op} \rightleftarrows \mathcal{S}^{\mathbf{C}^{op}}$$

defined by $\mathrm{Hom}_{\mathbf{C}}(-, {}_1R)$ and $\mathrm{Hom}_{\mathbf{C}^{op}}(-, R_r)$; many of the basic algebra-geometry dualities of mathematics

$$\mathrm{Alg}(\mathbf{C})^{op} \underset{\text{function alg}}{\overset{\text{spectrum}}{\rightleftarrows}} \mathrm{Geom}(\mathbf{C})$$

are just restrictions of exactly such an adjoint pair to subcategories of $(\mathcal{S}^{\mathbf{C}})^{op}$ and $\mathcal{S}^{\mathbf{C}^{op}}$ respectively, with Geom(**C**) even being a subtopos defined by a Grothendieck topology on **C**. A standard example of such R is $R(C', C) = \mathbf{C}(C', C)$, the "hom-functor" of **C**. However, in the case $\mathbf{C} = \mathbf{\Delta}_1$ we could even consider the trivial action on a set R; then any cylinder A yields a special kind of graph

$$R^A \rightleftarrows R^{A_0}$$

which in the case of a connected cylinder ($A_0 = 1$) is just the graph whose *points* are R, whose edges are all functions from A to R, and whose begin and finish vertex relations are given by evaluation

$$f\partial_0 = f(d_0)$$

$$f\partial_1 = f(d_1)$$

at the two points designated by the cylinder structure, for any $f \in R^A$.

As I have mentioned "singularity" several times, it may have occured to the reader that there is an objective way of measuring it. Indeed for any small category $\mathbf{C}$, there is a distinguished object $Eq = Eq_{\mathbf{C}}$ in the topos $S^{\mathbf{C}^{op}}$ such that for every object X there is a canonical map $X \xrightarrow{\sigma_X} Eq$ in $S^{\mathbf{C}^{op}}$ which does this. Namely the elements of sort C of Eq are just all the equivalence relations on C, where by an equivalence relation on C is meant the specification for every $D \in \mathbf{C}$ of a set of ordered pairs $D \rightrightarrows C$ of morphisms in $\mathbf{C}$ which is reflexive, symmetric, and transitive for each D and which is closed with respect to composition by arbitrary $D' \longrightarrow D$. If $C' \longrightarrow C$ in $\mathbf{C}$ and $E \in Eq(C)$, then $E \cdot \lambda \in Eq(C')$ is defined by taking for each D

$$\langle t_1, t_2 \rangle \in E \cdot \lambda \iff \langle \lambda t_1, \lambda t_2 \rangle \in E$$

thus making Eq into an object of $S^{\mathbf{C}^{op}}$. Of course it is more than just an object, having a natural intersection operation $Eq \times Eq \longrightarrow Eq$ and greatest point $1 \longrightarrow Eq$ making it into a semilattice object, so in particular into an ordered object. On the other hand, although the equality relation $\Delta_C \in Eq(C)$ for each C this is not natural, i.e. does not define a point $1 \overset{\Delta}{\dashrightarrow} Eq$ *unless* it happens that all morphisms in $\mathbf{C}$ are monomorphisms. The singularity measurement $X \xrightarrow{\sigma_X} Eq$ is defined, for each $C \xrightarrow{x} X$, by

$$\sigma(x) = \{\langle t_1, t_2 \rangle \mid x t_1 = x t_2\}$$

the "self-incidence"; then for $C' \xrightarrow{\lambda} C$ we have

$$\sigma(x\lambda) = \sigma(x)\lambda$$

since $(x\lambda)t_1 = (x\lambda)t_2$ iff $x(\lambda t_1) = x(\lambda t_2)$. The maps σ_X, although canonical, are not natural when X is varied, that is

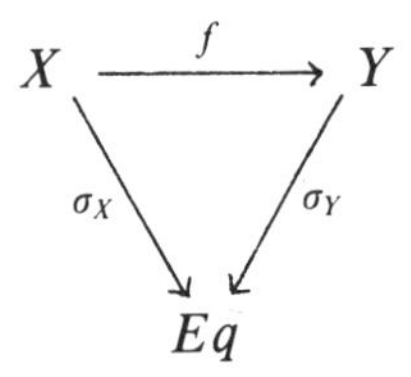

only commutes for "non-singular" f; of course

$$\sigma_X \subseteq \sigma_Y \circ f$$

for all f, so we could say that σ is natural in a suitable "2-categorical" sense. Some f might be "equisingular" in the sense that there exists an endomorphism $|f|$ of Eq so that a square commutes. For example with $\mathbf{C} = \mathbf{\Delta}_1$, $Eq_{\mathbf{C}}$ is the loop, and indeed in our generalization $\mathbf{M}(T)$ of $\mathbf{\Delta}_1$, Eq itself is very singular.

The reader may have also guessed that $\mathbf{\Delta}_1$ is the first of a sequence. The monoid $\mathbf{\Delta}_2$ of all order-preserving endomaps of the three-element ordered set $(0, 1, 2)$ has ten elements, three of which (the constants) are points z of the generic figure in that $z\lambda = z$ for all ten λ). The topos $S^{\mathbf{\Delta}_2^{op}}$ may be considered to consist of triangulated *surfaces,* wherein the triangles may be curled up or singular in other ways and are pasted together along edges or vertices in every conceivable way. A product $X \times Y$ is the 2-skeleton of the 4-dimensional object which one might imagine. The generic triangle has *nineteen* subobjects, so that the "nineteen values of superficial truth" Ω_2 form both a Heyting algebra and compatibly a triangulated surface in which only two of the singular triangles are degenerated all the way to points. If X is any such triangulated surface and $S \subset X$ any subsurface while x is any triangle of X, the statement "$x \in S$" thus has nineteen possible degrees of being false. After having built a model of the surface Ω_2 the reader might like to try $\Omega_3 \ldots$

III. I now will discuss three sequences of examples, assigning to each set T a topos $\mathcal{S}^{\mathbf{M}(T)^{op}}$ which will be gros for $T \geq 2$ and petit toposes $\mathcal{S}^{\mathbf{U}(T)^{op}}$, $\mathcal{S}^{\mathbf{P}(T)^{op}}$. Moreover $\mathcal{S}^{\mathbf{U}(T)^{op}}$ and $\mathcal{S}^{\mathbf{P}(T)^{op}}$ will turn out to be but two instances of a whole system $\mathcal{S}_T(B)$ of petit toposes attached to objects B of $\mathcal{S}^{\mathbf{M}(T)^{op}}$, namely those arising from the special choices $B = B_{\mathbf{U}}$ and $B = B_{\mathbf{P}}$. I will also point out some striking differences between even these simple $\mathcal{S}^{\mathbf{M}(T)^{op}}$ for $T = 1$ versus $T = 2$ versus $T > 2$ versus $T = \infty$. (It is surely relevant that many languages have *four* distinct "numbers": keine, eine, beide, viele.)

The category $\mathbf{U}(T)$ is simply the poset with $T+1$ elements, in which the added element is greater than all the elements of T and there are no other order relations. This poset indexes a basis of open sets in the topological space which has T isolated points and one "focal" point (whose only neighborhood is the whole space). The presheaf category $\mathcal{S}^{\mathbf{U}(T)^{op}}$ is then the category of *sheaves* on this space, for to specify a sheaf X is merely to specify a set of global sections $X(1)$, a set $X(U_t)$ of sections over each basic open U_t, and restriction maps $X(1) \longrightarrow X(U_t)$ for each $t \in T$. The sheaf condition in this simple case only applies to non-basic open sets

$$U_S = \bigcup_{t \in S} U_t \qquad \text{for } S \subseteq T$$

and then merely forces us to define

$$X(U_S) = \prod_{t \in S} X_t$$

since $U_{t1} \cap U_{t2} = 0$ for $t_1 \neq t_2$; this is the same answer we get if we consider that $U_S \hookrightarrow 1$ is a subobject of the terminal object 1 of $\mathcal{S}^{\mathbf{U}(T)^{op}}$ and define $X(U_S)$ to be the set of (natural, homogeneous) maps $U_S \longrightarrow X$ in $\mathcal{S}^{\mathbf{U}(T)^{op}}$. Everyone seems to agree that toposes constructed in this way from a poset are petit.

On the other hand, the category $\mathbf{P}(T)$, abstracting one simple idea of T parallel processes, is not a poset (unless $T \leq 1$): I define it to have only two objects U and $\mathbf{I}$ but T morphisms from U to $\mathbf{I}$ and no other non-identity morphisms. Thus, via the Yoneda-Grothendieck-Cayley-Dedekind embedding, the topos $\mathcal{S}^{\mathbf{P}(T)^{op}}$ has a full subcategory

$$U \rightrightarrows \mathbf{I}.$$
$$(T \text{ arrows})$$

Figures of type $U \longrightarrow X$ may be called the nodes of X and figures of type $\mathbf{I} \longrightarrow X$ T-gons, so that every T-gon in X has a t-th vertex node for each t in T (some of which may coincide); X is entirely specified by its set of nodes, its set of T-gons, and the t-th vertex relation to each t. For example, if $T = 2$, $\mathbf{P}(2) = \mathbf{P}$ previously alluded to and $\mathcal{S}^{\mathbf{P}(2)^{op}}$ consists of irreflexive graphs, wherein 2-gons are usually called edges and the two kinds of vertices are usually thought of as initial and terminal. Note that the *points* $1 \longrightarrow X$ in $\mathcal{S}^{\mathbf{P}(T)^{op}}$ are exactly the T-gons x whose t-th vertex $xt =$ the same node for all $t \in T$, i.e. generalized "loops".

Special interest attaches to the comma category at $\mathbf{I}$

$$\mathcal{S}^{\mathbf{P}(T)^{op}}/\mathbf{I} \approx \mathcal{S}^{\mathbf{P}(T)/\mathbf{I})^{op}}$$

since we can calculate that

$$\mathbf{P}(T)/\mathbf{I} \cong \mathbf{U}(T)$$

which is a poset ! The forgetful $\mathbf{P}(T)/\mathbf{I} \longrightarrow \mathbf{P}(T)$ is just the obvious

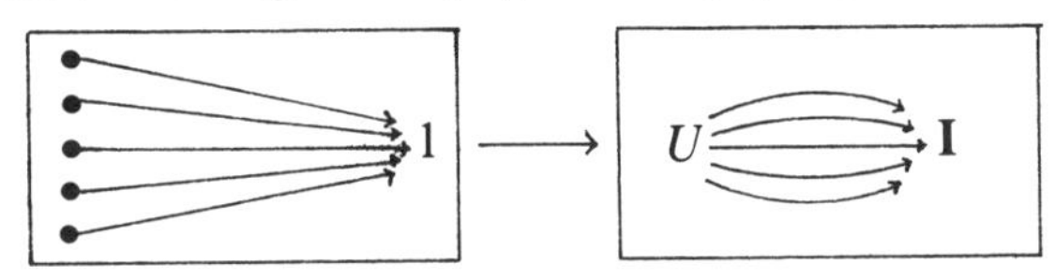

Now for any object $\mathbf{I}$ in any topos $\mathcal{E}$, there is a geometric morphism $\mathcal{E}/\mathbf{I} \xrightarrow{\pi} \mathcal{E}$, where the π correctly suggests both "projection" (from a total space to a base) and "product" (internal relative product = object of sections of any $E \longrightarrow \mathbf{I}$). π is just the right adjoint of the obvious functor which assigns

$$I^*F = F_{\mathbf{I}} \underset{\text{def}}{=} \begin{pmatrix} F \times \mathbf{I} \\ \downarrow \text{proj} \\ \mathbf{I} \end{pmatrix}$$

to any F in $\mathcal{E}$. Such a morphism $\mathcal{E}/I \longrightarrow \mathcal{E}$ is often considered to be a "local homeomorphism" since its inverse image functor $(\ )_{\mathbf{I}}$ preserves all the internal higher-order logic of $\mathcal{E}$:

$$(X \times Y)_{\mathbf{I}} = X_I \times Y_I = \text{product in } \mathcal{E}/\mathbf{I}$$
$$(Y^A)_{\mathbf{I}} = Y_{\mathbf{I}}^{A_{\mathbf{I}}} = \text{function space in the sense of } \mathcal{E}/\mathbf{I}$$
$$\Omega_{\mathbf{I}} = \text{the truth-value object of } \mathcal{E}/\mathbf{I} \text{ etc.}$$

If moreover the object $\mathbf{I}$ is a *covering* of the topos $\mathcal{E}$ in the sense that the unique map $\mathbf{I} \longrightarrow 1$ is an epimorphism of $\mathcal{E}$, then $\mathcal{E}/I \longrightarrow \mathcal{E}$ has the property of being "conservative", i.e. f is a monomorphism in $\mathcal{E}$ iff $f_{\mathbf{I}}$ is a monomorphism in $\mathcal{E}/\mathbf{I}$, f is a epimorphism in $\mathcal{E}$ iff $f_{\mathbf{I}}$ is an epimorphism in $\mathcal{E}/\mathbf{I}$, etc. Then one says that a topos $\mathcal{E}$ "locally" has a certain property if there exists a covering $\mathbf{I} \longrightarrow\!\!\triangleright\, 1$ of $\mathcal{E}$ such that $\mathcal{E}/\mathbf{I}$ actually has the property. It seems eminently reasonable that a topos which is locally petit should also be considered petit; at least, this seems to be implicit in Grothendieck's concept of *étendue* (which he now calls *topological multiplicity*): this refers to any topos $\mathcal{E}$ which is *locally* a subtopos of a presheaf topos $\mathcal{S}^{\mathbf{U}^{op}}$ where $\mathbf{U}$ is some *poset.* If $\mathbf{C}$ is any category in which every morphism is a monomorphism (for example any monoid satisfying the cancellation law $ax = ay \Rightarrow x = y$), then $\mathcal{S}^{\mathbf{C}^{op}}$ is an étendue, since it is locally presheaves on the poset $\mathbf{C}/\Sigma C$.

It is obvious that every morphism in $\mathbf{P}(T)$ is a monomorphism, since there are no non-trivial composites $ax = ay$ to check, and anyway we have already constructed the surjective local homeomorphism $\mathbf{U}(T) \longrightarrow \mathbf{P}(T)$ from a poset. Thus we are justified in considering that $\mathcal{S}^{\mathbf{P}(T)^{op}}$ is petit. On the other hand, it is not gros, for although the components functor $\mathcal{S}^{\mathbf{P}(T)^{op}} \longrightarrow \mathcal{S}$ takes Ω_T to 1 for $T \geq 2$, for the same T's it *fails* to preserve finite products.

The generic node U in $\mathcal{S}^{\mathbf{P}(T)^{op}}$ is not a covering, since in fact it is the unique non-trivial one among the three subobjects of 1 (thus Ω_T has three generalized "loops" $1 \longrightarrow \Omega_T$); $\mathcal{S}^{\mathbf{P}(T)^{op}}/U \cong \mathcal{S}$ since if $X \longrightarrow U$ then $(\mathbf{I}, X) = 0$ for $(\mathbf{I}, U) = 0$; thus X consists only of nodes $U \longrightarrow X$, each of which is a section of $X \longrightarrow U$; the map $X \longrightarrow U$ is unique if it exists since $U \hookrightarrow 1$; therefore $X = S \times U$ where $S = (U, X)$ is considered discrete. In particular the restriction of the object $\mathbf{I}$ to this open subtopos $\mathcal{S} \cong \mathcal{S}^{\mathbf{P}(T)^{op}}/U \hookrightarrow \mathcal{S}^{\mathbf{P}(T)^{op}}$ is the set T; in other words $U \times \mathbf{I} = T \times U$ with T considered discrete.

For the object $\mathbf{I}$ of $\mathcal{S}^{\mathbf{P}(T)^{op}}$, we have $\mathbf{I}^2 = \mathbf{I} + (T^2 - T) \times U$, where $T^2 - T$ is considered as a discrete object; for $(U, \mathbf{I}^2) = (U, \mathbf{I})^2 = T^2$ is the number of nodes of $\mathbf{I}^2$, whereas $(\mathbf{I}, \mathbf{I}^2) = (\mathbf{I}, \mathbf{I})^2 = 1^2 = 1$ is the number of T-gons of $\mathbf{I}^2$, and this unique element must be the *diagonal* $\mathbf{I} \longrightarrow \mathbf{I}^2$ which is nonsingular and hence have T distinct nodes, leaving $T^2 - T$ bare nodes in $\mathbf{I}^2$. Combining the above quadratic equation with the two equations $U^2 = U$, $U \times \mathbf{I} = T \times U$ we get a presentation of a small but significant part of the Burnside half-ring of the topos $\mathcal{S}^{\mathbf{P}(T)^{op}}$; significant because it generates the whole topos with the help of colimits, small because using only the half-ring operations, $\mathbf{I}$ and U don't generate any very complicated objects. This sub-half-ring has a very simple description after tensoring with the rationals $\mathbf{Q}$ or even with $\mathbf{Z}[1/T^2 - T]$, i.e. after adjoining negatives and sufficiently many denominators (here we assume $T > 1$ is finite, and treat $T^2 - T$ as the natural number which is its cardinality): For let x denote the element of this rationalized Burnside ring which corresponds to the generic figure $\mathbf{I}$, and similarly let u correspond to the generic node U; then the quadratic equation implies that the generic node can be expressed in terms of the generic figure

$$u = (x^2 - x)/(T^2 - T)$$

so that the subring in question is in fact generated (over the scalars $\mathbf{Z}[1/T^2 - T]$ by the single element x. However x satisfies the relations $xu = Tu$ where T is a whole number and also $u^2 = u$. Writing out the first of these we get

$$x^3 = (1+T)x^2 - Tx$$

and hence $x^4 = (1+T+T^2)x^2 - (1+T)Tx$. Then by multiplying out u^2 in terms of x we find that the second relation $u^2 = u$ actually follows from the first. On the other hand the cubic equation actually factors as

$$\boxed{x(x-1)(x-T) = 0}$$

Thus we have proved that

THEOREM. *The subring of the rationalized Burnside ring of* $\mathcal{S}^{\mathbf{P}(T)^{op}}$ *which is generated by the Yoneda embedding is actually the ring of functions on the three-point spectrum* $(0, 1, T) \subset \mathbf{Z}$.

This information helps to determine the points of the topos, $\mathcal{S} \longrightarrow \mathcal{S}^{\mathbf{P}(T)^{op}}$ whose inverse image functors are required to be left exact and cocontinuous, and hence correspond to *left* actions of the category $\mathbf{P}(T)$ which are "flat": namely the latter can only have the three possible cardinalities (for their set of "quantities" of type $\mathbf{I}$) $0, 1, T$. It would be interesting to find such an algebraic presentation of a portion including Ω of the (rationalized) Burnside ring of $\mathcal{S}^{\mathbf{P}(T)^{op}}$.

By $\mathbf{M}(T)$ I will mean the monoid which has $T+1$ elements, the added element being the identity element, with the multiplication law

$$\boxed{ts = t} \qquad t \in T, s \in \mathbf{M}(T).$$

Not only are all elements of $M(T)$ idempotent, but any pair of elements satisfies the following identity

$$\boxed{\alpha\beta\alpha = \alpha\beta}.$$

Such "graphic" monoids are relevant to the study of lists without repetitions and to "check-in" actions: if $\mathbf{M}$ is any graphic monoid and X is any right action of $\mathbf{M}$ on a set of states, then when α checks in, it may change the state, but if he tries later to check in a second time it is surely irrelevant. Not all graphic monoids $\mathbf{M}$ are of the simple form $\mathbf{M}(T)$, but there is a structure theorem for them saying that if we force commutativity $\mathbf{M} \xrightarrow{\sigma} \mathbf{S}$ we get a sup-semilattice with a support function

$$(\sigma(1) = 0, \sigma(\alpha\beta) = \sigma(\alpha) \cup \sigma(\beta))$$

with the property that if we consider $T_a = \{\alpha \mid \sigma(\alpha) = a\}$, the set of all elements with support equal to given $a \in \mathbf{S}$, then we get a homomorphic inclusion $\mathbf{M}(T_a) \longrightarrow \mathbf{M}$ for $a \neq 0$. Thus it is tempting to recast the structure theory in topos-theoretic terms by replacing the base topos $\mathcal{S}$ by another one constructed using the semilattice **S**, but I have not yet had time to work this out.

The idea to consider $\mathbf{M}(T)$ was reinforced by a remark of Professor Jorge Gracia in a discussion of the philosophical part/whole relation: the mere *idea* that T points cohere into a single whole could be considered totally abstractly as just another element adjoined to the abstract set T. Then much more complex wholes could be analyzed as being made of overlapping parts (possibly singular) of this simple kind. For example the idea of connecting two points ($T = 2$) may be expressed by the three element figure $\boxed{\bullet \longrightarrow \bullet}$. In a complex graph X, a given pair of points $2 \longrightarrow X$ (where $2 = 1 + 1$ is in itself discrete) may or may not be connectable

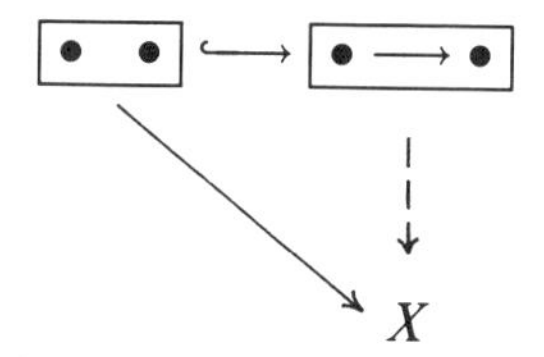

and if they are connectable it may be possible in many ways. So it is also for arbitrary T. Note that (because of the draconian multiplication table), $1 \in \overline{\mathbf{M}(T)}$, and indeed $\overline{\mathbf{M}(T)}$, the category obtained from the monoid by splitting idempotents, is equivalent to a category with only two objects, the terminal object and the one represented by the unique object of the monoid:

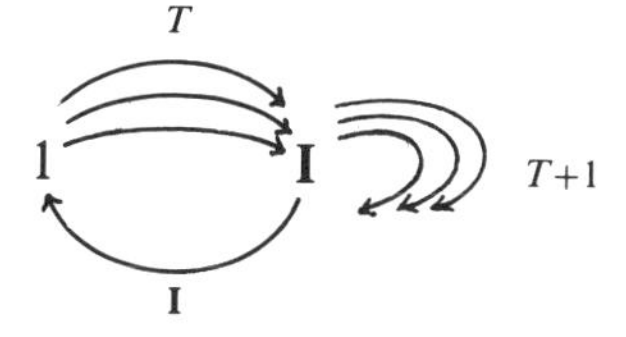

All endomaps of **I** except the identity $1_{\mathbf{I}}$ are constant; in other words, (for example $\mathbf{M}(2) = \mathbf{\Delta}_1$ as discussed in section II): The discrete object

$$T = \sum_T 1$$

is a subobject $T \subset \mathbf{I}$ in $\mathcal{S}^{\mathbf{M}(T)^{op}}$ which contains all points of **I**, but which is very different from **I**, for it lacks the breath of cohesiveness which **I** has and which makes all the difference although this restrictive method temporarily neglects further analysis of this unity. For any object X of $\mathcal{S}^{\mathbf{M}(T)^{op}}$ and any T-indexed family of points $1 \xrightarrow{x_t} X$ of X there is (by the universal mapping property of Σ) a single morphism $T \xrightarrow{x} X$ and hence a set (possibly empty) of (possibly singular) **I**-figures $\overline{x}$

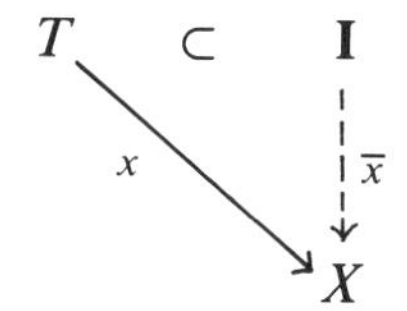

in X whose vertices agree $\overline{x}\cdot t = x_t$ for all t with the given points. The resulting myriads of incidence relations are preserved by any morphism $X \longrightarrow Y$.

By looking at the subobjects of $\mathbf{I}$ in $\mathcal{S}^{\mathbf{M}(T)^{op}}$ we see that there are precisely

$$2^T + 1$$

truth-values in the object Ω_T. The Heyting algebra thus resulting is also "simplest possible" in that the Boolean property

$$A = \neg\neg A$$

is violated for precisely one A, namely $A = T$, for $\neg\neg T = \mathbf{I} > T$, while it holds for $A = \mathbf{I}$ or for any $A \subseteq T$ with $A \neq T$.

For any T the components functor (left adjoint to the discrete inclusion) $\mathcal{S}^{\mathbf{M}(T)^{op}} \longrightarrow \mathcal{S}$ may be computed as the *reflexive* coequalizer

$$(T^2 \times \mathbf{I}, X) \rightrightarrows (1, X) \longrightarrow \pi_0(X)$$

and hence preserves finite cartesian products. For $T > 1$, there is a pair $1 \rightrightarrows \mathbf{I}$ of morphisms with empty intersection, thus (as was pointed out by Grothendieck), if we consider the map $\mathbf{I} \longrightarrow \Omega$ corresponding to one of them (which maps it to true), the others are mapped to false; this has the effect that all non-zero elements of Ω can be connected (by the action of $\mathbf{M}(T)$ on Ω, which is by inverse image) to zero and hence that $\pi_0(\Omega) = 1$. For any T, the points functor $\mathcal{S}^{\mathbf{M}(T)^{op}} \longrightarrow \mathcal{S}$ has the right adjoint sending any set S to S^T, providing a notion of chaotic object opposite to discrete. Thus for $T > 1$, $\mathcal{S}^{\mathbf{M}(T)^{op}}$ satisfies three of the distinctive properties of a "gros" topos.

For $T = 1$, $\mathbf{M}(T) = \{0, 1\}$ the multiplicative monoid with one idempotent $0 \neq 1$. We have that Ω has $2^1 + 1 = 3$ elements, but $\mathbf{M}(1)$ lacks sufficient action to connect T, $\mathbf{I}$ with 0, hence $\pi_0(\Omega) = 2$. Thus it seems that $\mathcal{S}^{\mathbf{M}(1)^{op}}$ is not gros. Probably it is not petit either; at least the $\mathbf{M}(1)/\mathbf{I}$ construction fails to produce a poset.

What possible use could the categories $\mathcal{S}^{\mathbf{M}(T)^{op}}$ for $T > 2$ have? As a theory of triangulated surfaces, $\mathcal{S}^{\mathbf{M}(3)^{op}}$ seems much poorer than the category $\mathcal{S}^{\Delta_2^{op}}$ discussed briefly in section II, since the six non-constant, non-identity operators in Δ_2 permit explicit calculation of the incidences of the *boundaries* of the triangles, not just of the vertices as here.

If T is a countably infinite set with a distinguished point ∞, then we can define a functor

$$\text{top} \longrightarrow \mathcal{S}^{\mathbf{M}(T)^{op}}$$

from the category of topological spaces and continuous maps, by using the distinguished point as follows: we topologize T by making every $t \neq \infty$ an open point and by making every *cofinite* subset containing ∞ into a neighborhood of ∞ resulting in a space denoted T_∞. Then for any space X, the set $\text{top}(T_\infty, X)$ of continuous maps $T_\infty \longrightarrow X$ is the set of *convergent sequences* in X, which has the obvious action of $\mathbf{M}(T)$ defined by $x \cdot t =$ the *constant* sequence which is constantly equal to the t-th term of the sequence x; in particular

$$\boxed{x_\infty = \lim_{t \neq \infty} x_t}.$$

Clearly

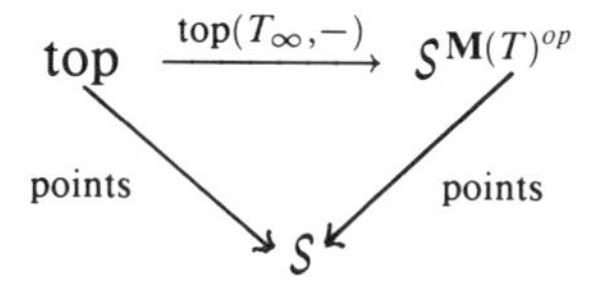

commutes, so that in particular our functor is faithful and for an object of $\mathcal{S}^{\mathbf{M}(T)^{op}}$ coming from top, the above limit condition is necessary and sufficient for a morphism x from the discrete T to extend to a (unique) cohesive figure

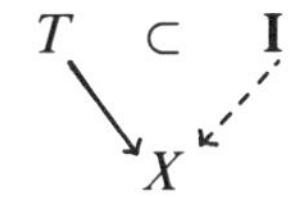

That is, cohesive means convergent for such objects. If we restrict our functor to the subcategory top_ω of sequentially determined spaces (which includes all metrizable spaces) our functor (which in any case preserves products and equalizers) becomes *full*; this is due to the fact that preservation of convergent sequences is sufficient to determine continuity in top_ω, whereas the naturality/homogeneity of the maps in $\mathcal{S}^{\mathbf{M}(T)^{op}}$ means precisely that convergent sequences are mapped to convergent sequence, preserving the evaluation at *each* t. The inclusion $\text{top}_\omega \hookrightarrow \mathcal{S}^{\mathbf{M}(T)^{op}}$ has a left adjoint, and preserves the function space construction (which exists in top_ω too). Thus we are justified in considering top_ω as a full subcategory of $\mathcal{S}^{\mathbf{M}(T)^{op}}$ (of course for a different choice of $\infty \in T$ we would get a different subcategory). Note that $\mathbf{I}$ is *not* in the subcategory; the space T_∞ determines a much bigger object

$$T \hookrightarrow \mathbf{I} \hookrightarrow T_\infty$$

(all with the same points) but of course for each X in $\mathcal{S}^{\mathbf{M}(T)^{op}}$ which comes from a space, every $\mathbf{I} \longrightarrow X$ uniquely extends to $T_\infty \longrightarrow X$. The Sierpinski space 2 (two points, three open sets) determines a Heyting algebra object $\mathbf{O} = \text{top}(T_\infty, 2)$ of $\mathcal{S}^{\mathbf{M}(T)^{op}}$ with a morphism $\mathbf{O} \longrightarrow \Omega$ given by restricting from T_∞ to $\mathbf{I}$; this suggests considering $\mathbf{O}^X$ as the "object of opens" for any object X of $\mathcal{S}^{\mathbf{M}(T)^{op}}$. However, we are not completely free to use $\mathcal{S}^{\mathbf{M}(T)^{op}}$ (with its super-simple definition) as a replacement (improvement in that it is a topos) for top_ω, because topological sums are not preserved by the embedding: there is the comparison map

$$\text{top}(T_\infty, X_1) + \text{top}(T_\infty, X_2) \longrightarrow \text{top}(T_\infty, X_1 + X_2)$$

(which is bijective on points) in the topos, but the "convergent sequences" in the presheaf-sum are either completely in one summand or completely in the other, whereas in the top_ω-sum on the right a convergent sequence may bounce back and forth any finite number of times before finally settling into either X_1 or X_2 where its limit is.

The fact that cohesiveness in $\mathcal{S}^{\mathbf{M}(T)^{op}}$ (for T countably infinite) is not necessarily of the "limit" kind is illustrated by a completely different cartesian-closed full reflective subcategory bor_ω obtained as follows. Bornological sets are usually defined to consist of a family of "bounded" subsets of a given set of points, subject to the axioms that all singletons are bounded, any subset of a bounded set is bounded, and the union of any finite collection of bounded sets is bounded. A morphism from one bornological set to another is any mapping of points for which the image of any bounded set in the domain is bounded in the codomain. Thus we get a (cartesian closed) category bor. A bornological set is discrete iff only the finite subsets are bounded, the opposite of "chaotic" in which all subsets are bounded; the structure of a bornological set is determined by knowing the chaotic figures in it. This seems to be a very loose structure, but in combination with algebraic structure it is quite important in functional analysis, where its covariant nature is much easier to deal with than the contrary "open set" determination: for example, the category ab(bor) of bornological abelian groups contains the usual categories of Frechet nuclear spaces, Banach spaces, etc. with continuous linear maps as *full* subcategories. Let T_{bor} be the codiscrete bornological set with T points. Then

$$\text{bor} \xrightarrow{\text{bor}(T_{\text{bor}},-)} \mathcal{S}^{\mathbf{M}(T)^{op}}$$

again preserves the natural point functors on both categories and has a left adjoint. If bor_ω is defined as the subcategory of those bornological sets for which any subset is bounded provided every countable subset of it is bounded, then the restriction of our functor to bor_ω is again full and has many good properties, but again fails to preserve sums. We get a disparate big enlargement of $\mathbf{I}$ with the same points

$$T \subset \mathbf{I} \begin{array}{l} \subset T_{\text{bor}} \\ \subset T_\infty \end{array}$$

in $\mathcal{S}^{\mathbf{M}(T)^{op}}$ such that $(T_{\text{bor}}, X) \xrightarrow{\sim} (\mathbf{I}, X)$ is a bijection for bornological objects X. This is the largest extension $\mathbf{I} \subset E$ of $\mathbf{I}$ which has the same points $T = (1, \mathbf{I}) \xrightarrow{\sim} (1, E)$ *and* which is separated by 1 in the sense that for any $\mathbf{I} \underset{g}{\overset{f}{\rightrightarrows}} E$, $f = g$ provided the induced mappings $(1, \mathbf{I}) \rightrightarrows (1, E)$ are equal, i.e. $\forall t \in T[ft = gt] \Rightarrow f = g$, i.e. $E \in \mathcal{Y}_{\mathbf{M}(T)}$.

It is even possible to get embeddings of topological categories into $\mathcal{S}^{\mathbf{M}(T)^{op}}$ which do preserve sums, provided we take T to be the power of the continuum and imagine it topologized as an interval or as a circle, since either of the latter is a *connected* space. For then

$$\text{top}(T_{\text{cts}}, \sum_i X_i) \xleftarrow{\sim} \sum_i \text{top}(T_{\text{cts}}, X_i)$$

in $\mathcal{S}^{\mathbf{M}(T)^{op}}$. Thus a cohesive figure is now interpreted to mean a continuous path, or a continuous loop, etc., depending on the meaning fixed for T_{cts}; the embedding will be *full* when restricted to locally arcwise-connected spaces. Such is the unexpected power of the action of constant maps.

Returning to the finite, let us consider briefly some of the remarkable properties of the apparently simple topos $\mathcal{S}^{\mathbf{M}(1)^{op}} = \mathcal{S}^{\{0,1\}^{op}}$ consisting of sets operated on by a single idempotent. Even though it is not gros, it is connected to any of our $\mathcal{S}^{\mathbf{M}(T)^{op}}$ through the homomorphism $\mathbf{M}(T) \longrightarrow \{0,1\}$ which collapses all T to 0, as well as by its T different sections $\{0,1\} \hookrightarrow \mathbf{M}(T)$. Restriction along the latter ones, and the left and right adjoints to restriction along the former one thus yield $T+2$ functors $\mathcal{S}^{\mathbf{M}(T)^{op}} \longrightarrow \mathcal{S}^{\{0,1\}^{op}}$ which may provide useful invariants for classifying the objects of $\mathcal{S}^{\mathbf{M}(T)^{op}}$, especially since most of these functors preserve both sums and products; especially for graphs ($T = 2$) one is always thirsty for systematic information. These invariants may be construed as quantitative in the Galileo-Cantor-Burnside-Grothendieck spirit as follows: recall that every X in $\mathcal{S}^{\{0,1\}^{op}}$ is uniquely expressible as a sum of connected components and that a connected object is just a pointed set. If we let B_n denote a standard pointed set with n elements, and $\hat{X}(n)$ denote the abstract set of components of type B_n, then we can write

$$X = \sum_n \hat{X}(n) \times B_n$$

(Incidentally $\sum_n \hat{X}(n)$ is the set of *points* of X, since B_n has a unique point). Now, since

$$B_n \times B_m \cong B_{n\times m}$$

we can compute cartesian products in terms of the above expansion as follows, using the fact that any topos is a distributive category:

$$\begin{aligned} X \times Y &= \Big(\sum_n \hat{X}(n) \times B_n\Big) \times \Big(\sum_m \hat{Y}(m) \times B_m\Big) \\ &= \sum_n \sum_m \hat{X}(n) \times \hat{Y}(m) \times (B_n \times B_m) \\ &= \sum_q \Big[\sum_{n\times m = q} \hat{X}(n) \times \hat{Y}(m)\Big] \times B_q \end{aligned}$$

This is just the rule for multiplying *formal Dirichlet series*, which are usually written with B_n replaced by $1/(n)^s$ where $1/(\ )^s$ is a formal character of the situation but for which one sometimes succeeds in evaluating $X(s)$ for some complex numbers s provided the coefficients $\hat{X}(n)$ comply. Now when the sets $\hat{X}(n)$ are finite, and non-zero only for finite n, we can via cardinality interpret the coefficients as whole numbers. This gives an explicit determination of the *Burnside ring* $\mathcal{R}(1)$ of the topos $\mathcal{S}^{\mathbf{M}(1)^{op}}$, which is defined to consist of all isomorphism types of objects satisfying the stated finiteness conditions, added and multiplied by categorical coproduct and product, with differences formally adjoined. (One of the main reasons for the finiteness conditions is to ensure that these "virtual" differences do not by their introduction produce collapse.) The Burnside rings were originally found useful in the case $\mathcal{S}^G$ of the topos of permutation representations of a group G. The $T+1$ ring homomorphisms and one linear operator

$$\mathcal{R}(T) \substack{\longrightarrow\\ \longrightarrow\\ \longrightarrow} \mathcal{R}(1)$$

thus assign to each $X \in \mathcal{S}^{\mathbf{M}(T)^{op}}$ (for example each graph when $T = 2$) a system of "ζ-functions" in the ring of formal Dirichlet series with $\mathbf{Z}$ coefficients. The classical Riemann ζ-function corresponds to the unique object of $\mathcal{S}^{\{0,1\}^{op}}$ which has exactly one component B_n of size n for every finite n.

To study a little more closely the differences between the toposes $\mathcal{S}^{\mathbf{M}(T)^{op}}$ for various T, consider the problem of computing the function space $\mathbf{I}^{\mathbf{I}}$, where $\mathbf{I}$ is the regular right representation. In general there is $1 \longrightarrow \mathbf{I}^{\mathbf{I}}$, the name of the identity map $\mathbf{I} \longrightarrow \mathbf{I}$, and also a map $\mathbf{I} \longrightarrow \mathbf{I}^{\mathbf{I}}$, the inclusion of constants coming as a special case of the $Y \longrightarrow Y^{\mathbf{I}}$ induced by the unique $\mathbf{I} \longrightarrow 1$. Putting these two together we have the map

$$1 + \mathbf{I} \longrightarrow \mathbf{I}^{\mathbf{I}},$$

as for any object in any topos. This map is sometimes an isomorphism, but certainly not in $\mathcal{S}^{\mathbf{M}(2)^{op}} = \mathcal{S}^{\Delta_1^{op}}$, where from our computation in section II, we see that $1 + \mathbf{I} \hookrightarrow \mathbf{I}^{\mathbf{I}}$ is the map whose internal picture is

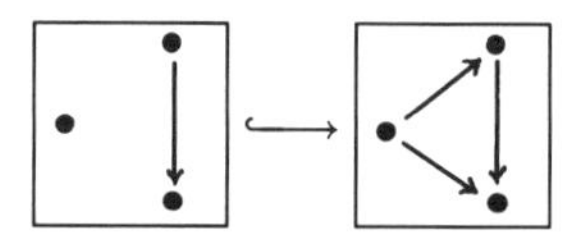

(The reader may wish as an exercise to compute the characteristic map $\mathbf{I}^{\mathbf{I}} \longrightarrow \Omega$ to the five-edge truth-value object for this inclusion.) On the other hand, for $T > 2$, the map $1 + \mathbf{I} \longrightarrow \mathbf{I}^{\mathbf{I}}$ *is* an isomorphism in $\mathcal{S}^{\mathbf{M}(T)^{op}}$; that is, all maps $\mathbf{I} \times \mathbf{I} \xrightarrow{f} \mathbf{I}$, (in other words all functions of two variables on the monoid to itself satisfying $f(ut, vt) = f(u,v)t$) are of a very simple form, as Steve Schanuel showed me.

Now $1 + x$ is a very simple polynomial, and we have said that for $T > 2$ the exponential $\mathbf{I}^{\mathbf{I}}$ is in fact that polynomial applied to $\mathbf{I}$. On the other hand in the category of graphs with $T = 2$, we have

$\mathbf{I}^{\mathbf{I}} =$

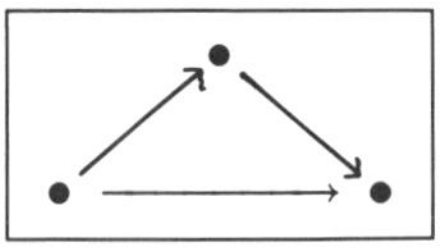

which is manifestly neither a sum nor a product and hence not any polynomial function of $\mathbf{I}$.

THEOREM. *For any $T \neq 2$, there is a polynomial Φ_T with* $\mathbf{N}$ *coefficients such that for the representable object* $\mathbf{I} \in \mathcal{S}^{\mathbf{M}(T)^{op}}$ *one has*

$$\mathbf{I}^{\mathbf{I}} = \Phi_T(\mathbf{I})$$

In fact $\Phi_T(x) = 1 + x$ *for all* $T > 2$, *whereas* $\Phi_0 = 1$ *and*

$$\Phi_1(x) = 2x^2.$$

To understand this further surprise from $\mathcal{S}^{\{0,1\}^{op}}$, note that for any connected B_n, $B_n^{\mathbf{I}}$ has (like any object) as many components as points, but its points correspond to morphisms $\mathbf{I} \longrightarrow B_n$, which are n in number. The number of general elements $\mathbf{I} \longrightarrow B_n^{\mathbf{I}}$ in $B_n^{\mathbf{I}}$ is the same as the number of morphisms $\mathbf{I}^2 \longrightarrow B_n$, so we need to consider $\mathbf{I}^2$; it has just one point, but $(\mathbf{I}, \mathbf{I}^2) = (\mathbf{I}, \mathbf{I})^2 = 2^2 = 4$ elements, so $\mathbf{I}^2 = B_4$. A map $B_4 \longrightarrow ?$ must map the point to a point but can map the other three elements arbitrarily. Hence $B_n^{\mathbf{I}}$ has n^3 elements, $n^3 - n$ of which are non-degenerate. If we could determine the apportionment of these $n^3 - n$ elements among the n components, we could essentially compute $Y^{\mathbf{I}}$ for any Y, since $\mathbf{I}$ is so strongly connected that

$$Y^{\mathbf{I}} = \Big(\sum_n \hat{Y}(n) \times B_n\Big)^{\mathbf{I}} = \sum_n \hat{Y}(n) \times B_n^{\mathbf{I}}$$

This apportionment for n a power of 2 follows from the case $B_2 = \mathbf{I}$. To determine how the $2^3 - 2 = 6$ nondegenerate elements of $\mathbf{I}^{\mathbf{I}}$ are apportioned among the two components, recall that the right action of T on these elements in their $\mathbf{I}^2 \xrightarrow{f} \mathbf{I}$ guise is by multiplying on the left only in the first variable

$$(ft)(s, x) = f(ts, x).$$

Hence for $T = \{0\}$ we get

$$\begin{aligned} (f \cdot 0)(1,1) &= f(0,1) \\ (f \cdot 0)(1,0) &= f(0,0) \\ (f \cdot 0)(0,1) &= f(0,1) \\ (f \cdot 0)(0,0) &= f(0,0) \end{aligned}$$

Thus, for any $\mathbf{I}^2 \xrightarrow{f} \mathbf{I}$, $f \cdot 0$ factors across the second projection $\mathbf{I}^2 \longrightarrow \mathbf{I}$, so is a *point* of $\mathbf{I}^{\mathbf{I}}$. The second projection itself corresponds to the point 1 of $\mathbf{I}^{\mathbf{I}}$, whereas the constantly 0 map $\mathbf{I}^2 \longrightarrow \mathbf{I}$ corresponds to the point 0 of $\mathbf{I}^{\mathbf{I}}$. Further consideration of cases shows that among the six f which do not factor across the second projection, three of them have $f \cdot 0 = 1$ and three of them have $f \cdot 0 = 0$. Hence the two components are of equal size $B_4 = \mathbf{I}^2$. In other words $\mathbf{I}^{\mathbf{I}} = 2\mathbf{I}^2$, as was to be shown.

[The exceptional behavior of $\mathcal{S}^{\mathbf{M}(T)^{op}}$ in the special case $T = 1$ may perhaps be explained by noting that since the monoid $\{0, 1\}$ is commutative, we could equally well consider it to be one of the sequence $\mathcal{S}^{\mathbf{M}(T)}$ of toposes of *left* actions, i.e. of "cylinders" which have T preferred sections instead of just two. All of these toposes (for $T \neq 0$) have *three* truth values Ω with two components, and all of them in fact have the components functor representable. The representing object for π_0 is T, the (*connected* this time) subobject of the generic object, which itself has no non-empty subobjects: $\pi_0 A = \mathcal{S}^{\mathbf{M}(T)}(T, A)$

for all objects A, so in particular T has only the identity endomorphism (as opposed to T^T in the case of right actions). The connected objects of $\mathcal{S}^{\mathbf{M}(T)}$ are parameterized by the category $T/\mathcal{S}$ of T-pointed sets, and are hence determined by the cardinality (of the set of figures, not "points") together with an equivalence relation on T. The Burnside ring is thus the tensor product of the formal Dirichlet series with the formal monoid ring of the monoid of all equivalence relations on T with intersection of equivalence relations as multiplication. The object A^A for A the generic object can be computed by a formula generalizing that for the case $T = 1$.]

The Theorem has the consequence that (for $T \neq 0$), the object $\mathbf{I}$ is *contractible* in $\mathcal{S}^{M(T)^{op}}$ iff $T = 2$.

Now it remains to discuss the relation between the gros toposes $\mathcal{S}^{\mathbf{M}(T)^{op}}$ (for $T \geq 2$) and the two sequences $\mathcal{S}^{\mathbf{U}(T)^{op}} \longrightarrow \mathcal{S}^{\mathbf{P}(T)^{op}}$ of petit toposes (actually étendue in these cases). I will do this by attaching to each object of the gros topos an associated petit topos by a reduction of the associated "comma category" or discrete fibration. That is, for suitable $\mathbf{C}$, (which if "big" enough will give a gros $\mathcal{S}^{\mathbf{C}^{op}}$) we will show that

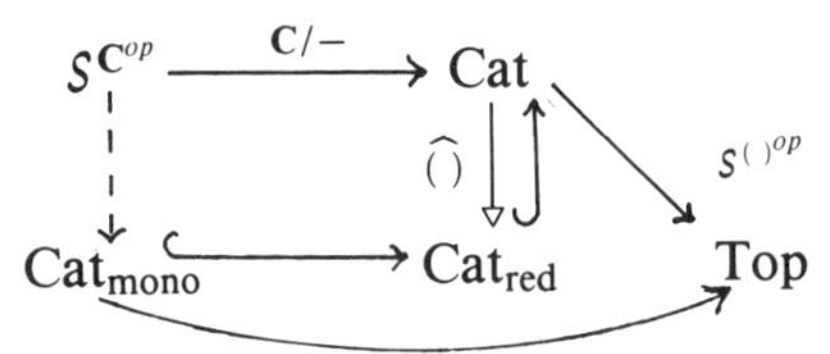

in other words that the reduction $\widehat{\mathbf{C}/B}$ of $\mathbf{C}/B$ for any $B \in \mathcal{S}^{\mathbf{C}^{op}}$ actually is a category whose topos of right actions is guaranteed to be petit, since $\widehat{\mathbf{C}/B}$ itself consists entirely of monomorphisms. For every $B_1 \xrightarrow{f} B_2$ in $\mathcal{S}^{\mathbf{C}^{op}}$, the resulting map of petit toposes is automatically "essential", that is we will always have the left Kan quantifier $f_!$ as well as the inverse image f^* and the right Kan quantifier f_*.

The meaning of "reduced" categories which will turn out to suffice for our limited purpose is

$$e \circlearrowleft C \ \& \ e^2 = e \Rightarrow e = 1$$

in other words there are no idempotents (except identities). Since most of our examples $\mathbf{C}$ consist "mainly" of idempotents and become $\mathbf{C}/B$ with the same feature, reducing $\mathbf{C}/B$ to $\widehat{\mathbf{C}/B}$ by collapsing the idempotents records principally the "change" in the "becoming".

Like any class of categories closed under arbitrary product and arbitrary subcategories, the inclusion $\mathrm{Cat}_{\mathrm{red}} \hookrightarrow \mathrm{Cat}$ has a left adjoint with surjective adjunction functors

$$\mathbf{C} \longrightarrow \widehat{\mathbf{C}}$$

which in this case are bijective on objects so determined by a categorical congruence relation, i.e. an equivalence relation on each set $\mathbf{C}(C', C)$, stable under composition. In our case the equivalence relation starts off being generated by all pairs $\langle e_1, e_2 \rangle \in \mathbf{C}(C, C)$ for which both e_1, e_2 are idempotents.

This indicates the meaning of my reduction process ($\hat{}$). It may be noted that $\widehat{\mathbf{C}}$ may also be described as the result of formally inverting all split epimorphisms (i.e. all p for which there exists a section s, $ps = id$) in case all idempotents split in $\mathbf{C}$. (In any case, this particular "fraction" construction, unlike most gives a *surjective* functor $\mathbf{C} \longrightarrow \widehat{\mathbf{C}}$). The reduction preserves finite products of categories.

Of course very strong conditions on $\mathbf{C}$ are required in order to conclude that all maps become monomorphisms in $\widehat{\mathbf{C}}$. Since I don't know a very explicit description in general of the congruence relation involved in $\mathbf{C} \longrightarrow \widehat{\mathbf{C}}$, I will need very strong conditions indeed. It would be very convenient (also in other applications such as simplicial sets) if one knew what conditions (if any) must be added to

> "Every map in $\mathbf{C}$ can be factored as a split epimorphism followed by a monomorphism"

in order to conclude that every map in $\widehat{\mathbf{C}}$ is a monomorphism; this comes down to whether $\mathbf{C} \longrightarrow \widehat{\mathbf{C}}$ *preserves* monomorphisms, since certainly any map in $\widehat{\mathbf{C}}$ is isomorphic to one of the form $[i]$, the congruence class of a map i which is a monomorphism in $\mathbf{C}$.

The condition in quotation marks above was carefully chosen to be stable under the passage from $\mathbf{C}$ to $\mathbf{C}/B$ where B is any object of $S^{\mathbf{C}^{op}}$, or equivalently to be stable under the passage from $\mathbf{C}$ to a category $\mathbf{E} \longrightarrow \mathbf{C}$ discretely fibered over $\mathbf{C}$; this stability would *not* hold if for example we had required the monomorphism to be split (i.e. admit retractions). Such stability smoothes our work, since anything proved for all $\mathbf{C}$ in a stable class can then be applied to all $\mathbf{C}/B$ for some $\mathbf{C}$ of particular interest. Since here we are interested in proving something about $\widehat{\mathbf{C}/B}$, this means that we can concentrate on studying the reduction process without at the same time combining that with a study of the particular "comma category" construction.

The stable class on which I will concentrate is the restricted one of "nodal" categories. By this I will mean any category $\mathbf{C}$ in which there exists some class $\mathcal{P}$ of objects such that

1. $C \underset{q}{\overset{p}{\rightrightarrows}} P \;\&\; P \in \mathcal{P} \Rightarrow p = q$;
2. Every map with codomain in $\mathcal{P}$ has a section;
3. Every map which does not have a section factors through $\mathcal{P}$.

It is then clear that every map in a nodal category can be factored as a split epimorphism followed by a monomorphism, since indeed *any* map with domain in $\mathcal{P}$ is (almost vacuously) a monomorphism. But this has been achieved in a very extreme way, since in fact

PROPOSITION. *In a category* $\mathbf{C}$ *nodal with respect to* $\mathcal{P}$, *every map is either an isomorphism or a* $\mathcal{P}$*-constant (in the sense of belonging to the bi-ideal of maps which factor through* $\mathcal{P}$*).*

For if p is not $\mathcal{P}$-constant, it has a section s; if s itself has a section, then that section is p and p is an isomorphism; alternatively s is $\mathcal{P}$-constant and so $s = iq$ where q has a section t because the codomain of q is in $\mathcal{P}$, but on the other hand q is a monomorphism since s is, so $tq = 1$. This shows that in the latter alternative the codomain of p is isomorphic to a $\mathcal{P}$-object, hence p itself is $\mathcal{P}$-constant.

Thus for a nodal category there is a third description of $\mathbf{C} \longrightarrow \widehat{\mathbf{C}}$: invert all maps which are both $\mathcal{P}$-constant and split epimorphisms. This smaller class of denominators may be thought of as "$\mathcal{P}$-degeneracies". Note that all $\mathcal{P}$-constant maps f really are constant in the sense that $fx_1 = fx_2$ whenever defined. Note also that if $C \longrightarrow P_1$ and $C \longrightarrow P_2$ are two degeneracies of the same C, then $P_1 \approx P_2$, that is that to a "degenerate object" there corresponds an essentially unique "$\mathcal{P}$-point".

For example the idempotent closure of the monoid of affine linear functions $ax + b : \mathbf{R} \longrightarrow \mathbf{R}$ is a nodal category.

A much smaller class which includes most of our examples are those nodal categories in which every *non-identity* map is $\mathcal{P}$-constant for a suitable $\mathcal{P}$. This remark should reinforce the reader's suspicion that much of the explicit part of this article may deserve the title "the theory of constant maps". However, the *actions* of constant maps apparently are not quite so trivial as might appear at first, somewhat as in continuum mechanics the "freezes" in configuration space often accompany serious activity in the state space. The category $\overline{\mathbf{\Delta}}_1 = \overline{\mathbf{M}(2)}$, involved in reflexive directed graphs belongs to this subclass of the nodal categories, but $\overline{\mathbf{F}}$, where $\mathbf{F}$ is the monoid of endomaps of a two-element set, is nodal though it has non-identity isomorphisms.

PROPOSITION. *If* $\mathbf{C}$ *is nodal and* $B \in \mathcal{S}^{\mathbf{C}^{op}}$, *then* $\mathbf{C}/B$ *is also nodal.*

PROOF. Let $\mathcal{P}_B$ be the class of all objects of $\mathbf{C}/B$ whose image under the forgetful "domain" functor $\mathbf{C}/B \longrightarrow \mathbf{C}$ is in $\mathcal{P}$, where $\mathcal{P}$ is chosen so that $\mathbf{C}$ is $\mathcal{P}$-nodal. Then $\mathcal{P}_B$ surely has the property 1) of nodality. (Note that many more objects of $\mathbf{C}/B$, namely those which are non-singular figures of B, will also have property 1); the letter $\mathcal{P}$ was chosen because in case $\mathbf{C} = \overline{\mathbf{M}(T)}$ with $\mathcal{P} = (1)$, $\mathcal{P}_B$ is the subcategory of *points* of B). As for property 2), it follows from the much more general fact that for any $\mathbf{C}$, if a map $\mathbf{C}/B$ lies over a split epimorphism in $\mathbf{C}$, then it is split in $\mathbf{C}/B$ by the "same" map. Property 3) is equally easy to lift: a map in $\mathbf{C}/B$ which is not an isomorphism factors in $\mathbf{C}$ through $\mathcal{P}$, but this immediately constructs an object in $\mathcal{P}_B$ through which both portions of the factorization continue to live in $\mathbf{C}/B$.

Thus to achieve our present goal it suffices to prove

THEOREM. *If* $\mathbf{C}$ *is a nodal category then every morphism in* $\widehat{\mathbf{C}}$ *is a monomorphism.*

But in fact that is true for a vacuous reason: Every non-isomorphism in $\widehat{\mathbf{C}}$ has domain P in $\mathcal{P}$, and in fact also has codomain *not* in $\mathcal{P}$, since any map between objects of $\mathcal{P}$ is already an isomorphism in $\mathbf{C}$. But any two maps $C \rightrightarrows P$ were already equal in $\mathbf{C}$, so certainly in $\widehat{\mathbf{C}}$, so any map with domain P is a monomorphism, even in $\widehat{\mathbf{C}}$.

Now in applying the above to our two object example $\overline{\mathbf{M}(T)}$, note that the objects of the category $\overline{\mathbf{M}(T)/B}$ are of two kinds, points $\mathcal{P}_B$ and "stars", and that for any star there are T maps to it from points expressing the vertex relations; but there are also, for the *degenerate* stars maps with label I degenerating them to actual vertices, with the resulting T idempotents from each degenerate star to itself, and it is these latter which collapse to identities in $(\overline{\mathbf{M}(T)}/B)^\wedge$.

Since the degenerate stars have been now identified with their associated points and hence may be omitted, up to equivalence the latter category may thus be pictured in two levels with all maps going down and in particular no non-identity maps to any *point*

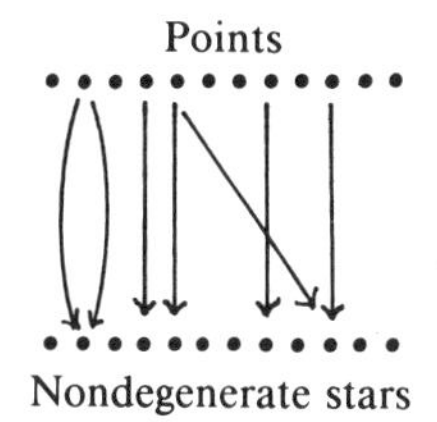

This category will itself be a poset if every star is non-singular, and in any case is "locally" a poset as required.

Now for given T, let $B_{\mathbf{U}}$ be the object I of $\mathcal{S}^{\mathbf{M}(T)^{op}}$. Then there is just one star, and it is non-singular, so

$$\overline{(\mathbf{M}(T)}/B_{\mathbf{U}})^\wedge = \mathbf{U}(T)$$

On the other hand, let $B_{\mathbf{P}}$ be the right $\mathbf{M}(T)$-set which has just one star and also just one point, so that every $t \in T$ must act the same; in other words the star is completely singular but not degenerate (so $B_{\mathbf{P}}$ = the generic loop in the case $T = 2$ of reflexive graphs). Then

$$\overline{(\mathbf{M}(T)}/B_{\mathbf{P}})^\wedge = \mathbf{P}(T)$$

the "parallel process". In the case $T = 2$, this means that irreflexive graphs may be identified with reflexive graphs equipped with a map with discrete fiber to the loop.

Thus we have shown that for each T, each of the petit toposes $\mathcal{S}^{\mathbf{U}(T)^{op}}$, $\mathcal{S}^{\mathbf{P}(T)^{op}}$ are but two examples, corresponding to two particular objects B of the gros topos $\mathcal{S}^{\mathbf{M}(T)^{op}}$, of a family of petit toposes associated to all the objects B. Each such petit topos is actually equivalent to a subcategory

$$\mathcal{S}^{\mathbf{M}(T)^{op}}/B \overset{\longrightarrow}{\underset{--\rightarrow}{\longleftarrow}} \mathcal{S}(B)$$

which is both reflective and coreflective, with the reflection being just the left Kan extension of $\overline{\mathbf{M}(T)}/B \longrightarrow \overline{(\mathbf{M}(T)}/B)^\wedge$. Which subcategory is it? In general if $\mathbf{C} \longrightarrow \mathbf{D}$ is surjective or a fraction construction, then $\mathcal{S}^{\mathbf{C}^{op}} \hookleftarrow \mathcal{S}^{\mathbf{D}^{op}}$ is the subcategory in which **D**-identified actors in **C** act the same or **D**-inverted actors in **C** act bijectively. Since in our example $\mathcal{P} = \{1\}$ and $\mathcal{P}_B$ is the inverse image in $\overline{\mathbf{M}(T)}/B$, the condition that $X \longrightarrow B$, an object of $\mathcal{S}^{\mathbf{M}(T)^{op}}/B$ belong to $\mathcal{S}(B)$ is just that it be "orthogonal" to $\mathbf{I} \longrightarrow 1$.

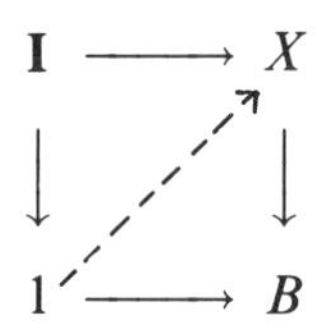

in other words, that the fibers be discrete.

We could invert still more maps, for example those labeled by a fixed subset of T (as we did in the case $T = 2$ to obtain the free categories $\mathcal{F}$) to obtain a finer notion of the petit topos associated to the objects B. Further, there is sometimes a Grothendieck topology functorially associated to the objects B, giving a system of petit *sheaf* toposes. For example if our gros determiner is $\mathbf{C} = \mathbf{M}(2)/C$ where C is the loop, and hence the B's are the simplest kind of *labeled* graph, we can define a notion of covering in $\mathcal{F}(B)$ by declaring that all maps of *positive* label are coverings. (Here we picture the basic label values as $C =$ [loop diagram: 0, +].) The resulting notion of petit topos assigns to each C-labeled graph B a generalized Jónsson-Tarski topos $\mathrm{sh}(B)$, the original example of the latter arising from $B =$ [two-loop diagram], two loops with the non-degenerate labeling. Every such $\mathrm{sh}(B)$ is an étendue locally homeomorphic to a generalized Cantor space $\mathcal{X}(B)$.

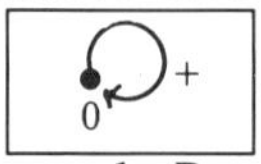

Finally, why should one be interested in petit toposes when we may as well study $X \longrightarrow B$ without the above restriction, and gros toposes seem to do better at englobing the study of geometrical objects? One reason is that we often want to consider actions as processes. Once a process f is *specified,* its parts must have been specified too. Thus if $X_0 \xrightarrow{f} X_1$ can be factored into stages

$$X_0 \xrightarrow{f_0} \bullet \underset{w}{\overset{u}{\rightrightarrows}} \bullet \xrightarrow{f_1} X_1$$

so that $uf_0 = wf_0$ and $f_1u = f_1w$, then $u = w$; this seems to be inherent in the notion of a *given* process. It will be guaranteed if all maps in the category of states and processes are monomorphisms, as Grassmann 1844 §8 pp. 40–41 seems to imply. On the other hand, the (self-dual) more general condition just stated *also* defines an epireflective subcategory of Cat; a topos defined by a site satisfying it is locally definable by a poset site which lives (not in $\mathcal{S}$ but) in the atomic Boolean topos $\mathcal{S}'$ of "combinatorial functors" on the category of finite sets and monomorphisms studied by Myhill and Schanuel. Because of the latter result (proved by Johnstone in his work on "QD" toposes), the Myhill-Schanuel topos $\mathcal{S}'$ seems destined to play a role in such attempts to further broaden the concept of "petit" (i.e. to generalize the notion of space while retaining some space-like features): while internally the typical objects of $\mathcal{S}'$ are natural combinatorial constructions such as binomial coefficients, externally $\mathcal{S}'$ can be characterized by the model-theoretic job it does among all $\mathcal{S}$-toposes, which is to classify infinite decidable objects; many of its remarkable properties in both these roles can be deduced from Schanuel's discovery that it is atomic in the sense of Barr, which means that $\mathcal{S}' \longrightarrow \mathcal{S}$ is a generalized local homeomorphism in the sense that the inverse image preserves all higher order logic. $\mathcal{S}'$ is petit in our generalized sense for the opposite reason from all our examples: it is defined by a site $\mathbf{C}$ (= the opposite of the category of finite sets and monomorphisms) in which all maps are *epimorphisms* in $\mathbf{C}$.

REFERENCES

M. Barr, Atomic toposes. Journal of Pure and Applied Algebra, **17** (1980) no. 1, pp 1–24.

J. Boman, Differentiability of a function and of its compositions with functions of one variable, Math. Scand. **20** (1967), pp 249–268.

A. H. Clifford, *The algebraic theory of semigroups*, Vol. 1, American Mathematical Society, 1961.

J. Duskin, Free groupoids, trees, and free groups, to appear in a special volume of the Journal of Pure and Applied Algebra in honor of Bernhard Banachewski.

S. M. Gersten, On fixed points of automorphisms of finitely generated free groups. Bulletin of the American Mathematical Society **8** (1983) 451–454.

H. Grassmann, *Die Ausdehnungslehre von 1844*, pp 40–41, Chelsea Publishing Co. 1969.

A. Grothendieck, *Théorie des topos et cohomologie étale des schémas*, Lectures Notes in Math. Vols. 269–270, Springer-Verlag, 1972.

A. Grothendieck, *Pursuing stacks*, 400pp. manuscript, 1983.

A. Grothendieck, *Récoltes et semailles*, Vols. 1–4, Université des Sciences et Techniques du Languedoc, Montpellier, 1985.

P. T. Johnstone, Quotients of decidable objects in a topos, Math. Proc. Cambridge Philos. Soc. **93** (1983) 409–419.

P. T. Johnstone, *Topos theory*, Academic Press, 1977.

P. T. Johnstone and G. C. Wraith, Algebraic theories in toposes, in *Indexed categories and their applications*, Lecture Notes in Math. Vol. 661, 1978, 141–242.

F. W. Lawvere, Taking categories seriously, Revista Colombiana de Matemáticas, **20** (1986) 147–178.

F. W. Lawvere, Categories of spaces may not be generalized spaces, as exemplified by directed graphs, Revista Colombiana de Matemáticas, **20** (1986) 179–185.

J. P. Serre, *Trees*, Springer-Verlag, 1980.

DEPARTMENT OF MATHEMATICS, UNIVERSITY AT BUFFALO, STATE UNIVERSITY OF NEW YORK, BUFFALO, NEW YORK 14214-3093

Contemporary Mathematics
Volume **92**, 1989

Typed Lambda Models and Cartesian Closed Categories (preliminary version)

John C. Mitchell[*]
Department of Computer Science
Stanford University
Stanford, CA 94305

Philip J. Scott[†]
Department of Mathematics
University of Ottawa
Ottawa, Ontario K1N 6N5

September 16, 1988

Abstract

The interpretations of typed lambda calculi in Henkin models and cartesian closed categories (ccc's) are summarized. We discuss general and specialized completeness theorems, and relationships between interpretation functions. We also point out that ccc's satisfy the non-equational property of extensionality, provided one is content with the internal logic of the structure.

1 Introduction

This paper reviews the interpretation of typed lambda calculi in set-theoretic structures called Henkin models [Hen50] and cartesian closed categories (ccc's). To emphasize the parallels between the two frameworks, we use a version of typed lambda calculus that corresponds naturally to ccc's. Specifically, we assume that types include a terminal type **1**, products $\sigma \times \tau$, and exponentials $\sigma \rightarrow \tau$ (function spaces). In addition, we allow equations between types, and "nonlogical" term formation rules.

The primary points of comparison are the interpretation functions and soundness and completeness theorems associated with each kind of structure. Many of the technical results are known to experts in either lambda calculus or category theory. However, there are some subtle points that we believe are sometimes misunderstood. In particular, there seem to be a multitude of opinions regarding the proper "meaning function" for cartesian closed categories, and the status of extensionality (axiom (η) and rule (ξ)) over ccc's. We view the interpretation of typed lambda calculus in arbitrary ccc's as a natural generalization of the interpretation in Henkin models, and do not see any technical reason for emphasizing so-called "concrete" categories, or categories in which **1** is a generator.

[*]Supported by an NSY PYI Award.
[†]Partially supported by a grant from the Natural Sciences and Engineering Research Council of Canada.

2 Syntax of Terms and Equations

2.1 Types

We will consider typed lambda calculi over any "algebra" of types. Specifically, the collection *Typ* of types may be any collection containing $\mathbf{1}$ and closed under the binary operations $\times$ and $\rightarrow$. This includes the special case that type expressions are freely generated over $\mathbf{1}$ and given "ground" types $b_1, b_2, \ldots$ using $\times$ and $\rightarrow$, but also allows types such as $a \rightarrow b$ and $a \times b$ to be identified. While freely generated type expressions are sufficient for most purposes, allowing identifications between types gives us a better correspondence between lambda calculus and cartesian closed categories.

2.2 Terms

Given a collection *Typ* of types, the well-formed terms over *Typ* are characterized using the subsidiary notion of type assignment. A *type assignment* Γ is a finite set of formulas $x{:}\,\sigma$ associating types $\sigma \in \mathit{Typ}$ to variables, with no variable x occurring twice. The formula $x{:}\,\sigma$ may be read "the variable x has type σ." We write $\Gamma, x{:}\,\sigma$ for the type assignment

$$\Gamma, x{:}\,\sigma = \Gamma \cup \{x{:}\,\sigma\},$$

where, in writing this, we assume that x does not appear in Γ. Terms will be written in the form $\Gamma \triangleright M{:}\,\sigma$, which may be read, "$M$ has type σ relative to Γ."

In addition to pure terms, we will allow lambda terms over "nonlogical" constants and "nonlogical" term formation rules. The well-typed terms are assumed to be freely generated from the following axioms, logical formation rules and any additional nonlogical rules. We will always take $*$ as a constant symbol of type $\mathbf{1}$.

$$(var) \qquad x{:}\,\sigma \triangleright x{:}\,\sigma$$

$$(cst) \qquad \emptyset \triangleright c{:}\,\sigma, \quad c \text{ a constant of type } \sigma$$

$$(\rightarrow E) \qquad \frac{\Gamma \triangleright M{:}\,\sigma \rightarrow \tau, \ \Gamma \triangleright N{:}\,\sigma}{\Gamma \triangleright M \cdot_\sigma N{:}\,\tau}$$

$$(\rightarrow I) \qquad \frac{\Gamma, x{:}\,\sigma \ \triangleright M{:}\,\tau}{\Gamma \triangleright \lambda x{:}\,\sigma.M{:}\,\sigma \rightarrow \tau}$$

$$(\times E) \qquad \frac{\Gamma \triangleright M{:}\,\sigma \times \tau}{\Gamma \triangleright \mathbf{Proj}_1^{\sigma,\tau} M{:}\,\sigma, \ \Gamma \triangleright \mathbf{Proj}_2^{\sigma,\tau} M{:}\,\tau}$$

$$(\times I) \qquad \frac{\Gamma \triangleright M{:}\,\sigma, \ \Gamma \triangleright N{:}\,\tau}{\Gamma \triangleright \langle M, N\rangle{:}\,\sigma \times \tau}$$

$$(add\ hyp) \qquad \frac{\Gamma \triangleright M{:}\,\tau}{\Gamma, x{:}\,\sigma \ \triangleright M{:}\,\tau}$$

Application is written using the symbol "$\cdot_\sigma$" so that the type of the argument is explicit. (This is also done in [BM85].) We will often omit "$\cdot_\sigma$", writing, *e.g.*, $\Gamma \triangleright MN{:}\,\tau$, when the type σ is either unimportant or determined by context.

A *language* is a set of well-typed terms over some set of typed constants, closed under the above term formation rules, and any additional formation rules desired.

2.3 Equations

Given our formulation of terms, it is natural to write equations in the form

$$\Gamma \triangleright M = N : \tau$$

where we assume that $\Gamma \triangleright M\colon \tau$ and $\Gamma \triangleright N\colon \tau$. Since we include type assignments explicitly, we have the structural rule

(*add hyp*) $$\frac{\Gamma \triangleright M = N : \sigma}{\Gamma, x\colon \tau \triangleright M = N : \sigma}$$

which lets us add additional typing hypotheses to equations.

Provable equality will be an equivalence relation, and a congruence with respect to the term formation operations. Thus we have the axiom and rules

(*ref*) $$\Gamma \triangleright M = M : \sigma$$

(*sym*) $$\frac{\Gamma \triangleright M = N : \sigma}{\Gamma \triangleright N = M : \sigma}$$

(*trans*) $$\frac{\Gamma \triangleright M = N : \sigma, \ \Gamma \triangleright N = P : \sigma}{\Gamma \triangleright M = P : \sigma}$$

(ξ) $$\frac{\Gamma, x\colon \sigma \triangleright M = N : \tau}{\Gamma \triangleright \lambda x\colon \sigma.M = \lambda x\colon \sigma.N : \sigma \to \tau}.$$

(μ) $$\frac{\Gamma \triangleright M_1 = M_2 : \sigma \to \tau, \ \Gamma \triangleright N_1 = N_2 : \sigma}{\Gamma \triangleright M_1 N_1 = M_2 N_2 : \tau}$$

The basic properties of **1**, pairing and lambda abstraction are formalized by the following axioms.

(*one*) $$\Gamma \triangleright x = * : \mathbf{1}$$

($\mathbf{Proj}_1$) $$\Gamma \triangleright \mathbf{Proj}_1 \langle M, N \rangle = M : \sigma$$

($\mathbf{Proj}_2$) $$\Gamma \triangleright \mathbf{Proj}_2 \langle M, N \rangle = N : \tau$$

(**Pair**) $$\Gamma \triangleright \langle \mathbf{Proj}_1\, M, \mathbf{Proj}_2\, M \rangle = M : \sigma \times \tau$$

(α) $$\Gamma \triangleright \lambda x\colon \sigma.M = \lambda y\colon \sigma.[y/x]M : \sigma \to \tau, \text{ provided } y \notin FV(M)$$

(β) $$\Gamma \triangleright (\lambda x\colon \sigma.M) \cdot_\sigma N = [N/x]M : \tau$$

(η) $$\Gamma \triangleright \lambda x\colon \sigma.(M \cdot_\sigma x) = M : \sigma \to \tau, \text{ provided } x \notin FV(M)$$

A *typed lambda theory over a language* **L** is a set of well-typed equations between **L**-terms that includes all instances of the axioms and is closed under the inference rules. We will often use the phrase "typed lambda theory" to refer to a language, together with a theory for this language. Consequently, our typed lambda theories are equivalent to the "arbitrary typed lambda calculi" of [LS86].

3 Henkin models for typed lambda calculus

3.1 Overview

One accepted model theory for typed lambda calculus is based on Henkin models, first described in [Hen50]. There are several ways of defining these models precisely. The definition we will use has three parts. We first define typed applicative structures, and then specify two additional conditions that are required of models. A typed applicative structure consists of families of sets and mappings indexed by types, or pairs of types, of the form necessary to give meanings to terms. To be a model, an applicative structure must be extensional, which guarantees that the meaning of a term is unique, and there must be enough elements so that every well-typed term in fact has a meaning. One way of specifying that there are enough elements is called the "environment model" condition. We will also discuss the equivalent "combinatory model" condition briefly.

The equational proof system is easily seen to be sound for all Henkin models, but additional rules are needed for deductive completeness. With a relatively minor addition, we have soundness and completeness over models without empty types. General completeness without the non-emptiness assumption is also possible, but requires a more substantial modification of the proof system.

3.2 Applicative structures, extensionality and frames

A *typed applicative structure* $\mathcal{A}$ is a tuple

$$\langle A^\sigma, \mathbf{App}^{\sigma,\tau}, \mathbf{Proj}_1^{\sigma,\tau}, \mathbf{Proj}_2^{\sigma,\tau}, *\rangle$$

of families of sets and mappings indexed by types. Specifically, for all σ and τ we assume the following conditions.

- A^σ is some set,
- $\mathbf{App}^{\sigma,\tau}: A^{(\sigma\to\tau)\times\sigma} \longrightarrow A^\tau$ is a set-theoretic map from $A^{(\sigma\to\tau)\times\sigma}$ to A^τ,
- $\mathbf{Proj}_1^{\sigma,\tau}: A^{\sigma\times\tau} \longrightarrow A^\sigma$,
- $\mathbf{Proj}_2^{\sigma,\tau}: A^{\sigma\times\tau} \longrightarrow A^\tau$,
- $* \in A^1$.

Note that the symbol "$\to$" is used in two ways here, first as a symbol in type expressions, as in $A^{(\sigma\to\tau)\times\sigma}$, and second as a meta symbol in writing the domain and range of set-theoretic functions. To help distinguish these uses, we will use a slightly longer arrow for the meta symbol.

An applicative structure is *extensional* if it satisfies the following conditions.

- $\forall x: A^1.\, x = *$
- $\forall p, q: A^{\sigma\times\tau}.\, (\mathbf{Proj}_1 p = \mathbf{Proj}_1 q \wedge \mathbf{Proj}_2 p = \mathbf{Proj}_2 q) \supset p = q$
- $\forall f, g: A^{\sigma\to\tau}.$
  $(\forall \langle f, d\rangle, \langle g, d\rangle: A^{(\sigma\to\tau)\times\sigma}.\, \mathbf{App}\langle f, d\rangle = \mathbf{App}\langle g, d\rangle) \supset f = g$

Strictly speaking, the third condition is extensionality. However, it is convenient to group the three conditions together under the same name. The second condition guarantees that for every pair of elements $a \in A^\sigma$ and $b \in A^\tau$, there is at most one "pair" $p \in A^{\sigma\times\tau}$ whose first and second projections yield a and b, respectively. When the pair p exists, we write $\langle a, b\rangle$ for p. Assuming the extensionality condition for pairs, it is easy to see that $\langle \mathbf{Proj}_1 p, \mathbf{Proj}_2 p\rangle$ must equal p, and so every element of $A^{\sigma\times\tau}$ may be written as a pair. Thus, the the quantification over all pairs $\langle f, d\rangle, \langle g, d\rangle: A^{(\sigma\to\tau)\times\sigma}$ in the application clause may be regarded as a shorthand for quantifying over all $p, q: A^{(\sigma\to\tau)\times\sigma}$ such that $\mathbf{Proj}_1 p = f, \mathbf{Proj}_1 q = g$ and $\mathbf{Proj}_2 p = \mathbf{Proj}_2 q$.

In place of applicative structures, many authors use a definition that assumes extensionality. This is relatively natural when types are freely generated. Instead of letting $A^{\sigma\to\tau}$ be any set, and requiring an application map to make elements of $A^{\sigma\to\tau}$ behave like functions, we could require that $A^{\sigma\to\tau}$ actually be some set of functions from A^σ to A^τ. (However, we do not want to assume that $A^{\sigma\to\tau}$ contains all functions.) Similarly, we may assume $A^{\sigma\times\tau}$ is a set of pairs and A^1 a singleton set. To be precise, a *type frame* is an applicative structure $\langle A^\sigma, \mathbf{App}^{\sigma,\tau}, \mathbf{Proj}_1^{\sigma,\tau}, \mathbf{Proj}_2^{\sigma,\tau}, *\rangle$ such that

- A^1 has only one element
- $*$ is the only element of A^1.
- $A^{\sigma\times\tau} \subseteq A^\sigma \times A^\tau$ is a set of ordered pairs
- $\mathbf{Proj}_1^{\sigma,\tau}\langle x, y\rangle = x$ and $\mathbf{Proj}_2^{\sigma,\tau}\langle x, y\rangle = y$

- $A^{\sigma\to\tau} \subseteq (A^\tau)^{(A^\sigma)}$ is a set of functions
- $\mathbf{App}^{\sigma,\tau}\langle f, x\rangle = f(x)$

It is easy to see that when types are freely generated, type frames are equivalent to extensional applicative structures. However, type identifications like $\sigma\to\tau = \sigma\times\tau$ may make it impossible to construct a type frame from an extensional applicative structure.

Lemma 3.1 *Let $\mathcal{A}$ be an applicative structure for a freely-generated collection of types. Then $\mathcal{A}$ is extensional iff there is a type frame $\mathcal{B}$ with $\mathcal{A} \cong \mathcal{B}$.*

Proof. It is easy to see that every type frame is extensional, and extensionality is invariant under isomorphism. For the converse, we construct $\mathcal{B}$ and isomorphism $i: \mathcal{A} \longrightarrow \mathcal{B}$ by induction on types. ∎

3.3 Environment and combinatory model conditions

An *environment* η for applicative structure $\mathcal{A}$ is a mapping from variables to the union of all A^σ. If Γ is a type assignment, then we say η *satisfies* Γ, written $\eta \models \Gamma$, if $\eta(x) \in A^\sigma$ for every $x{:}\sigma \in \Gamma$.

An extensional applicative structure is an *environment model* if the clauses below define a total meaning function $[\![\cdot]\!]\cdot$ on terms $\Gamma \triangleright M{:}\sigma$ and environments η such that $\eta \models \Gamma$. In other words, we require that for every well-typed term $\Gamma \triangleright M{:}\sigma$, and every environment $\eta \models \Gamma$, the meaning $[\![\Gamma \triangleright M{:}\sigma]\!]\eta \in A^\sigma$ exists as defined below.

$$
\begin{array}{lcl}
[\![\Gamma \triangleright x{:}\sigma]\!]\eta & = & \eta(x) \\
[\![\Gamma \triangleright *{:}\mathbf{1}]\!]\eta & = & * \in A^{1} \\
[\![\Gamma \triangleright c{:}\sigma]\!]\eta & = & \hat{c} \in A^\sigma,\ \text{which is assumed given} \\
[\![\Gamma \triangleright \mathbf{Proj}_1^{\sigma,\tau} M{:}\sigma]\!]\eta & = & \mathbf{Proj}_1^{\sigma,\tau}\ [\![\Gamma \triangleright M{:}\sigma\times\tau]\!]\eta \\
[\![\Gamma \triangleright \mathbf{Proj}_2^{\sigma,\tau} M{:}\tau]\!]\eta & = & \mathbf{Proj}_2^{\sigma,\tau}\ [\![\Gamma \triangleright M{:}\sigma\times\tau]\!]\eta \\
[\![\Gamma \triangleright \langle M,N\rangle{:}\sigma\times\tau]\!]\eta & = & \text{the unique } p \in A^{\sigma\times\tau} \text{ such that} \\
 & & \mathbf{Proj}_1^{\sigma,\tau}(p) = [\![\Gamma \triangleright M{:}\sigma]\!]\eta \text{ and} \\
 & & \mathbf{Proj}_2^{\sigma,\tau}(p) = [\![\Gamma \triangleright N{:}\tau]\!]\eta \\
[\![\Gamma \triangleright M \cdot_\sigma N{:}\tau]\!]\eta & = & \mathbf{App}^{\sigma,\tau}\langle [\![\Gamma \triangleright M{:}\sigma\to\tau]\!]\eta, [\![\Gamma \triangleright N{:}\sigma]\!]\eta\rangle \\
[\![\Gamma \triangleright \lambda x{:}\sigma.M{:}\sigma\to\tau]\!]\eta & = & \text{the unique } f \in A^{\sigma\to\tau} \text{ such that} \\
 & & \forall d{:}A^\sigma.\mathbf{App}\, f d = [\![\Gamma, x{:}\sigma \triangleright M{:}\tau]\!]\eta[d/x] \\
[\![\Gamma, x{:}\sigma \triangleright M{:}\tau]\!]\eta & = & [\![\Gamma \triangleright M{:}\tau]\!]\eta,\ x \text{ not free in } M
\end{array}
$$

It is well-known that the condition that every term has a meaning is equivalent to the existence of combinators, and that each combinator is characterized by an equational axiom. We say an applicative structure $\mathcal{A}$ *has combinators* if, for all ρ, σ, τ, there exist elements

$$
\begin{array}{lcl}
K_{\sigma,\tau} & \in & A^{\sigma\to(\tau\to\sigma)} \\
S_{\rho,\sigma,\tau} & \in & A^{(\rho\to\sigma\to\tau)\to(\rho\to\sigma)\to\rho\to\tau} \\
Pair^{\sigma,\tau} & \in & A^{\sigma\to\tau\to\sigma\times\tau} \\
Proj_1^{\sigma,\tau} & \in & A^{(\sigma\times\tau)\to\sigma} \\
Proj_2^{\sigma,\tau} & \in & A^{(\sigma\times\tau)\to\tau}
\end{array}
$$

satisfying the equational conditions

$$
\begin{array}{lcl}
K_{\sigma,\tau}xy & = & x \\
S_{\rho,\sigma,\tau}xyz & = & (xz)(yz)
\end{array}
$$

for all x, y, z of the appropriate types, and the obvious equations involving $Pair^{\sigma,\tau}$, $Proj_1^{\sigma,\tau}$ and $Proj_2^{\sigma,\tau}$.

Lemma 3.2 *An extensional applicative structure $\mathcal{A}$ is an environment model iff $\mathcal{A}$ has combinators.*

Since each combinator is defined by an equational axiom, environment models may be characterized as the extensional applicative structures which satisfy the equational combinator axioms.

This is often called the *combinatory model definition,* since combinators play an important role. In this terminology, Lemma 3.2 says that the environment and combinatory model definitions are equivalent.

We will use the neutral and historical term *Henkin model* for both environment and combinatory models.

In logical terms, a typed applicative structure is any model of a many-sorted first-order signature with one sort for each type, and function symbols $\mathbf{App}^{\sigma,\tau}$, and so on. Extensionality is a first-order property (expressible over the same signature), while the requirement that every term have a meaning may be stated equationally. Thus Henkin models may be regarded as models of the "first-order theory of typed lambda calculus." Since extensionality is not an equational property, Henkin models do not form an algebraic variety.

3.4 Soundness and completeness for Henkin models

The proof system is sound for deriving consequences of any set of well-typed equations between lambda terms. However, as shown in [MMMS87,MM87], additional rules are required to achieve deductive completeness.

Theorem 3.3 (Soundness) *If* $\mathcal{E} \vdash \Gamma \triangleright M = N : \tau$, *for any set* $\mathcal{E}$ *of typed equations, then every Henkin model satisfying* $\mathcal{E}$ *also satisfies* $\Gamma \triangleright M = N : \tau$.

Although the converse of Theorem 3.3 fails, the additional rule

$$(nonempty) \qquad \frac{\Gamma, x\colon \sigma \triangleright M = N : \tau}{\Gamma \triangleright M = N : \tau} \; x \text{ not free in } M, N.$$

gives us completeness over models without empty types. If a type σ is empty (*i.e.*, $A^\sigma = \emptyset$), then rule *(nonempty)* is unsound, since $\Gamma, x\colon \sigma \triangleright M = N : \tau$ may hold solely because no environment can give x a value of type σ.

Theorem 3.4 (Completeness without empty types) *Let* $\mathcal{E}$ *be any lambda theory closed under the rule (nonempty). Then there is an Henkin model* $\mathcal{A}$, *with no* $A^\sigma = \emptyset$, *satisfying precisely the equations belonging to* $\mathcal{E}$.

Proof. (Sketch) Let $\mathcal{H}$ be any infinite set of variable typings $x\colon \sigma$ such that each type is given infinitely many variables, and no variable appears twice. Define the equivalence class $[M]_\mathcal{E}$ of M by

$$[M]_\mathcal{E} = \{\, N \mid \mathcal{E} \vdash \Gamma \triangleright M = N : \tau, \text{ some finite } \Gamma \subseteq \mathcal{H} \,\}$$

and let A^τ be the collection of all $[M]_\mathcal{E}$ with $\Gamma \triangleright M\colon \tau$ some $\Gamma \subseteq \mathcal{H}$. Application and other operations are straightforward. ■

It is worth mentioning that if all types are freely generated from one type constant b, then any Henkin model with a nontrivial equational theory must satisfy *(nonempty)*. The reason is that if the model is nontrivial, some type must have at least two elements. But this can only happen if b has more than one element. Since all constant functions are lambda definable (over elements of the model), it follows that every type will be nonempty. For this reason, when only one type constant is used, and types are freely generated, it is common to write equations without specifying the set of free variables explicitly.

Another special case involving *(nonempty)* is the pure theory with no nonlogical axioms. It may be shown that this theory is actually closed under *(nonempty)*, and so there is a Henkin model without empty types satisfying precisely the equations provable without nonlogical axioms. In addition, Friedman [Fri75] has shown that the full hierarchy of all set-theoretic functions over any infinite set satisfies precisely these equations. This theorem is also described in [Sta85].

In general, we are interested in typed lambda calculi with an arbitrary collection of ground types. Since some of these may not have any definable elements, or may naturally be considered empty (as may arise when types are given by specifications), it is important to be able to reason about terms over possibly empty types (see [MMMS87,MM87] for further discussion).

For Henkin models in which some types may be empty, we may achieve completeness using additional rules presented in [MMMS87]. The main purpose of these additional rules is to capture reasoning of the form "if $M = N$ whenever σ is empty, and $M = N$ whenever σ is nonempty, then we must have $M = N$". To facilitate reasoning about empty types, it is convenient to add assumptions of the form $empty(\sigma)$ to type assignments. An *extended equation* will be a formula $\Gamma \triangleright M = N : \sigma$ with Γ the union of a type assignment Γ_1 and a set Γ_2 of formulas $empty(\sigma)$. We require that $\Gamma_1 \triangleright M : \tau$ and $\Gamma_1 \triangleright N : \tau$, so that emptiness assertions do not affect the syntactic types of terms.

The proof system for reasoning about empty types uses an axiom scheme for introducing equations that use emptiness assertions

$$(empty\ I) \qquad \Gamma, empty(\sigma), x{:}\sigma \triangleright M = N : \tau$$

and an inference rule which lets us use emptiness assertions to reason by cases

$$(empty\ E)\ \frac{\Gamma, x{:}\sigma \triangleright M = N : \tau, \quad \Gamma, empty(\sigma) \triangleright M = N : \tau}{\Gamma \triangleright M = N : \tau}\, x \notin FV(M, N)$$

Technically speaking, the side condition in rule (*empty E*) is redundant, since the second equation in the antecedent cannot be well-formed unless x is not free in M or N. We will write $\vdash^{(emptyI,E)}$ for provability using $\vdash$ and the axiom and inference rule for empty types.

Theorem 3.5 (Completeness for Set-Theoretic Models [MMMS87]) *Let $\mathcal{E}$ be a set of extended equations, possibly containing emptiness assertions, and $\Gamma \triangleright M = N : \sigma$ an extended equation. Then $\mathcal{E} \vdash^{(emptyI,E)} \Gamma \triangleright M = N : \sigma$ iff every Henkin model satisfying $\mathcal{E}$ also satisfies $\Gamma \triangleright M = N : \sigma$.*

Proof. (Sketch) The completeness proof uses an infinite set $\mathcal{H}$ of typings $x{:}\sigma$ and emptiness assertions $empty(\sigma)$. If $\Gamma \triangleright M_0 = N_0 : \sigma_0$ is a chosen equation not provable from $\mathcal{E}$, then $\mathcal{H}$ is constructed so that (i) $\Gamma \subseteq \mathcal{H}$, (ii) for every σ, either $x{:}\sigma \in \mathcal{H}$ for infinitely many x, or $empty(\sigma) \in \mathcal{H}$, (iii) for every finite $\Gamma' \subset \mathcal{H}$, the equation $\Gamma' \triangleright M_0 = N_0 : \sigma_0$ is not provable from $\mathcal{E}$. The construction of $\mathcal{H}$ proceeds in stages, using an enumeration of all types. The remainder of the proof is similar to the proof of Theorem 3.4. ∎

Since (*empty E*) is not an equational inference, the rules for reasoning about empty types have a different flavor from the proof system given in the last section, and we give up the property that every set of equations closed under semantic implication is the theory of a single "minimal" model (see [MM87] for further discussion).

3.5 Combinatory and lambda algebras

Two nonextensional structures are occasionally of interest. Combinatory algebras are typed applicative structures that have combinators (as described above), but are not necessarily extensional. In a sense, every term can be given a meaning in a combinatory algebra, since every term can be put in combinatory form, and every applicative combination of combinators has a straightforward interpretation. However, many natural equations between lambda terms may fail. For example, $SKK = SKS$ holds in every lambda model (provided these are given the same type), since these functions are extensionally equal. However, both are in combinatory normal form, and so may have distinct interpretations in a combinatory algebra. Consequently, combinatory algebras are not models of the equational theory of typed lambda calculus.

There exists an equationally axiomatized class of combinatory algebras satisfying the pure equational theory of typed lambda calculus (β, η-conversion). These structures are called *lambda algebras,* and the unmemorable axioms may be found in [Bar84,Mey82], for example.

The main difference between lambda algebras and models is that lambda algebras do not satisfy (ξ) in the standard sense. Specifically, we would generally interpret

$$(\xi) \qquad \frac{\Gamma, x{:}\,\sigma \triangleright M = N : \tau}{\Gamma \triangleright \lambda x{:}\,\sigma.M = \lambda x{:}\,\sigma.N : \sigma \rightarrow \tau}.$$

as saying that whenever M and N have the same meaning for all values of x, we have $\lambda x{:}\,\sigma.M = \lambda x{:}\,\sigma.N$. This is guaranteed sound by the extensionality condition, but this reading of (ξ) fails in lambda algebras.

Combinatory algebras may be distinguished from Henkin models using the multi-sorted first-order language described in Section 3.3. The reason for using a first-order language is that the extensionality axiom

$$(ext) \qquad (\forall x{:}\,\sigma)(fx = gx) \supset f = g$$

describing function equality requires a quantifier (cf. [Bar84, Definition 5.2.19]). It seems natural to call first-order theories containing *(ext)* and some form of combinator axioms "first-order lambda theories." As mentioned in Section 3.3, Henkin models are precisely the models of these theories.

3.6 Reformulation of meaning function

The comparison of Henkin and categorical interpretations of lambda terms is simplified using a slight reformulation of the meaning function. A minor change of notation will make this easier. Below, we will write a type assignment $\{x_1{:}\,\sigma_1, \ldots, x_k{:}\,\sigma_k\}$ in the form $\bar{x}{:}\,\bar{\sigma}$, where $\bar{x} = x_1, \ldots, x_k$ and $\bar{\sigma} = \sigma_1, \ldots, \sigma_k$. We will also abuse notation and write $\bar{\sigma}$ for the type $\sigma_1 \times \ldots \times \sigma_k$, with $\times$ associated to the right. If $\bar{\sigma}$ is the empty sequence, then the corresponding type is $\bar{\sigma} = \mathbf{1}$. In addition, for every type $\bar{\sigma}$ and integer $i < k$, we will write $\mathbf{Proj}^{\bar{\sigma}}_{i,k}$ for the combination of projection functions such that

$$\mathbf{Proj}^{\bar{\sigma}}_{i,k}{:}\, A^{\bar{\sigma}} \longrightarrow A^{\sigma_i}$$

gives the i-th component of the cascaded sequence $\langle x_1, \langle x_2, \ldots, x_k\rangle\rangle{:}\, A^{\bar{\sigma}}$.

The meaning $[\![\bar{x}{:}\,\bar{\sigma} \triangleright M{:}\,\tau]\!]$ of a well-typed term will now be a function

$$[\![\bar{x}{:}\,\bar{\sigma} \triangleright M{:}\,\tau]\!]{:}\, A^{\bar{\sigma}} \longrightarrow A^{\tau}$$

The following definition may be proved equivalent to the meaning defined earlier, in a straightforward precise sense.

$$\begin{array}{lcl}
[\![\bar{x}{:}\,\bar{\sigma} \triangleright x_i{:}\,\sigma_i]\!] & = & \mathbf{Proj}^{\bar{\sigma}}_{i,k} \\
[\![\bar{x}{:}\,\bar{\sigma} \triangleright *{:}\,\mathbf{1}]\!] & = & \text{the unique } \mathcal{O}^{\bar{\sigma}}{:}\, A^{\bar{\sigma}} \longrightarrow A^{1} \\
 & & \quad \text{such that } \forall d{:}\, A^{\bar{\sigma}}.\mathcal{O}^{\bar{\sigma}}(d) = * \\
[\![\bar{x}{:}\,\bar{\sigma} \triangleright c{:}\,\tau]\!] & = & \hat{c} \circ \mathcal{O}^{\bar{\sigma}}, \text{ where } \hat{c}{:}\, \mathbf{1} \rightarrow \tau \text{ is assumed given} \\
[\![\bar{x}{:}\,\bar{\sigma} \triangleright \mathbf{Proj}^{\sigma,\tau}_1 M{:}\,\sigma]\!] & = & \mathbf{Proj}^{\sigma,\tau}_1 \circ [\![\bar{x}{:}\,\bar{\sigma} \triangleright M{:}\,\sigma \times \tau]\!] \\
[\![\bar{x}{:}\,\bar{\sigma} \triangleright \mathbf{Proj}^{\sigma,\tau}_2 M{:}\,\tau]\!] & = & \mathbf{Proj}^{\sigma,\tau}_2 \circ [\![\bar{x}{:}\,\bar{\sigma} \triangleright M{:}\,\sigma \times \tau]\!] \\
[\![\bar{x}{:}\,\bar{\sigma} \triangleright \langle M, N\rangle{:}\,\rho \times \tau]\!] & = & \text{the unique } p{:}\, A^{\bar{\sigma}} \longrightarrow A^{\rho \times \tau} \text{ such that} \\
 & & \mathbf{Proj}^{\sigma,\tau}_1 \circ p = [\![\bar{x}{:}\,\bar{\sigma} \triangleright M{:}\,\sigma]\!] \text{ and} \\
 & & \mathbf{Proj}^{\sigma,\tau}_2 \circ p = [\![\bar{x}{:}\,\bar{\sigma} \triangleright N{:}\,\tau]\!] \\
[\![\bar{x}{:}\,\bar{\sigma} \triangleright M \cdot_{\sigma} N{:}\,\tau]\!] & = & \mathbf{App}^{\sigma,\tau} \circ \langle [\![\bar{x}{:}\,\bar{\sigma} \triangleright M{:}\,\sigma \rightarrow \tau]\!], [\![\bar{x}{:}\,\bar{\sigma} \triangleright N{:}\,\sigma]\!]\rangle \\
[\![\bar{x}{:}\,\bar{\sigma} \triangleright \lambda y{:}\,\rho.M{:}\,\rho \rightarrow \tau]\!] & = & \text{the unique } f{:}\, A^{\bar{\sigma}} \longrightarrow A^{\rho \rightarrow \tau} \text{ such that } \forall d{:}\, A^{\bar{\sigma}} \longrightarrow A^{\rho}. \\
 & & \mathbf{App} \circ \langle f, d\rangle = [\![\bar{x}, y{:}\,\bar{\sigma}, \rho \triangleright M{:}\,\tau]\!] \circ \langle \mathbf{Id}^{\sigma}, d\rangle
\end{array}$$

In the application and lambda abstraction cases, for functions $f{:}\, A^{\sigma} \longrightarrow A^{\rho}$ and $g{:}\, A^{\sigma} \longrightarrow A^{\tau}$, we have written $\langle f, g\rangle$ for the mapping $x \mapsto \langle f(x), g(x)\rangle$. Note that the order of $\bar{x}$ and $\bar{\sigma}$ affect the meaning of $\bar{x}{:}\,\bar{\sigma} \triangleright M{:}\,\tau$. However, this is a minor point and we will try to ignore it as much as possible.

4 Categorical Models

4.1 Introduction

Cartesian closed categories (ccc's) play a fundamental role in the study of typed lambda calculus. Indeed, from the viewpoint of categorical logic, syntax and semantics are interchangeable. This is demonstrated by a categorical equivalence.

A *cartesian closed category* is a category with *specified* terminal object, products and exponentials. This means that each ccc has a specified

- object $\mathbf{1}$ with unique $\mathcal{O}^\sigma : \sigma \longrightarrow \mathbf{1}$ for each σ,
- binary object map $\times$ with specified arrows $\mathbf{Proj}_1^{\sigma,\tau}, \mathbf{Proj}_2^{\sigma,\tau}$, and map $\langle\ ,\ \rangle : Hom(\sigma,\tau) \times Hom(\sigma,\rho) \longrightarrow Hom(\sigma,\tau \times \rho)$ for each σ,τ,ρ,
- binary object map $\rightarrow$ with specified arrow $\mathbf{App}^{\sigma,\tau}$ and map $\mathbf{Curry}^{\sigma,\tau,\rho}: Hom(\sigma \times \tau, \rho) \longrightarrow Hom(\sigma, \tau \rightarrow \rho)$ for each σ,τ,ρ

forming an adjoint situation in each case. (See [LS86,Mac71] for further information.) With $\mathbf{1}, \times$ and $\rightarrow$ specified as part of each ccc, it is straightforward to interpret any type expression, once we have chosen an object for each type constant. Since the interpretation of type expressions in ccc's is entirely straightforward, we will often ignore the distinction between syntax and semantics, writing type expressions for objects of a given ccc.

The category *CCC* is the category whose objects are cartesian closed categories and morphisms are functors preserving cartesian closed structure. The morphisms of *CCC* are called *cartesian closed functors,* or *cc-functors*. Since $\mathbf{1}$, $\times$, $\rightarrow$, and associated maps are specified as part of each ccc, we require that cc-functors preserve these. The category λ*−Calc* is the category whose objects are typed lambda theories (*i.e.*, languages equipped with specified theories), and morphisms are language translations preserving the structure of types, terms, and equations. Full definitions are given in [LS86].

From any cartesian closed category $\mathbf{C}$, we can construct a typed language and lambda theory by introducing variables (indeterminates), as in [LS86]. This calculus, $\mathcal{L}(\mathbf{C})$, is called the *internal language* of $\mathbf{C}$. Conversely, any typed language and lambda theory $\mathbf{T}$ determine a category $\mathcal{C}(\mathbf{T})$, called the *category generated by* $\mathbf{T}$. The objects of $\mathcal{C}(\mathbf{T})$ are the types of $\mathbf{T}$, and the arrows are equivalence classes of terms, modulo equality of $\mathbf{T}$. To eliminate problems having to do with the names of free variables, it is simplest to choose one variable of each type, and define the arrows from σ to τ using terms over the chosen free variable of type σ. We will write $[z{:}\sigma \triangleright M{:}\tau]_{\mathbf{T}}$ for the arrow of $\mathcal{C}(\mathbf{T})$ given by the term $z{:}\sigma \triangleright M{:}\tau$, modulo $\mathbf{T}$. The theory $\mathbf{T}$ will be omitted when this is clear from context. The definition of the unique arrow $\mathcal{O}^\sigma : \sigma \longrightarrow \mathbf{1}$, and operations $\mathbf{Proj}_1, \mathbf{Proj}_2, \langle\ \rangle, \mathbf{App}, \mathbf{Curry}$ are straightforward. For example, $\langle\ \rangle$ in $\mathcal{C}(\mathbf{T})$ is defined by

$$\langle [x{:}\sigma \triangleright M{:}\rho], [x{:}\sigma \triangleright N{:}\tau] \rangle = [x{:}\sigma \triangleright \langle M,N \rangle{:}\rho \times \tau].$$

Since the category $\mathcal{C}(\mathbf{T})$ is defined using terms, it might also be called the "term model of $\mathbf{T}$." As shown in [LS86], the functors $\mathcal{L}$ and $\mathcal{C}$ are inverses (up to isomorphism), giving us an equivalence between cartesian closed categories and typed lambda theories.

Theorem 4.1 *[LS86, p.79–80] The functors $\mathcal{L} : CCC \longrightarrow \lambda{-}Calc$ and $\mathcal{C} : \lambda{-}Calc \longrightarrow CCC$ between the category CCC of cartesian closed categories and the category λ−Calc of typed lambda calculi give a categorical equivalence. In other words, $\mathcal{C} \circ \mathcal{L} \cong Id$ and $\mathcal{L} \circ \mathcal{C} \cong Id$.*

Theorem 4.1 suggests the mathematical importance of typed λ-calculi with product types, surjective pairing, and possibly empty types: they correspond exactly to ccc's. Of course for such a general theorem we pay a price: we must allow "applied" lambda theories whose types need not be freely generated from "atomic" types. Moreover, there may be unexpected equations between types and the types may even form a proper class.

4.2 Interpreting lambda terms in categories

We can use the meaning function defined in Section 3.6 to interpret lambda terms as arrows in a ccc. Since this definition requires that meanings of constants be given, we must choose an arrow of the appropriate type for each constant of the language. In addition, if terms are built using "non-logical" formation rules, we must choose appropriate maps on arrows for these cases. It is easy to see that each clause in the definition of Section 3.6 makes sense in any ccc, except possibly those that require uniquely determined functions. By regarding conditions such as "the unique $f: A^{\bar{\sigma}} \longrightarrow A^{\tau}$ such that ..." as meaning, "the unique $f \in Hom(\bar{\sigma}, \tau)$ such that ...," we can show that these conditions do make sense and determine the meaning of each term uniquely. In fact, the definition of cartesian closed category is precisely what is required to verify this.

With a more general meaning function, we can generalize soundness to arbitrary ccc's. Let $[\![\cdot]\!]$ be a meaning function as described in Section 3.6, mapping terms of a language $\mathbf{L}$ into a ccc $\mathbf{D}$. For any well-typed equation $\bar{x}: \bar{\sigma} \rhd M = N : \sigma$ of $\mathbf{L}$, we say $\mathbf{D}$ *satisfies* $\bar{x}: \bar{\sigma} \rhd M = N : \sigma$, written

$$\mathbf{D} \models \bar{x}: \bar{\sigma} \rhd M = N : \tau$$

if $[\![\bar{x}: \bar{\sigma} \rhd M: \tau]\!]$ and $[\![\bar{x}: \bar{\sigma} \rhd N: \tau]\!]$ are the same arrow of $\mathbf{D}$.

Theorem 4.2 (Soundness for CCC's) *If $\mathcal{E} \vdash \Gamma \rhd M = N : \tau$, for any set $\mathcal{E}$ of typed equations, then every cartesian closed category satisfying $\mathcal{E}$ also satisfies $\Gamma \rhd M = N : \tau$.*

If $\mathbf{D}$ is any cartesian closed category, and $[\![\cdot]\!]$ is a meaning function into $\mathbf{D}$, determined as in Section 3.6 by choosing interpretations for the constants and non-logical term formation rules of language $\mathbf{L}$, then we call the set of equations satisfied by $\mathbf{D}$ with respect to $[\![\cdot]\!]$ the $\mathbf{L}$-*theory induced by* $\mathbf{D}$.

There are several interesting connections between the meaning function $[\![\cdot]\!]$ and the categories λ-*Calc* and *CCC* of lambda theories and cartesian closed categories. A meaning function $[\![\cdot]\!]: \mathbf{L} \to \mathbf{D}$ induces a lambda theory $\mathbf{T}$, which in turn generates a category $\mathcal{C}(\mathbf{T})$. From $[\![\cdot]\!]$, we can construct a functor $F_{[\![\cdot]\!]}: \mathcal{C}(\mathbf{T}) \longrightarrow \mathbf{D}$ as follows. For types, the objects of $\mathcal{C}(\mathbf{T})$, we let $F_{[\![\cdot]\!]}(\sigma)$ be the interpretation of σ in $\mathbf{D}$, and for arrows we let

$$F_{[\![\cdot]\!]}([x: \sigma \rhd M: \tau]_{\mathbf{T}}) = [\![x: \sigma \rhd M: \tau]\!].$$

Using Theorem 4.2, we can show that $F_{[\![\cdot]\!]}$ is well-defined, and, using the properties of $[\![\cdot]\!]$, a cartesian closed functor.

Lemma 4.3 *Let $[\![\cdot]\!]: \mathbf{L} \longrightarrow \mathbf{D}$ be a meaning function from language $\mathbf{L}$ into ccc $\mathbf{D}$, and let $\mathbf{T}$ be the induced lambda theory over $\mathbf{L}$. Then $F_{[\![\cdot]\!]}: \mathcal{C}(\mathbf{T}) \longrightarrow \mathbf{D}$ is a cartesian closed functor from the category generated by $\mathbf{T}$ into $\mathbf{D}$.*

Basically, $[\![\cdot]\!]$ is the composition of $F_{[\![\cdot]\!]}$ and the process of collapsing terms modulo $\mathbf{T}$. However, there is a minor complication having to do with the difference between terms of $\mathbf{L}$ and arrows of $\mathcal{C}(\mathbf{T})$. Suppose $[\![\cdot]\!]$ is a meaning function from $\mathbf{L}$ into $\mathbf{D}$, and let $\mathbf{T}$ be the induced theory. By substituting projections of z for variables $x_1, \ldots, x_k$, we can transform any $\mathbf{L}$-term $\bar{x}: \bar{\sigma} \rhd M: \tau$ into a term $z: \bar{\sigma} \rhd M': \tau$ of the same type, with one free variable, and such that

$$[\![\bar{x}: \bar{\sigma} \rhd M: \tau]\!] = [\![z: \bar{\sigma} \rhd M': \tau]\!].$$

Since the name of the free variable (z above) may be chosen arbitrarily, we can define a map $i_{\mathbf{L}}$ on $\mathbf{L}$-terms mapping each term to a term with the one free variable of the appropriate type used to construct $\mathcal{C}(\mathbf{T})$. We can then map the term $i_{\mathbf{L}}(\bar{x}: \bar{\sigma} \rhd M: \tau) = z: \bar{\sigma} \rhd M': \tau$ to its equivalence class, modulo $\mathbf{T}$. Let $j_{\mathbf{T}}$ be the map from terms to equivalence classes taking $\bar{x}: \bar{\sigma} \rhd M: \tau$ to the equivalence class $[i_{\mathbf{L}}(\bar{x}: \bar{\sigma} \rhd M: \tau)]_{\mathbf{T}}$. Then we have the following factorization of $[\![\cdot]\!]$.

Theorem 4.4 *Let $[\![\cdot]\!]: \mathbf{L} \longrightarrow \mathbf{D}$ be a meaning function from $\mathbf{L}$ into a ccc $\mathbf{D}$, and $\mathbf{T}$ the induced lambda theory. Then $[\![\cdot]\!]$ factors as the composition*

$$[\![\cdot]\!] = F_{[\![\cdot]\!]} \circ j_{\mathbf{T}}$$

of a map taking terms to equivalence classes, and a cc-functor from $\mathcal{C}(\mathbf{T})$ into $\mathbf{D}$.

Conversely, we can show that every cartesian closed functor $F : \mathcal{C}(\mathbf{T}) \longrightarrow \mathbf{D}$ determines a meaning function as defined in Section 3.6. Cartesian closedness of a functor F means that $\mathbf{1}, \times$ and $\rightarrow$ are preserved, so that $F(\mathbf{1}_{\mathcal{C}(\mathbf{T})}) = \mathbf{1}_{\mathbf{D}}$, $F(\sigma \times_{\mathcal{C}(\mathbf{T})} \tau)) = F(\sigma) \times_{\mathbf{D}} F(\tau)$, and similarly for function spaces. In addition, the term formation operations $\mathbf{Proj}_1, \mathbf{Proj}_2, \langle\ \rangle, \mathbf{App}$ and **Curry** must be preserved. For **App**, for example, this means that

$$\begin{aligned} F([z{:}\bar{\sigma} \triangleright M \cdot_\rho N{:}\tau]_{\mathbf{T}}) &= F(\mathbf{App} \circ \langle [z{:}\bar{\sigma} \triangleright M{:}\rho{\rightarrow}\tau]_{\mathbf{T}}, [z{:}\bar{\sigma} \triangleright N{:}\rho]_{\mathbf{T}} \rangle) \\ &= \mathbf{App} \circ \langle F([z{:}\bar{\sigma} \triangleright M{:}\rho{\rightarrow}\tau]_{\mathbf{T}}), F([z{:}\bar{\sigma} \triangleright N{:}\rho]_{\mathbf{T}}) \rangle \end{aligned}$$

Writing $[\![\ ]\!]$ for the composition of F and j, we have as a consequence

$$[\![\bar{x}{:}\bar{\sigma} \triangleright M \cdot_\rho N{:}\tau]\!] = \mathbf{App} \circ \langle [\![\bar{x}{:}\bar{\sigma} \triangleright M{:}\rho{\rightarrow}\tau]\!], [\![\bar{x}{:}\bar{\sigma} \triangleright N{:}\rho]\!] \rangle.$$

The other conditions work out similarly, giving us the following theorem.

Theorem 4.5 *Let $F : \mathcal{C}(\mathbf{T}) \longrightarrow \mathbf{D}$ be a cartesian closed functor from the category generated by theory $\mathbf{T}$ over $\mathbf{L}$ into a ccc $\mathbf{D}$. Then the function $[\![\cdot]\!]{:}\, \mathbf{L} \longrightarrow \mathbf{D}$ given by composing F with the map $j_{\mathbf{T}}$ from terms to equivalence classes modulo $\mathbf{T}$ is a meaning function, as specified in Section 3.6. Furthermore, the theory $\mathbf{T}'$ induced by $[\![\cdot]\!]$ is an extension of $\mathbf{T}$, with $\mathbf{T}' = \mathbf{T}$ if F is one-to-one.*

Since the identity functor on $\mathcal{C}(\mathbf{T})$ is one-to-one, we now have an easy completeness theorem.

Corollary 4.6 *(Easy Completeness) Let $\mathbf{T}$ be an equational typed lambda theory. Then $\mathbf{T} \vdash \bar{x}{:}\bar{\sigma} \triangleright M = N : \tau$ iff the interpretation determined by the functor $Id{:}\,\mathcal{C}(\mathbf{T}) \longrightarrow \mathcal{C}(\mathbf{T})$ satisfies $\bar{x}{:}\bar{\sigma} \triangleright M = N : \tau$.*

For a related result for toposes, cf. [LS86, p.212].

Since the categories λ*-Calc* and *CCC* are equivalent, there is a bijective correspondence between cartesian closed functors $F{:}\,\mathcal{C}(\mathbf{T}) \longrightarrow \mathbf{D}$ and language translations $M{:}\,\mathbf{T} \longrightarrow \mathcal{L}(\mathbf{D})$. Consequently, every language translation determines a meaning function, and vice versa. In summary, meaning functions, language translations, and cartesian closed functors are essentially equivalent.

4.3 Extensionality in cartesian closed categories

Several authors have described connections between untyped lambda algebras and reflexive objects in cartesian closed categories [Bar84,Koy82,HS86,OW78]. A similar connection between typed lambda algebras and cartesian closed categories is described in [BM85]. Generally, these comparisons begin with a locally small ccc and construct an applicative structure by taking the set of arrows $\mathbf{1} \longrightarrow \sigma$ as elements of A^σ. Since an arrow $f{:}\,\sigma \longrightarrow \tau$ determines an extensional function only if $\mathbf{1}$ is a generator[1], one generally obtains a lambda algebra. This correspondence between ccc's and lambda algebras has been interpreted as meaning that cartesian closed categories are not "extensional," since rule (ξ) has a dubious status over lambda algebras. However, there is a subtle point here, and the situation merits further investigation.

It follows from Theorem 4.2 that rule (ξ) *is* sound for the interpretation of terms in cartesian closed categories. In particular, $\mathbf{Curry} f = \mathbf{Curry} g$ iff $f = g$, or, put another way, cartesian closedness guarantees that

$$[\![\bar{x}{:}\bar{\sigma} \triangleright \lambda y{:}\,\rho.M{:}\rho{\rightarrow}\tau]\!] = [\![\bar{x}{:}\bar{\sigma} \triangleright \lambda y{:}\,\rho.N{:}\rho{\rightarrow}\tau]\!]$$

$$\textit{iff} \quad [\![\bar{x}, y{:}\,\bar{\sigma}, \rho \triangleright M{:}\tau]\!] = [\![\bar{x}, y{:}\,\bar{\sigma}, \rho \triangleright N{:}\tau]\!]$$

[1] If $\mathbf{1}$ is a generator, then C is concrete with representation $Hom(\mathbf{1}, -){:}\, C \longrightarrow Set$. In some of the computer science literature, the term "concrete" is used for the property of having one as a generator. See [Mac71,Sco80] for further discussion.

To emphasize the point that regardless of how cartesian closed categories are compared with Henkin structures, ccc's always *behave* as if they are extensional models, we will show that every ccc is extensional in a standard *internal* sense. Thus, according to the internal logic of any category with sufficient structure to interpret quantification, ccc's are models of first-order typed lambda theories. In other words, contrary to the suggestions of [Bar84,BM85,Koy82], there is no technical reason to focus on ccc's with 1 as a generator.

Since extensionality is a quantified formula, the argument that ccc's satisfy extensionality involves extra machinery beyond cartesian closedness. To make the point in a fairly general way, we will interpret universal quantification using subobjects (see [MR77]). When a ccc supports quantification, extensionality is easily shown to be true. Moreover, as pointed out to us by Peter Freyd, one doesn't even need to explicitly mention universal quantification to understand extensionality. Thus, to the degree that extensionality may be expressed in any ccc, ccc's are always extensional.

To see how extensionality holds in ccc's, we begin with a brief review of universal quantification on subobjects. Given a ccc $\mathbf{C}$, and objects C, A, suppose the projection $p: C \times A \longrightarrow C$ induces a functor $p^*: Sub(C) \longrightarrow Sub(C \times A)$ by pulling back along p, and that p^* has right adjoint $\forall_A$. This situation is summarized symbolically as follows.

$$Sub(C \times A) \underset{\forall_A}{\overset{p^*}{\rightleftarrows}} Sub(C) \qquad\qquad p^* \dashv \forall_A$$

Since adjointness is equivalent to natural isomorphism of hom-sets, the functor $\forall_A$ is characterized by the fact that for any subobjects $S \leq C$ and $R \leq C \times A$, we have

$$(*) \qquad S \times A \leq R \quad \textit{iff} \quad S \leq \forall_A R.$$

In other words, condition $(*)$ may be taken as the defining property of $\forall_A$. We will be concerned with quantification $\forall_A$ over A with respect to $C = B^A \times B^A$.

Consider the subobject of $B^A \times B^A \times A$ given by the pullback

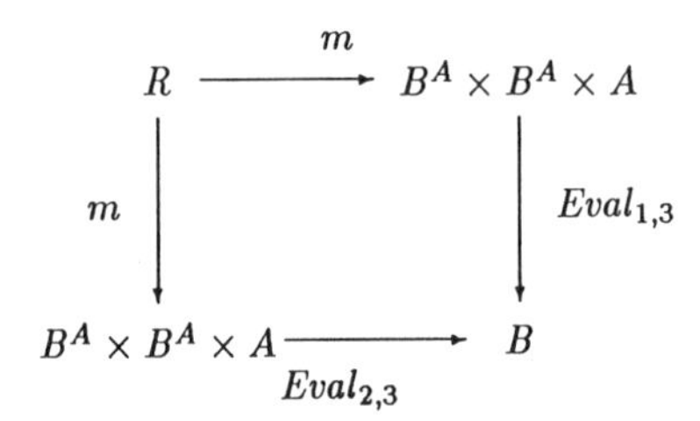

where $Eval_{i,j}$ means "evaluate the i-th function on the j-th argument." Intuitively, the object R is the collection of triples $\langle f, g, d \rangle$ with $f, g \in B^A$ and $fd = gd$, so the subobject $\forall_A R \rightarrowtail B^A \times B^A$ corresponds to the collection

$$\forall_A R \;\approx\; \{ \langle f, g \rangle \in B^A \times B^A \mid \forall x{:}\, A.\, fx = gx \}.$$

Thus extensionality may be expressed by saying that $\forall_A R$ is also the collection $\{ \langle f, g \rangle \mid f = g \}$. Put in terms of subobjects, extensionality is the equation

$$(ext) \qquad \forall_A R \rightarrowtail B^A \times B^A \;=\; B^A \overset{\Delta}{\rightarrowtail} B^A \times B^A$$

where $\Delta = \langle 1, 1 \rangle$ is the diagonal.

As pointed out by Peter Freyd, the straightforward verification of extensionality does not actually depend on the existence of $\forall_A R$. Essentially, $\forall_A R$ is the maximal subobject $S \leq B^A \times B^A$ such that $S \times A \leq R$, while the diagonal is the minimal such subobject. To show that these are equal, it suffices to show the stronger property that there is only one subobject $S \leq B^A \times B^A$ such that $S \times A \leq R$. This stronger statement does not involve quantification, and so we will prove it directly.

Let S be any subobject $S \leq B^A \times B^A$ with $S \times A \leq R$. Then the following diagram commutes.

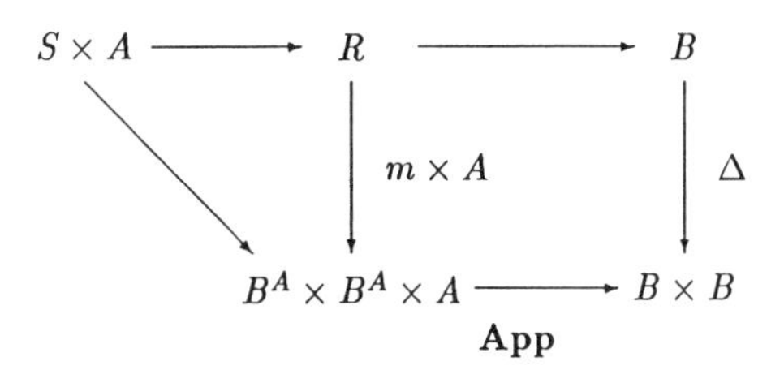

Here $\mathbf{App} = \mathbf{App}^{A, B \times B}$ uses the isomorphism $B^A \times B^A \cong (B \times B)^A$.

Now apply the functor $(\cdot)^A$ to the "outer" square with R eliminated. This gives us the right half of the commutative diagram below. The left half is straightforward.

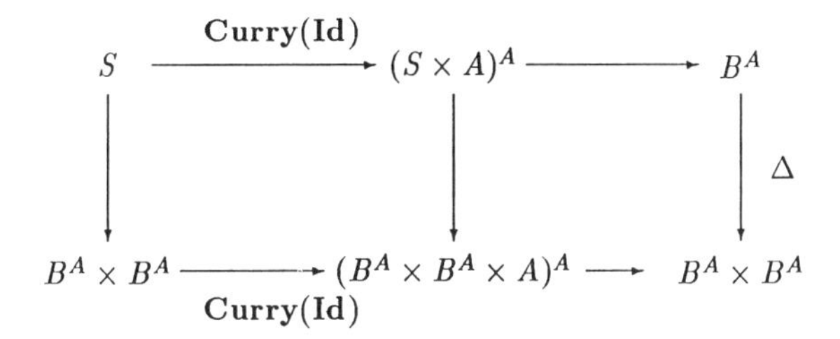

Since the composition along the bottom of the diagram may be seen to be the identity, this give us the inclusion

$$S \rightarrowtail B^A \times B^A \quad \leq \quad B^A \overset{\Delta}{\rightarrowtail} B^A \times B^A.$$

The converse is straightforward, proving the desired identity of subobjects.

This shows that once we have the necessary equalizer or pullback to have R, we can see that any ccc satisfies extensionality. In addition, one may also paraphrase the use of equalizers. (We leave this as an exercise for the interested reader.) Thus we may conclude that whenever one has enough machinery to state extensionality internally, it is simply true.

An alternative view of extensionality for ccc's is described in [Sco80]. The Yoneda embedding $Y: C \hookrightarrow Set^{C^{op}}$ may be used to represent any ccc as a full cartesian closed subcategory of a functor category. Since $Set^{C^{op}}$ is a topos, extensionality of C may always be expressed in the internal logic of $Set^{C^{op}}$. Since the subobject lattice interpretation of extensionality (described above) coincides with the internal topos interpretation of extensionality in $Set^{C^{op}}$, extensionality of C always holds in the internal logic of $Set^{C^{op}}$. Thus any ccc may be embedded in an interpretation of higher-order logic, and extensionality of the given ccc always holds in this interpretation. It is well-known that in fact the internal (higher-order, intuitionistic) logic of $S^{C^{op}}$ is conservative over the (first-order) typed lambda calculus (with extensionality axiom) of C. In addition, Pitts' extension of the Yoneda lemma [Pit87, Lemma 4.5] covers still more general first order (lambda) theories, so there is actually quite a lot that can be said about first-order theories in ccc's, even though ccc's most naively correspond only to equational lambda theories.

5 Completeness for Categorical Interpretations

For typed lambda theories, there are two forms of completeness relating a class $\mathcal{T}$ of theories and a class $\mathcal{M}$ of models (or interpretations). One is that for every theory $\mathbf{T}$ from $\mathcal{T}$, there is a *minimal model* from $\mathcal{M}$ satisfying precisely the equations which belong to $\mathbf{T}$. The weaker form of completeness is that every $\mathbf{T}$ is closed under semantic implication with respect to $\mathcal{M}$, meaning that for every well-formed equation e not belonging to $\mathbf{T}$, there is a *counter model* from $\mathcal{M}$ satisfying $\mathbf{T}$ but not e. Both forms are important to our understanding of completeness for Henkin models. We have minimal models without empty types for theories closed under rule *(nonempty)*, and countermodels (possibly with empty types) for the class of theories closed under $\vdash^{(emptyI,E)}$. However, we do have any form of completeness unless we either add rule *(nonempty)*, or add the axiom and inference rule for possibly empty types.

The situation for arbitrary categories is somewhat different. By Corollary 4.6, every theory $\mathbf{T}$ has a minimal categorical model $\mathcal{C}(\mathbf{T})$. However, the equivalence between $\lambda-Calc$ and CCC seems so elementary that this is not a very satisfying completeness theorem. In fact, categories are often regarded as theories. Therefore, we prefer completeness theorems for special classes of ccc's that seem more "semantic" in some way or another.

The standard completeness theorems, beyond Corollary 4.6, all involve functor categories $Set^{\mathcal{C}^{op}}$. The easiest such result, as remarked earlier, is to use the Yoneda embedding $Y: \mathcal{C}(\mathbf{T}) \hookrightarrow Set^{\mathcal{C}(\mathbf{T})^{op}}$. Since Y is full and faithful, we have

$$\mathbf{T} \vdash \Gamma \triangleright M = N : \sigma \ \text{ iff } \ Set^{\mathcal{C}(\mathbf{T})^{op}} \models \Gamma \triangleright M = N : \sigma.$$

To view the Henkin completeness theorems in similar terms, it is helpful to use the internal logic of $Set^{\mathcal{C}^{op}}$. As pointed out earlier, the definition of extensional applicative structure may be formalized in multi-sorted first-order logic, with one sort for each type. In addition, the existence of combinators may be put equationally, so Henkin models may be formalized in first-order logic. Since we can easily interpret the first-order definition of lambda model in the internal logic of any topos, and $Set^{\mathcal{C}^{op}}$ is always a topos, we are led to the notion of *lambda model in* $Set^{\mathcal{C}^{op}}$. It is easy to show that a lambda model *in* any topos must be a ccc. Since $Set = Set^{\{*\}}$ for a one-point category $\{*\}$, we can describe the Henkin completeness theorems as completeness for various classes of lambda models in $Set^{\{*\}}$.

For general lambda theories, the completeness theorem described in [MM87] says it suffices to consider lambda models in toposes of the form $Set^{P^{op}}$, where P is any poset. Since a topos $Set^{P^{op}}$ is often called a Kripke model, it makes sense to call a lambda model in $Set^{P^{op}}$ a *Kripke lambda model.* A direct completeness proof showing that every lambda theory $\mathbf{T}$ has a minimal Kripke lambda model is given in [MM87]. The completeness theorem may also be proved using the composition of two functors

$$\mathcal{C}(\mathbf{T}) \xrightarrow{Y} Set^{\mathcal{C}(\mathbf{T})^{op}} \xrightarrow{D} Set^{P^{op}}$$

where D, the "Diaconescu cover" [Joh77], preserves first order logic, and hence the definition of lambda model. However, it is worth emphasizing that the Diaconescu cover does not preserve exponentials, so that the exponentials in $\mathcal{C}(\mathbf{T})$ will not necessarily be interpreted as exponentials of $Set^{P^{op}}$. (This is similar to situation with Henkin models, where $A^{\sigma\to\tau}$ is not necessarily isomorphic to the set of all functions from A^{σ} to A^{τ}.) The poset P used in the Diaconescu cover is the poset of strings of $\mathcal{C}(\mathbf{T})$ maps ordered by $u \leq v$ iff v is an initial segment of u. This construction is analogous to the two-step completeness proof for type theories, presented in [LS86, Theorems 19.1, 19.2]. For type theories, every unprovable formula fails in a Kripke model $Set^{P^{op}}$, where P is a poset of pairs $\langle X, F\rangle$, with X a type assignment and F a filter of propositions. However, for the less expressive language of lambda calculus, filters are not required.

Another form of completeness theorem may be given using internal categories. Recent work by A. Pitts on second- and higher-order polymorphic lambda calculi uses internal ccc's in topoi [Pit87].

It is worth mentioning that these completeness theorems do not necessarily carry over easily to typed lambda calculi with a natural numbers type (*i.e.*, with iteration). For, unlike the topos case, the functors above do not preserve natural numbers. In particular, the Yoneda embedding does not, although $Y(N)$ does, in certain cases, support various strong forms of induction (*c.f.* [MR87].

To describe the notion of "lambda models *in* a topos" categorically, and to give general accounts of completeness (in the form of representation theorems for ccc's, with additional structure like natural numbers objects) we feel that additional research is required.

6 Applications to untyped lambda calculus

Although we will not take the time to discuss this in detail, it is worth noting that untyped lambda theories may be regarded as a particular class of typed lambda theories. Specifically, as described in [Sco80], an untyped theory is essentially a typed theory satisfying the equation $\Phi \circ \Psi = Id$ for constants with types $\Phi: b \to b \to b$ and $\Psi: (b \to b) \to b$, for some b. Consequently, the general discussion of soundness and completeness theorems for both Henkin models and ccc's pertains to untyped lambda calculus as well. (See [Bar84] for further discussion.) However, with constants Φ and Ψ, the type b cannot be empty, so there is no need to consider Henkin models with empty types.

Acknowledgements

We thank Andre Scedrov for encouraging us to write this paper, and Peter Freyd for helpful observations on the interpretation of extensionality in ccc's. Albert Meyer and a referee also provided useful comments and suggestions.

P.J. Scott wishes to thank the University of Pennsylvania Department of Mathematics for great hospitality during his sabbatical, when we began writing this paper.

References

[Bar84] H.P. Barendregt. *The Lambda Calculus: Its Syntax and Semantics.* North Holland, 1984. (revised edition).

[BM85] V. Breazu-Tannen and A.R. Meyer. Lambda calculus with constrained types. In *Logics of Programs*, pages 23–40, Springer LNCS 193, June 1985.

[Fri75] H. Friedman. Equality between functionals. In R. Parikh, editor, *Logic Colloquium*, pages 22–37, Springer-Verlag, 1975.

[Hen50] L. Henkin. Completeness in the theory of types. *Journal of Symbolic Logic*, 15(2), June 1950. pages 81-91.

[HS86] W.S. Hatcher and P.J. Scott. Lambda algebras and c-monoids. *Zeitschr. f. math. Logic und Grundlagen d. Math.*, 32:415 – 430, 1986.

[Joh77] P. Johnstone. *Topos Theory.* Academic Press, 1977.

[Koy82] C.P.J. Koymans. Models of the lambda calculus. *Information and Control*, 52(3):306–323, 1982.

[LS86] J. Lambek and P.J. Scott. *Introduction to Higher-Order Categorical Logic.* Cambridge studies in advanced mathematics 7, 1986.

[Mac71] S. MacLane. *Categories for the Working Mathematician.* Volume 5 of *Graduate Texts in Mathematics*, Springer-Verlag, 1971.

[Mey82] A.R. Meyer. What is a model of the lambda calculus ? *Information and Control*, 52(1):87–122, 1982.

[MM87] J.C. Mitchell and E. Moggi. Kripke-style models for typed lambda calculus. In *IEEE Symp. Logic in Computer Science*, pages 303–314, June 1987.

[MMMS87] A. R. Meyer, J. C. Mitchell, E. Moggi, and R. Statman. Empty types in polymorphic lambda calculus. In *Proc. 14-th ACM Symp. on Principles of Programming Languages*, pages 253–262, January 1987.

[MR77] M. Makkai and G.E. Reyes. *First-order categorical logic.* Springer Lecture Nores in Math. 611, 1977.

[MR87] I. Moerdijk and G.E. Reyes. A smooth version of the zariski topos. *Advances in Mathematics*, 65(3):229–253, 1987.

[OW78] A. Obtulowicz and A. Wiwiger. *Categorical, functorial and algebraic aspects of the type free lambda calculus.* Technical Report Preprint 164, Institute of Mathematics, Polish Academy of Sciences, Sniadeckich 8, Skr. Poczt. 137, 00-950, Warszaw, Poland, 1978.

[Pit87] A.M. Pitts. Polymorphism is set-theoretic, constructively. In *Proceedings Summer Conf. on Category Theory and Computer Science*, Springer LNCS, 1987. To appear.

[Sco80] D.S. Scott. Relating theories of the lambda calculus. In *To H.B. Curry: Essays on Combinatory Logic, Lambda Calculus and Formalism*, pages 403–450, Academic Press, 1980.

[Sta85] R. Statman. Equality between functionals, revisited. In *Harvey Friedman's Research on the Foundations of Mathematics*, pages 331–338, North-Holland, 1985.

Contemporary Mathematics
Volume 92, 1989

SOME CONNECTIONS BETWEEN MODELS OF COMPUTATION*

Philip S. Mulry[†]

Introduction

Considerable attention has been directed towards defining a general environment in which basic constructs in computation can be described and also carried out. This has led to different notions of an effective space or structure. Further these structures have been collected inside categories which might be considered suitable universes for the discussion of computability and effectivity. Among the constructions by now well known are the enumerated sets and the effective f_o spaces of Ersov, the effective domains of Scott, the effective topos and the recursive topos. Various other categories such as the categories of partial equivalence relations and generalized enumerated sets have also been studied. These different categories reflect to some extent the roles topology, effectivity, computability, and separability play in the study of computation.

In this paper we investigate some of the connections that exist between the different categories mentioned above. Much has already been written about these categories individually; considerably less has been written about their interrelationships. Here we focus our attention on one idea, namely the role the natural number object (both standard and non-standard) plays in these concerns. In the first section we see how the existence of a natural number object (n.n.o.) plays a role in determining what categorical structure can exist on various subcategories of the effective and recursive toposes. In the second section we observe how the lack of n.n.o.'s in semantic categories can be overcome in helping describe fixed point

*1980 Mathematics Subject Classification (1985 Revision) 68Q05, 18B20, 03D45.

[†]Partially supported by NSF Grant CCR8706333.

results. Also the n.n.o. is used throughout to describe and explain functorial connections that exist between these various categorical models of computation.

Section 1

We begin by recalling the effective and recursive toposes denoted by *Eff* and *Rec* respectively. By now several references exist for both *Eff* and *Rec* (see [H] and [M1] and [R]) so we will limit our discussion to several remarks.

The construction of *Eff* was motivated by Kleene realizability and in a sense that can be made precise *Eff* constitutes a universe of effective mathematics. In particular, objects of *Eff* consist of pairs $(A, =)$ where it is realized $=: A \times A \to P\,\mathbf{N}$ satisfies symmetry and transitivity. Maps, which are equivalence classes of functional relations, can be thought of as generalized effective maps. *Eff* is not a Grothendieck topos but rather arises from the tripos formed from the standard partial applicative structure on $\mathbf{N}$ ($n(m) \approx \phi_n(m)$ where ϕ_n is the nth partial recursive function). A geometric morphism $Set \stackrel{\Delta}{\to}$ *Eff* exists whose left adjoint is the global sections functor.

Rec on the other hand is constructed by taking canonical sheaves on the monoid R of computable functions (thought of as the one object site $N \circlearrowleft$). In this context an object X in *Rec* can be thought of as a set of paths (i.e. an element x of X is the path $N \stackrel{x}{\to} X$ by Yoneda) acted on by computable functions. So for example if x is a path in X and f is a computable function, then $x \circ f$ is a new path in X. These paths satisfy a further glueing (sheaf) condition. For example, if x and y are paths defined on the even and odd natural numbers respectively, then there exists a unique path $N \to X$ which extends x and y.

It might well be suspected that partial (as opposed to total) computable functions would be a more natural starting point. We note however that for site R, $sh(R) \cong sh(R^\sharp)$ where $R^\sharp$ is the idempotent splitting of R. In this case, $R^\sharp$ consists of r.e. sets and partial computable functions. *Rec* can be thought of as a universe of computable mathematics where the maps, which now are natural transformations, can be thought of as generalized computable maps. Like *Eff* there exists a geometric morphism from Set to *Rec* which is now essential, i.e., the global sections functor has both left and right adjoints $\Gamma_! \dashv \Gamma \dashv \Gamma^!$.

Example. If X is an arbitrary set then $\Delta(X)$ is the object in *Eff* with $P\,\mathbf{N}$-valued equality defined by $\mid x = y \mid = \mathbf{N}$ if $x = y$; ϕ otherwise. $\Gamma^!(X)$, on the other hand, is the set of all (arbitrary) sequences $\{x_n\}$ in X. If f is total computable then $\{x_n\} \circ f = \{x_{fn}\}$. We note for future reference that $Hom_{Eff}(\mathrm{N}, \Delta X) \cong \Gamma^!(X)$.

The notion of a natural number object (n.n.o.) plays an important role in both toposes. In *Eff*, the n.n.o. is $\mathrm{N} = (\mathbf{N}, =)$ where $\mid n = m \mid = \{n\}$ if $n = m$, ϕ otherwise. In *Rec*, discrete N, i.e. $\Gamma_!(\mathbf{N})$ is the n.n.o. There is also another, nonstandard, candidate for a natural number object, namely the representable functor N. Note in particular that Γ N $= \Gamma N = \mathbf{N}$. We will use the notation $\mathbf{N}$, N, N consistently to denote the n.n.o. in *Set*, the n.n.o. in an arbitrary topos, and the non-standard n.n.o. in *Rec* respectively.

It is now well known that N in *Eff* and N in *Rec* satisfy a number of important properties such as Church's Thesis, Markov's Principle, Countable Choice, and restricted induction. Motivated by this, a general notion of a nonstandard natural number object (NNO*) in a topos has been defined in [R]. Both the n.n.o. N in *Eff* and the nonstandard N in *Rec* are examples of NNO* but the n.n.o. N in *Rec* is not.

Another category which plays a role in our discussion is the category of enumerated sets, *En*. The objects of *En* are pairs $< S, v >$ where S is a set and v is an onto enumeration $\mathbf{N} \rightarrow$ S. Arrows $< S, v > \stackrel{G}{\rightarrow} < S', v' >$ are set maps G for which there exists a total computable f making the diagram commute

$$\begin{array}{ccc} \mathbf{N} & \stackrel{f}{\longrightarrow} & \mathbf{N} \\ \downarrow v & & \downarrow v' \\ S & \stackrel{G}{\longrightarrow} & S' \end{array}$$

Many details can be found in [E2].

Theorem. *En* is fully embedded inside *Rec* taking the n.n.o. of *En* to the nonstandard N in *Rec*. Similarly, *En* is fully embedded inside *Eff* preserving n.n.o.'s.

Proof. See [M1] and [R] for the first and second statements respectively.

The natural number object in *En* is just the identity enumeration $\mathbf{N} \rightarrow \mathbf{N}$. For the first embedding an enumerated set $< S, v >$ is taken to the sheaf of paths in S where a path is of the form $v \circ f$ for f total computable. Such paths correspond to arrows from the n.n.o. to $< S, v >$ in *En* so the embedding can be alternately viewed as a means of making the n.n.o. $< \mathbf{N}, id >$ act like a representable, i.e., a generator. Simply extending this embedding in the natural way from *En* to *Eff* generates a functor, namely Hom (N, –), *Eff* $\rightarrow$ *Rec* which once again takes the n.n.o. to the nonstandard N and which forces N in *Eff* to act representably. This functor is denoted K in [R] where it is pointed out K is no longer full or faithful but does satisfy (i) $\Gamma K \cong \Gamma$ and (ii) $K\Delta \cong \Gamma^!$.

An interesting question naturally arises. *En* while embedded inside both *Eff* and *Rec* is not cartesian closed and thus is not a particularly satisfactory universe for discussing certain aspects of computation. Can we do better, i.e., can we find a full cartesian closed subcategory of *Eff* and *Rec* which properly includes *En*? The answer is no, as the following theorem shows.

Theorem. Let $\mathbf{D}$ be fully embedded inside both *Rec* and *Eff.* If $\mathbf{D}$ is cartesian closed then the embedding of *En* inside *Rec* and *Eff* cannot factor through $\mathbf{D}$.

Proof. *En* has a n.n.o. $<\mathbf{N}, \mathrm{id}>$. If the embedding of *En* factors through $\mathbf{D}$ then $\mathbf{D}$ contains N, the image of $<\mathbf{N}, id>$ via the embedding, as well as maps $\mathrm{N}^{\mathrm{N}} \to \mathrm{N}$ since it is a cartesian closed category. These would coincide by fullness exactly to the corresponding maps in *Rec* and *Eff* which in turn correspond to Banach-Mazur functionals and the effective operations respectively. By a classical result of Friedberg, however, these are known not to agree. See [M2] and [Ro].

Example. Let *Mod* denote the category of modest sets. The objects of *Mod* are sets A equipped with an enumeration $X \overset{v_A}{\to} A$ of A where X is a subset of N. Equivalently A can be thought of as a quotient of X, with the enumeration v_A as the quotient map. Maps $< A, v_A > \to < B, v_B >$ are set functions $A \overset{G}{\to} B$ for which a partial recursive function ϕ exists (with domain $\phi \supseteq X$) so that the diagram commutes

$$\begin{array}{ccc} X & \xrightarrow{\phi} & Y \\ \downarrow v_A & & \downarrow v_B \\ A & \xrightarrow{G} & B \end{array}$$

En is easily seen to be a full subcategory of *Mod* which in turn can be included in both *Eff* and *Rec.* The inclusion into *Eff* which takes $< A, v_A >$ to the set X with the obvious =-relation is full since each ϕ has a corresponding index. The inclusion into *Rec* takes $< A, v_A >$ to the set of all paths in A of the form $v_A \circ f$ where f is a total recursive function and the range of f is a subset of X. This inclusion into *Rec*, however, is not full by the theorem.

Example. Let Rec_1 denote the quasitopos of $\neg\neg$-separated objects in *Rec.* It is obviously full in *Rec* and the inclusion of *En* in *Rec* factors through Rec_1. Equivalently Rec_1 can also be described as consisting of the objects of *Rec* weakly generated by 1. Rec_1 has also been investigated in a different context in [L,M] where the authors call it the category of generalized enumerated sets. The coincidences of the two categories is pointed out in [R]. Now, however, by the theorem, Rec_1 cannot be fully included inside *Eff.*

The proof of the theorem actually proves a bit more than the theorem states. We have the following corollary.

Corollary. There does not exist a cartesian closed category possessing a n.n.o. N which is fully embedded inside both *Rec* and *Eff* taking N to N and N respectively.

What the above theorem indicates is that we are severely restricted in creating a categorical environment which reflects both recursivity and effectivity, at least when a n.n.o.

is present. From this point of view, *En*, which possesses a n.n.o., had no hope of being cartesian closed. Of course, there are cartesian closed full subcategories of both *Rec* and *Eff*. We consider these in the next section.

Section 2

We begin this section by retreating from the question raised in section 1. We look in the other direction, namely at a subcategory of *En* and concentrate once more on natural number objects.

We assume the reader is familiar with the notion of a Scott domain.

Definition. A Scott domain D is effective iff its neighborhood system has a computable presentation, the compatibility relation and sup operation are effective on finite elements, and every element is the recursively enumerable sup of its finite approximations. The category of effective domains (e-domains) *EDom* has arrows $D \rightarrow E$ corresponding to computable mappings, i.e., approximable mappings f for which the relation on neighborhoods $X_n f Y_m$ is recursively enumerable in n and m. See [S] for details.

It is convenient, in fact, to think of D as possessing a computable enumeration of its elements and the arrows $D \rightarrow E$ as actual functions between elements of D and E subject to certain restraints. This leads to the following result.

Theorem. EDom fully embeds inside *Rec* and *Eff*.

Proof. In the previous section we observed that *En* is full inside both *Rec* and *Eff* so it suffices to show *EDom* fully embeds inside *En*. If D is an e-domain then there exists an effective enumeration v, of the elements of D. A computable map $D \xrightarrow{f} D'$ between e-domains generates a corresponding set function between the elements. When such an f exists it is easy to show there exists a total computable $h : N \rightarrow N$ for which $Fv(n) = v'h(n)$ for all $n \epsilon N$. Thus *EDom* is included in *En*. That the inclusion is also full was observed by Ersov (in the context of constructively complete f_o spaces) [E2].

The above result indicates that if we restrict our attention to effective domains, it matters not whether we describe our maps as computable in *EDom* (i.e. approximable plus r.e. restraints), effective in *Eff* or computable in *Rec* (i.e. a natural transformation). This explains why many of the usual functionals and operators of classical recursion theory, which act on particular e-domains, have several different characterizations.

The corollary of the previous section indicates that *EDom* has no n.n.o. Since *EDom* can be fully embedded into *Rec* and *Eff*, however, e-domains have a n.n.o. inside each of the larger topoi. In the case of *Rec*, however, the n.n.o. is discrete and not as effective as N, the nonstandard n.n.o. The following corollary then is interesting.

Corollary. N acts like the n.n.o. (in *Rec*) on effective domains.

Proof. In fact, the result easily holds for enumerated sets as indicated in [M1] but we provide a simple argument here. Given data $1 \xrightarrow{d} D \xrightarrow{F} D$ where D is an effective domain, we know by the theorem there exists a corresponding total computable h and an $n \epsilon N$ so that $v_D(n) = d$. This provides new data $1 \xrightarrow{n} N \xrightarrow{h} N$, but this in turn defines a computable $g : N \to N$ so that the diagram commutes

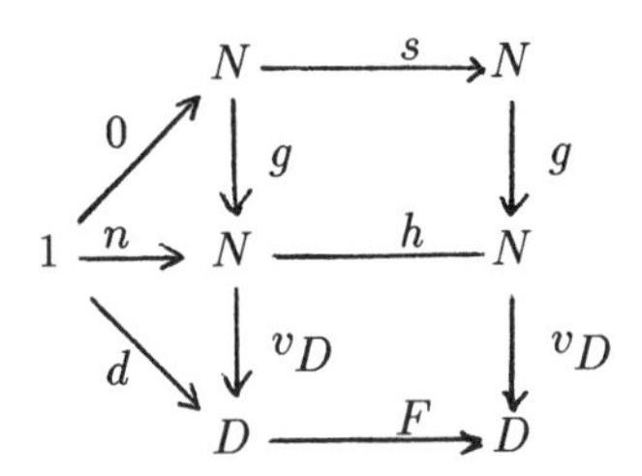

Although neither h nor n is unique, the composite $v_D \circ g$ is.

Example. The proof of the last corollary is reminiscent of the proof of the first recursion theorem. Let $PR \xrightarrow{F} PR$ be a recursive operator on the effective domain of partial recursive functions. Since N is a n.n.o. for objects of *EDom*, the data $1 \xrightarrow{\uparrow} PR \xrightarrow{F} PR$ (where $\uparrow$ refers to the everywhere divergent function) generates a unique path $N \xrightarrow{g} PR$ in *Rec*. In fact g generates the effective sequence of functions g(0), g(1), ... whose sup is the least fixed point of F.

This same construction works, in fact, more generally for any operator $D \xrightarrow{F} D$ in *EDom*, generating a sequence of elements in D whose sup is the least fixed point of F. This observation warrants further study and will be the subject of a forthcoming paper.

We now consider fixed point results from a different point of view than the last example. The connection with previous results is the role the n.n.o. once again plays in the examples. Let C be a category with 1.

Definition. (i) $X \epsilon C$ has the *fixed point property* if every morphism $X \xrightarrow{t} X$ in C has a global point $(1 \to X)$ fixed by t.

(ii) Let $X \xrightarrow{G} Y$ be a morphism in C. X has the *fixed point property* relative to G if for every map $X \xrightarrow{t} X$ there exists a global point x of X so that Gtx = Gx.

(iii) A map $X \xrightarrow{f} Y$ is *A-path surjective* if any map $A \to Y$ factors through f. The map is *point surjective* if it is 1-path surjective.

The notion of a path surjective or principal map appears in [M1] where it refers to N-path surjectivity in our present notation. Standard enumerations of the partial recursive functions, for example, are path surjective (giving an s-m-n theorem). The principal map notation appears again in [L-M] where it refers to the specific case of A-path surjective maps of the form $A \to B$ in *En*. These maps are in fact N-path surjective in *Rec* by an elementary argument.

Theorem. Suppose C is a cartesian closed category. If there exists a map $A \to X^A$ which is point surjective then X has the fixed point property. If in addition $A \cong A \times A$ then A has the fixed point property relative to any A-path surjective map $A \xrightarrow{G} X$.

Proof. The first statement can be found in [L], the second in [M2].

Corollary. First and second recursion theorems hold for any e-domain D in *Eff* or *Rec*.

Proof. Let A be N or N in *Eff* or *Rec* respectively. If D is an e-domain then both point surjective $A \to D^A$ and A-path surjective maps exist and clearly $A \cong A \times A$. By the theorem, any map $D \to D$ has a fixed point and N or N has the fixed point property relative to the path surjective enumeration of the elements of D in *Eff* or *Rec* respectively. These last statements correspond to first and second recursion theorems when D is PR.

We end by observing that the above results suggest it may be useful to characterize just when N can act like a standard n.n.o. For example, since K(N) = N, N is standard not only for elements of *EDom* and *En* but also for those elements K(X), where X is in *Mod* or more generally *Eff*, for which K is full (i.e. endomaps of KX are of the form Kf). From this point of view K can be seen as not only forcing the representability of N but also the standardness of N. *Set* itself would recognize N as standard since $\Gamma^! : Set \to Rec$ is an inclusion and $\Gamma^! \simeq K\Delta$ while other objects such as the n.n.o. N in *Rec* would not. Classical results in recursion, however, as indicated in section 1, may place severe restrictions on what is possible.

BIBLIOGRAPHY

[E1] Ersov, Ju. La theorie des enumerations, in: *Actes du Congres International des Mathematiciens 1970*, vol. 1 (Gauthier-Villars, 1971).

[E2] Ersov, Ju. Model C of Partial Continuous Functions, in *Logic Colloquium 76* (North Holland, Amsterdam, 1977).

[H] Hyland, J. M. E. The effective topos, in: A. S. Troelstra and D. Van Dalen, eds., *L. E. J. Brouwer, Centenary Symposium* (North Holland, Amsterdam, 1982).

[LM] Longo, G. and E. Moggi, Cartesian closed categories for effective type structures, in: *Lecture Notes in Computer Science* 173 (Springer, Berlin, 1984).

[M1] Mulry, P. S. Generalized Banach-Mazur Functionals in the Topos of Recursive Sets, *Journal of Pure and Applied Algebra* 26 (1982), 71-83.

[M2] Mulry, P. S. A Categorical Approach to the Theory of Computation, to appear in *Annals of Pure and Applied Logic.*

[Ro] Rogers, H. *Theory of Recursive Functions and Effective Computability* (New York: McGraw-Hill, 1967).

[R] Rosolini, G. Continuity and Effectiveness in Topoi, Ph.D. Thesis. Oxford, 1986.

[S] Scott, D. *Lectures on a Mathematical Theory of Computation.* Technical Monograph PRG-19. Oxford University, 1981.

DEPARTMENT OF COMPUTER SCIENCE
COLGATE UNIVERSITY
HAMILTON, NY 13346

Contemporary Mathematics
Volume **92**, 1989

Some Applications of Categorical Model Theory

Robert Paré

July 1, 1988

Introduction

Taking a somewhat extreme point of view, we might say that categorical model theory is the study of the 2-category of accessible categories, accessible functors, and natural transformations. The theory of accessible categories presents itself in two different forms. One is that accessible categories are precisely the categories of models of theories expressible in infinitary first order logic, i.e. in $L_{\infty,\infty}$. The morphisms are functions preserving all relevant structure. The word "relevant" is made precise by specifying a certain fragment of $L_{\infty,\infty}$, and requiring that morphisms preserve the interpretations of all formulas in it. In this form, many natural questions, arising from model theory, present themselves. Below, we show that two categorical notions of theory, that of triple (or monad) and that of **Prop**, give rise to accessible categories when their models are taken in an accessible category.

The other form that the theory takes is purely categorical. Accessible categories form a large class of nice categories with good stability properties. Most of the applications come from the fact that accessible categories have small dense subcategories on the one hand, and that there are good tools for showing that categories are accessible (such as the Limit theorem) on the other. Some examples are given below.

The main results come from the opposition between the abstract and the concrete. Thus we see a number of very general 2-categorical theorems giving results about particular categories of models.

Most of the results presented here were developed in conversations with Michael Makkai while working on our joint paper [**M/P**], to which the reader is referred for more details. In fact, this paper may be considered an elaboration of some points in it. Conversations with Barry Jay and Richard Wood were also useful.

1 Review of accessible categories

Let κ be an infinite regular cardinal. Recall from [G/U] that a small category $\mathbf{I}$ is *κ-filtered* if every diagram $\mathbf{J} \to \mathbf{I}$, where $\mathbf{J}$ has $< \kappa$ morphisms, has a cocone. An object A of a category $\mathbf{A}$, is *κ-presentable* if the hom functor $\mathbf{A}(A, -) : \mathbf{A} \to \mathbf{Set}$ preserves κ-filtered colimits. This notion is reasonable only if $\mathbf{A}$ has κ-filtered colimits, in fact only if $\mathbf{A}$ is in some sense a category of models, e.g. accessible.

Definition 1 ([M/P]) *A category* $\mathbf{A}$ *is* accessible *if there exists an infinite regular cardinal* κ *such that*
(i) $\mathbf{A}$ *has* κ*-filtered colimits,*
(ii) the full subcategory $\mathbf{A}_\kappa$ *of* κ*-presentable objects is essentially small,*
(iii) every object of $\mathbf{A}$ *is a* κ*-filtered colimit of* κ*-presentable objects.*

Examples of accessible categories include all algebraic categories with rank, or more generally all locally presentable categories in the sense of [G/U], all small categories with split idempotents, the category of infinite sets and monomorphisms, but not $\mathbf{Set}^{op}$ or the category of topological spaces and continuous maps.

An *accessible functor*, $F : \mathbf{A} \to \mathbf{B}$, between accessible categories is one that preserves κ-filtered colimits for some κ. These are sometimes called functors with rank in the literature. There is no restriction on the natural transformations. All of this gives a 2-category, $\mathcal{A}cc$, which is a sub-2-category of $\mathcal{CAT}$, the 2-category of (large) categories.

The accessible categories are precisely the categories of models of theories in $L_{\infty,\infty}$, which we briefly describe (more details can be found in [M/P]).

We start with a language L (which is, very roughly, a sort of graph). L has a set of *sorts*, $A, B, C, \ldots$, a set of *function symbols*, $f, g, \ldots$, which have arities (families $\langle A_i \rangle_{i \in I}$ of sorts, I a set) and value sorts (think $f : \prod A_i \to B$), and *predicate symbols*, $P, Q, \ldots$, also with arities (think $P \subseteq \prod A_i$).

The set of formulas, $L_{\infty,\infty}$, is built by induction. First we define *terms* by putting variables into function symbols and substituting these into other function symbols (e.g. $f(x, g(x, y), h(z))$). The *atomic formulas* are then $t_1 = t_2$ and $P(\langle t_i \rangle)$ where the t_i are terms. Then $L_{\infty,\infty}$ consists of all *formulas* that can be built in the usual way, out of atomic formulas using $\neg$ (negation), $\to$ (implication), $\bigvee$ (disjunction of a (possibly infinite) *set* of formulas), $\bigwedge$ (conjunction of a set of formulas), $\exists_X$, $\forall_X$ (quantifiers over a *set* X of variables).

An *L-structure* M assigns to each sort A of L a set $M(A)$, to each function symbol $f : \prod A_i \to B$, a function $M(f) : \prod M(A_i) \to M(B)$, and to each predicate symbol P, a relation $M(P) \subseteq \prod M(A_i)$. Then all formulas Φ have an interpretation in M,

$$\|\Phi\| \subseteq \prod_{x \in X} M(A_x)$$

where X is a set of variables containing the free variables of Φ, and A_x is the sort of the variable x.

A theory will be a set of sentences of $L_{\infty,\infty}$, and models will be L-structures that satisfy these sentences. But we want a *category* of models, and so we must have the proper notion of morphism of models. Morphisms must preserve some, but not necessarily all, formulas used in the theory. Thus we first start by choosing a fragment of $L_{\infty,\infty}$. Its purpose is to determine the morphisms we will use.

A *fragment* $\mathcal{F}$ is a set of formulas of $L_{\infty,\infty}$ such that
(i) all atomic formulas are in $\mathcal{F}$,
(ii) $\mathcal{F}$ is closed under substitution of terms for free variables,
(iii) $\mathcal{F}$ is closed under subformulas,
(iv) if $\Phi \to \Psi \in \mathcal{F}$ then $\neg\Phi \vee \Psi \in \mathcal{F}$,
(v) if $\forall_X \Phi \in \mathcal{F}$ then $\neg\exists_X \neg\Phi \in \mathcal{F}$.

If M, N are two L-structures, an *$\mathcal{F}$-elementary morphism* $t : M \to N$ is an assignment of a function $t(A) : M(A) \to N(A)$ to every sort A of L, such that for every $\Phi \in \mathcal{F}$, there is a restriction

$$\begin{array}{ccc} \|\Phi\| & \subseteq & \prod M(A_x) \\ \downarrow & & \downarrow \prod t(A_x) \\ \|\Phi\| & \subseteq & \prod N(A_x), \end{array}$$

i.e. t preserves the interpretation of all formulas in $\mathcal{F}$.

A *theory* $\mathbf{T}$ consists of a fragment $\mathcal{F}$ and a set of sentences of the form

$$(\forall)(\Phi \to \Psi)$$

where Φ and Ψ are built from the formulas in $\mathcal{F}$ using only $\bigwedge, \bigvee, \exists_X$ ($(\forall)$ means quantification over all free variables, i.e. the universal closure). An L-structure M satisfies $(\forall)(\Phi \to \Psi)$ if in M, $\|\Phi\| \subseteq \|\Psi\|$. $Mod(\mathbf{T})$ is then the category of models of $\mathbf{T}$, i.e. of L-structures satisfying all formulas in $\mathbf{T}$, with $\mathcal{F}$-elementary morphisms.

The example below should help to clarify these concepts.

A basic theorem in the subject, due essentially to Lair **[L]**, is the following:

Theorem 1 *A category is accessible if and only if it is of the form $Mod(\mathbf{T})$ for $\mathbf{T}$ a theory in $L_{\infty,\infty}$ for some language L.*

We give an example to illustrate these concepts. We shall construct a theory in a language L whose category of models is the category of coalgebras over a given commutative ring R.

An *R-coalgebra* is an R-module C with linear maps $\epsilon : C \to R$ and $\delta : C \to C \otimes C$ such that

$$\begin{array}{ccc} C & \xrightarrow{\delta} & C \otimes C \\ & {\scriptstyle\rho}\searrow \quad \swarrow{\scriptstyle C\otimes\epsilon} & \\ & C \otimes R & \end{array} \qquad \begin{array}{ccc} C & \xrightarrow{\delta} & C \otimes C \\ & {\scriptstyle\lambda}\searrow \quad \swarrow{\scriptstyle \epsilon\otimes C} & \\ & R \otimes C & \end{array}$$

$$
\begin{array}{ccccc}
C & \xrightarrow{\delta} & C\otimes C & \xrightarrow{C\otimes\delta} & C\otimes(C\otimes C) \\
\downarrow{\scriptstyle\delta} & & & & \uparrow{\scriptstyle\alpha} \\
C\otimes C & \xrightarrow[\delta\otimes C]{} & (C\otimes C)\otimes C & &
\end{array}
$$

commute, where $\otimes$ denotes the tensor product of R-modules, and ρ,λ,α are the canonical isomorphisms. A homomorphism of R-coalgebras $\phi : (C, \epsilon, \delta) \to (D, \epsilon, \delta)$ is a linear map $\phi : C \to D$ such that

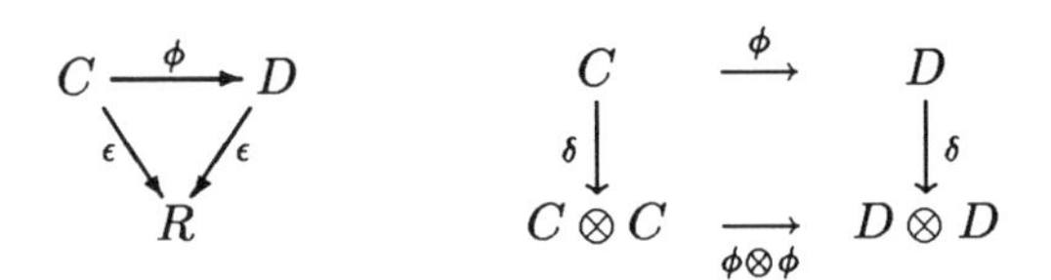

commute.

The sorts of the language L will be C, $C \otimes C$, $C \otimes (C \otimes C)$,$(C \otimes C) \otimes C$, $\bar{R}$, $C \otimes \bar{R}$, $\bar{R} \otimes C$. Thus there are seven sorts. The names are merely symbols, chosen suggestively, but could have been taken to be $A_1, \ldots, A_7$. There are no relation symbols in L. There are however many function symbols which we break into four groups:

(1) If A is any one of the sorts of L, we have

$+ : A \times A \to A$

$0 : 1 \to A$ (the arity of 0 is the empty family of sorts)

$r \cdot (\ ) : A \to A$ (one for each $r \in R$)

(2) five binary function symbols

$\otimes : C \times C \to C \otimes C$

$\otimes : C \times (C \otimes C) \to C \otimes (C \otimes C)$

$\otimes : (C \otimes C) \times C \to (C \otimes C) \otimes C$

$\otimes : C \times \bar{R} \to C \otimes \bar{R}$

$\otimes : \bar{R} \times C \to \bar{R} \otimes C$

(3) one nullary one

$\bar{1} : 1 \to \bar{R}$

(4) and nine unary function symbols

$\epsilon : C \to \bar{R}$

$\delta : C \to C \otimes C$

$C \otimes \epsilon : C \otimes C \to C \otimes \bar{R}$

$\epsilon \otimes C : C \otimes C \to \bar{R} \otimes C$

$C \otimes \delta : C \otimes C \to C \otimes (C \otimes C)$

$\delta \otimes C : C \otimes C \to (C \otimes C) \otimes C$

$\rho : C \to C \otimes \bar{R}$

$\lambda : C \to \bar{R} \otimes C$

$\alpha : (C \otimes C) \otimes C \to C \otimes (C \otimes C).$

This describes the language.

The fragment consists of all atomic formulas. The axioms of the theory **T** fall into four corresponding groups:

(1) For every sort A of L, axioms that say A is an R-module

- (i) $(x+y)+z = x+(y+z)$ (To put this formula into the form given above, write $\forall x, y, z(\text{true} \to (x+y)+z = x+(y+z))$ where "true" is the conjunction of the empty set of formulas.)
- (ii) $x+y = y+x$
- (iii) $0x = 0$
- (iv) $(r+s)x = rx+sx$ (this is a whole family of axioms, one for each pair $r, s \in R$: $\forall x(\text{true} \to (r+s)x = rx+sx)$)
- (v) $r(x+y) = rx+ry$
- (vi) $1x = x$

(2) For every function symbol of type (2) above, $\otimes : A \times B \to A \otimes B$, axioms that express the fact that $A \otimes B$ will become the tensor product in any model. (i)– (iii) say that $\otimes$ is bilinear, (iv) says that $A \otimes B$ is spanned by elements of the form $x \otimes y$, and (v), a slight variation on Proposition I.8.8 of **[Stn]**, says that no more gets identified in $A \otimes B$ than is absolutely necessary (tuples of elements of A and B are represented by row and column vectors respectively).

- (i) $(x_1 + x_2) \otimes y = x_1 \otimes y + x_2 \otimes y$
- (ii) $x \otimes (y_1 + y_2) = x \otimes y_1 + x \otimes y_2$
- (iii) $(rx) \otimes y = r(x \otimes y) = x \otimes (ry)$
- (iv) $\forall z : A \otimes B(\bigvee_{n \in N} \exists x_1 \ldots x_n, y_1 \ldots y_n(z = \sum_{i=1}^{n} x_i \otimes y_i))$ (an example of a countable disjunction). (We write $\forall z : A \otimes B$ to mean $\forall z$ and to indicate that z is a variable of type $A \otimes B$.)
- (v) $\sum_{i=1}^{n} x_i \otimes y_i = 0 \to \bigvee_{(l,m) \in N^2} \exists \mathbf{u}_{:A^l} \exists \mathbf{z}_{:B^m} \bigvee_{C \in R^{n \times l}} \bigvee_{D \in R^{n \times m}} (\mathbf{x} = \mathbf{u}C \wedge C\mathbf{y} = D\mathbf{z} \wedge \mathbf{u}D = \mathbf{0})$

(3) (i) for each $r \neq s$ in R we have the axiom $r\bar{1} \neq s\bar{1}$, i.e. $r\bar{1} = s\bar{1} \to$ false, where false is $\bigvee \emptyset$.

- (ii) $\forall x : \bar{R}(\bigvee_{r \in R} x = r\bar{1})$

(4) (i) axioms expressing the fact that all function symbols in (4) are R-linear, e.g. $\epsilon(x+y) = \epsilon(x) + \epsilon(y)$, $\epsilon(rx) = r\epsilon(x)$, etc.

- (ii) Coalgebra axioms:

$C \otimes \epsilon(\delta(x)) = \rho(x)$

$\epsilon \otimes C(\delta(x)) = \lambda(x)$

$\alpha(\delta \otimes C(\delta(x))) = C \otimes \delta(\delta(x))$

(iii)
$$\begin{aligned} &C \otimes \epsilon(x \otimes y) = x \otimes \epsilon(y) \\ &\epsilon \otimes C(x \otimes y) = \epsilon(x) \otimes y \\ &C \otimes \delta(x \otimes y) = x \otimes \delta(y) \\ &\delta \otimes C(x \otimes y) = \delta(x) \otimes y \\ &\rho(x) = x \otimes \bar{1} \\ &\lambda(x) = \bar{1} \otimes x \\ &\alpha((x \otimes y) \otimes z) = x \otimes (y \otimes z) \end{aligned}$$

Once the intent of the above axioms is understood, it becomes clear that a model is nothing but an R-coalgebra. Furthermore, since the fragment contains only atomic formulas, the morphisms of models are exactly coalgebra homomorphisms.

We remark that although infinite disjunctions were used, all conjunctions and all quantifiers were finite.

Although the use of theories in $L_{\infty,\infty}$ to describe usual categories of models is probably the most flexible, when it comes to working with these categories a more categorical presentation is desirable. The main tool used for this in **[M/P]** is the notion of a sketch and its models, introduced by Ehresmann **[E]** and studied extensively by his group **[L]**, **[G/L]**, ...

A *sketch* $\mathcal{S}$ is a quadruple $(\mathbf{G}, \mathbf{D}, \mathbf{L}, \mathbf{C})$ where $\mathbf{G}$ is a graph (for us a graph is always a directed multigraph, i.e. a "category without multiplication or identities"), $\mathbf{D}$ is a class of diagrams of the form

$$\begin{array}{ccccccccc} & \nearrow & G_1' & \to & G_2' & \to & G_3' & \cdots & G_n' & \searrow & \\ G & & & & & & & & & & G' \\ & \searrow & G_1 & \to & G_2 & \to & G_3 & \cdots & G_m & \nearrow & \end{array}$$

in $\mathbf{G}$. (We allow empty paths.) $\mathbf{L}$ is a class of cones in $\mathbf{G}$, and $\mathbf{C}$ is a class of cocones in $\mathbf{G}$. A cone consists of a small graph $\mathbf{I}$ and a graph morphism

$$\mathbf{I}^- \to \mathbf{G}$$

where $\mathbf{I}^-$ is the graph obtained by adding a new object $-\infty$ to $\mathbf{I}$ and an arrow $-\infty \to v$ for every vertex v of $\mathbf{G}$. Cocones are similar.

A *morphism of sketches* $\mathcal{S} \to \mathcal{S}'$ is a graph morphism $\mathbf{G} \to \mathbf{G}'$ that takes diagrams in $\mathbf{D}$ to diagrams in $\mathbf{D}'$, cones and cocones of $\mathcal{S}$ to cones and cocones of $\mathcal{S}'$. Every category $\mathbf{A}$ has an underlying sketch: the graph consists of all objects and arrows of $\mathbf{A}$, the diagrams of all commutative diagrams of $\mathbf{A}$, the cones and cocones of all limit cones and colimit cocones respectively. A *model* of a sketch $\mathcal{S}$ is a sketch morphism from $\mathcal{S}$ into the underlying sketch of **Set**, the category of sets. (Models with values in any category make sense, at least if the category is complete and cocomplete.) A *morphism of models* is a natural transformation (a notion which makes sense for morphisms from a graph into a category).

The fact that this notion is so simple makes the following theorem, also due to Lair **[L]**, all the more surprising.

Theorem 2 *The categories of models of small sketches are the same as categories of models of theories in* $L_{\infty,\infty}$.

Although its proof is not deep, this theorem is basic to our theory.

An example of an easily sketchable category is the category of connected graphs. As a sketch we can take

$$A \rightrightarrows B \to 1$$

where 1 is to be a terminal object and the whole diagram a coequalizer. It is clear which sketch we mean, although it is stated somewhat imprecisely. Explicitly, let

(i) $\mathbf{G}$ be the graph

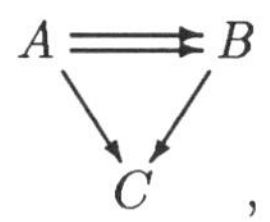

(ii) $\mathbf{D} = \emptyset$

(iii) $\mathbf{L} = \{\bar{C} : \emptyset^- \to \mathbf{G}\}$

(iv) $\mathbf{C} = \{\mathbf{I}^+ \cong \mathbf{G} \xrightarrow{1_G} \mathbf{G}\}$ where $\mathbf{I}$ is the graph $A \rightrightarrows B$.

The presentation of a Grothendieck topos as sheaves on a site gives another example of sketchable category. Let $(\mathbf{C}, \mathcal{T})$ be a site. We build a sketch $\mathcal{S}$ as follows. The graph of $\mathcal{S}$ is the underlying graph of $\mathbf{C}^{op}$ and the diagrams of $\mathcal{S}$ are all diagrams that commute in $\mathbf{C}^{op}$ (so that a model of $\mathcal{S}$ will be, in particular, a presheaf on $\mathbf{C}$). $\mathcal{S}$ has no cocones. It has one cone for each covering sieve $R \hookrightarrow \mathbf{C}(-, C)$. The projection $P : El(R) \to \mathbf{C}$ has a canonical cocone to C, $P' : El(R)^+ \to \mathbf{C}$. The cone we want is $P'^{op} : El(R)^{+op} = El(R)^{op-} \to \mathbf{C}^{op}$. Then $Mod(\mathcal{S}) \cong Sh(\mathbf{C}, \mathcal{T})$.

2 The Limit Theorem

Theorem 3 (Limit Theorem) *The 2-category $\mathcal{A}cc$ is closed in $\mathcal{CAT}$ under weighted bilimits.* □

A proof of this theorem is given in [M/P]. Here we will give some of its applications. Most of the applications revolve around the fact that an accessible category has generators, a useful fact that is sometimes difficult to prove in practice.

However, before the applications, we will discuss the notion of weighted bilimit. It is one of the notions of limit appropriate to a 2-category (or bicategory) such as $\mathcal{A}cc$ or $\mathcal{CAT}$. Some care must be taken as not all limits are appropriate. For example, pullbacks are not since they are not "invariant under equivalence." The problem is illustrated by the following example. Let $\mathbf{A} \hookrightarrow \mathbf{Set}^{\mathbf{C}}$ be any full subcategory. Define functors $F_1, F_2 : \mathbf{Set}^{\mathbf{C}} \to \mathbf{Set}^{\mathbf{C}}$, both equivalent to the identity functor, by

$$F_i(\Phi) = \begin{cases} \Phi & \text{if } \Phi \text{ is in } \mathbf{A} \\ \Phi \times \mathrm{Const}(\{i\}) & \text{otherwise.} \end{cases}$$

($\mathrm{Const}(\{i\})$ is the constant functor with value $\{i\}$.) Then the pullback of F_1 and F_2 is $\mathbf{A}$, but of course, not every such $\mathbf{A}$ is accessible. One feels that there is something unnatural about this pullback in $\mathcal{CAT}$ and that the pullback should be "taken up to isomorphism." This will be an example of a bilimit.

The notion of *weighted bilimit* was introduced by Street [S] under the name "indexed bilimit." Since we are only interested in weighted bilimits in $\mathcal{CAT}$, we shall give a concrete description of them in this situation (for a general discussion the reader is referred to [S] or [**M/P**]).

There are several ways of generalizing to $\mathcal{CAT}$ the construction of limits in **Set** as compatible families of elements. One such way is to consider families of objects compatible up to isomorphisms satisfying an appropriate coherence condition. If the universal property is expressed as an isomorphism of categories, we get pseudo-limits; if it is expressed as an equivalence of categories, we get bilimits. Bilimits are only unique up to equivalence. If the pseudo-limit exists, as it does in $\mathcal{CAT}$, it is a bilimit and every bilimit is equivalent to it. Thus we first describe pseudo-limits.

In fact, we need *weighted* pseudo-limits. Let $\mathcal{I}$ be a small 2-category and $W : \mathcal{I} \to \mathcal{C}at$ a 2-functor into the 2-category of small categories. This is the *weight* 2-functor. The case of ordinary pseudo-limits is given by $W = \mathrm{Const}(1)$. If $\Phi : \mathcal{I} \to \mathcal{CAT}$ is another 2-functor, then the *pseudo-limit of Φ weighted by W* (or the *W-pseudo-limit of Φ*) is the category $\mathrm{Lim}_W \Phi$ whose objects are pseudo-natural transformations

$$W \to \Phi$$

and whose morphisms are modifications.

A *pseudo-natural transformation* $\theta : W \to \Phi$ is given by:

(i) for every object I in $\mathcal{I}$, a functor

$$\theta(I) : W(I) \to \Phi(I)$$

(ii) for every morphism $i : I \to J$ in $\mathcal{I}$, a natural isomorphism

$$\begin{array}{ccc} W(I) & \xrightarrow{\theta(I)} & \Phi(I) \\ {\scriptstyle W(i)}\downarrow & \Downarrow \theta(i) & \downarrow{\scriptstyle \Phi(i)} \\ W(J) & \xrightarrow[\theta(J)]{} & \Phi(J). \end{array}$$

These must satisfy:

(iii) $\theta(1_I) = 1_{\theta(I)}$ for every I,

(iv) for $I \xrightarrow{i} J \xrightarrow{j} K$ in $\mathcal{I}$

$$\begin{array}{ccc} W(I) & \xrightarrow{\theta(I)} & \Phi(I) \\ {\scriptstyle W(i)}\downarrow & \Downarrow \theta(i) & \downarrow{\scriptstyle \Phi(i)} \\ W(J) & \xrightarrow{\theta(J)} & \Phi(J) \\ {\scriptstyle W(j)}\downarrow & \Downarrow \theta(j) & \downarrow{\scriptstyle \Phi(j)} \\ W(K) & \xrightarrow[\theta(K)]{} & \Phi(K) \end{array} \quad = \quad \begin{array}{ccc} W(I) & \xrightarrow{\theta(I)} & \Phi(I) \\ {\scriptstyle W(ji)}\downarrow & \Downarrow \theta(ji) & \downarrow{\scriptstyle \Phi(ji)} \\ W(K) & \xrightarrow[\theta(K)]{} & \Phi(K), \end{array}$$

(v) for every 2-cell $\alpha : i \to i' : I \to J$ of $\mathcal{I}$,

$$\begin{array}{ccc} \Phi(i)\theta(I) & \xrightarrow{\theta(i)} & \theta(J)W(i) \\ {\scriptstyle \Phi(\alpha)\theta(I)}\downarrow & & \downarrow{\scriptstyle \theta(J)W(\alpha)} \\ \Phi(i')\theta(I) & \xrightarrow[\theta(i')]{} & \theta(J)W(i') \end{array}$$

commutes.

A *modification* of pseudo-natural transformations $m : \theta \to \phi$ is given by:

(i) for every I in $\mathcal{I}$, a natural transformation

$$m(I) : \theta(I) \to \phi(I)$$

such that

(ii) for every morphism $i : I \to J$ in $\mathcal{I}$

$$\begin{array}{ccc} \Phi(i)\theta(I) & \xrightarrow{\theta(i)} & \theta(J)W(i) \\ {\scriptstyle \Phi(i)m(I)}\downarrow & & \downarrow{\scriptstyle m(J)W(i)} \\ \Phi(i)\phi(I) & \xrightarrow[\phi(i)]{} & \phi(J)W(i) \end{array}$$

commutes. (There are no conditions on 2-cells.)

A category is a *(weighted) bilimit* of a diagram if it is equivalent to the (weighted) pseudo-limit.

Some examples will help clarify these concepts.

Example 1: Let

$$\begin{array}{ccc} & & \mathbf{A} \\ & & \downarrow{\scriptstyle F} \\ \mathbf{B} & \xrightarrow[G]{} & \mathbf{C} \end{array}$$

be a diagram in $\mathcal{CAT}$. The pseudo-pullback is the category whose objects are quintuples (A, B, C, c, c') where A, B, C are objects of $\mathbf{A}$, $\mathbf{B}$, $\mathbf{C}$ respectively, and $c : FA \to C$ and $c' : GB \to C$ are isomorphisms in $\mathbf{C}$. The morphisms are the obvious ones, i.e. triples of maps commuting with the isomorphisms.

This category is equivalent to the category whose objects are triples (A, B, c) where A and B are objects of $\mathbf{A}$ and $\mathbf{B}$ respectively and $c : FA \to GB$ is an isomorphism in $\mathbf{C}$, with the obvious morphisms. So this last category (which is somewhat simpler) is a bipullback but not a pseudo-pullback.

Suppose G has the following property: for every B in $\mathbf{B}$ and isomorphism $c : C \to GB$ in $\mathbf{C}$, there exist B' in $\mathbf{B}$ and an isomorphism $b : B' \to B$ such that $G(b) = c$. This happens, for example, if G is a fibration or a cofibration, or if $\mathbf{B}$ is a *replete* full subcategory of $\mathbf{C}$. In this case, the actual pullback is equivalent to the above bipullback and so it is a bipullback too.

Example 2: Take the same diagram in $\mathcal{CAT}$ but now give it the weight

$$\begin{array}{ccc} & & \mathbf{1} \\ & & \downarrow{\scriptstyle \bar{0}} \\ \mathbf{1} & \xrightarrow[\bar{1}]{} & \mathbf{2} \end{array}$$

(the weight in example 1 was implicitly the constant diagram **1**). Then the pseudo-limit has as objects seventuples

$$(A, B, C_1, C_2, c_1, c_2, c)$$

where $c_1 : FA \to C_1$ and $c_2 : GB \to C_2$ are isomorphisms and $c : C_1 \to C_2$ is a morphism of $\mathbf{C}$, and has the obvious morphisms. This is easily seen to be equivalent to the comma category (F, G), so the comma category appears as a weighted bilimit.

Example 3: To illustrate what the weight does, consider the case where W is the constant functor with value $\mathbf{C}$, $\bar{\mathbf{C}} : \mathbf{1} \to \mathcal{C}at$, and $\Phi : \mathbf{1} \to \mathcal{CAT}$ given by $\mathbf{A}$. Then the pseudo-limit is the functor category $\mathbf{A}^{\mathbf{C}}$.

Example 4: Let $F : \mathbf{A} \to \mathbf{A}$ be a functor such that $F^2 = F$ and consider the category fix(F) of fixed points of F, i.e. the subcategory of $\mathbf{A}$ with objects A such that $FA = A$ (equal!) and morphisms a such that $Fa = a$. On the surface, this does not seem like a wholesome sort of construction. However, the pseudo-limit of the diagram, with $\mathcal{I}$ the monoid $\{1, e \mid e^2 = e\}$, has as objects (A, a) where $a : FA \to A$ is an isomorphism and $Fa = 1$, and as morphisms the obvious ones. And it is easily seen that this category is equivalent to fix(F), so fix(F) *is* a bilimit.

Example 5: The (weighted) lax limit of a diagram is described in the same way as the pseudo-limit except that the natural transformations $\theta(i)$ are not required to be isomorphisms anymore. The reader is referred to **[K]** for more details. In particular it is mentioned there that there is a construction of a weight $W^\dagger$ from a given one $W : \mathcal{I} \to \mathcal{C}at$ such that the 2-limit weighted by $W^\dagger$ is the same as the lax limit weighted by W. In fact, this $W^\dagger$ also serves to show that the W-lax limit is a $W^\dagger$-bilimit, thus weighted lax limits are also appropriate for the Limit theorem.

The oplax limit is defined in the same way as the lax limit except that the $\theta(i)$ go in the opposite direction. Thus oplax $\lim(\Phi) \cong (\text{lax} \lim(\Phi^{op}))^{op}$ and since the opposite of a bilimit is also a bilimit, we see that oplax limits are bilimits too.

These examples should suffice for now. Below, we give some applications where the indexing 2-category is not merely a category.

The first application is taken from **[M/P]**. We mention it here as motivation for the problem below.

Proposition 1 *An accessible category* $\mathbf{A}$ *with cokernel pairs is well-copowered.*

Proof: The Limit theorem shows that the functor categories $\mathbf{A}^{\mathbf{2}}$ and $\mathbf{A}^{\mathbf{E}}$ are accessible ($\mathbf{E}$ is the category $\cdot \rightrightarrows \cdot$). For any object A in $\mathbf{A}$, the comma category $(A, \mathbf{A})$ is also accessible by the Limit theorem. The functor coker : $(A, \mathbf{A}) \to \mathbf{A}^{\mathbf{E}}$, which takes the cokernel pair of a map, is easily seen to be accessible, as is the diagonal functor $\Delta : \mathbf{A}^{\mathbf{2}} \to \mathbf{A}^{\mathbf{E}}$ given by

$$\Delta(B \xrightarrow{f} C) = B \overset{f}{\underset{f}{\rightrightarrows}} C.$$

Furthermore, Δ makes $\mathbf{A}^{\mathbf{2}}$ into a replete full subcategory of $\mathbf{A}^{\mathbf{E}}$, so the pullback

$$\begin{array}{ccc} (A,\mathbf{A}) & \xrightarrow{\text{coker}} & \mathbf{A}^{\mathbf{E}} \\ \uparrow & & \uparrow \Delta \\ \mathrm{Epi}(A) & \longrightarrow & \mathbf{A}^{\mathbf{2}} \end{array}$$

is accessible. But $\mathrm{Epi}(A)$ is the poset of epimorphisms out of A and an accessible poset must be small (every object is a sup of objects in a small subcategory of κ-presentables). Thus $\mathbf{A}$ is well-copowered. □

If $\mathbf{A}$ has an accessible completion $\mathbf{B}$ (i.e. if it can be fully embedded in a complete and cocomplete accessible category such that the inclusion preserves all existing limits and colimits) then the above proof extends to show that $\mathbf{A}$ is well-powered. (Simply replace both $\mathbf{A}$s on the right in the above square by $\mathbf{B}$s.) This happens, for example, if there exist arbitrarily large compact cardinals or if $\mathbf{A}$ is the category of models of a theory in $L_{\omega,\omega}$, usual (finitary) first order logic.

In fact, there are a number of results that hold for theories in $L_{\omega,\omega}$ that do not extend to the more general $L_{\infty,\infty}$. This suggests the following:

Problem: Find a categorical characterization of categories of models of theories in $L_{\omega,\omega}$.

It is well-known [**B/B**] that a triple $\mathbf{T} = (T, \eta, \mu)$ on a category $\mathbf{A}$ gives rise to a simplicial object in $\mathcal{CAT}(\mathbf{A}, \mathbf{A})$

$$1 \xrightarrow{\eta} T \underset{\eta T}{\overset{T\eta}{\underset{\longleftarrow}{\overset{\mu}{\rightrightarrows}}}} T^2 \;\equiv\; T^3 \;\cdots$$

If $\mathcal{I}$ is the 2-category with one object $*$ and $\mathcal{I}(*,*)$ the strict monoidal category $\mathbf{\Delta}$ of finite ordinals, then a triple gives a 2-functor $\Phi : \mathcal{I} \to \mathcal{CAT}$ by letting $\Phi(*) = \mathbf{A}$, $\Phi([n]) = T^n$, and for 2-cells as suggested above. Then a calculation shows that the lax limit of Φ weighted by $\mathbf{1}$ is the Eilenberg-Moore category $\mathbf{A}^{\mathbf{T}}$ of $\mathbf{T}$. Thus if $\mathbf{A}$ and T are accessible, the Limit theorem says that $\mathbf{A}^{\mathbf{T}}$ is too.

Dually, the Eilenberg-Moore category of a cotriple is an oplax limit. Thus if $\mathbf{G} = (G, \epsilon, \delta)$ is a cotriple on an accessible category $\mathbf{A}$ with G also accessible, then $\mathbf{A}_{\mathbf{G}}$ is accessible too.

It has long been known (see e.g. [**K/W**]) that the Eilenberg-Moore category of a left exact cotriple on a topos is again a topos. It is also known that, if the topos we start from is Grothendieck, then the constructed one is Grothendieck if and only if the cotriple has rank. Ieke Moerdijk has pointed out that this fact follows easily from our theory. First of all, any Grothendieck topos is accessible, as mentioned above. Any functor with a left or right adjoint between accessible categories is itself accessible ([**M/P**], Proposition 2.4.8), so if the Eilenberg-Moore category is Grothendieck then the cotriple is accessible, i.e. has rank.

Conversely, if the cotriple has rank, i.e. is accessible, then the Eilenberg-Moore category is accessible so it has generators. Since we already know that it is a cocomplete topos, it is Grothendieck.

Our final application is to categories like the category of coalgebras over a ring. In section 1 we gave a theory whose models are R-coalgebras so that category is accessible. It follows that it has generators. This was the main result of **[B]**. There are other categories similar to R-coalgebras, such as R-Hopf algebras, graded coalgebras, etc. In **[MacL]**, MacLane invented the notion of a **Prop** in order to model such structures.

A **Prop**, $\mathbf{P}$, is a commutative strict monoidal category whose objects are the natural numbers on which $\otimes$ is addition. If $\mathbf{V}$ is a symmetric monoidal category, a *model* of $\mathbf{P}$ in $\mathbf{V}$ is a symmetric monoidal functor $\mathbf{P} \to \mathbf{V}$. For example, the **Prop** for coalgebras has morphisms $\epsilon : [1] \to [0]$ and $\delta : [1] \to [2]$ and all consequences (such as $\epsilon \otimes 1 : [2] \to [1]$, $1 \otimes \delta : [2] \to [3]$) subject to equations making $([1], \epsilon, \delta)$ a comonoid (e.g. $(\epsilon \otimes 1)\delta = 1$). In fact it is the free symmetric strict monoidal category with comonoid, generated by the empty graph. (The reader is referred to **[MacL]** for more details.) In **[F]**, Fox generalized the results of Barr to categories of models of a **Prop** in a locally presentable monoidal category. Here we generalize Fox's results to show that models of a **Prop** in an accessible monoidal category is again accessible. In fact, it is not necessary to restrict our attention to **Props**. One can easily generalize the notion of **Prop** to include multisorted models such as pairs (C, A) where C is a coalgebra and A is a C-comodule. It is not necessary to assume commutativity of the tensor, but as most of the applications use a commutative tensor we treat only this case. The modifications necessary to get the noncommutative version of the theorem below are trivial. Thus we shall show that the category of monoidal functors from a small symmetric monoidal category into an accessible monoidal category is again accessible.

Definition 2 *A symmetric monoidal category* $(\mathbf{V}, I, \otimes, \alpha, \rho, \lambda, \sigma)$ *is called* accessible *if* $\mathbf{V}$ *and* $\otimes : \mathbf{V} \times \mathbf{V} \to \mathbf{V}$ *are both accessible.*

Theorem 4 *Let* $\mathbf{P}$ *and* $\mathbf{V}$ *be symmetric monoidal categories with* $\mathbf{P}$ *small and* $\mathbf{V}$ *accessible. Then the category* $Mon(\mathbf{P}, \mathbf{V})$ *of monoidal functors and monoidal natural transformations from* $\mathbf{P}$ *to* $\mathbf{V}$ *is accessible.*

Proof: We shall show that $Mon(\mathbf{P}, \mathbf{V})$ is a weighted bilimit of a diagram of accessible categories. For this we use a 2-category which is a slight variation on Barry Jay's "2-**Prop** of a pseudo-monoid" **[J]**. We start with the Lawvere theory of algebras with one constant i and one binary operation t (no equations). We make this into a 2-category $\mathcal{I}$ by adding the following 2-cells freely:

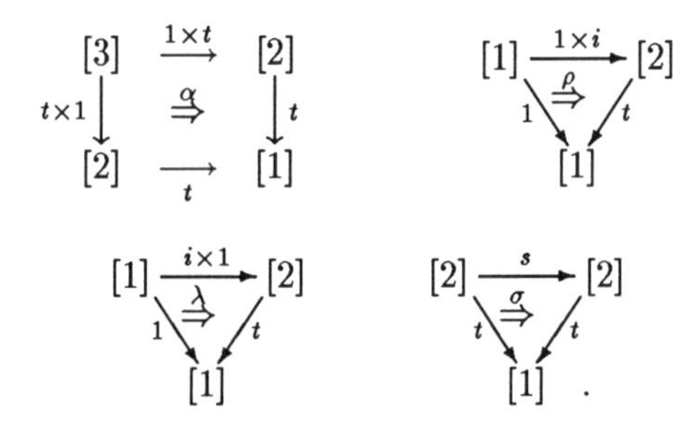

We don't even require that they be invertible, nor that the products $[m] \times [n] \cong [m+n]$ be 2-products.

If $\mathbf{V}$ is a monoidal category, we get a 2-functor $\Phi : \mathcal{I} \to \mathcal{CAT}$ by considering $(\mathbf{V}, I, \otimes)$ as an algebra for the above theory and then taking the 2-cells $\alpha,\rho,\lambda,\sigma$ to the natural isomorphisms of the same names in $\mathbf{V}$. Thus $\Phi([n]) = \mathbf{V}^n$, $\Phi(i) = \bar{I} : \mathbf{V}^0 \to \mathbf{V}$ and $\Phi(t) = \otimes : \mathbf{V}^2 \to \mathbf{V}$.

Let Φ be as just described and $W : \mathcal{I} \to \mathcal{C}at$ be a similar 2-functor corresponding to $\mathbf{P}$. We claim that the W-weighted pseudo-limit of Φ is equivalent to the category $Mon(\mathbf{P}, \mathbf{V})$.

Let $\theta : W \to \Phi$ be a pseudo-natural transformation. Let $\theta([1]) = F : \mathbf{P} \to \mathbf{V}$. For any $[n]$ we have projections $p_k : [n] \to [1]$ in $\mathcal{I}$, which give isomorphisms

$$\begin{array}{ccc} \mathbf{P}^n & \xrightarrow{\theta([n])} & \mathbf{V}^n \\ {\scriptstyle proj_k}\downarrow & \Downarrow \theta(p_k) & \downarrow{\scriptstyle proj_k} \\ \mathbf{P} & \xrightarrow[F]{} & \mathbf{V}. \end{array}$$

Thus $\langle \theta(p_k) \rangle : \theta([n]) \to F^n$ is a natural isomorphism. But in any pseudo-natural transformation we can always replace each $\theta(I)$ by an isomorphic one $\theta'(I)$ $(\tau_I : \theta(I) \xrightarrow{\cong} \theta'(I))$ provided we replace the isomorphisms $\theta(i)$ by $\theta'(i) = \tau_J \theta(i) \tau_I^{-1}$, and thus get an isomorphic pseudo-natural transformation. So, in the above, we can assume that $\theta([n])$ is F^n and that all $\theta(p_k)$ are identities. Corresponding to the binary operation $t : [2] \to [1]$ we have

$$\begin{array}{ccc} \mathbf{P}^2 & \xrightarrow{F^2} & \mathbf{V}^2 \\ \otimes\downarrow & \Downarrow \theta(t) & \downarrow\otimes \\ \mathbf{P} & \xrightarrow[F]{} & \mathbf{V} \end{array}$$

which we call $\tilde{F} : \otimes F^2 \to F\otimes$, and corresponding to the nullary operation $i : [0] \to [1]$ we have

$$\begin{array}{ccc} \mathbf{1} & \longrightarrow & \mathbf{1} \\ {\scriptstyle I}\downarrow & \Downarrow \theta(i) & \downarrow{\scriptstyle I} \\ \mathbf{P} & \xrightarrow[F]{} & \mathbf{V} \end{array}$$

which is called $F^0 : I \to F(I)$. Conditions (iv) and (v) in the definition of a pseudo-natural transformation, applied to α,ρ,λ and σ, show that $(F, F^0, \tilde{F})$ is the data for a monoidal functor. For example, (iv) shows that the top and bottom of the diagram

$$\begin{array}{ccccc} (FA \otimes FB) \otimes FC & \xrightarrow{\tilde{F}\otimes 1} & F(A \otimes B) \otimes FC & \xrightarrow{\tilde{F}} & F((A \otimes B) \otimes C) \\ {\scriptstyle \alpha}\downarrow & & & & \downarrow{\scriptstyle F\alpha} \\ FA \otimes (FB \otimes FC) & \xrightarrow[1\otimes\tilde{F}]{} & FA \otimes F(B \otimes C) & \xrightarrow[\tilde{F}]{} & F(A \otimes (B \otimes C)) \end{array}$$

are $\theta(t \cdot t \times 1)$ and $\theta(t \cdot 1 \times t)$ respectively evaluated at (A, B, C), and its commutativity follows from (v) for α in $\mathcal{I}$.

Some calculations show that this establishes an equivalence of categories between the weighted pseudo-limit of Φ and the category $Mon(\mathbf{P}, \mathbf{V})$. If $\mathbf{V}$ and $\otimes$ are accessible, Φ maps into $\mathcal{A}cc$ and thus $Mon(\mathbf{P}, \mathbf{V})$ is accessible. □

Examples of accessible monoidal categories are given by R-modules in a Grothendieck topos $\mathbf{E}$, where R is a commutative ring in $\mathbf{E}$, and also the graded and differential graded versions of these. Some examples of categories which appear as $Mon(\mathbf{P}, \mathbf{V})$ are the categories of (commutative) algebras or coalgebras or Hopf algebras in the above monoidal categories.

An application is given by the following observations: an accessible category is complete if and only if it is cocomplete and an accessible functor between complete accessible categories has a left (right) adjoint if and only if it preserves limits (resp. colimits). Take $\mathbf{V}$ to be R-mod($\mathbf{E}$) as above. Then R-coalg($\mathbf{E}$) is easily seen to be cocomplete, so it is complete too. Knowing this, R-Hopf($\mathbf{E}$) is easily seen to be complete and thus it is cocomplete. (But neither limits nor colimits are obvious for R-Hopf algebras.) The forgetful functor $U : R$-Hopf($\mathbf{E}$) $\to R$-coalg($\mathbf{E}$) clearly preserves limits and thus has a left adjoint, i.e. in a Grothendieck topos every coalgebra generates a free Hopf algebra.

References

[B] M. Barr, Coalgebras over a Commutative Ring, J. Alg. 32(1974) 600-610.

[B/B] M. Barr and J. Beck, Homology and Standard Constructions, Springer Lecture Notes 80(1969) 245-335.

[E] C. Ehresmann, Esquisses et types de structures algébriques, Bull. Instit. Polit., Iasi, XIV(1968),1-14.

[F] T. Fox, Purity in Locally-Presentable Monoidal Categories, J. Pure Appl. Alg. 8(1976) 261-265.

[G/L] R. Guitart et C. Lair, Limites et co-limites pour représenter les formules, Diagrammes 7(1982).

[G/U] P. Gabriel et F. Ulmer, Lokal prasentierbare Kategorien, Springer Lecture Notes 221(1971).

[J] C. B. Jay, Languages for monoidal categories, preprint.

[K] G. M. Kelly, Elementary Observations on 2-Categorical Limits, preprint.

[K/W] A. Kock and G. C. Wraith, Elementary Toposes, Aarhus Lecture Notes Series 30(1971).

[L] C. Lair, Catégories modelables et catégories esquissables, Diagrammes 6(1981).

[MacL] S. MacLane, Categorical Algebra, Bull. A.M.S. 71(1965) 40-106.

[M/P] M. Makkai and R. Paré, Accessible Categories: The Foundations of Categorical Model Theory, to appear in Contemporary Mathematics.

[Stn] B. Stentröm, Rings of Quotients, Springer-Verlag, New York Heidelberg Berlin, 1975.

[S] R. Street, Fibrations in bicategories, Cahiers de Topologie et Géom. Diff. 21(1980), 111-160.

Contemporary Mathematics
Volume 92, 1989

COHERENCE FOR BICATEGORIES WITH FINITE BILIMITS I

A.J. Power[1]

ABSTRACT. After a brief discussion on the importance of coherence questions to the working Category Theorist, we define three classes of limits appropriate to the study of 2-categories: bilimits, pseudo-limits, and flexible limits. We then discuss the Yoneda functor for 2-categories, and establish results analogous to those for the ordinary Yoneda functor. Using these results, we resolve a long-standing coherence question: every bicategory with finite bilimits is biequivalent to a 2-category with finite flexible limits. This, together with a companion result relating homomorphisms that preserve finite bilimits to 2-functors that preserve finite flexible limits, gives a complete theory of coherence for "two-dimensional" essentially algebraic theories.

1. INTRODUCTION. The following situation is increasingly common in Category Theory: one has a theorem, with a proof; but the proof is not quite complete, because it has not been verified that some naturally arising isomorphisms fit together in the way that the formal theory demands. The verification is almost always routine, and one's intuition is almost always vindicated; but to check the detail is often a very tedious job. Of course, one should still do it. Sometimes, that results in enormous diagrams going into print, [6] and [7]; other times, having checked the detail, one morely asserts the result and omits the detail, [9] and [15]. Neither solution is entirely satisfactory. This problem may be called the problem of coherence, cf. [8].

There are three basic approaches to coherence questions: ignore them; check the coherence of each collection of isomorphisms as it arises; or prove a coherence theorem, enabling one to resolve a class of such instances with one technical result. The first approach can be dangerous, as illustrated in

1980 Mathematics Subject Classification (1985 Revision). 18D05
1 Supported by National Science Foundation of Australia.

[2], because on rare occasions, one's intuition fails; the second approach is often unavoidable; but clearly, the third approach, when possible, is preferable. In fact, there have been several important coherence results proved over the past thirty years: one of the early fundamental ones appeared in Mac Lane's article [12]; and the theory of clubs was developed precisely in order to study coherence [8]. At present, Kelly, other colleagues, and myself are writing a series of articles, commencing with [4], that study the problem of coherence of "two-dimensional" monad theory and essentially algebraic theories. This paper forms part of that series.

In this paper, we prove a coherence theorem: every bicategory with finite bilimits is biequivalent to a 2-category with finite flexible limits. As early as 1980, [15], Street observed the absence of such a theorem, and remarked that such an absence forced him to prove results in the "bi-" case. It has long been known that every bicategory is biequivalent to a 2-category, but coherence for bilimits has proved to be of substantial difficulty. In a later article, we shall exhibit a companion result relating homomorphisms that preserve finite bilimits to functors that preserve finite flexible limits. Then, by work analogous to that in [4], we will have a complete theory of coherence for "two-dimensional" essentially algebraic theories.

Section 2 herein revises the definitions of bilimits, pseudo-limits, and flexible limits. Section 3 discusses the Yoneda functor for 2-categories. Section 4 gives the coherence result.

I offer warm thanks to Max Kelly for his advice and encouragement, and more generally to the Sydney Category Seminar and its many visitors.

2. BILIMITS, PSEUDO-LIMITS AND FLEXIBLE LIMITS. We assume the reader is acquainted with the notions of bicategory, homomorphism (or psuedo-functor), strong transformation, and modification, as defined for example in [1,15]. Following [15], we call **Hom(K,Cat)** the 2-category of homomorphisms, strong transformations, and modifications from a bicategory K to **Cat**. If **K** is in fact a 2-category, we call **Ps(K,Cat)** the full sub-2-category of **Hom**(K,**Cat**) determined by the 2-functors from **K** to **Cat**; and we call [K,**Cat**] the functor 2-category.

2.1. DEFINITIONS (i) [15] Given bicategories **D** and K, and homomorphisms f: D $\to$ **Cat** and g: D $\to$ K, an f-*weighted bilimit of* g is a birepresenting object for **Hom(D,Cat)** (f,K(-,g)): $K^{op} \to$ **Cat**, i.e. an object **bilim**(f,g) of K such that K(-,**bilim**(f,g)) $\simeq$ **Hom**(D,**Cat**)(f,**K**(-,g)) in the 2-category **Hom**(K,**Cat**).

(ii) [10] If, in (i), D and K are 2-categories, and f and g are 2-functors, an f-*weighted pseudo-limit of* g is a representing object for $\mathbf{Ps}(D,\mathbf{Cat})(f,K(-,g)):K^{op} \to \mathbf{Cat}$, i.e. an object $\mathbf{pslim}(f,g)$ of K such that $K(-,\mathbf{pslim}(f,g)) \cong \mathbf{Ps}(D,\mathbf{Cat})(f,K(-,g))$ in the 2-category $[K,\mathbf{Cat}]$.

It is immediate that any pseudo-limit acts as a bilimit. However, it is common for a bilimit to exist, where f and g are functors, without the existence of a corresponding pseudo-limit. For instance, **Groth Top** has all bilimits but in general does not have pseudo-limits. The same is true for all 2-categories of the form $T\text{-}\mathbf{Alg}^{op}$ as defined in [4]: some such **T-Alg**'s are **Lex**, **Cart**, and **Strong Mon**. Of course, a bilimit is unique only up to equivalence.

At this level, as illustrated in [9], it is essential to consider weighted limits, not just conical limits, in order to determine completeness properties. Moreover, the definition of flexible limits, to be given below, cannot even be properly formulated if attention is restricted to conical limits. We move towards the definition of flexible limits as follows:

2.2 DEFINITION [5] Given 2-categories D and K, and 2-functors $f: D \to \mathbf{Cat}$ and $g: D \to K$, an f-*weighted limit of* g is a representing object for $[D,\mathbf{Cat}](f,K(-,g)):K^{op} \to \mathbf{Cat}$, i.e. an object $\mathbf{lim}(f,g)$ such that $K(-,\mathbf{lim}(f,g)) \cong [D,\mathbf{Cat}](f,K(-,g))$ in $[K,\mathbf{Cat}]$.

2.3 PROPOSITION [4] Given a small 2-category D, the inclusion 2-functor $J: [D,\mathbf{Cat}] \to \mathbf{Ps}(D,\mathbf{Cat})$ has a left adjoint $(\ )'$.

Proposition 2.3 is an instance of the main theorem of [4]; an explicit construction of the left adjoint appears in [3]. It tells us that every small pseudo-limit is a **Cat**-enriched limit, since $\mathbf{pslim}(f,g) \cong \mathbf{lim}(f',g)$, either side existing if the other does. Observe that conical pseudo-limits, i.e. those pseudo-limits weighted by $\Delta 1: D \to \mathbf{Cat}$, are represented in general as non-conical Cat-enriched limits. Henceforth, assume D is small.

We are now in a position to define flexible limits.

2.4. DEFINITIONS [3,4] (i) A 2-functor $f: D \to \mathbf{Cat}$ is *flexible* if f is a retract of f' in $[D,\mathbf{Cat}]$.

(ii) If $f: D \to \mathbf{Cat}$ is flexible, say $\mathbf{lim}(f,g)$ is a *flexible limit*.

In [4], Definition 2.4(i) is made in a more general context, and it is shown that there are several equivalent statements. For instance, f is flexible if and only if f is a retract of some g' in $[D,\mathbf{Cat}]$; and f is flexible if and only if f is retract-equivalent to f', i.e. there exist $i: f \Rightarrow f'$ and $r: f' \Rightarrow f$ such that $ri = 1$ and $ir \cong 1$. In particular, if f is flexible, it follows that f is equivalent to f'. Applying the remarks after Proposition 2.3, this shows us:

2.5. REMARK. If f is flexible, $\mathbf{lim}(f,g) \simeq \mathbf{pslim}(f,g)$ (if these limits exist).

Flexible limits include pseudo-limits, since f' is trivially a retract of g' for some g. They also include most of the **Cat**-enriched limits defined in [10]: lax limits, inserters, iso-inserters, equifiers, comma objects, products, cotensors, and Eilenberg-Moore objects. They do not include equalizers or identifiers. It is shown in [3] that a 2-category **K** admits flexible limits if and only if it admits pseudo-limits and splitting of idempotents, i.e. given a 1-cell e in K such that $e^2 = e$, there exist r and i with $ri = 1$ and $ir = e$. Some 2-categories that admit flexible limits are **Lex**, **Cart**, and $\mathbf{Groth\ Top}^{op}$.

3. THE YONEDA FUNCTOR FOR 2-CATEGORIES. By a mild abuse of notation, given a small 2-category K, we regard the Yoneda (2-)functor as having codomain $\mathbf{Hom}(K^{op},\mathbf{Cat})$; so we have $Y: K \to \mathbf{Hom}(K^{op},\mathbf{Cat})$, $YK = K(-,K)$.

3.1. PROPOSITION [15] Given a small 2-category **K**, the Yoneda functor $Y: K \to \mathbf{Hom}(K^{op},\mathbf{Cat})$ is locally an equivalence, i.e. for each K and L in **K**, $Y_{KL}: K(K,L) \to \mathbf{Hom}(K^{op},\mathbf{Cat})(K(-,K), K(-,L))$ is an equivalence of categories.

3.2. PROPOSITION [13] Every bicategory is biequivalent to a 2-category, i.e. for every bicategory **K**, there is a 2-category **K'**, and there are homomorphisms $F: K \to \mathbf{K}'$ and $G: \mathbf{K}' \to K$ such that $FG \simeq 1_{K'}$ and $GF \simeq 1_K$.

3.3. LEMMA. Given small bicategories K and **L** and homomorphisms $f: K \to \mathbf{Cat}$, $g: \mathbf{L} \to \mathbf{Cat}$, and $h: K \times L \to \mathbf{Cat}$,

$$\mathbf{Hom}(K,\mathbf{Cat})(f-,\mathbf{Hom}(L,\mathbf{Cat})(g?,h(-,?))) \simeq \mathbf{Hom}(L,\mathbf{Cat})(g?,\mathbf{Hom}(K,\mathbf{Cat})(f-,h(-,?))).$$

PROOF By Proposition 3.2, it suffices to assume K and **L** are 2-categories. Moreover, by an example in [14], every homomorphism from a 2-category C into **Cat** is equivalent to a functor from C into **Cat**. So, we may assume f, g and h are functors. So, we must show that

$$\mathrm{Ps}(K,\mathbf{Cat})(f-,\mathrm{Ps}(L,\mathbf{Cat})(g?,h(-,?))) \simeq \mathrm{Ps}(L,\mathbf{Cat})(g?,\mathrm{Ps}(K,\mathbf{Cat})(f-,h(-,?))).$$

In fact, applying Proposition 2.3 to a standard result of enriched category theory ([9](3.7) and (3.19)), the above follows immediately.

3.4. REMARK. It is easy to verify that the equivalence of Lemma 3.3 as constructed therein is natural in f, in the sense that it lifts to an equivalence in **Hom**(**Hom**(K,**Cat**),**Cat**).

3.5. PROPOSITION. For any small 2-category **K**, the Yoneda functor $Y: K \to \mathbf{Hom}(K^{op},\mathbf{Cat})$ preserves whatever bilimits exist in **K**.

PROOF. Given f: **D** → K, g: **D** → **Cat** homomorphisms such that **bil im**(f,g) exists in **K**. By Proposition 3.2, we may assume **D** is a 2-category. Then,

$$\begin{aligned} \mathbf{Hom}(K^{op},\mathbf{Cat})(-,Y\,\mathbf{bilim}(f,g)) &= \mathbf{Hom}(K^{op},\mathbf{Cat})(-,K(-,\mathbf{bilim}(f,g))) \\ &\simeq \mathbf{Hom}(K^{op},\mathbf{Cat})(-,\mathbf{Hom}(D,\mathbf{Cat})(f?,K(-,g?))) \\ &\simeq \mathbf{Hom}(D,\mathbf{Cat})(f?,\mathbf{Hom}(K^{op},\mathbf{Cat})(-,K(-,g?))) \\ &= \mathbf{Hom}(D,\mathbf{Cat})(f?,\mathbf{Hom}(K^{op},\mathbf{Cat})(-,Yg?)); \end{aligned}$$

so Y **bilim**(f,g) acts as a bilimit **bilim**(f,Yg).

3.6. PROPOSITION. For any small 2-category **K**, **Hom**(K^{op},**Cat**) has all flexible limits.

PROOF. It is illustrated in [14] that **Hom**(K^{op},**Cat**) is of the form **Ps-T-Alg** for a 2-monad **T** on [|**K**|,**Cat**], where |K| is the underlying set of objects of K. It is an elementary but tedious calculation to show that every such 2-category has flexible limits: they are given as in the base 2-category [|K|,**Cat**]. First one proves the existence of pseudo-limits: the algebra structure is given by the universal property of the limits in the base. Then one splits idempotents, again by doing so in the base, then lifting the splitting to **Ps-T-Alg**. More detail will appear in [11].

4. THE MAIN COHERENCE RESULT

4.1. THEOREM. Every bicategory with finite bilimits is biequivalent to a 2-category with finite flexible limits.

PROOF. Let K be a bicategory with finite bilimits. By Proposition 3.2, we may assume that **K** is a 2-category. Consider Y: K → **Hom**(K^{op},**Cat**). Let $\bar{K}$ be the closure of ob(Y**K**) in **Hom**(K^{op},**Cat**) under equivalences, and let Y factor as $K \xrightarrow{i} \bar{K} \xrightarrow{j} \mathbf{Hom}(K^{op},\mathbf{Cat})$. Since Y is locally an equivalence by Proposition 3.1, it follows that the inclusion i: K → $\bar{K}$ is a biequivalence. Let k: $\bar{K}$ → K be a biequivalence inverse of i. Given a 2-category **D** and functors f: **D** → **Cat** and g: **D** → $\bar{K}$ such that f is a finite flexible weight, **bilim**(f,kg) exists in K, and so Y preserves it by Proposition 3.5.

$$\begin{aligned} \text{So} \quad \lim(f,jg) &\simeq \mathrm{pslim}(f,jg) && \text{by Remark 2.5} \\ &= \mathrm{bilim}(f,Ykg) && \text{since } Yk = jik \simeq j \\ &\simeq Y\,\mathrm{bilim}(f,kg); \end{aligned}$$

so $\mathbf{lim}(f,jg)$ lies in $\bar{K}$, i.e. $\bar{K}$ is closed in $\mathbf{Hom}(K^{op},\mathbf{Cat})$ under finite flexible limits.

For reasons of size, $\bar{K}$ may not be a legitimate bicategory; so let $\hat{K}$ be the closure of ob(iK) in $\bar{K}$ under finite flexible limits. By the general considerations of [9], $\hat{K}$ is legitimate. It is immediate that $\hat{K}$ is biequivalent to K.

3.2. REMARKS. (i) It is clear that this description of $\hat{K}$, and the proof that it has the desired properties, does not involve the Axiom of Choice in any substantial way.

(ii) The "finiteness" condition can obviously be replaced by any smallness condition that distinguishes between the size of K and the size of the limits it is to possess.

BIBLIOGRAPHY

1. J. Bénabou, Introduction to bicategories, Lect. Notes in Math. 47 (Springer, Berlin, 1967), 1-77.

2. J. Bénabou,"Fibred categories and the foundations of naive category theory", J. Symb. Logic, 50 (1985), 10-37.

3. G. Bird, "Limits of locally-presentable categories", Ph.D. thesis, Univ. of Sydney (1984).

4. R. Blackwell, G.M. Kelly, and A.J. Power, "Two-dimensional monad theory" (submitted).

5. F. Borceux, G.M. Kelly, "A notion of limit for enriched categories", Bull. Aust. Math. Soc. 12 (1975), 49-72.

6. M. Bunge, "Coherent extensions and relational algebras", Trans Amer. Math. Soc. 197 (1974)), 355-390.

7. J. W. Gray, "Formal category theory: adjointness for 2-categories", Lect. Notes in Math. 391 (Springer, Berlin, 1974).

8. G.M. Kelly, "On clubs and doctrines", Lect. Notes in Math. 420 (Springer, Berlin, 1974), 181-256.

9. G.M.Kelly, Basic Concepts of Enriched Category Theory, (Cambridge University Press, Cambridge, 1982).

10. G.M. Kelly, "Elementary observations on 2-categorical limits"(submitted).

11. G.M. Kelly and A.J. Power, "Pseudo-algebras and flexibility for a monad" (in preparation).

12. S. Mac Lane,"Natural associativity and commutativity", Rice University Studies 49, (1963), 28-46.

13. S. Mac Lane, R. Paré, "Coherence for bicategories and indexed categories", J. Pure Appl. Algebra, 37 (1985), 59-80.

14. A.J. Power, "A general coherence result" (to appear).

15. R. Street, "Fibrations in bicategories", Cahiers de Top. et Geom. Diff. 21, (1980), 111-160.

DEPARTMENT OF PURE MATHEMATICS
UNIVERSITY OF SYDNEY
N.S.W., 2006
AUSTRALIA

Contemporary Mathematics
Volume **92**, 1989

On Partial Cartesian Closed Categories.

Leopoldo Román

Introduction.

Recently the study of partial categorical notions has been subject by several people ([Cu.Ob.], [Di.He.], [Ro.]). The main interest in this paper is to clarify the relationship between Partial Cartesian Categories, Partial Cartesian Closed Categories, and Cartesian, Cartesian Closed Categories. We will follow the spirit of [Cu.Ob.] and we will try to answer some natural questions which arise after reading that paper.

I want to express my sincere thanks to M. Makkai and R. Paré for his encouragement and advice on several aspects of this work.

This paper was written while the author was a guest of McGill University and of the Groupe Interuniversitaire en études catégoriques in Montréal.

1. Categories with partial morphisms.

1.1 Definition [Cu.Ob.]. A category with partial morphisms or $p\underline{C}$ is a category $\underline{C}$ such that for every homset $hom_{\underline{C}}(A, B)$ we have a partial order structure and has a distinguished subset of maximal arrows called total; composition is monotonic.

Total arrows satisfy the following: Identity arrows are total, total arrows are closed under composition.

1.2 Examples.

Every pointed category [Di.He.] is a $p\underline{C}$-category. Every category with finite limits is a $p\underline{C}$-category.

For every $p\underline{C}$- category $\underline{C}$, we can consider the subcategory $\underline{C}_T$ of $\underline{C}$ whose objects are the same as $\underline{C}$ and whose morphisms are the total arrows. We are interested in sudying the relationship between the notions of partial product, partial exponentiation, for $\underline{C}$ and the corresponding notions (products, exponentiation) for $\underline{C}_T$.

1.3 Definition [Cu.Ob.]. A $p\underline{C}$ category $\underline{C}$ has liftings if for every $\underline{C}$-object A there is an object $\tilde{A}$ called the lifting of A, such that for every object B the is a bijection $\sigma_{B,A} : hom_{\underline{C}}(B, A) \to hom_{\underline{C}_T}(B, \tilde{A})$ such that for all $f : C \to B$ and $g : B \to A$ in $\underline{C}$ we have:

$$\sigma_{B,A}(g)f \leq \sigma_{C,A}(gf).$$

Perhaps, liftings are motivated by the following example. If we start with any topos $\underline{E}$ and consider the category of partial morphisms then $\tilde{A}$ is the partial map classifier of A.

Notice that when a $p\underline{C}$-category $\underline{C}$ has liftings then for every $\underline{C}$-object A we have an inclusion

$$\eta_A : A \to \tilde{A}$$

where $\eta_A = \sigma_{A,A}(1_A)$; in fact, this arrow is a $\underline{C}$-section via $ex_A = \sigma^{-1}_{\tilde{A},A}(1_A)$; Moreover,$\eta_A$ is always a total morphism but ex_A is ot necessaraly a total morphism.We will say that A is a complete object when in fact there is a total map

$$p : \tilde{A} \to A$$

such that $p\eta_A = 1_A$.

Curien and Obtulowicz study in detail complete objects; for instance,

$$\tilde{:}\underline{C} \longrightarrow \underline{C}_T$$

is a functor and it is right adjoint to the inclusion functor $i : \underline{C}_T \longrightarrow \underline{C}$. In particular, the following proposition will be very useful.

1.4 Proposition [Cu.Ob.] An object A is complete if and only if, for every object B and every morphism $f : B \to A$, there is a total morphism $g : B \to A$ such that $f \leq g$. In particular, $\tilde{A}$ is complete for every object A.

2.Partial Cartesian and Cartesian Closed Categories.

We introduce now the notion of partial cartesian category.

2.1 Definition [Cu.Ob.] A $p\underline{C}$-category $\underline{C}$ is called partially cartesian or $p\underline{CC}$ when it has the following structure:

I.-A partial terminal object 1 and for every object A, an arrow $!_A : A \to 1$ satisfying

i. $!_A$ is maximum in $hom(A,1)$ for every object A.

ii. The total morphisms are exactly those $f : A \to B$ such that $!_B f = !_A$.

iii. If $f, f', g : A \to B$ are such that $f, f' \leq g$ and $!_B f \leq !_B f'$ then $f \leq f'$.

iv. If $h, h' : A \to B$ are such that $h \leq h'$ then for any $g : B \to 1$ we have $gh = (gh') \cap (gh)$.

II.- For every object A, B a partial product given by an object $A \times B$, partial projections $\pi_{A,B} : A \times B \to A$, $\pi'_{A,B} : A \times B \to B$, and such that for every pair of morphisms $f : C \to A, g : C \to B$ there is a morphism $\langle f, g\rangle : C \to A \times B$, satisfying:

v. $\pi_{A,B}, \pi'_{A,B}$ are total morphisms.

vi. $\langle -, -\rangle$ is monotonic in both arguments.

vii. For any $h : C \to A \times B$, $h = \langle \pi_{A,B}h, \pi'_{A,B}h\rangle$

viii. $\pi_{A,B}\langle f, g,\rangle \leq f$ and $\pi'_{A,B}\langle f, g\rangle \leq g$.

ix. For all $k : D \to C, \langle f, g\rangle k = \langle fk, gk\rangle$.

x. $1_{A\times B}\langle fg,\rangle = (!_B f) \cap (!_B g)$.

2.2 Example.

Every category with finite limits is a partial cartesian category.

An interesting consequence of this definition, as is pointed out in [Cu.Ob.], is that $\underline{C}_T$ is actually a Cartesian Category.

We are now ready to introduce the notion of partial cartesian closed category.

2.3 Definition [Cu.Ob.]. A partial cartesian category $\underline{C}$ is partially cartesian closed or $p\underline{CCC}$ if for any objects A, B there is an object A^B such that for every object C there is a bijection

$$\lambda : hom_{\underline{C}}(C \times B, A) \to hom_{\underline{C_T}}(C, A^B)$$

satisfying the following condition:

For all $f : D \to C, g : C \times B \to A$

$$\lambda(g)f \leq \lambda(g(f \times 1_B))$$

where $f \times 1_B \equiv \langle f\pi_{D,B}, \pi'_{D,B}\rangle$.

2.4 Example. If $\underline{E}$ is any topos and we consider the category of partial morphisms $\underline{E}_p$ then A^B is nothing but $(\tilde{A})^B$. This suggests that every $p\underline{CCC}$-category has liftings.

2.5 Proposition [Cu.Ob.] Every $p\underline{CCC}$-category $\underline{C}$ has liftings; namely, for every $\underline{C}$-object A, we have $A^1 = \tilde{A}$.

In fact we can exploit more the definition of a $p\underline{CCC}$-category. Notice that for every object A, $1 \times A$ is isomorphic to A. Therefore we have the following isomorphisms:

$$hom_{\underline{C_T}}(C \times B, \tilde{A}) \simeq hom_{\underline{C}}(1 \times (C \times B), A)$$

$$\simeq hom_{\underline{C}}(C \times B, A) \simeq hom_{\underline{C_T}}(C, A^B).$$

Complete objects in a $p\underline{CCC}$-category $\underline{C}$ satisfy the following:

2.6 Lemma. 1 is a complete object. Complete objects are closed under products and for every $\underline{C}$-objects A, B A^B is a complete object.

Proof. We will use the characterization of complete objects.

Indeed, 1 is a complete object because $!_B : B \to 1$ is maximum in $hom_{\underline{C}}(B, 1)$. Suppose now, A, B are complete objects and take any $\underline{C}$-morphism $f : C \to A \times B$ then making $f_A = \pi_{A,B}f : C \to A, f_B = \pi'_{A,B}f : C \to B$; by the hypothesis, there are total morphisms f_A^-, f_B^- such that $f_A \leq f_A^-$ and $f_B \leq f_B^-$. Notice that

$$f = \langle f_A, f_B\rangle$$

and $$f \leq \langle f_A^-, f_B^-\rangle.$$

Clearly, $\langle f_A^-, f_B^-\rangle$ is a total morphism.

Finally, if A, B are arbitrary objects of $\underline{C}$ and $f : C \to A^B$ is any $\underline{C}$-morphism, then we want a total morphism $f^- : C \to A^B$; i.e., a $\underline{C}$-morphism $g : C \times B \to A$.

From $f : C \to A^B$, we have: $f \times 1_B : C \times B \to A^B \times B$.Call

$$ev_A = \lambda^{-1}(1_{A^B}) : A^B \times B \to A$$

then we get $g = ev_A(f \times 1_B) : C \times B \to A$. Define, $f^- = \lambda(g) : C \to A^B$; then, we have

$$f \leq \lambda(g)$$

simply because $\lambda(g) = \lambda(ev_A(f \times 1_B)) \leq \lambda(ev_A)f = f$.

In [Cu.Ob.], it was stated that the full subcetegory $\underline{C}_{TL}$ of $\underline{C}_T$ which has as objects 1 and $\tilde{A}$ for all $\underline{C}$-object A, is Cartesian Closed. The problem is that there is no good reason for $\tilde{A}^{\tilde{B}}$ be of the form $\tilde{C}$ for some $\underline{C}$-object C.

2.7 Proposition. If $\underline{C}$- is a partial cartesian closed category then the full subcategory $\underline{CC}_T$ of $\underline{C}_T$ whose objects are complete objects of $\underline{C}$, is a Weak cartesian closed category.

In [L.S.] (see exercise 2, p.97) , the notion of a weak $\underline{C}$-monoid is introduced. In order to be more clear we will define a weak cartesian closed category.

2.8 Definition. A weak cartesian closed category $\underline{C}$, is a category with the following structure:

i. A terminal object 1

ii. For every objects A, B, an object $A \times B$ together with morphisms $\pi_{A,B} : A \times B \to B, \pi'_{A,B} : A \times B \to B$.

iii. For every objects A, B, an object A^B, and an arrow $ev_{A,B} : A^B \times B \to A$.

These data satisfy the following conditions:

a. For all $f : C \to A, g : C \to B$, there is an arrow

$$\langle f, g, \rangle : C \to A \times B.$$

b. For any $k : C \times B \to A$ there is an arrow $k^* : C \to A^B$.

satisfying the following equations.

1. $\pi_{A,B}\langle f, g\rangle = f$.
2. $\pi'_{A,B}\langle f, g\rangle = g$.
3. For all $h : D \to C, \langle f, g\rangle h = \langle fh, gh\rangle$.
4. $ev_{A,B}\langle k^* f, g\rangle = k\langle f, g\rangle$.
5. $k^* f = (k\langle f\pi_{A,B}, \pi'_{A,B}\rangle)^*$.

We will prove now the proposition 2.7.

Proof of 2.7 .

The only thing we need to show is the weak exponentiation. If A, B are arbitrary $\underline{CC}_T$-objects, then by the lemma 2.6, A^B is a complete object. We define first the evaluation morphism.

We already have a $\underline{C}$-arrow $ev_A : A^B \times B \to A$: since A is a complete object, there is a total arrow

$$ev_{A,B} : A^B \times B \to A$$

such that $ev_A \leq ev_{A,B}$.

Given any CC_T-morphism $f : C \times B \to A$, we get a CC_T-morphism $f^* = \lambda(f) : C \to A^B$; we will show that

$$ev_{A,B}\langle f^*\pi_{C,B}, \pi'_{C,B}\rangle = f.$$

Since $\lambda(f), \pi_{C,B}, \pi'_{C,B}$, are total morphisms we have: $\lambda(ev_A\langle f^*\pi_{C,B}, \pi'_{C,B}\rangle)f^* = \lambda(ev_A)f^* = \lambda(f)$. Therefore $f = ev_A\langle f^*\pi_{C,B}, \pi'_{C,B}\rangle$. Using $ev_A \leq ev_{A,B}$ we get $f = ev_A\langle f^*\pi_{C,B}, \pi'_{C,B}\rangle \leq ev_{A,B}\langle f^*\pi_{C,B}, \pi'_{C,B}\rangle$; since f is total we get:

$$ev_{A,B}\langle f^*\pi_{C,B}, \pi'_{C,B}\rangle = f.$$

From this equality, 2.8.4 and 2.8.5 follow easily, hence $\underline{CC}_T$ is a weak cartesian closed category.

How can we get a cartesian closed category from $\underline{CC}_T$? Using the Karoubi envelope of a category (see [L.S.], p.100), we can prove that the Karoubi envelope of $\underline{CC}_T$, denoted by $K(\underline{CC}_T)$, is a cartesian closed category. In fact we can state a more general result; if $\underline{C}$ is a weak cartesian closed category then $K(\underline{C})$ is cartesian closed. This result is stated for weak $\underline{C}$-monoids in [L.S.] (see exercise 3 p.100).

2.9 Proposition The Karoubi envelope of a weak cartesian closed category $\underline{C}$ is cartesian closed.

Proof. Recall that the objects of $K(\underline{C})$ are idempotents $f_A : A \to A$ and the morphisms are $\underline{C}$-arrows $\phi : A \to B$ such that $f_B\phi f_A = \phi$. We will show first that $K(\underline{C})$ is cartesian.

Clearly, $K(\underline{C})$ has a terminal object. Now, if f_A, f_B are to arbitrary objects of $K(\underline{C})$ the the product is:

$$f_A \times f_B = \langle f_A\pi_{A,B}, f_B\pi'_{A,B}\rangle.$$

Using 2.8 it is easy to see that $f_A \times f_B$ is an idempotent. The projections are $f_A\pi_{A,B} : f_A \times f_B \to f_A$ and $f_B\pi'_{A,B} : f_A \times f_B \to f_B$. We check, for instance that $f_A\pi_{A,B}$ is a morphism.

$(f_A\pi_{A,B})(f_A \times f_B) = f_Af_A\pi_{A,B} = f_A\pi_{A,B}$.

If $\psi : f_C \to f_A$ and $\phi : f_C \to f_B$ are two $K(\underline{C})$-morphisms then $\langle\psi, \pi\rangle : f_C \to f_A \times f_B$ is a $K(\underline{C})$-morphism because

$$\langle\psi, \phi\rangle f_C = \langle\psi f_C, \phi f_C\rangle$$
$$= \langle\psi, \phi\rangle.$$

Using 2.8, it is easy to see that $\langle\psi, \phi\rangle$ is unique.

Finally, $K(\underline{C})$ is cartesian closed. If f_A, f_B are two arbitrary $K(\underline{C})$-objects $f_A^{f_B}$ is an idempotent and $f_Aev_{A,B}(1_{A^B} \times f_B) : f_A^{f_B} \times f_B \to f_A$ is a morphism (using again 2.8). Given $h : f_C \times f_B \to f_A$, $h^* : f_C \to f_A^{f_B}$ satisfies:

i. $f_A^{f_B}h^*f_C$
$=(f_Aev_{C,B}\langle h^*\pi_{C,B}, f_B\pi'_{C,B}\rangle))^*f_C$
$=(f_Ah(1_C \times f_B))^*f_C$.
$(f_Ah(1_C \times f_B)(f_C \times 1_B))^*$.
h^*.

hence, h^* is a $K(\underline{C})$-morphism.

ii. $f_Aev_{A,B}(1 \times f_B)(h^* \times 1_B)(f_C \times f_B)$.
$=f_Aev_{A,B}(h^* \times 1_B)(f_C \times f_B)$.
$f_Ah(f_C \times f_B) = h$.
iii. h^* is unique.

If $k : f_C \to f_A^{f_B}$ satisfies also $f_A ev_{A,B}(1 \times f_B)(k \times 1_B)(f_C \times f_B) = h$. then since $k = f_A^{f_B} k f_C$ we have

$$k = (f_A ev_{A,B}(kf_A \times f_B))^* = h^*.$$

Therefore $K(\underline{C})$ is a cartesian closed category.

3. When $\underline{C}_T$ is cartesian closed ?

If $\underline{C}$ is a $p\underline{CCC}$-category, when $\underline{C}_T$, the category of total maps is cartesian closed? We will give an answer to this but first we will make the following observation.

3.1 Lemma. If $\underline{C}$ is a partial category with liftings then the diagram

$$A \xrightarrow{\eta_A} \tilde{A} \underset{\tilde{\eta}_A}{\overset{\eta_{\tilde{A}}}{\rightrightarrows}} \tilde{\tilde{A}}$$

is a $\underline{C}_T$-equalizer.

Proof. The diagram commutes because

$$\eta_{\tilde{A}}\eta_A = \sigma_{\tilde{A},A}(1_{\tilde{A}})\eta_A$$

$$\tilde{\eta}_A\eta_A = \sigma_{\tilde{A},\tilde{A}}(\eta_A ex_A)\eta_A =$$

$$= \sigma_{A,\tilde{A}}(\eta_A).$$

hence $\eta_{\tilde{A}}\eta_A = \tilde{\eta}_A\eta_A$.

If $f : C \to \tilde{A}$ is a $\underline{C}_T$-morphism such that $\eta_{\tilde{A}} f = \tilde{\eta}_A f$ then

$$\eta_{\tilde{A}} f = \sigma_{\tilde{A},\tilde{A}}(1_{\tilde{A}})f = \sigma_{C,\tilde{A}}(f)$$

and $\tilde{\eta}_A f = \sigma_{\tilde{A},\tilde{A}}(\eta_A ex_A)f = \sigma_{C,\tilde{A}}(\eta_A ex_A f)$; hence $f = \eta_A(ex_A f)$. Since f is a total morphism then $ex_A f$ is also a total morphism.

By the remark after 2.5, we already know that for any objects $\tilde{A}, B, (\tilde{A})^B$ exists in $\underline{C}_T$. Using the lemma 3.1 we can prove the following:

3.2 Proposition If $\underline{C}_T$ has equalizers then $\underline{C}_T$ is a cartesian category.

Proof. Let A, B any objects of $\underline{C}_T$ considerthe following two $\underline{C}$-arrows:

$$\{\eta_{\tilde{A}}\}^B : (\tilde{A})^B \to (\tilde{\tilde{A}})^B.$$

$$\{\tilde{\eta}_A\}^B : (\tilde{A})^B \to (\tilde{\tilde{A}})^B.$$

Define A^B as the equalizer of these two arrows:

$$A^B \xrightarrow{\varepsilon} (\tilde{A})^B \underset{(\tilde{\eta}_A)^B}{\overset{(\eta_{\tilde{A}})^B}{\rightrightarrows}} (\tilde{\tilde{A}})^B$$

We can form the following commutative diagram:

Then we have an arrow $ev_{A,B} : A^B \times B \to A$. Now, given any morphism $f : C \times B \to A$ in $\underline{C}_T$, we get a $\underline{C}_T$-morphism $f^\dagger : C \times B \to \tilde{A}$ and therefore $(f^\dagger)^\star : C \to (\tilde{A})^B$ equalizes $(\eta_{\tilde{A}})^B$ and $(\tilde{\eta}_A)^B$ because:

$$ev_{\tilde{\tilde{A}},B}((\eta_{\tilde{A}})^B)((f^\dagger)^\star \times 1_B)$$

$$= \eta_{\tilde{A}} ev_{\tilde{A},B}((f^\dagger)^\star \times 1_B) = \eta_{\tilde{A}} f^\dagger.$$

and

$$ev_{\tilde{\tilde{A}},B}((\tilde{\eta}_A)^B(f^\dagger)^\star \times 1_B)) == \tilde{\eta}_A ev_{\tilde{A},B}((f^\dagger)^\star \times 1_B) = \eta_{\tilde{A}} f^\dagger.$$

it is easy to see that $\eta_{\tilde{A}} f^\dagger = \tilde{\eta}_A f^\dagger$ using $ex_A f^\dagger = f$. Therefore there is an arrow $f^\star : C \to A^B$ making the diagram

$$\begin{array}{ccccc} & & C & & \\ & \swarrow^{f^*} & \downarrow (\bar{f})^* & & \\ A^B & \xrightarrow{\epsilon} & (\tilde{A})^B & \rightrightarrows & (\tilde{\tilde{A}})^B \end{array}$$

commute.

Clearly, $ev_{A,B}(f^\star \times B) = f$, because we can form the following commutative diagram

$$\begin{array}{ccccc} & & C \times B & & \\ & \swarrow^{f^* \times 1_B} & \downarrow (\bar{f})^* \times 1_B & & \\ A^B \times B & \rightarrowtail & (\tilde{A})^B \times B & \rightrightarrows & (\tilde{\tilde{A}})^B \times B \\ \downarrow & \swarrow^{f} & \downarrow ev_{\tilde{A},B} & & \downarrow ev_{\tilde{\tilde{A}},B} \\ A & \underset{\eta_A}{\rightarrowtail} & \tilde{A} & \underset{\tilde{\eta}_A}{\overset{\eta_{\tilde{A}}}{\rightrightarrows}} & \tilde{\tilde{A}} \end{array}$$

Now,given any $h : C \to A^B$ in $\underline{C}_T$, $(ev_{A,B}(h \times 1_B))^\star = h$ if and only if $\eta_{A^B}(ev_{A,B}(h \times 1_B))^\star = \eta_{A^B} h$; Where the arrow $\eta_{A^B}(ev_{A,B}(h \times 1_B))^\star$ is given by:

$\overline{(ev_{A,B}(h \times 1_B))^\star}$ and $\overline{ev_{A,B}(h \times 1_B)} = \overline{ev_{A,B}}(h \times 1_B)$; Moreover, $\overline{ev_{A,B}} = \eta_A ev_{A,B}$ as is easily seen; hence

$$\eta_{A^B}(ev_{A,B}(h \times 1_B))^\star = \eta_A ev_{A,B}(h \times 1_B)$$

$$= ev_{\tilde{A},B}(\eta_{A^B} \times 1_B)(h \times 1_B) = ev_{\tilde{A},B}(\eta_{A^B} h \times 1_B)$$

Therefore, $(ev_{A,B}(h \times 1_B))^\star = (ev_{\tilde{A},B}(\eta_{A^B} h \times 1_B))^\star = \eta_{A^B} h$ and we prove the assertion.

REFERENCES .

[Cu.Ob.] P.L.Curien and A.Obtulowicz. Partial and Cartesian Closedness, preprint 1986.

[Di.He.] R.DiPaola and A.Heller, Dominical Categories,J. Symb. Logic ,1988.

[L,S] J.Lambek and P.J.Scott, Intoduction to Higher Order Categorical Logic, Cambridge Studies in Advanced Mathematics 7, Cambridge University Press, Cambridge England,1986.

[R.] G.Rosolini, Continuity and Effectiveness in Topoi, Ph.D. Thesis, Carnegie Mellon University, 1986.

Department of Mathematics.
McGill University.
Monttreal Quebec , H3A 2K6.

Current Adress:
Instituto de Matemáticas
Universidad Nacional Autónoma de México.
México D.F.; C.P. 04510.

Contemporary Mathematics
Volume 92, 1989

Normalization Revisited

ANDRE SCEDROV

University of Pennsylvania
Departments of Mathematics and Computer Science

ARPANET: Andre@cis.upenn.edu

ABSTRACT: We give a semantic proof of normalization for second-order polymorphic lambda calculus, that every polymorphic lambda term may be reduced to an irreducible one in finitely many steps.

INTRODUCTION

Anyone interested in mastering polymorphic lambda calculus should try to come up with one's own proof of normalization and see oneself make a mistake in the process. Then try giving a *correct* proof.

The normalization theorem was originally proved by Girard [71, 72] by proof-theoretic means. This theorem is a theoretical basis for program verification mechanisms such as type checking in various functional programming languages that feature parameterized abstract data types, see Reynolds [74, 83] , Huet [86, 87] , Bruce et al. [88]. Normalization is also a fundamental theorem in logic, in particular in proof theory, where its consequences include cut-elimination, the existence and disjunction properties, and the consistency of second-order arithmetic (see Girard [87], Takeuti [87], Troelstra [73]).

In this largely expository paper we present a completely self-contained semantic proof of normalization. Our argument came about from the study of partial equivalence relations (PER) semantics and the related

realizability interpretations in Freyd and Scedrov [87] and in Carboni et al. [88], but it eventually boiled down to a rather simple proof, which is given in section 2. Section 3 contains some comments on our normalization proof.
We would like to thank Henk Barendregt, Peter Freyd, Jean-Yves Girard, John Mitchell, Dana Scott, Phil Scott, and Rick Statman for very helpful comments and discussions. This work was partially supported by NSF grant CCR-8705596.

1. SECOND-ORDER POLYMORPHIC LAMBDA CALCULUS

Second-order polymorphic types are built inductively from type variables:

$$p \mid A \Rightarrow B \mid \forall p.A .$$

Free occurrences of type variables in types are defined as usual in logic. In particular, p is bound in $\forall p.A$ and hence p does not occur free in $\forall p.A$. We use A , B , C , ... to denote second-order polymorphic types, or *types* in short. We identify types that differ only in their bound type variables. $A[p{:=}B]$ is the result of substituting type B for the free occurrences of type variable p in type A , where bound type variables in A may be renamed so as to be distinct from the free type variables in B .

We assume another countably infinite alphabet of (ordinary) variables, written x , y , z , ... A *context* Γ is a finite list of expressions $x{:}A$, where x is an ordinary variable and A is a type, and no x appears twice. An expression $x{:}A$ may be read "variable x has type A ". In writing a list Γ , $x{:}A$ we always assume that x does not appear in Γ . *Second-order polymorphic lambda terms* will be the terms given by the following inductive definition of the *typing judgments* "In context Γ term t has type A " , written $\Gamma \vdash t{:}A$. The definition is given by deduction rules for deriving typing judgments. (One simultaneously defines the notion of a *free* occurrence of an ordinary variable in a polymorphic lambda term. We assume that the free occurrences are inherited in a lower line of a rule unless a restrictive comment is made.) The rules are:

$\Gamma \vdash x{:}A$ *if* $x{:}A$ *appears in* Γ ,

$$\frac{\Gamma,\ x{:}A \ \vdash\ t : B}{\Gamma \ \vdash\ \lambda x{:}A.\, t : A \Rightarrow B}$$ (then x does not occur free in $\lambda x{:}A.\,t$)

$$\frac{\Gamma \vdash t : A \Rightarrow B \qquad \Gamma \vdash u{:}A}{\Gamma \vdash tu : B}$$

$$\frac{\Gamma \vdash t{:}A}{\Gamma \vdash \Lambda p.\, t : \forall p.\, A}$$ *provided* p *not free in any* A *with* x:A *in* Γ *and* x *free in* t ,

$$\frac{\Gamma \vdash t : \forall p.\, A}{\Gamma \vdash t\{B\} : A[p{:=}B]}$$

$$\frac{\Gamma \vdash t{:}A}{\Gamma' \vdash t{:}A}$$ *where* Γ' *is a permutation of* Γ

$$\frac{\Gamma \vdash t{:}A}{\Gamma,\ x{:}B \ \vdash t{:}A}$$

Let t and u be polymorphic lambda terms of type B and A , respectively (in the same context). We write t[x:=u] for the result of substituting u for all free occurrences of the variable x of type A in t , where the bound (i.e., not free) variables in t may be renamed if necessary to prevent clashes (as with types, we identify terms up to renaming of bound variables, i.e. up to alpha conversion).

We define:

$(\lambda x{:}A.\,t)u$ is *immediately reducible* to $t[x{:=}u]$,

$(\Lambda p.\,w)\{C\}$ is *immediately reducible* to $w[p{:=}C]$,

and let *reduction* be the least reflexive transitive relation containing immediate reduction and compatible with term formation rules. Reduction relation is written as v ->> w and pronounced " v is reducible to w " , or: " w is a reduct of v ". (Seely [87b] considers reduction in polymorphic lambda calculus in the setting of indexed 2-categories.)

It may be shown by induction on derivation of typing judgments that reduction in second-order polymorphic lambda calculus has the *Church-Rosser property*:

THEOREM. Any two reducts of the same term have a common reduct. ▮

This theorem was first proved in Girard [72: I.4]; for exposition see also Scedrov [88].

A polymorphic lambda term is in *normal form* if none of its subterms has an immediate reduct. w is a normal form of v if v ->> w and w is in normal form. By the Church-Rosser property, the normal form of a term must be unique if it exists. Here we concentrate on the question of existence.

Our proof of normalization deals with *untyped lambda terms*, which, as the reader will recall, are defined inductively:

a) Assume a countably infinite collection of variables, each of which is an untyped lambda term (term, in short),
b) If a is a term and x is a variable, then $\lambda x.a$ is a term,
c) If a and b are terms, then ab is a term.

Free occurences of a variable in a term, substitution of terms, and reduction are defined in the same manner as above (but without restrictions due to types). In particular, we identify terms up to renaming of bound variables (alpha conversion). Reduction is often called *beta reduction*. *Beta conversion* on untyped lambda terms is the least equivalence relation containing beta reduction (on untyped lambda terms) and compatible with the formation rules a) - c).

Beta reduction enjoys the Church-Rosser property and will again be denoted as a ->> b . Normalization fails for untyped lambda terms. For example, $(\lambda x.xx)(\lambda x.xx)$ is its only reduct but it is not in normal form.

Bibliographical comments: The reader is referred to Barendregt [84] and Hindley & Seldin [86] for a detailed study of the untyped lambda calculus and the lambda calculus with simple types. Connections with computer science and with category theory are discussed in Huet [87], Cousineau et al. [86] , Curien [86]. The relationship between cartesian closed categories and lambda calculus with simple types is discussed in detail in Lambek & Scott [86].)

It shall be convenient to consider beta reduction on untyped lambda terms as the transitive closure of *simple reduction*, which is defined by the following inductive clauses:

(1) $a \to a$,
(2) If $a \to b$, then $\lambda x.a \to \lambda x.b$,
(3) If $a \to b$ and $c \to d$, then $ac \to bd$,
(4) If $a \to b$ and $c \to d$, then $(\lambda x.a)c \to b[x{:=}d]$.

We will write t , u , v , ... for polymorphic typed lambda terms and a , b , c , ... for untyped lambda terms.

2. A SEMANTIC PROOF OF NORMALIZATION

Consider the collection of normalizable untyped lambda terms, i.e., those untyped lambda terms which may be reduced to a normal form in finitely many steps. (Notice that the normalization property is of existential nature.) In the definition of our model and in the proof of Lemma 1 we will identify beta convertible untyped terms. This causes no harm because the class of normalizable terms is closed under beta conversion.

A *type set* is any set of (beta conversion equivalence classes of) normalizable untyped lambda terms that contains all terms of the form $xt_1 \ldots t_n$, where x is any variable and applications are associated to the left. Polymorphic types are interpreted as type sets:

i) $\|p\|$ is an arbitrary type set for any type variable p ,
ii) $\|A \Rightarrow B\| = \{c\colon c$ is normalizable and for any $a \in \|A\|$, $ca \in \|B\|\}$,
iii) $\|\forall p.\, A\| = \bigcap \{\, \|A\| : \|p\|$ any type set$\}$.

Note that the set defined in ii) is a type set. (In fact, it is this preservation property that suggests our definition of type sets once one is convinced that type sets must contain variables; see proof of Lemma 1 below.)

Any context Γ is interpreted as the (finite) list of type sets that interpret the types occurring in Γ.

Polymorphic terms are interpreted simply by erasing types. More precisely:

$$(\lambda x{:}A.\, t)^- = \lambda x.\, t^- ,$$
$$(tu)^- = t^- u^- ,$$
$$(\Lambda p.\, t)^- = t^- ,$$
$$(t\{A\})^- = t^- .$$

We now show that this interpretation is sound. If $\vec{x}$ is a list of variables $x_1 \ldots x_n$ and a is a list of (beta conversion equivalence classes of) normalizable untyped lambda terms, then we write $t^-[x{:=}a]$ for the result of substituting a_k for x_k in t^-, $1 \leq k \leq n$. We remind the reader that the empty conjunction is true.

LEMMA 1. Let $\Gamma \vdash t{:}A$, $\vec{a} \in \|\Gamma\|$. Then $t^-[\vec{x}{:=}\vec{a}] \in \|A\|$.

PROOF: By induction on the length of derivation of typing judgments (see the inductive definition of derivable typing judgments in section 1). The only problematic case is first-order lambda abstraction:

$$\frac{\Gamma,\ x{:}A \ \vdash\ t{:}B}{\Gamma\ \vdash\ \lambda x{:}A.\, t \,:\, A\Rightarrow B}\ .$$

Let Γ be $y_1{:}C_1, \ldots, y_n{:}C_n$. By inductive hypothesis, for any $c_k \in \|C_k\|$, $1\leq k\leq n$, and any $a \in \|A\|$, $t^-[\vec{y}, x := \vec{c}, a] \in \|B\|$ and in particular the untyped lambda term $t^-[\vec{y}, x := \vec{c}, a]$ has a normal form. Because $\|A\|$ is a type set it must contain the beta conversion equivalence class of the variable x. Hence $t^-[\vec{y}{:=}\vec{c}] \in \|B\|$ and thus $t^-[\vec{y}{:=}\vec{c}]$ must have a normal form, say d. Therefore $(\lambda x.\, t^-)[\vec{y}{:=}\vec{c}]$ has a normal form, namely $\lambda x.\, d$. Of course, $\lambda x.\, t^-$ is $(\lambda x{:}A.\, t)^-$. It remains to be shown that $(\lambda x{:}A.\, t)^-[\vec{y}{:=}\vec{c}]$ belongs to the type set $\|A\Rightarrow B\|$ for any $\vec{c} \in \|\Gamma\|$. But given any $a \in \|A\|$, $((\lambda x.\, t^-)a)[\vec{y}{:=}\vec{c}]$ is beta convertible to $t^-[\vec{y}, x := \vec{c}, a]$, which belongs to $\|B\|$ by inductive hypothesis. ∎

COROLLARY 1. The type erasure of every second-order polymorphic lambda term has a normal form in the untyped lambda calculus. ∎

Now first observe that:

PROPOSITION 1. If t is immediately reducible to u in polymorphic lambda calculus by a reduction step on a type application (Λp.w){A} then the number of occurrences of the symbol Λ in u is one less than in t. ∎

We finish the proof of normalization for second-order polymorphic lambda calculus by appealing to a kind of stability of type-erasure under reduction, which was first observed in Girard [72: I.A]:

LEMMA 2. If t is reducible to u in polymorphic lambda calculus, then t^- is reducible to u^- in untyped lambda calculus. More precisely, if t is reducible to u in polymorphic lambda calculus by a finite sequence that includes an immediate reduction on a subterm (λx:A.w)v, then t^- is reducible to u^- in untyped lambda calculus by a finite sequence that includes at least one step (λx.a)c -> a[x:=c]. Furthermore, if t^- is reducible to an untyped term c in untyped lambda calculus, then in polymorphic lambda calculus t is reducible to some polymorphic term u whose type-erasure is c.

Proof of Lemma 2 relies on:

PROPOSITION 2. Let $\Gamma \vdash t : A \Rightarrow B$ and let t^- be λx.c. Then t is reducible to some polymorphic term λx:A.u such that u^- is c.

PROOF OF PROPOSITION 2: By induction on the number of type applications that occur in t. *Base:* If no type applications occur, then t itself must be of the form λx:A.u, where c is u^-. *Step:* It suffices to consider t of the form $(\Lambda p.w)\{B_1\}\ldots\{B_n\}$, $n \geq 1$. Then t is reducible to $w[p:=B_1]\{B_2\}\ldots\{B_n\}$, whose type erasure is still λx.c and we may use the induction hypothesis. ∎

PROOF OF LEMMA 2: The first part is clear. Regarding the second part, we may assume that t^- is simply-reducible to an untyped lambda term c. Proof will be by double induction. The principal

induction is along the definition of simple reducibility in the untyped lambda calculus; the auxiliary induction is on the number of type abstractions and type applications that occur in t .
BASE: $t^- \to c$ by rule (1) . Then t^- is c , hence u may be t .
PRINCIPAL STEP: $t^- \to c$ by one of the rules (2) , (3) , or (4). Use the auxiliary induction. *Base:* t is vw or $\lambda x{:}A.v$. If $t^- \to c$ by rules (2) or (3), then already $v^- \to a$ and $w^- \to b$, so by the principal induction hypothesis there exist polymorphic lambda terms r and s such that $v \to\!\!> r$, $w \to\!\!> s$, r^- is a , and s^- is b . Then let u be rs or $\lambda x{:}A.r$, respectively. If $t^- \to c$ by rule (4), then v^- is $\lambda x.a$ and we already have $a \to b$ and $w^- \to d$. By the principal induction hypothesis there exist polymorphic lambda terms r and s such that $v \to\!\!> r$, $w \to\!\!> s$, r^- is $\lambda x.b$, and $s^- \to d$. By Proposition 2 there is a polymorphic lambda term u' such that $r \to\!\!> \lambda x{:}A.u'$ and $(u')^-$ is b . Then let u be $u'[x{:=}s]$. *Auxiliary step:* t^- is $\Lambda p.v$ or $v\{A\}$. In either case t^- is v^- , so we have $v^- \to c$. By the auxiliary induction hypothesis there exists a polymorphic lambda term w such that $v \to\!\!> w$ and w^- is c . Then let u be $\Lambda p.w$ or $w\{A\}$, respectively. ∎

Therefore:

THEOREM. Every second-order polymorphic lambda term has a normal form.

PROOF: Corollary 1 , Proposition 1 , and Lemma 2. ∎

3. SOME COMMENTS ON THE PROOF

Specialists will note that our proof is a simplified version of the proof given in Tait [75] and Mitchell [86]. On the other hand, that proof obtains more, namely that *every* reduction sequence of a second-order polymorphic lambda term must terminate in finitely many steps in in a unique normal form (*strong normalization*), originally proved by Girard [72]. In comparison with our approach, note that beta conversion of untyped terms does not preserve strong normalizability. Thus Girard, Tait, and Mitchell use a more involved definition of type sets. (In fact, Girard uses a notion of type set of polymorphic terms, candidat du réductibilité, rather than using type erasure.)

We would also like to point out that no cartesian closed categories arise in our approach. In section 2 we were not interested in eta conversion of polymorphic lambda terms. Thus type sets were enough and there was no need to consider partial equivalence relations. In category-theoretic sense, this means that we were dealing with a weakening of a cartesian closed structure and with weak products with respect to certain uniform families of morphisms. Let us briefly discuss the relevant category.

Let $\mathbb{L}$ be the collection of all normalizable untyped lambda terms modulo beta conversion. We shall be interested in partial endofunctions f on $\mathbb{L}$ for which there exists $a \in \mathbb{L}$ such that $f(b) \simeq ab$ for all $b \in \mathbb{L}$ (the Kleene symbol " $\simeq$ " means that whenever one side is defined so is the other and in that case they are both equal). Such functions are *partial* because, e.g., although $\lambda x.xx$ is in normal form, $(\lambda x.xx)(\lambda x.xx)$ is not normalizable.

Consider the category whose objects are type sets and whose morphisms $X \longrightarrow Y$ are partial endofunctions described above and defined on X. In other words, a morphism $X \longrightarrow Y$ is named by some $a \in \mathbb{L}$ such that for every $b \in X$ it is the case that $ab \in Y$ (and hence that ab is normalizable because type sets are nonempty.) Equality of morphisms is equality of functions, so a morphism may have many names.

a, b, c, ... will denote elements of $\mathbb{L}$. Composition in order of execution will be denoted by semicolon.

Recall that applications are associated to the left. Let $\langle a,b\rangle =$ $(\lambda x.\lambda y.\lambda z.zxy)ab = \lambda z.zab$, $\mathfrak{l}(c) \simeq (\lambda z.z(\lambda x.\lambda y.x))c \simeq c(\lambda x.\lambda y.x)$, $\mathfrak{r}(c) \simeq (\lambda z.z(\lambda x.\lambda y.y))c \simeq c(\lambda x.\lambda y.y)$. Clearly $\mathfrak{l}\langle a,b\rangle = a$ and $\mathfrak{r}\langle a,b\rangle = b$ for all $a,b \in \mathbb{L}$. Let L denote $\lambda x.\lambda y.x$ and let R denote $\lambda x.\lambda y.y$.

Given type sets X and Y, let $a \in X \cdot Y$ iff aL and aR are normalizable and $aL \in X$, $aR \in Y$. It is easily seen that $X \cdot Y$ is a type set and that $\mathfrak{l}\colon X \cdot Y \longrightarrow X$ and $\mathfrak{r}\colon X \cdot Y \longrightarrow Y$ are morphisms such that for any two morphisms $f\colon Z \longrightarrow X$ and $g\colon Z \longrightarrow Y$ there exists a morphism $h\colon Z \longrightarrow X \cdot Y$ for which $h;\mathfrak{l} = f$ and $h;\mathfrak{r} = g$. Moreover, h is given canonically by the pairing described above.

We abuse the notation and write $\langle f, g \rangle$ for h. Note that we are not claiming the uniqueness of h, i.e. one has only a canonically given weak cartesian product of X and Y.

Further, given type sets X and Y, let $a \in X \Rightarrow Y$ iff a names a morphism $X \longrightarrow Y$, i.e. for any $b \in X$ it is the case that $ab \in Y$ The reader may verify that $X \Rightarrow Y$ is again a type set and that the normal form $\lambda z.((zL)(zR))$ names a morphism $e: X \Rightarrow Y \cdot X \longrightarrow Y$ such that for any morphism $f: Z \cdot X \longrightarrow Y$ there is a morphism $g: Z \longrightarrow X \Rightarrow Y$ for which $\langle \ell, r \rangle ; f = \langle (\ell ; g), r \rangle ; e$. Note that we are claiming neither the uniqueness of g nor that f can be expressed via g and e. In other words, one has only a rather weak cartesian closed structure, albeit given canonically. We reiterate that this weak structure suffices for the semantics used in our normalization proof in section 2.

Finally, let us consider the semantic equivalent of type abstraction in this setting. A family of morphisms $f_\alpha : X_\alpha \longrightarrow Y_\alpha$ is called *realizable* iff all f_α's are named by a single normalizable untyped lambda term. (We are interested in the cases when the α's range over all type sets and the domains X_α are all the same, say some fixed type set X.) Given any family of type sets Y_α indexed by a set A, let $\Pi_{\alpha \in A} Y_\alpha$ be the intersection of all Y_α's. Note that $\lambda x.x$ names a realizable family of morphisms $\pi_\tau : \Pi_{\alpha \in A} Y_\alpha \longrightarrow Y_\tau$, $\tau \in A$, such that for any realizable family $f_\tau : X \longrightarrow Y_\tau$, $\tau \in A$, there exists a morphism $h: X \longrightarrow \Pi_{\alpha \in A} Y_\alpha$ for which $h ; \pi_\tau = f_\tau$, each $\tau \in A$. In other words, the category of has canonically given weak products w.r.t. realizable families.

Once polymorphic types have been interpreted as type sets, nonempty contexts may also be interpreted as type sets by considering canonical finite products (say, associated to the left). Then Lemma 1 says that every derived typing judgment $\Gamma \vdash t : A$, the appropriate untyped lambda abstraction of the type erasure t^- names a morphism $\|\Gamma\| \longrightarrow \|A\|$ (i.e., an element of $\|A\|$ when Γ is the empty list).

It is to be hoped that the properties of the category just described, when put in a proper 2-categorical setting, may play a role in operational semantics similar to the role that cartesian closed categories play in denotational semantics. A related development is discussed in Seely [87b].

REFERENCES

Barendregt, H. [84], *"The Lambda Calculus. Its Syntax and Semantics"*, Revised Edition. North-Holland, Amsterdam.

Breazu-Tannen, V., Coquand, T. [87], *Extensional Polymorphism.* Proc. TAPSOFT '87 - CFLP, Pisa. Springer LNCS 250 . Expanded version to appear in Theor. Comp. Sci.

Bruce, K.B., Meyer, A.R., Mitchell, J.C. [88], *The Semantics of Second-Order Lambda Calculus.* Information and Computation, to appear.

Carboni, A., Freyd, P.J., Scedrov, A. [8], *A Categorical Approach to Realizability and Polymorphic Types.* In: "Third ACM Workshop on Math. Foundations of Prog. Language Semantics, Proceedings New Orleans, Louisiana, April 1987", ed by M. Main et al., Springer LNCS 298 , pp. 23-42.

Cousineau, G., Curien, P.L., Robinet, B. (eds.) [86], *"Combinators and Functional Programming Languages".* Springer LNCS 242 .

Curien, P.L. [86], *"Categorical Combinators, Sequential Algorithms, and Functional Programming".* Research Notes in Theoretical Computer Science, Pitman.

Girard, J.Y. [71], *Une Extension de l'Interprétation de Gödel ...* In: "Second Scandinavian Logic Symposium", ed. by J.E. Fenstad, North-Holland, Amsterdam.

Girard, J.Y. [72], *"Interprétation Fonctionelle et Élimination des Coupures ...".* Thèse de Doctorat d'Etat, Université Paris VII.

Girard, J.Y. [87], *"Proof Theory and Logical Complexity"*. Bibliopolis, Napoli.

Freyd, P.J., Scedrov, A. [87], *Some Semantics Aspects of Polymorphic Lambda Calculus".* In: 2nd IEEE Symposium on Logic in Computer Science, Ithaca, NY, pp. 315-319.

Freyd, P. J., Scedrov, A. [89], *"Categories, Allegories: Volume One of Geometric Logic"*. North-Holland, to appear.

Hindley, J. R., Seldin, J. P. [86], *"Introduction to Combinators and Lambda Calculus"*. Cambridge University Press.

Hyland, J. M. E. [87], *A Small Complete Category*. Preprint.

Hyland, J. M. E., Robinson, E. P., Rosolini, G. [87], *The Discrete Objects in the Effective Topos*. Preprint.

Huet, G. [86], *Deduction and Computation*. In: "Fundamentals in Artificial Intelligence", eds. W. Bibel and P. Jorrand, Springer LNCS 232.

Huet, G. (ed.) [87], *"Logical Foundations of Functional Programming, Proceedings Austin, Texas, June 1987"*. Addison-Wesley, to appear.

Lambek, J., Scott, P. J. [86], *"Introduction to Higher-Order Categorical Logic"*. Cambridge University Press.

Mitchell, J. C. [86], *A Type-Inference Approach to Reduction Properties and Semantics of Polymorphic Expressions*. In: Proc. 1986 ACM Symposium on Lisp and Functional Programming, pp. 308-319.

Reynolds, J. C. [74], *Towards a Theory of Type Structure*. Springer LNCS 19 , pp. 408-425.

Reynolds, J. C. [83], *Types, Abstraction, and Parametric Polymorphism*. In: "Information Processing '83", ed. by R. E. A. Mason. North-Holland, Amsterdam, pp. 513-523.

Rosolini, G. [86] *About Modest Sets*. Preprint.

Scedrov, A. [88] *A Guide to Polymorphic Types*. In: "Logic and Computer Science, Proceedings C. I. M. E. Summer School, Montecatini Terme, June 1988", ed. by P. Odifreddi, Springer LNCS, to appear.

Seely, R. A. G. [87a] *Categorical Semantics for Higher-Order Polymorphic Lambda Calculus*. J. Symbolic Logic 52 , pp. 969-989.

Seely, R. A. G. [87b] *Modelling Computations: A 2-Categorical Framework.* 2nd IEEE Symposium on Logic in Computer Science, Ithaca, NY, pp. 65-71.

Tait, W. W. [75] *A Realizability Interpretation of the Theory of Species.* In: Springer LNM 453 .

Takeuti, G. [87] *"Proof Theory"*, Second Edition. North-Holland, Amsterdam, 1987.

Troelstra, A. S. [73] *"Metamathematical Investigation of Intuitionistic Arithmetic and Analysis"*. Springer LNM 344 .

Contemporary Mathematics
Volume 92, 1989

LINEAR LOGIC, $*$-AUTONOMOUS CATEGORIES AND COFREE COALGEBRAS

R.A.G. SEELY

ABSTRACT. A brief outline of the categorical characterisation of Girard's linear logic is given, analagous to the relationship between cartesian closed categories and typed λ-calculus. The linear structure amounts to a $*$-autonomous category: a closed symmetric monoidal category $\mathbf{G}$ with finite products and a closed involution. Girard's exponential operator, ! , is a cotriple on $\mathbf{G}$ which carries the canonical comonoid structure on A with respect to cartesian product to a comonoid structure on $!A$ with respect to tensor product. This makes the Kleisli category for ! cartesian closed.

0. INTRODUCTION. In "Linear logic" [1987], Jean-Yves Girard introduced a logical system he described as "a logic behind logic". Linear logic was a consequence of his analysis of the structure of qualitative domains (GIRARD [1986]): he noticed that the interpretation of the usual conditional "$\Rightarrow$" could be decomposed into two more primitive notions, a linear conditional "$\multimap$" and a unary operator "!" (called "of course"), which is formally rather like an interior operator:

$$X \Rightarrow Y = !X \multimap Y \tag{1}$$

The purpose of this note is to answer two questions (and perhaps pose some others.) First, if "linear category" means the structure making valid the proportion

linear logic : linear category = typed λ-calculus : cartesian closed category

then what is a linear category? This question is quite easy, and in true categorical spirit, one finds that it was answered long before being put, namely by BARR [1979]. Our intent here is mainly to supply a few details to make the matter more precise (though we leave many more details to the reader), to point out some similarities with work of LAMBEK [1987] (see these proceedings), and to appeal for a change in some of the notation of GIRARD [1987].

---

[1]1980 *Mathematics Subject Classification* (1985 *Revision*). 03G30, 18A15.
Partially supported by grants from Le Fonds F.C.A.R., Quebec.

Second, what is the meaning of Girard's exponential operator ! ? Since Girard has in fact offered several variants of ! in [1987], and another in GIRARD and LAFONT [1987], one cannot be too dogmatic here, but some certainty as to the minimal demands ! makes is possible — in particular we show that ! ought to be a cotriple, and its Kleisli category ought to be cartesian closed, in order to capture the initial motivation of the exponential. (This is already implicit in equation (1).)

ACKNOWLEDGEMENT. This note should be regarded as a "gloss" on GIRARD [1987], providing the categorical context and terminology for that work; I think the categorical setting provides a genuine improvement, and in particular, indicates how the notation may be made clearer. Others have come to similar conclusions: elsewhere in this volume DE PAIVA [1987] considers these matters, giving a fuller discussion of the interpretation of "$!A$" as "the cofree commutative comonoid over A", in the context of Dialectica categories. I would like to thank Michael Barr for pointing out that he had considered the essence of linear categories in [1979], thus giving further evidence of "the unreasonable influence of category theory in mathematics".

1. LINEAR LOGIC. There are several variations in the style Girard uses to present linear logic: *eg.* one sided sequents in [1987] and traditional sequents in GIRARD and LAFONT [1987]. I think the essence of the structure, especially its symmetry, is clearest when sequents in the style of Szabo's and Lambek's polycategories (SZABO[1975]) are used; here a sequent has the form

$$A_1, A_2, \ldots, A_n \rightarrow B_1, B_2, \ldots, B_m$$

(Of course, formally this is just an ordered pair of finite sequences — actually sets would do — of formulas.) The commas on the left should be thought of as some kind of conjunction, those on the right disjunction. (Better, think of the A_i on the left as data each to be used exactly once, and of the B_j on the right as possible alternate responses.)

1.1 DEFINITION. A (propositional) linear logic consists of formulas and sequents. Formulas are generated by the binary connectives $\otimes, \odot, \times, +$, and $\multimap$, and by the unary operation $\neg$, from a set of constants including I, ϕ, 1, and 0, and from variables.

Sequents consist of ordered pairs of finite sequences of formulas, as above; actually, finite sets of formulas would be better, in view of (*perm*) below, but let us pass over this point. The sequents are generated by the following rules from "initial sequents" (*i.e.* axioms), which include the following. (Greek capitals represent finite sequences (sets) of formulas.)

AXIOMS.

$$(id_A) : A \rightarrow A$$

$$(IR) : \quad \rightarrow I \qquad (\phi L) : \quad \phi \rightarrow$$

$$(1R) : \Gamma \rightarrow 1, \Delta \qquad (0L) : \quad \Gamma, 0 \rightarrow \Delta$$

$$(d) : A \rightarrow \neg\neg A \qquad (d^{-1}) : \neg\neg A \rightarrow A$$

RULES.

$$(perm) : \frac{\Gamma \rightarrow \Delta}{\sigma\Gamma \rightarrow \tau\Delta} \quad \text{for any permutations } \sigma, \tau.$$

$$(cut) : \frac{\Gamma \rightarrow A, \Delta \qquad A, \Theta \rightarrow \Psi}{\Gamma, \Theta \rightarrow \Psi, \Delta}$$

$$(\neg var): \frac{\Gamma, A \to B, \Delta}{\Gamma, \neg B \to \neg A, \Delta}$$

$$(IL): \frac{\Gamma \to \Delta}{\Gamma, I \to \Delta} \qquad (\phi R): \frac{\Gamma \to \Delta}{\Gamma \to \phi, \Delta}$$

$$(\otimes L): \frac{\Gamma, A, B \to \Delta}{\Gamma, A \otimes B \to \Delta} \qquad (\otimes R): \frac{\Gamma \to A, \Delta \qquad \Theta \to B, \Psi}{\Gamma, \Theta \to A \otimes B, \Delta, \Psi}$$

$$(\odot L): \frac{\Gamma, A \to \Delta \qquad \Theta, B \to \Psi}{\Gamma, \Theta, A \odot B \to \Delta, \Psi} \qquad (\odot R): \frac{\Gamma \to A, B, \Delta}{\Gamma \to A \odot B, \Delta}$$

$$(\multimap L): \frac{\Gamma \to A, \Delta \qquad \Theta, B \to \Psi}{\Gamma, \Theta, A \multimap B \to \Psi, \Delta} \qquad (\multimap R): \frac{\Gamma, A \to B, \Delta}{\Gamma \to A \multimap B, \Delta}$$

$$(\times L): \frac{\Gamma, A \to \Delta}{\Gamma, A \times B \to \Delta} \qquad \frac{\Gamma, B \to \Delta}{\Gamma, A \times B \to \Delta} \qquad (\times R): \frac{\Gamma \to A, \Delta \qquad \Gamma \to B, \Delta}{\Gamma \to A \times B, \Delta}$$

$$(+L): \frac{\Gamma, A \to \Delta \qquad \Gamma, B \to \Delta}{\Gamma, A + B \to \Delta} \qquad (+R): \frac{\Gamma \to A, \Delta}{\Gamma \to A + B, \Delta} \qquad \frac{\Gamma \to B, \Delta}{\Gamma \to A + B, \Delta}$$

1.2 REMARKS. (1) Concerning notation: In GIRARD [1987] a somewhat different notation is used. I have made changes so as to use wherever possible notation that is standard from a categorical viewpoint. This table summarises the changes:

Girard	:	$A^{\perp}$	1	$\perp$	$\top$	⅋ or $\sqcup$	& or $\sqcap$	$\oplus$
Here	:	$\neg A$	I	ϕ	1	$\odot$	$\times$	$+$

(Symbols not changed: $0, \otimes, \multimap$.)

I believe it is more important to pair the connectives by de Morgan duality ($\otimes$with $\odot$, $\times$ with $+$) than by "distributivity considerations" (as would justify Girard's $\sqcup$ with $\sqcap$.) Furthermore, $\times$ and $+$ seem to really be cartesian product and categorical sum, so those symbols seem more appropriate than Girard's (particularly his $\oplus$.) I must confess to being unable to find an entirely satisfactory notation for the de Morgan dual to tensor product, either in words ("dual tensor" seems preferable to "cotensor" or "tensor sum", or to Girard's "par") or in symbols ($\odot$ has been chosen for its neutrality; $\oplus$ might have been better were it not already so widely in use elsewhere.)

(2) The following sequents may be derived:

$(m_{AB}) : A, B \to A \otimes B$ $\qquad (\omega_{AB}) : A \odot B \to A, B$

$(e_{AB}) : A, A \multimap B \to B$

$(\pi^1_{AB}) : A \times B \to A$ $\qquad (\varpi^1_{AB}) : A \to A + B$

$(\pi^2_{AB}) : A \times B \to B$ $\qquad (\varpi^2_{AB}) : B \to A + B$

$(s_{AB}) : A \otimes B \to B \otimes A$

$(a_{ABC}) : (A \otimes B) \otimes C \to A \otimes (B \otimes C)$ $\qquad (a^{-1}_{ABC}) : A \otimes (B \otimes C) \to (A \otimes B) \otimes C$

(and similar sequents s', a' for $\odot$,)

$(\neg_{AB}) : A \multimap B \to \neg B \multimap \neg A$

It is easy to see that (m_{AB}) is equivalent to $(\otimes R)$, (w_{AB}) to $(\odot L)$, (e_{AB}) to $(\multimap L)$, (π_{AB}) to $(\times L)$, (ϖ_{AB}) to $(+R)$, **in the presence of** (cut). (Indeed, the rules amount to building in the required amount of (cut) to allow cut-elimination to go through.) As for symmetry and associativity, these follow from the rule $(perm)$ and the (implicit) associativity of concatenation. We give (a_{ABC}) as an illustration:

$$\dfrac{\dfrac{\dfrac{\dfrac{\dfrac{B, C \xrightarrow{(m)} B \otimes C \qquad A, B \otimes C \xrightarrow{(m)} A \otimes (B \otimes C)}{A, B, C \longrightarrow A \otimes (B \otimes C)}\ (cut)}{C, A, B \longrightarrow A \otimes (B \otimes C)}\ (perm)}{C, A \otimes B \longrightarrow A \otimes (B \otimes C)}\ (\otimes L)}{A \otimes B, C \to A \otimes (B \otimes C)}\ (perm)}{(A \otimes B) \otimes C \xrightarrow{(a)} A \otimes (B \otimes C)}\ (\otimes L)$$

As for $(\neg_{AB})$, it is given by

$$\dfrac{\dfrac{A, A \multimap B \xrightarrow{(e)} B}{\neg B, A \multimap B \to \neg A}\ (\neg var,\ perm)}{A \multimap B \to \neg B \multimap \neg A}\ (\multimap R, perm)$$

1.3 If we are to characterize the notion of a "linear category", we must complete the description of linear logic as a "deductive system" (in the sense of LAMBEK and SCOTT [1986]). First we must add the equations between derivations of sequents needed to get the structure of a polycategory (SZABO [1975]); these equations essentially make (cut) into a "polycomposition" of "polyarrows" which is associative, "partially commutative", and has units (id_A). (Analagous equations for multicategories may be found in this volume in LAMBEK [1987]; for this reason I will not go into detail here for these or the remaining equations.) Next, we must account for the monoidal structure of I, $\otimes$ (and their duals $\phi, \odot$) by adding equations which make sequents $A_1, \ldots, A_n \to B_1, \ldots, B_m$ equivalent to sequents $A_1 \otimes \cdots \otimes A_n \to B_1 \odot \cdots \odot B_m$. (Clearly there are maps, using the evident "hom" notation

$$[A_1, \ldots, A_n; B_1, \ldots, B_m] \leftrightarrow [A_1 \otimes \cdots \otimes A_n; B_1 \odot \cdots \odot B_m]$$

given by the rules $(\otimes L, R)$, $(\odot L, R)$, (cut); the point is that these maps be isomorphisms and inverse to each other.)

Similarly, it is likely that we want the structure to be symmetric monoidal, closed, and have finite products and coproducts — each of these adds to the list of equations in the evident way. For instance, (a) and (a^{-1}) must be inverse, as must (s_{AB}) and (s_{BA}). (This last could be weakened, if we only want a braided monoidal category, as in JOYAL and STREET [1986]. However, this would complicate the rest of the structure, so we shall not pursue this further.) Moreover, $(\multimap R)$ should give a bijection

$$[\Gamma, A; B, \Delta] \xrightarrow{\sim} [\Gamma; A \multimap B, \Delta]$$

whose inverse is

$$\frac{\Gamma \to A \multimap B, \Delta \qquad A, A \multimap B \xrightarrow{(e)} B}{\Gamma, A \to B, \Delta} \quad (cut,perm).$$

Finally, it seems that $\neg$ is a contravariant functor (in view of $(\neg var)$), that it is strong (in view of $(\neg_{AB})$), and that it is an involution (in view of (d) — which thus must be inverse to (d^{-1}).) These yield further equations, including the following, (if we are to have a $*$-autonomous category, as defined in BARR [1979]): for any A, B, these derivations of the sequent $A \multimap B \to \neg\neg A \multimap \neg\neg B$ are equal: $(d^{-1} \multimap d) = (\neg_{\neg B \neg A})(\neg_{AB})$. Here $(d^{-1} \multimap d)$ is a case of the schema

$$\frac{C \xrightarrow{(f)} A \qquad B \xrightarrow{(g)} D}{A \multimap B \xrightarrow{(f \multimap g)} C \multimap D}$$

given by

$$\frac{C \xrightarrow{(f)} A \qquad \dfrac{A, A \multimap B \xrightarrow{(e)} B \qquad B \xrightarrow{(g)} D}{A, A \multimap B \to D}\ (cut)}{\dfrac{C, A \multimap B \to D}{A \multimap B \to C \multimap D}\ (\multimap R)}\ (cut)$$

and $(\neg_{\neg B \neg A})(\neg_{AB})$ is a case of *(cut), viz.* in general:

$$\frac{A \xrightarrow{(f)} B \quad B \xrightarrow{(g)} C}{A \xrightarrow{(g)(f)} C} \quad (cut)$$

Since the required equations may be easily generated from the above recipe (and are in essence to be found in the references given, for the most part), and since this process is familiar (for instance, to that of LAMBEK and SCOTT [1986] for λ-calculus), I shall avoid the messy notational baggage needed to make all the details explicit, by stating boldly and without discussion:

1.4 DEFINITION. A linear category **G** is a $*$-autonomous category with finite products.

REMARKS. For a fuller discussion of $*$-autonomous categories, see BARR [1979]. Here just let me say that **G** is a closed symmetric monoidal category **G** with an involution $\neg\colon \mathbf{G}^{op} \to \mathbf{G}$ given by a dualising object ϕ: in our notation this means $\neg A = A \multimap \phi$ and the canonical arrow $A \to ((A \multimap \phi) \multimap \phi)$ is an isomorphism. (Barr uses $*$ for our $\neg$.) In such a category the existence of finite coproducts follows from finite products by de Morgan duality.

1.5 PROPOSITION. Given any linear logic $\mathcal{L}$, a linear category $\mathbf{G}(\mathcal{L})$ may be constructed (whose objects are formulas and whose morphisms $A \to B$ are equivalence classes of derivations of sequents $A \to B$); given any linear category $\mathbf{G}$, a linear logic $\mathcal{L}(\mathbf{G})$ may be constructed (whose constants are the objects of $\mathbf{G}$ and whose axioms are the morphisms of $\mathbf{G}$). Furthermore $\mathbf{G} \simeq \mathbf{G}(\mathcal{L}(\mathbf{G}))$ and (in a suitable sense) $\mathcal{L}$ is equivalent to $\mathcal{L}(\mathbf{G}(\mathcal{L}))$.

2. THE EXPONENTIAL OPERATOR ! .

In GIRARD [1987], these rules are given (in our notation) for the operator ! :

$$(der): \quad \frac{\Gamma, A \to \Delta}{\Gamma, !A \to \Delta} \qquad (\text{"dereliction"})$$

$$(thin): \quad \frac{\Gamma \to \Delta}{\Gamma, !A \to \Delta} \qquad (\text{"thinning" or "weakening"})$$

$$(contr): \quad \frac{\Gamma, !A, !A \to \Delta}{\Gamma, !A \to \Delta} \qquad (\text{"contraction"})$$

$$(!): \quad \frac{!\Gamma \to A}{!\Gamma \to !A}$$

In (!), $!\Gamma$ means $!A_1, !A_2, \ldots, !A_n$. Girard actually gives the rules for the de Morgan dual ? ; we shall not discuss ? .

2.1 It is perhaps worth simplifying these rules:

PROPOSITION. In the presence of linear logic:

(1) (der) is equivalent to the scheme $(\epsilon_A) : !A \to A$,
(2) $(thin)$ is equivalent to the scheme $(\epsilon'_A) : !A \to I$,
(3) $(contr)$ is equivalent to the scheme $(\delta'_A) : !A \to !A \otimes !A$,
(4) if $(der), (thin), (contr)$, then (!) is equivalent to:

$$(\delta_A) : !A \to !!A \qquad \text{and } (fun) : \frac{A \to B}{!A \to !B}$$

$$(\Delta\ iso) : !A \otimes !B \leftrightarrow !(A \times B) : (\Delta\ iso)^{-1}$$

$$(i\ iso) : I \leftrightarrow !1 : (i\ iso)^{-1}$$

REMARKS. (1) $(\delta_A), (fun)$ arise from the case $n = 1$ of (!), (Δiso) from the $n > 1$ case, and $(i\ iso)$ from the $n = 0$ case.

(2) Notice these rules seem to imply that we should regard ! as a functor (by (fun)), indeed a cotriple (or comonad) (by $(\epsilon_A), (\delta_A)$), and each $!A$ seems to be a comonoid (with respect to the monoidal structure $I, \otimes$), in view of $(\epsilon'_A), (\delta'_A)$. Furthermore, this comonoid structure seems to be the image under ! of the canonical comonoidal structure $(1 \leftarrow A \to A \times A)$ with respect to the cartesian structure $1, \times$, in view of $(\Delta\ \ iso)$, $(i\ \ iso)$. (These comments will take us straight to Definition 2.2.)

PROOF OF 2.1:

(1) For (ϵ_A), apply (der) to (id_A).For (der), apply (cut) to (ϵ_A).
(2) For (ϵ'_A), apply $(thin)$ to (IR). For $(thin)$, use (IL) and (cut) with (ϵ'_A).
(3) For (δ'_A), apply $(contr)$ to $(m_{!A!A})$. For $(contr)$, use $(\otimes L)$ and (cut) with (δ'_A).
(4) Given $(!)$, (δ_A) is $(!)$ applied to $(id_{!A})$.(fun) is the derived rule

$$\frac{\dfrac{A \to B}{!A \to B}\ (der)}{!A \to !B}\ (!)$$

$(\Delta\ iso)^{-1}$ is

$$\frac{\dfrac{\dfrac{A \times B \xrightarrow{(\pi^1)} A}{!(A \times B) \to !A}\ (fun) \qquad \dfrac{A \times B \xrightarrow{(\pi^2)} B}{!(A \times B) \to !B}\ (fun)}{!(A \times B), !(A \times B) \to !A \otimes !B}\ (\otimes R)}{!(A \times B) \to !A \otimes !B}\ (contr)$$

$(\Delta\ iso)$ is

$$\frac{\dfrac{\dfrac{\dfrac{\dfrac{!A \to A}{!A, !B \to A}\ (thin) \qquad \dfrac{!B \to B}{!A, !B \to B}\ (thin, perm)}{!A, !B \to A \times B}\ (\times R)}{!A, !B \to !(A \times B)}\ (!)}{!A \otimes !B \to !(A \times B)}\ (\otimes L)}{}$$

Conversely, given (δ_A), (fun), and the Δ, i iso's, we derive $(!)$ as follows.

First treat the $n = 0$ case as $I \to A$, *i.e.* as a special case of the $n = 1$ case. This is possible because of $(i\ iso)$; $I \to A$ may be thought of as $!1 \to A$. Then for $n = 1$, $(!)$ becomes the rule

$$\frac{!A_1 \to A}{!A_1 \to !A}$$

given by

$$\frac{!A_1 \xrightarrow{(\delta)} !!A_1 \qquad \dfrac{!A_1 \to A}{!!A_1 \to !A}\ (fun)}{!A_1 \to !A}\ (cut)$$

$(\Delta\ iso)$ allows the case when $n > 1$ to be reduced to the $n = 1$ case in view of the evident induced bijections (using the hom notation of 1.3):

$$[!A_1, \ldots, !A_n; A] \cong [!A_1 \otimes \cdots \otimes !A_n; A] \cong [!(A_1 \times \cdots \times A_n); A].$$

2.2 If we impose the appropriate equations on derivations, it is clear that we shall arrive at the following structure.

DEFINITION. A Girard category consists of a linear category $\mathbf{G}$ together with a cotriple $!:\mathbf{G} \to \mathbf{G}$ satisfying the following:

(i) for each A of $\mathbf{G}$, $!A$ is a comonoid with respect to the tensor structure: $I \xleftarrow{\epsilon'_A} !A \xrightarrow{\delta'_A} !A \otimes !A$;
(ii) there are natural isomorphisms $!A \otimes !B \xrightarrow{\sim} !(A \times B), I \xrightarrow{\sim} !1$; moreover ! takes the comonoid structure $(1 \xleftarrow{A} A \xrightarrow{\Delta} A \times A)$ with respect to the cartesian structure, to the comonoid structure in (i): *i.e.* these diagrams commute:

$$\begin{array}{ccc} !A & \xrightarrow{\delta'} & !A \otimes !A \\ \| & & \downarrow\wr \\ !A & \xrightarrow{!\Delta} & !(A \times A) \end{array} \qquad \begin{array}{ccc} !A & \xrightarrow{\epsilon'} & I \\ \| & & \downarrow\wr \\ !A & \xrightarrow{!A} & !1 \end{array}$$

REMARK. In fact it is easy to note that (i) follows from (ii), the diagrams defining ϵ' and δ'. However, in view of the "uncertainty" surrounding (!), it seems best to keep all the rules seperate.

2.3 As before, we claim without further ado:

PROPOSITION. Given a linear logic $\mathcal{L}$ with exponential operator !, a Girard category may be constructed on $\mathbf{G}(\mathcal{L})$; given a Girard category $\mathbf{G}$, !, the linear logic $\mathcal{L}(\mathbf{G})$ can be equipped with an exponential operator !. These constructions extend the equivalences of Proposition 1.5 in the evident way.

2.4 The essence of Girard's translation of intuitionistic logic into linear logic is the following:

PROPOSITION. If $\mathbf{G}$, ! is a Girard category, then the Kleisli category $\mathbf{K}(\mathbf{G})$ is cartesian closed.

PROOF. (For a basic reference on the categorical notions of cotriple, comonoid, and Kleisli category, see MAC LANE [1971].) Recall that the objects of $\mathbf{K}(\mathbf{G})$ are those of $\mathbf{G}$ itself, while the morphisms are given by

$$Hom_{\mathbf{K}(\mathbf{G})}(A, B) = Hom_{\mathbf{G}}(!A, B).$$

Writing $A \Rightarrow B$ for exponentiation B^A in $\mathbf{K}(\mathbf{G})$, $X \multimap Y$ for the internal hom in $\mathbf{G}$, it seems likely that

$$A \Rightarrow B = !A \multimap B$$

will do the trick. It is an easy matter to verify that

(i) the terminal object of $\mathbf{K}(\mathbf{G})$ is 1, the terminal object of $\mathbf{G}$;
(ii) the product $A \times B$ of $\mathbf{K}(\mathbf{G})$ is the same as in $\mathbf{G}$;
(iii) $A \Rightarrow B$ is $!A \multimap B$.

Appropriate bijections are given by

$$\frac{\dfrac{\dfrac{C \to A \times B \text{ in } \mathbf{K}(\mathbf{G})}{!C \to A \times B \text{ in } \mathbf{G}}}{!C \to A, \quad !C \to B \text{ in } \mathbf{G}}}{C \to A, \quad C \to B \text{ in } \mathbf{K}(\mathbf{G})}
\qquad
\frac{\dfrac{\dfrac{\dfrac{C \to (A \Rightarrow B) \text{ in } \mathbf{K}(\mathbf{G})}{!C \to (!A \multimap B) \text{ in } \mathbf{G}}}{!A \otimes !C \to B \text{ in } \mathbf{G}}}{!(A \times C) \to B \text{ in } \mathbf{G}}}{A \times C \to B \text{ in } \mathbf{K}(\mathbf{G})}$$

2.5 REMARKS. (1) In general $\mathbf{K}(\mathbf{G})$ does not have coproducts; Girard's interpretation of disjunction

$$A \vee B = !A + !B$$

is mysterious from this point of view, for the appropriate maps do not lie in $\mathbf{K}(\mathbf{G})$, though we do have a glimmer of the correct coproduct structure — *viz.* the bijections

$$\frac{\dfrac{!A + !B \to C \quad \text{in } \mathbf{G}}{!A \to C \quad !B \to C \text{ in } \mathbf{G}}}{A \to C \quad B \to C \text{ in } \mathbf{K}(\mathbf{G})}$$

(2) In GIRARD and LAFONT [1987], a stronger structure is considered for ! , with the intention that $!A = I \times A \times (!A \otimes !A)$. What Girard and Lafont seem to require (again, since they do not give a deductive system, but only a logic, we are left to supply appropriate equations between derivations) amounts to $!A$ being the cofree commutative comonoid over A. This condition implies (and is stronger than) that $\mathbf{G}$, ! is a Girard category; the question is whether it is too strong. (The structure of coherent spaces and linear maps, in the next section, does not have this extra property, for instance, nor is the structure of DE PAIVA [1987] an example since it is not $*$-autonomous.) It seems to me that what is really wanted is that $\mathbf{K}(\mathbf{G})$ be cartesian closed, so the question is: what is the minimal condition on ! that guarantees this — *i.e.* can we axiomatise this condition satisfactorily? (L. Román asked a related question: what condition on ! makes the Eilenberg-Moore category cartesian closed?)

(3) In GIRARD [1987], the propositional system is extended to a predicate linear logic by the addition of free variables of appropriate types and quantifiers $\bigwedge, \bigvee$, subject to the rules

$$(\textstyle\bigwedge L) \quad \frac{\Gamma, A[x := t] \to \Delta}{\Gamma, \bigwedge x A \to \Delta} \qquad\qquad (\textstyle\bigwedge R) \quad \frac{\Gamma \to A, \Delta}{\Gamma \to \bigwedge x A, \Delta}$$

(with the usual restriction on $(\bigwedge R)$: x not free in Γ, Δ).

(The rules for $\bigvee$ are given by de Morgan duality, as is $\bigvee$ itself.)

Girard is not specific about the nature of the types here — we may suppose, for example, that the above amounts to the following categorical structure:

An indexed linear category consists of a category $\mathbf{S}$ with finite products, and an indexed category $\mathbf{G}$ over $\mathbf{S}$; for each S of $\mathbf{S}$, the fibre $\mathbf{G}^S$ is a linear category, whose structure is preserved by t^*, t any morphism of $\mathbf{S}$; furthermore, each π^* has both adjoints $\bigvee_\pi \dashv \pi^* \dashv \bigwedge_\pi$, where π is a projection morphism of $\mathbf{S}$. The idea here, of course, is that $\mathbf{G}^S$ consists of the linear formulas with free variable of type S and (equivalence classes of) derivations of such formulas. (To be certain the logic is properly fibred in this way we ought to add conditions to the rules of inference to ensure that in any derivation of propositional linear logic, the same variables appear throughout, and in the quantifier rules, the only variables lost are those explicitly indicated — such restrictions are analogous to those of SEELY [1983] for first order intuitionistic logic and [1987] for polymorphic λ–calculus, and cause no loss of expressive power, (with a liberal use of dummy free variables.)

As with the logic, the adjoints $\bigvee_\pi, \bigwedge_\pi$ are dual, and so one only need assume one exists. (This is analogous to the situation for cartesian product and sum in $*$-autonomous categories.)

In this context, we would define an indexed Girard category as an indexed linear category $\mathbf{G}$ over $\mathbf{S}$ so that each $\mathbf{G}^S$ was a Girard category (i.e., had a ! cotriple with the usual properties), and that each t^* preserved this structure also. For such an indexed cotriple, one can define the indexed Kleisli category $\mathbf{K}(\mathbf{G})$ over $\mathbf{S}$ ($\mathbf{K}(\mathbf{G})^S$ will be $\mathbf{K}(\mathbf{G}^S)$); we already know $\mathbf{K}(\mathbf{G})$ will be (indexed) cartesian closed, and a similar analysis will easily show that in $\mathbf{K}(\mathbf{G})$, each π^* (π a projection of $\mathbf{S}$) will have a right adjoint $\pi^* \dashv \Pi_\pi$ (given, on objects, by $\bigwedge_\pi$.)

In general, we won't have a $\sum_\pi \dashv \pi^*$; the situation is similar to that for coproducts, as the following bijections show:

$$\frac{\dfrac{A \to \pi^* B \text{ in } \mathbf{K}(\mathbf{G})^{S\times T}}{!A \to \pi^* B \text{ in } \mathbf{G}^{S\times T}}}{\bigvee_\pi !A \to B \text{ in } \mathbf{G}^S}$$

(Girard set $\sum_\pi A = \bigvee_\pi !A$.)

The corresponding bijections for Π show why $\Pi_\pi A = \bigwedge_\pi A$ works:

$$\begin{array}{c} \pi^* B \to A \text{ in } \mathbf{K}(\mathbf{G})^{S\times T} \\ \hline !\pi^* B \to A \text{ in } \mathbf{G}^{S\times T} \\ \hline \pi^* !B \to A \text{ in } \mathbf{G}^{S\times T} \\ \hline !B \to \bigwedge_\pi A \text{ in } \mathbf{G}^S \\ \hline B \to \bigwedge_\pi A \text{ in } \mathbf{K}(\mathbf{G})^S \end{array}$$

3. AN EXAMPLE: COHERENT SPACES. In [1987] GIRARD gives an example of a model of linear logic; I shall briefly summarise how that example may be presented in this set-up. (This section will not be self-contained; I assume the reader has a copy of GIRARD [1987] in front of her.)

A coherent space is an atomic Scott domain closed under sups of families of pairwise compatible (or consistent) elements. Such a space X may be represented as a subdomain of the powerset $\mathcal{P}(|X|)$, where $|X|=$ the set of atoms of X; by this representation, the atoms are singletons. In fact, the structure of X is entirely given by the graph on $|X|$ defined by compatability: $x \frown y (mod X)$iff $x \vee y \in X$ (iff x, y are compatible in X.)

A linear map $f: X \to Y$ of coherent spaces must preserve sups of families of pairwise compatible elements and binary infs of compatible elements, as well as the order. Such a map is entirely determined by its trace: $\{< x, y >| \; x$ an atom of X, y an atom of $Y, y \leq f(x)\}$, since, for $a \in X$,

$$f(a) = \bigvee_{x \leq a} \bigvee_{y \leq f(x)} y.$$

The category **COHL** of coherent spaces and linear maps is a linear category; this is essentially proven in GIRARD [1987]. Furthermore, !: **COHL** $\to$ **COHL** makes **COHL** a Girard category; this is implicit in GIRARD [1987], but some of the details might be useful. Given a coherent space X, $!X$ is given by $|!X| = \mathcal{P}_{cfin}(|X|) =$ compatible finite subsets of $|X|$, with compatibility in $!X$ canonically induced by compatibility in X. (Viewing a space X as a subdomain of $\mathcal{P}(|X|)$, this would be written $|!X| = X_{fin} =$ finite elements of X.) Given a linear map $f: X \to Y$, $!f: !X \to !Y$ is characterised by the following: given an atom $\{x_1, \ldots, x_n\}$ of $!X$, an atom $\{y_1, \ldots, y_m\}$ of $!Y$, then $\{y_1, \ldots, y_m\} \leq (!f)(\{x_1, \ldots, x_n\})$ iff $n = m$ and $y_i \leq f(x_i)$ for each $i \leq n$. (!f is the "direct image made linear".)

For X in **COHL**, the map $\epsilon_X: !X \to X$ is given by: for a an atom of $!X$, x an atom of X, $x \leq \epsilon_X(a)$ iff $x \in a$, (*i.e.* $\{x\} \leq a$ in $!X$.) $\delta_X: !X \to !!X$ is given by: for a an atom of $!X$, b an atom of $!!X, b \leq \delta_X(a)$ iff $\bigvee b \leq a$.

The trace of $\epsilon'_X: !X \to I$ is the singleton $\{< \emptyset, 1 >\}$, where 1 is the unique atom of I. And $\delta'_X : !X \to !X \otimes !X$ is given by: for a, b, c atoms of $!X$, so that $< b, c >$ is an atom of $!X \otimes !X, < b, c > \leq \delta'_X(a)$ iff $b \vee c \leq a$.

It is a matter of straightforward calculation to show that these maps satisfy the equations for a cotriple and comonoid, and that the natural isomorphisms of Definition 2.2(ii) have the stated properties. The Kleisli category **K(COHL)** is **COHS**, the category of coherent spaces and *stable* maps, originally introduced (as "binary qualitative domains and stable maps") in GIRARD [1986].

COHS is well known to be cartesian closed, and does not have finite coproducts. (Girard discusses his treatment of sums in [1986].) Furthermore, **COHL** does not model the stronger axiom for ! in GIRARD and LAFONT [1987] $!A \cong I \times A \times (!A \otimes !A)$, nor does ! create cofree comonoids. (GIRARD [1987] mentions varying the notion of coherent space, using trees, in order to model this situation.)

BIBLIOGRAPHY

M. Barr, *-autonomous categories, Springer Lecture Notes in Mathematics 752, Berlin, 1979.

J.-Y. Girard, "The system F of variable types, fifteen years later," J. Theoretical Computer Science 45 (1986), 159 - 192.

J.-Y. Girard, "Linear logic," J. Theoretical Computer Science 50 (1987), 1 - 102.

J.-Y. Girard and Y. Lafont, "Linear logic and lazy computation", preprint, 1987.

A. Joyal and R. Street, "Braided monoidal categories," Macquarie Mathematics Report (No. 860081), Macquarie University, 1986.

S. Mac Lane, Categories for the working mathematician, Springer, New York, 1971.

J. Lambek, "Deductive systems and categories II," Springer Lecture Notes in Mathematics 86, pp. 76 - 122, Berlin, 1969.

J. Lambek, "Multicategories revisited," Proc. A.M.S. Conf. on Categories in Computer Science and Logic, 1987.

J. Lambek and P.J. Scott, Introduction to higher order categorical logic, Cambridge studies in advanced mathematics 7, Cambridge University Press, Cambridge, 1986.

C.V. de Paiva, "The Dialectica interpretation category," Proc. A.M.S. Conf. on Categories in Computer Science and Logic, 1987.

R.A.G. Seely, "Hyperdoctrines, natural deduction, and the Beck condition," Zeitschr. f. math. Logik und Grundlagen d. Math. 29(1983), 505 - 542.

R.A.G. Seely, "Categorical semantics for higher order polymorphic lambda calculus," J. Symbolic Logic 52(1987), 969 - 989.

M.E. Szabo, "Polycategories," Comm. Alg. 3 (1975), 663 - 689.

Department of Mathematics and Computer Science
John Abbott College
P.O. Box 2000
Ste. Anne de Bellevue
Quebec H9X 3L9

ABCDEFGHIJ — 89